Springer Finance

Springer
Berlin
Heidelberg
New York
Barcelona
Hong Kong
London
Milan
Paris
Tokyo

Helyette Geman Dilip Madan
Stanley R. Pliska Ton Vorst (Editors)

Mathematical Finance – Bachelier Congress 2000

Selected Papers
from the First World Congress
of the Bachelier Finance Society,
Paris, June 29–July 1, 2000

Springer

Hélyette Geman
CEREMADE, UMR CNRS 7534, Université Paris IX - Dauphine
Place du Maréchal De Lattre De Tassigny, 75775 Paris Cedex 16, France
and
Groupe ESSEC, Av. Bernhard-Hirsch, B.P. 50105, 95021 Cergy-Pontoise Cedex, France
e-mail: geman@dauphine.fr

Dilip Madan
University of Maryland, Robert H. Smith School of Business,
4427 Van Munching Hall, College Park, MD 20742, USA
e-mail: dmadan@rhsmith.umd.edu

Stanley R. Pliska
University of Illinois, Department of Finance,
242 University Hall, 601 South Morgan Street, Chicago, IL 60607-7, USA
e-mail: srpliska@uic.edu

Ton Vorst
Erasmus Universiteit Rotterdam, Postbus 1738, 3000 DR Rotterdam, The Netherlands
e-mail: vorst@opres.few.eur.nl

Mathematics Subject Classification (2000):
91-06, 91B28, 91B30, 91Bxx, 60-xx, 58J65, 60Gxx, 60G40, 65C05, 65T50, 93A30

Library of Congress Cataloging-in-Publication Data

Bachelier Finance Society. World Congress (1st : 2000 : Paris, France)
Mathematical finance--Bachelier Congress, 2000 : selected papers from the First World Congress of the Bachelier Finance Society, Paris, June 29-July 1, 2000 / Helyette Geman ... [et al.] (editors).
p. cm. --(Springer finance) Includes bibliografical references and index.

1. Financial futures--Congresses. 2. Speculation--Congresses. 3. Brownian motion processes--Congress. I. Title. II. Series
HG60024.3 .B33 2000 332.64'5--dc21 2001049334

ISBN 978-3-642-08729-5

Springer-Verlag is a part of Springer Science+Business Media

springeronline.com

Printed in Germany

Cover design: *design & production*, Heidelberg

Printed on acid-free paper

Preface

"La probabilité qui est une abstraction rayonne comme un petit soleil"
Le Jeu, la Chance et le Hasard by Louis Bachelier.

Even though it has only recently emerged as a scientific field of its own, mathematical finance came to existence in 1900, with the doctoral dissertation of the French mathematician Louis Bachelier entitled "Theorie de la Spéculation".

Researchers in finance will find there the first definition of options (it seems that options on tulips were traded in the Netherlands in the seventeenth century but it is not possible to trace any document which contains the precise description of the contract) as well as the first pricing model ever formalized in finance. Assuming that the underlying stock price dynamics are driven by a random walk, Bachelier *derives* from it the price of the option. At the same time, Bachelier supports his novel theoretical analysis with a sophisticated study of the French capital markets which were at the turn of the century a significant trading place and the main one worldwide for perpetual bonds.

Bachelier's work remained mostly unknown to financial economists until it was rediscovered in the 1950s by two brilliant minds. Jimmie Savage was sending postcards to several theorists in the field asking whether any of them "knew of a French guy named Bachelier who had written a little book on speculation". The answer was a definite yes on the part of Samuelson who had heard of him in the late 1930s from the Polish-American mathematician Stan Ulam, and had also kept in mind a footnote reference to Bachelier in Volume I of *Probability Theory and its Applications* published by Feller in 1950. In this footnote, Feller states that "Credit for discovering the connections between random walks and diffusions is due principally to L. Bachelier. Kolmogorov's theory of stochastic processes of the Markov type is based largely on Bachelier's ideas".

Hence, when Paul Cootner decided to produce a 1960s anthology of finance memoirs, Paul Samuelson urged him to commission an English translation of Bachelier; he also gave in 1964 a vibrant tribute to Bachelier's work: "So outstanding is his work that we can say that the study of speculative prices has its moment of glory at its moment of conception". Accordingly, Paul Samuelson decides to adopt Bachelier's model while correcting for the possibly negative values implied by Gaussian distributions and incompatible with the limited-liability feature of common stocks. And he "pragmatically replaces the Absolute Gaussians by log-normal probabilities".

The stochastic differential equation which became the key assumption in the Black-Scholes-Merton and many other pricing formulas first appeared in the 1965 paper by Paul Samuelson *Rational Theory of Warrant Pricing.* To report a modest anecdote of mine, while I was driving Professor Samuelson back to his hotel on June 29, after the magnificent inaugural ceremony in the Amphithéâtre Marguerite de Navarre at the Collège de France, marked by the talks given by Paul Samuelson, Henry McKean and Robert Merton to an audience comprising an impressive number of great names in mathematics and economics, I mentioned to him that many people were unaware that he was responsible for the fundamental equation. Professor Samuelson replied: 'Yes, I had the equation but "they" got the formula...' As we know, the number of times the equation and the formula have been restated and used in the last 35 years is beyond counting.

The 1965 paper of Paul Samuelson has another remarkable trait: it contains the unique (to my knowledge) piece of work Henry McKean ever dedicated to finance: in the Appendix, the explicit solution of the American option problem is provided when the underlying stock follows a geometric Brownian motion and the maturity is infinite. As of today, the exact solution for a finite maturity in the same setting has not yet been obtained. However, we understand better why the problem was "easier" for an infinite maturity (or for a maturity which would be an exponential time independent of the Brownian motion) thanks to recent pieces of work on functionals of Brownian motion, by Marc Yor in particular.

Coming to mathematicians, they recognize in Bachelier's pioneering work the development of the properties of Brownian motion that Brown had started to exhibit; the first expression of the Markov property which was only made fully explicit in 1905 (and which remains today a key assumption in most of the reference models in finance); and the introduction of the beautiful concept of trajectories at a time when the classical probabilistic representation was a sequence of heads and tails in coin tossing. Even years later, the great Kolmogorov was more interested in the analytical objects attached to stochastic processes than in their trajectories. Louis Bachelier paved the way to the work by Wolfgang Doeblin in the late thirties and to the profound study of Brownian excursions by Paul Lévy. Returning to finance, trajectories of diffusions, jump-diffusions or pure jump processes have become familiar tools, indispensable for the trader placing his orders in commodity or equity markets on the basis of charts, for the fundamental analyst relating stock price changes to earning announcements or news arrival and for the risk-manager simulating trajectories to compute the Value at Risk or the economic capital attached to a position or a portfolio.

The first World Congress of the Bachelier Finance Society, taking place in the home country of Louis Bachelier one hundred years after the defence of his PhD dissertation, and during the year 2000 which had been declared World Mathematical Year by the International Mathematical Union, had to be an exceptional manifestation and indeed it was. Paul Samuelson crossed the ocean to talk to us about "Finance Theory Within One Lifetime", a tale that by definition, he only was in the position to recount. Robert Merton who often honoured our country with his presence, was this time in the company of his PhD adviser and current peer in the exclusive group of Nobel Laureates. Henry McKean covered nine blackboards of the amphitheatre with his elegant formulas and figures. S.R.S. Varadhan gave one of his brilliant talks and yet, more research needs to be done to analyze the applications of large deviations to finance. Albert Shiryaev, Hans Föllmer and David Heath, experts in – among other topics – potential theory and stochastic processes, proved that they had also fully captured the major subtleties in financial economics. Last but not least, Steve Ross and Eduardo Schwartz were representatives of the field of mathematical finance via excellence in financial economics from arbitrage to interest rate models, from information theory to option pricing.

As shown (only partially) by this volume, the quality of the audience was as impressive as the list of Invited Speakers; the streets between the Collège de France, Ecole Normale Supérieure and Institut Henri Poincaré were humming for four days with animated discussions and we all left with the emotion of having been part of a unique scientific event.

Paris, July 2000

Hélyette Geman
President of the Bachelier Finance Society

Table of Contents

Bachelier and His Times: A Conversation with Bernard Bru *1*
M.S. Taqqu

Modern Finance Theory Within One Lifetime *41*
P. Samuelson

Future Possibilities in Finance Theory and Finance Practice *47*
R.C. Merton

Brownian Motion and the General Diffusion: Scale & Clock *75*
H.P. McKean

Rare Events, Large Deviations *85*
S.R.S. Varadhan

Conquering the Greeks in Monte Carlo: Efficient Calculation of the Market Sensitivities and Hedge-Ratios of Financial Assets by Direct Numerical Simulation 93
M. Avellaneda, R. Gamba

On the Term Structure of Futures and Forward Prices 111
T. Björk, C. Landén

Displaced and Mixture Diffusions for Analytically-Tractable Smile Models 151
D. Brigo, F. Mercurio

The Theory of Good-Deal Pricing in Financial Markets 175
A. Černý, S. Hodges

Spread Option Valuation and the Fast Fourier Transform 203
M.A.H. Dempster, S.S.G. Hong

The Law of Geometric Brownian Motion and its Integral, Revisited; Application to Conditional Moments 221
C. Donati-Martin, H. Matsumoto, M. Yor

The Generalized Hyperbolic Model: Financial Derivatives and Risk Measures 245
E. Eberlein, K. Prause

Using the Hull and White Two-Factor Model in Bank Treasury Risk Management 269
R.J. Elliott, J. Van der Hoek

Default Risk and Hazard Process 281
M. Jeanblanc, M. Rutkowski

Utility-Based Derivative Pricing in Incomplete Markets 313
J. Kallsen

Pricing Credit Derivatives in Credit Classes Frameworks 339
F. Moraux, P. Navatte

An Autoregressive Conditional Binomial Option Pricing Model 353
J.-L. Prigent, O. Renault, O. Scaillet

Markov Chains and the Potential Approach to Modelling Interest Rates and Exchange Rates 375
L.C.G. Rogers, F.A. Yousaf

Theory and Calibration of HJM with Shape Factors 407
A. Roncoroni, P. Guiotto

Optimal Investment in Incomplete Financial Markets 427
W. Schachermayer

Evaluating Investments in Disruptive Technologies 463
E. Schwartz, C. Zozaya-Gorostiza

Quickest Detection Problems in the Technical Analysis of the Financial Data 487
A. Shiryaev

Bachelier and His Times: A Conversation with Bernard Bru*†‡

Murad S. Taqqu

Department of Mathematics
Boston University
111 Cummington St.
Boston, MA 02215, USA
murad@math.bu.edu

Abstract. Louis Bachelier defended his thesis "Theory of Speculation" in 1900. He used Brownian motion as a model for stock exchange performance. This conversation with Bernard Bru illustrates the scientific climate of his times and the conditions under which Bachelier made his discoveries. It indicates that Bachelier was indeed the right person at the right time. He was involved with the Paris stock exchange, was self-taught but also took courses in probability and on the theory of heat. Not being a part of the "scientific establishment," he had the opportunity to develop an area that was not of interest to the mathematicians of the period. He was the first to apply the trajectories of Brownian motion, and his theories prefigure modern mathematical finance. What follows is an edited and expanded version of the original conversation with Bernard Bru.

Bernard Bru is the author, most recently, of *Borel, Lévy, Neyman, Pearson et les autres* [38]. He is a professor at the University of Paris V where he teaches mathematics and statistics. With Marc Barbut and Ernest Coumet, he founded the seminars on the history of Probability at the EHESS (École des Hautes Études en Sciences Sociales), which bring together researchers in mathematics, philosophy and the humanities.

M.T. : It took nearly a century for the importance of Louis Bachelier's contributions to be recognized. Even today, he is an enigmatic figure. Little is known about his life and the conditions under which he worked. Let's begin with his youth. What do we know about it?

B.B. : Not much. Bachelier was born in Le Havre to a well-to-do family on March 11, 1870. His father, Alphonse Bachelier, was a wine dealer at Le Havre and his mother Cécile Fort-Meu, was a banker's daughter. But he lost his parents in 1889 and was then forced to abandon his studies in order to earn his livelihood. He may have entered the family business, but he seems

* This article first appeared in *Finance and Stochastics* [119]. This is a slightly expanded version. It appears in French in [120].

† *AMS 1991 subject classifications:* 01A55, 01A60, 01A65, 01A70.

‡ The work was partially supported by the NSF Grant ANI-9805623 at Boston University.

to have left Le Havre for Paris after his military service around 1892 and to have worked in some capacity at the Paris Stock Exchange. We know that he registered at the Sorbonne in 1892 and his thesis "Theory of Speculation" [5] of 1900 shows that he knew the financial techniques of the end of the 19th century perfectly.

M.T. : How important was the Paris Stock Exchange at that time?

B.B. : The Paris Stock Exchange, had become by 1850, the world market for the *rentes*, which are perpetual government bonds. They are fixed-return securities. When the government wished to contract a loan, it went through the Paris Exchange. The bond's stability was guaranteed by the state and the value of the gold franc. There was hardly any inflation until 1914. The rate ranged between 3 and 5%. The securities had a nominal value, in general 100 francs, but once a bond was issued, its price fluctuated. The sums that went through Paris were absolutely enormous. Among the French, the bonds remained in families through generations. A wealthy Frenchman was a "rentier", a person of independent means, who lived on the products of his bonds.

M.T. : I thought that a "rentier" is someone who lives off his land holdings.

B.B. : That's also true but an important part, that which was liquid because easy to transfer, came from financial bonds. It all began with "the emigrants' billion" (*le milliard des émigrés*). During the French Revolution, the nobility left and their holdings were sold as national property. When they returned in 1815, it was necessary to make restitution. The French state took a loan of a billion francs at the time, which was a considerable sum. The state paid the interest on it but never repaid the capital. It is what was called a "perpetual bond", and the success of the original offering led to subsequent new issues. In 1900 the nominal capital of this public debt was some 26 billion francs (on a France's annual budget of 4 billion). The international loans (from Russia, Germany, etc.) brought the total to 70 billion gold francs. All of the commercial houses had part of their funds invested in bonds. The state guaranteed that every year interest would be paid to the holders at fixed rates. This continued until the war of 1914, when the franc collapsed.

M.T. : Could the bonds be sold?

B.B. : They were sold for cash or as forward contracts or options, through stockbrokers. There was an official market on the exchange and a parallel market. It's quite complicated, but it required a large workforce, for there were no phones, so there were assistants who carried out the transactions. Many of the financial products we know today existed then. There were many ways to sell bonds. If you read Bachelier's thesis, he explains the workings of the system briefly.

M.T. : Why did people sell their perpetual bonds?

B.B. : For purposes of transfer or for speculation. It was, however, a speculation that was tolerated since it was not particularly risky. The bonds prices fluctuated markedly only during the great French political crises of 1830, 1848, and 1870.

M.T. : Was there fear of default?

B.B. : Yes. Considerable fortunes were then made and lost. These extreme fluctuations were not addressed by Bachelier in his thesis, he was merely concerned with the ordinary day-by-day fluctuations.

M.T. : Where did Bachelier work?

B.B. : I've searched, but I've been unable to locate the firm where Bachelier worked. It remains a mystery. But what is indisputable is that he loved science. As soon as he was able to set aside some funds, he returned to his studies. He earned his degree in mathematics at the Sorbonne in 1895 where he studied under professors such as Paul Appell, Émile Picard and Joseph Boussinesq, a mathematical physicist. There were two important areas in mathematics at the end of the 19th century: mathematical physics (that is, mechanics) and geometry. Those were the things one studied at that time. He therefore learned the theory of heat (diffusion equation) with Boussinesq [35], and also, he had Henri Poincaré. It was prior to Poincaré's change of chair.

M.T. : At the Sorbonne?

B.B. : Yes, where Poincaré occupied the chair in mathematical physics and probability between 1886 and 1896. Poincaré then transferred to a chair in celestial mechanics.

M.T. : So Bachelier almost missed studying under Poincaré?

B.B. : He would no doubt have followed his courses on celestial mechanics, since Poincaré was idolized at the time. Poincaré's courses were difficult to follow; they were also very innovative and without exams. The math degree[1] required taking exams in mechanics, differential and integral calculus, and astronomy. Bachelier finally succeeded in passing these. He also took Poincaré's exam in mathematical physics in 1897[2]. So Bachelier and Poincaré did meet.

M.T. : Was it an oral exam?

B.B. : Yes. It was probably there that Bachelier got the idea of continuing his studies. At the time, it was an honor, since the next degree was the thesis[3].

---

[1] equivalent to a Bachelor/Master of Arts.

[2] This course had been offered since 1834, but there were no exams because the course used to be elective. Bachelier was the first to pass the examination after the rules changed.

[3] In fact, there were two theses, an original one and a second one, which is an oral examination and whose purpose is to test the breadth and teaching abilities of the candidate. Bachelier's second thesis was about Boussinesq's work on fluid mechanics. The subject involved the motion of a sphere in a liquid.

After the thesis, it was necessary to find a university position, and these were rare. At the universities in the provinces, there were probably about fifty positions in mathematics. There were two at each university. To teach at a university required a thesis, but that was not enough, for there were almost no positions.

M.T. : The subject of Bachelier's thesis was out of the ordinary.

B.B. : In fact, it was exceptional. On the other hand, Bachelier was the right man at the right time, first because of his experience in the stock exchange. Secondly, he knew the theory of heat (this was the height of classical mathematical physics). Third, he was introduced to probability by Poincaré and he also had the probability lecture notes [27] of Joseph Bertrand, which served him well. If you look at Bertrand's chapter on gambling losses, you will see that it was useful to Bachelier. But the idea of following trajectories is attributable to Bachelier alone. It's what he observed at the Stock Exchange.

M.T. : Bachelier does seem to have been the right man at the right time.

B.B. : He was undoubtedly the only one who could have done it. Even Poincaré couldn't have done it. It had to happen in Paris, the center of speculation in bonds. It required a mathematical background, but not too extensive, since the mathematics of the time was not about that: it was about the theory of functions, especially functions of complex variables. The thesis of Émile Borel, that of Jacques Hadamard, were on the theory of functions. Bachelier was incapable of reading that. Moreover, Bachelier's thesis did not receive the distinction that he needed to open the doors of the university. It required getting the grade "very honorable". He only received the grade "honorable".

M.T. : Were there two possible grades?

B.B. : There was "adjourn", which indicated that the thesis was not worthy of being considered. And there were three grades: "passable", which was never given; "honorable", which meant "that's very good, sir, so long", and the "very honorable" grade, which offered the possibility of a university career, although not automatically.

M.T. : Why do you believe that he received only the grade "honorable"?

B.B. : It was a subject that was utterly esoteric compared to the subjects that were dealt with during that period, generally the theses of mechanics, which is to say partial differential equations. The big theses of the era were theses on the theory of functions (Borel, Baire, Lebesgue). Therefore, it was not an acceptable thesis topic. If we look, moreover, at the grades Bachelier earned in his degree exams, which are preserved in the national archives, they were very mediocre. He had a written exam in analysis, mechanics and astronomy. He had a great deal of difficulty. He tried many times before finally succeeding, and when he did succeed, it was just barely. He was last or next-to-last. That was still very good, since there were relatively few successes. The exams were difficult, and he was self-taught.

M.T. : Why?

B.B. : He did not go to a *lycée* following his *baccalauréat.* He had to take a job right away. The baccalaureate was the exam that opened the doors of the university. But in fact, all of the students followed two years of "special mathematics" in a *lycée* in order to gain entrance to the great scientific schools (such as the École Polytechnique or the École Normale Supérieure). The fundamentals of science were acquired at the *lycée* level. Bachelier must have studied on his own, which explains his difficulties on examinations. Thus Bachelier never had a chance to obtain a university chair. In the end, the quality of his thesis, the fact that it was appreciated by Poincaré, the greatest French intellect of the time, did not change the fact that Bachelier lacked the "necessary" distinction.

M.T. : Was he already working?

B.B. : He was working and studying at the same time. He occasionally took courses and also examinations. He was employed, I don't know where, perhaps in a commercial firm. Since his thesis was not enough for him to gain employment at a university, it is likely that he continued to work.

M.T. : Were there any errors in his thesis?

B.B. : No, absolutely not, there were no errors. The thesis was written rather in the language of a physicist. Fundamentally, this was not the problem. At that time, Poincaré would have pointed out a true error, had there been one. Poincaré's way of reasoning was similar: he left the details aside, he assumed them justified and didn't dwell on them. Bourbaki came only later. As for the question of "errors", that was something else. It came after the war of 1914. The thesis was in 1900. He was not awarded a position because he was not "distinguished" enough. What's more, Probability did not start to gain recognition in France until the 1930's. This was also the case in Germany.

M.T. : Who were the great probabilists in 1900?

B.B. : There were none. Probability as a mathematical discipline dates from after 1925. There was a Laplace period until 1830, then it's the crossing of the desert – mathematicians took no interest in those things – their interest was rekindled only much later. Let's take Paris, for example. Bachelier's thesis was 1900. We'd have to wait another twenty years for Deltheil, Francis Perrin and especially the end of the 30's with Dugué, Doeblin, Ville, Malécot, Fortet, Loève.

M.T. : Was Bachelier's thesis considered a probability thesis?

B.B. : No. It was a mathematical physics thesis, but since it was not physics, it was about the Stock Exchange, it was not a recognized subject.

M.T. : Wasn't there some notion of Brownian motion at the time?

B.B. : Bachelier doesn't refer to it at all. He learned of this much later, for there were to be many popularized publications on the subject. But in 1900, zero. The translation of Boltzmann[4] [28] in France was done in 1902 and 1905. And Boussinesq was a mathematician doing mechanics and hydrodynamics. For him, mathematical physics was differential equations.

M.T. : Why did Bachelier introduce Brownian motion?

B.B. : To price options. (The options considered by Bachelier were somewhat different from the ones we know today.) He uses the increments of Brownian motion to model "absolute" price changes, whereas today, one prefers to use them to model "relative" price changes (see Samuelson [113–115][5]).

---

[4] Brownian motion is named after Robert Brown [36], the Scottish botanist who noticed in 1827 that grains of pollen suspended in water had a rapid oscillatory motion when viewed under a microscope. The original experiment and its re-enactment are described in [55]. The kinetic theory of matter, which relates temperature to the average kinetic energy, was developed later in the century, in particular by Ludwig Boltzmann, and it is the basis of Einstein's explanation of Brownian motion [51] in 1905.

[5] The idea of modeling the logarithm of prices by independent and normally distributed random variables was also suggested by Osborne [96] in 1959. Osborne was a physicist working at the Naval Research Laboratory in Washington, D.C. At the time, he knew apparently of neither Bachelier nor Samuelson (see also [2] and [26]) He later wrote an interesting book [98] which are his lecture notes at the University of California at Berkeley. In his 1959 article [96], Osborne does not mention Bachelier but, following a letter by A. G. Laurent [82] in the same volume, Osborne provided a reply [97], where he quotes Bachelier. He starts [97] by indicating that after the publication of his 1959 article [96], many people drew his attention to earlier references, and then he gives the following nice summary of Bachelier's thesis (the reference numbers in the text below are ours):

> *I believe the pioneer work on randomness in economic time series, and yet most modern in viewpoint, is that of Bachelier [5] also described in less mathematical detail in reference [15]. As reference [5] is rather inaccessible (it is available in the Library of Congress rare book room), it might be well to summarize it here. In it Bachelier proceeds, by quite elegant mathematical methods, directly from the assumption that the expected gain (in francs) at any instant on the Bourse is zero, to a normal distribution of price changes, with dispersion increasing as the square root of the time, in accordance with the Fourier equation of heat diffusion. The theory is applied to speculation on rente, an interest-bearing obligation which appeared to be the principle vehicle of speculation at the time, but no attempt was made to analyze the variation of prices into components except for the market discounting of future coupons, or interest payments. The theory was fitted to observations on rente for the years 1894-98. There is a considerable quantitative discussion of the expectations from the use of options (puts and calls). He also remarked that the theory was equally applicable to other types of speculation, in stock, commodities, and merchandise. To him is due credit for major priority on this problem.*

M.T. : Is it Poincaré who wrote the report on the thesis?

B.B. : Yes, that's how it was done at that time. There were three people in the jury but only one reported. The other two members of the jury were Appell and Boussinesq. They probably read nothing, as opposed to Poincaré, who read everything. When there was a thesis that no one wanted to read, on any subject, applied physics, experimental physics, it was directed to Poincaré. I've seen some Poincaré reports on some incredible works. He had an unbelievably quick intelligence.

M.T. : Is that why he was asked to report on Bachelier's thesis?

B.B. : Perhaps. But it's also because he knew Bachelier.

M.T. : Bachelier had indeed taken his course. But in those courses, did one speak to the professor?

B.B. : Never. It was unthinkable to question a professor. Even after the course. In the biography of Jerzy Neyman[6] by Constance Reid [112], Neyman recounts that, when he was a Rockefeller fellow in Paris, he followed Borel's course in probability[7]. He once approached Borel to ask him some questions. Borel answered, "You are probably under the impression that our relationships with people who attend our courses are similar here to what they are elsewhere. I am sorry. This is not the case. Yes, it would be a pleasure to talk to you, but it would be more convenient if you would come this summer to Brittany where I will be vacationing"[8]. This was in 1926. Neyman was at the still young age of 32.

M.T. : Where did you find Poincaré's thesis report?

B.B. : At the National Archives[9], where things remain for eternity. Here's the beginning of the report[10]:

---

[6] This is the Neyman (1894-1981) of the celebrated Neyman-Pearson Lemma in hypotheses testing.

[7] Émile Borel (1871-1956) founded the French school of the theory of functions (Baire, Lebesgue, Denjoy). In his 1898 book [29], he introduces his measure as the unique countably additive extension of the length of intervals; it became the basis of modern measure and integration theory. Borel sets are now named after him. Starting in 1905, Borel focused on probability and its applications and developed properties related to the notion of almost sure convergence. See [56] for the story of his life.

[8] See [112], p. 66.

[9] The original document of Poincaré's thesis report is held at the *Registre des thèses de la Faculté des Sciences de Paris*, at the *Archives nationales*, 11 rue des Quatre-Fils, 75003 Paris, classification AJ/16/5537. It is dated March, 29, 1900, the day of the defense.

[10] The full text, translated into English, by Selime Baftiri-Balazoski and Ulrich Hausmann, can be found in [44]. The French text of the report is given below, as well as the short defense report, signed by Paul Appell.

The subject chosen by Mr. Bachelier is somewhat removed from those which are normally dealt with by our applicants. His thesis is entitled "Theory of Speculation" and focuses on the application of probability to the stock market. First, one may fear that the author had exaggerated the applicability of probability as is often done. Fortunately, this is not the case. In his introduction and further in the paragraph entitled "Probability in Stock Exchange Operations", he strives to set limits within which one can legitimately apply this type of reasoning. He does not exaggerate the range of his results, and I do not think that he is deceived by his formulas.

M.T. : Poincaré does not seem convinced of the applicability of probability to the stock market.

B.B. : It must be said that Poincaré was very doubtful that probability could be applied to anything in real life. He took a different view in 1906 after the articles of Émile Borel. But prior to this, there was the Dreyfus Affair.

M.T. : What is the connection between Poincaré and the Dreyfus Affair?

B.B. : Dreyfus was accused of dissimulating his writings in a compromising document. The question was then to determine whether this document was written in a natural way, or whether it was constrained writing, in other words, "forged," a typical problem in hypotheses testing. Poincaré was called by the defense to testify in writing on the actual value of the probabilistic argument. Poincaré began by saying that the expert witness for the prosecution, Alphonse Bertillon, had committed "colossal" computational errors and that, in any case, probability could not be applied to the human sciences (*sciences morales*)[11]. If you look at Poincaré's course on probability, you will see that he is skeptical with regard to its applications.

M.T. : What especially interested Poincaré in Bachelier's thesis?

B.B. : It's the connection to the heat equation. Yet this connection was already commented upon by Rayleigh in England. Rayleigh (1842-1919) was a great physicist, the successor of Maxwell at Cambridge and a specialist in random vibrations. He received the Nobel Prize in 1904. Rayleigh had made the connection between the problem of random phase and the heat equation [106,107]. You are adding n oscillations together. The simplest version of this

[11] The transcript appeared in the newspaper *Le Figaro* on September 4, 1899. Poincaré's letter, concerning Bertillon's way of reasoning, was addressed to Painlevé who was a defense witness. Painlevé read it in court. Here is what Poincaré writes around the end of his letter: *None of this is scientific and I do not understand why you are worried. I do not know whether the defendant will be found guilty, but if he is, it will be on the basis of other proofs. It is not possible that such arguments make any impression on people who are unbiased and have a solid mathematical education.* [Translation by M.T.].

is coin tossing. One of Bachelier's proofs (he had a number of different arguments) is a bit like that. On the other hand, what Rayleigh did not see at all, and what Bachelier saw, and Poincaré understood and appreciated, was the exploitation of symmetries, the reflection principle, which leads to the law of the maximum. It's something that probably comes from Bertrand [27]. Poincaré was undoubtedly the only one capable of quickly understanding the relevance of Bachelier's method to the operations of the Stock Exchange because, as of 1890, he had introduced in celestial mechanics a method, called the *chemins conséquents*, which involves trajectories.

M.T. : Is the reflection principle attributable to Bertrand?

B.B. : For coin tossing, yes. The purely combinatorial aspect of the reflection principle is due to Désiré André, a student of Bertrand. Désiré André was a mathematician, professor in a parisian *lycée* . He had written his thesis, but was never able to obtain a position at the University of Paris. He did some very fine work in combinatorics (1870-1880). The reflection principle in gambling losses can already be found in Bertrand [27], but especially in Émile Borel. But the continuous time version is not obvious. Evidently, Bachelier obtained it in a heuristic fashion, but this is nonetheless remarkable.

M.T. : Désiré André discovered the reflection principle. Wasn't he then the first to see trajectories since the reflection principle is based on them?

B.B. : The argument in Désiré André involves combinatorial symmetry but not time or trajectory, but he is obviously not far away. Trajectories are implicit in the work of almost all the classical probabilists, but they do not take the ultimate step of making them explicit. Things would have been different, had they done so. For them, these are combinatorial formulas. Today our view is distorted. In coin tossing, we see the trajectories rise and fall. At that time, this was not the case.

M.T. : Bachelier learned probability in Poincaré's course. Do the lecture notes still exist?

B.B. : Yes, they do (see reference [102]). There are two editions, the first is from 1896, the second from 1912, the year of Poincaré's death. The 1912 edition is very interesting. The one of 1896, which Bachelier must have read, is less so. Bachelier referred primarily to Bertrand's book [27], which appeared in 1888. Bertrand is a controversial figure. He gave us "the Bertrand series", "the Bertrand curves", etc. He died in 1900, the year of Bachelier's thesis. He was professor of mathematical physics at the Collège de France. He taught a course on probability all his life, for he was jointly professor at the École Polytechnique, and his book is very brilliant.

M.T. : Did Poincaré know of Rayleigh's results?

B.B. : Not at all. Rayleigh's works on random vibrations began in 1880 and ended the year of his death in 1919. (The second edition of his book

[106], dated 1894, contains many results on the subject.) Rayleigh's articles were published in English journals, which were not read in France. At that time, the French did not read English. French physics then was in a state of slumber. It's Pólya [104], then in Zürich, Switzerland, who in 1930 made Rayleigh's results known in Paris. Pólya read widely. He became interested in geometric probability in 1917, and in road networks during the 20s.

M.T. : But I suppose that after Einstein, one made the connection with what Rayleigh did.

B.B. : These were different fields. Their synthesis occurred when probability was being revived in the 1930s. One then realized that all this was somewhat similar but belonging to different scientific cultures.

M.T. : After his thesis, did Bachelier want to do something else?

B.B. : No, not at all. When he discovered diffusion, it was a revelation, a fascination that never left him. These were ideas that had been around since Laplace (1749-1827). Laplace went from differential equations to partial derivatives. He had no problem with that. It was only analysis with a combinatorial perspective. Bachelier was of a physical mind set, very concrete. He could see the stock fluctuations. They were right before his eyes. And that changed his point of view. He was in an original, unique position. Rayleigh did not have this vision. He saw vibrations. Bachelier saw trajectories. From that moment on, Bachelier committed all his energies to the subject, as far as we can determine. This can be seen by looking at the manuscripts that are in the Archives of the Academy of Science. The formulas are calligraphed as though they were works of art (while the proofs are slapdashed). He was never to cease until his death in 1946. As soon as he defended his thesis, he published an article [6] in 1901, where he revised all of the classic results on games with his technique of approximation by a diffusion (as it is now called). He corrected Bertrand's book in large part, and he completely rewrote everything while adopting as he said, a "hyperasymptotic" view. For according to Bachelier, Laplace clearly saw the asymptotic approach, but never did what he, Bachelier, had done.

M.T. : The asymptotic approach deals with the Gaussian limit. The hyperasymptotic one concerns limits of trajectories, which is continuity perceived from a distance.

B.B. : He did it in a very clumsy manner, for he wasn't a true mathematician. But Kolmogorov [76] in 1931[12] and Khinchine [75] in 1933[13]and the

---

[12] See below.

[13] This is what Khinchine [75] writes (page 8):

> *This new approach differs from the former, in that it involves a direct search for the distribution function of the continuous limiting process. As a consequence, the solution appears as a proper distribution law (and not, as before,*

post-war probabilists understood the richness of the approximation-diffusion point of view.

M.T. : But these techniques did not exist at the time of Bachelier.

B.B. : No, but there is a freshness in the point of view and enthusiasm. He therefore continued to work, and he tried to obtain some grants. There were some research grants in France during that period, an invention attributable to the bond holders. A few among them had no descendants and bequeathed their bonds to the university. The first research grants date back to 1902. Before that, they did not exist. That's why research in France was strictly marginal. It was only at the Université de Paris that research was done, and even there not that much.

M.T. : Did Bachelier have any forerunners at the Exchange?

B.B. : There was Jules Regnault who published a book [111] in 1863 (see [70]). Forty years before Bachelier, he saw that the square-root law applied, namely that the mean deviation[14] is expressed in terms of square-root of time. It's a book on the philosophy of the Exchange that is quite rare. I know only of one copy, at the *Bibliothèque Nationale*[15].

M.T. : To find that law without an available mathematical structure means that it must have been observed empirically.

B.B. : The reason that Regnault gave is curious (the radius of a circle where time corresponds to the surface...)[16]. But he verified the square-root law on stock prices. How he found it, I don't know. Regnault is obviously not someone who studied advanced mathematics. I tried to see whether he got his *baccalauréat*, but I could not find this. No doubt he studied alone, probably the works of Quetelet and perhaps Cournot[17]. We still know nothing of this

---

as a limit of distribution laws). Bachelier [5,12] was the first to take this new approach, albeit with mathematically inadequate means. The recent extensive development and generalisation of this approach by Kolmogoroff [76,77] and de Finetti [46,45] constitute one the most beautiful chapters dealing with probability theory ...

[Translated from the German. The reference numbers are ours.]

[14] L'*écart moyen* in French. Regnault does not provide a formal definition but the term seems to refer to the average of the absolute deviations of prices between two time periods. It was translated incorrectly as "standard deviation" in [119].

[15] There is also one copy at the *Library of Congress* in Washigton D.C. The card catalogue indicates that Jules Regnault died in 1866.

[16] Excerpts are given below.

[17] Adolphe Quetelet (1796-1874) was influenced by Laplace and Fourier. He used the normal curve in settings different from that of the error law [105]. Antoine Augustin Cournot (1801-1877) wrote [43] but also [42], where he discusses supply and demand functions.

Regnault, who would have been the Kepler of the Exchange just as Bachelier would have been its Newton (relatively speaking).

M.T. : Who published Regnault's book – the Exchange?

B.B. : There is a gigantic body of literature on the Exchange. But these are not interesting books ("How to Make a Fortune", etc.). There's Regnault's book which is unique, and which we know about. Émile Dormoy, an important French actuary, quotes it[18] in 1873 in reference to the square root law (see [49]). The stockbrokers took Regnault's book into account and if you look at the finance courses of the end of the 19th century, they do refer to the square-root law.

M.T. : So Bachelier must have been familiar with that law.

B.B. : Certainly – in the same way that Bachelier knew Lefèvre's diagrams, which represent the concrete operations of the Exchange[19]. One could buy and sell the same product at the same time in different ways. There is a graphic means of representing this. Bachelier's first observations are based on these diagrams.

M.T. : Does all of this apply only to bonds?

B.B. : Yes.

M.T. : Bonds must then have been issued on a regular basis?

---

[18] Dormoy writes ([49], page 53):

> *In order to get an idea of the real premium on each transaction, one must estimate the mean deviation of prices in a given time interval. But following the observations made and summarized a long time ago by Mr. Jules Regnault in his book titled Philosophie de la Bourse, the 30 day mean deviation is 1.55 francs for the rente. For time intervals that are either longer or shorter than a month, the mean deviation of prices is proportional to the square root of the number of days.*

[Translation by M.T.].

[19] Henri Lefèvre was born in Châteaudun in 1827. He obtained a university degree in the natural sciences in 1848. Not finding a teaching position, he worked as an economics correspondent for several newspapers. He later became the chief editor of *El eco hispano-americo*, a newspaper with focus on South America. Lefèvre in 1869, was one of the founders of *l'Agence centrale de l'Union financière* and his books on the stockmarket [83,85] date from that period. He was well acquainted with the economic life of the time and his diagrams are quite clever (see [69]). These diagrams were rediscovered independently by Léon Pochet [101], a graduate from the *École Polytechnique*, but Lefèvre complains and claims priority [84]. Lefèvre then became a full member of the society of actuaries and worked at the *Union*, one of the most important insurance companies in Paris.

B.B. : For example, the Germans financed the war of 1870 by issuing loans in Paris and the French paid "reparations" to the Germans after the war by a loan of five billion francs underwritten at the Paris Exchange. The large networks of railroads were financed by loans underwritten in Paris, etc.

M.T. : Where did Bachelier publish?

B.B. : Until 1912 Bachelier published his works thanks to the support of Poincaré, for it was necessary that someone recommend them to the *Annales de l'École Normale Supérieure* or to the *Journal de Mathématiques Pures et Appliquées*. These were important journals. But Bachelier's articles were not read. And though Poincaré in the end clearly did not read them, he encouraged him

M.T. : Was Bachelier's thesis published?

B.B. : It was published in the *Annales de l'École Normale Supérieure* [5] in 1900.

M.T. : It was also translated into English and reprinted in 1964 in the book, *The Random Character of Stock Market Prices* [41].

B.B. : What is curious is that Émile Borel, who was a prominent mathematician and who was part of the establishment, never took an interest in Bachelier. His interest was in statistical physics, in conjunction with the theory of kinetics and the paradox of irreversibility. Borel published his first works on probability [30] in 1905.

M.T. : Was he younger than Bachelier?

B.B. : No, they were about the same age. Borel born in 1871, Bachelier in 1870. Borel surely was very interested in probability, but not in Bachelier. Borel occasionally had to report on Bachelier's requests for grants. He always wrote favorable reports, for Bachelier had little money, but without ever taking any interest in his works (as far as I know).

M.T. : But Bachelier worked at the Exchange?

B.B. : Perhaps, but he must have made a very modest salary. Borel had a prominent position on the Council of the Faculty of Sciences. Each time that Bachelier submitted a request, Borel wrote a favorable report. These were small sums of money. I believe he received 2000 francs four times. This was in gold francs, but it was a small sum. So Bachelier, beginning in 1906-1907, obtained small grants three or four times like that. It was then that he must have written his enormous treatise on probability, published at his own expense [12]. But, in that book, he only went over his articles.

M.T. : He wrote an article on diffusions after his thesis. Was it interesting?

B.B. : Yes, it's an article published in 1906 entitled "On continuous probability" (cf. [7]). It's an extraordinary article. He had two major accomplishments, his thesis and this.

M.T. : Was Bachelier rather isolated before the First World War?

B.B. : De Montessus[20] [47] published a book in 1908 on probability and its applications, which contains a chapter on finance based on Bachelier's thesis. Bachelier's arguments can also be found in the 1908 book of André Barriol[21] [25] on financial transactions. And there is also a popularizing book on the stock market by Gherardt [60], where Regnault and Bachelier are quoted[22]. But yes, Bachelier was essentially isolated. In those years he remained in Paris. He seemed to have no interactions with anyone.

M.T. : But how is it that Émile Borel had so much power to award grants? Wasn't he also very young?

B.B. : Borel defended his doctorate in 1894 at the age of 23. He was exceptional. He was appointed to the Sorbonne at 25, something unprecedented, since most appointments to the Sorbonne took place after one turned fifty. Borel was first in everything. He married the daughter of Paul Appell, dean of the Faculté des Sciences de Paris.

M.T. : Appell of polynomial fame?

B.B. : Yes. Appell was an important mathematician. Borel wrote extensively, but he doesn't seem to have paid attention to Bachelier. Borel took a great interest in Probability. In 1912 (cf. [33]), he wrote that he wanted to

[20] Robert de Montessus (1870-1937) was professor at the *Faculté Catholique des Sciences* of Lille and at the *Office National Météorologique*. In 1905 he wrote a thesis on continuous algebraic functions, which was awarded the "Grand Prix des Sciences Mathématiques" in 1906.

[21] Alfred Barriol (1873-1959) graduated from the *École Polytechnique* in 1892 and became an economist and actuary. He was the first professor of finance at the *Institut de Statistique* of the University of Paris and financial advisor to several french governments. Whereas the book of de Montessus [47] did not have much success, the one by Barriol [25] was used by generations of students in finance and insurance.

[22] Maurice Gherardt did not belong to a scientific organization. He wrote books entitled *Vers la fortune par les courses, guide pratique du parieur aux courses de chevaux...exposé théorique et pratique d'une méthode rationnelle et inédite de paris par mises égales permettant de gagner 4000frs par an avec 500frs de capital* (Paris: Amat, 1906); *La vie facile par le jeu à la roulette et au trente-et quarante* (Paris: Amat, 1908); *Le gain mathématique à la Bourse; la spéculation de bourse considérée comme un jeu de pur hasard, théorie mathématique de la probabilité en matière de cours, écarts et équilibres, conjectures alternantes, tableaux et graphiques à l'usage des spéculateurs, exposé théorique d'une méthode de spéculation assurant un bénefice considérable et continuel* (Paris: Amat, 1910), which is [60].

dedicate all of his energy to the development of applications of probability, and he succeeded. He viewed probability as a general philosophy, an approach to understanding the sciences, in particular, physics. But Bachelier's appeared to him to have little importance, because this business of the Stock Exchange was not too serious. And this business of hyperasymptotic diffusion, just did not interest Borel who was a brilliant thinker. He undoubtedly judged it pointless, since Stirling's formula sufficed for games. But Borel directed Francis Perrin's thesis on Brownian motion and its applications to physics[23]. It's a remarkable thesis published in 1928. Borel is somewhat paradoxical. He was a powerful mathematician and a founder of the modern theory of functions. On the other hand, Borel was very elitist. Do you understand what "elitist" means within the French context? It means that Bachelier was unimportant.

M.T. : Why did Bachelier write a book?

B.B. : It was his lecture notes [12]. Bachelier was allowed to teach an open but unpaid course on probability at the University of Paris from 1909 until 1914[24]. He also wrote another book which appeared in 1914, entitled *Game, Chance and Randomness* [15], which proved very popular. In any case, the war in 1914 stopped all these scientific activities.

M.T. : Was he drafted?

B.B. : Yes, he served through the entire war and was promoted to lieutenant. In a manner of speaking he had a "good war". The war killed many young mathematicians. This presented new career opportunities for Bachelier. From 1919, Bachelier was lecturing at the universities of Besançon (1919-1922), Dijon (1922-1925) and Rennes (1925-1927). The position of *chargé de cours* (lecturer) was without tenure but it was paid and relatively stable. The lecturer replaces a professor who is away or whose position is temporarily vacant.

M.T. : Did Bachelier apply for a permanent position?

B.B. : René Baire's chair in differential calculus in Dijon became available in 1926 and Bachelier applied for it, at the age of 56. In the provincial

[23] Francis Perrin (1901-1992), the son of the Nobel prize laureate Jean Perrin, did not receive the usual schooling. Together with the children of Marie Curie and those of Paul Langevin, he was tutored privately by the best scientists of the time. Émile Borel taught him Mathematics (Borel was a close friend of his father since their days at the École Normale Supérieure). After his theses, one in Mathematics, the other in Physics, Francis Perrin became a professor at the Sorbonne and then at the Collège de France. As high commissioner of atomic energy, he played a major role in designing the French nuclear policy of the 50s and 60s.

[24] Borel taught a probability course [32] twice in 1908 and 1909 and it is likely that this is the course that Bachelier took over. After the First World War, in 1919, Borel taught the course again after transferring from the chair in function theory that he had held since 1908 to the chair in probability and mathematical physics, then held by Boussinesq.

universities, there were two chairs in mathematics: a differential calculus chair and a mechanics chair. Those were the two required courses for the degree. The mechanics chair in Dijon was occupied by a well known mathematician, Maurice Gevrey[25], a specialist in partial differential equations. He was to write a report on Bachelier. He must have gone over Bachelier's writings very quickly since it was not his own theory and it looked strange. Bachelier, in fact, often took shortcuts, not paying much attention to questions of normalization and of convergence.

M.T. : This was undoubtedly a matter of simplification.

B.B. : Yes, indeed. Reading Bachelier, one occasionally gets the impression that he considers that Brownian motion is differentiable though it is not. Gevrey had the 1913 article published in the *Annales de l'École Normale Supérieure* [13], where Bachelier asks the following: "A geometric point M is moving at a speed v whose velocity is constant but where direction keeps varying randomly. The position of M is projected on the three rectangular axes centered at its initial position. What is the probability that at time t, the point M will have given coordinates x, y, z ?". The answer is that the point M moves according to Bachelier's Brownian motion, but this is not possible if the speed is constant and finite, as Bachelier seems to suppose. Indeed, if we place ourselves in dimension 1, the speed of Bachelier's point M is at every instant either $+v$ or $-v$, with probability 1/2 each. Its position at time t is $\sum \pm vdt$. Therefore the mean of its position is 0 and the variance of its position is $\mathrm{Var}(\sum \pm vdt) = (v\,dt)^2 t/dt$, of the order of dt. Since dt is infinitesimal, the variance is negligeable and there is no motion. The point M can never leave its original position. In order that there be motion, one must normalize v by $1/\sqrt{dt}$, and therefore give to M an infinite speed, which will allow it to move. Normalizing v by $1/\sqrt{dt}$ means setting $v = v_0/\sqrt{dt}$, where $0 < v_0 < \infty$, and thus replacing the increments vdt by $(v_0/\sqrt{dt})dt = v_0\sqrt{dt}$. This gives $\mathrm{Var}(\sum \pm vdt) = \mathrm{Var}(\sum \pm v_0\sqrt{dt}) = (v_0^2 dt)t/dt = v_0^2 t$, a finite and non-zero quantity. That's what Bachelier had done in his thesis, within the context of coin tossing, but he did not reproduce this reasoning in 1913.

M.T. : But did Gevrey know that?

B.B. : No, he had no idea, but he must have read this page and gone through the roof. For Bachelier, it was his usual way of talking.

M.T. : It was a true misfortune then.

B.B. : It fell to the wrong referee. He wrote a devastating report. But since he was not competent in probability, he sent it to Paul Lévy[26]. Lévy,

[25] Maurice Gevrey (1884-1957) was an important mathematician working on parabolic partial differential equations, following Hadamard [64]. The existence and uniqueness theorem of Markov processes in Feller [53] is based on the theory of Hadamard and Gevrey. His collected works can be found in [59].

[26] Together with Kolmogorov and Émile Borel, Paul Lévy (1886-1971) is one of the most important probabilists of the first half of the twentieth century. He

at that time (1926), had just published an important work on probability (cf. [86]). Gevrey knew him very well, for they were both students of Jacques Hadamard. Hadamard was professor at the Collège de France and was surrounded by many brilliant students who formed a type of caste. Obviously, Gevrey wanted nothing to do with Bachelier. Gevrey sent Lévy the incriminating page asking him (I'm paraphrasing) "What do you think of this?" Lévy answered, "You're right, it doesn't work," having read nothing but this famous page. One can imagine that Bachelier's goal in his 1913 article was to show that his modeling of stock market performance is equally applicable to the Brownian motions whose importance had just been pointed out by Jean Perrin in the context of the motion of molecules. Indeed, in 1913, Jean Perrin published "The Atoms" (cf. [100]), aimed at a popular audience, in which he talks about his experience with Brownian motion. One could just as well imagine that this is also why Poincaré, who had read Bachelier's thesis, recommended an article of this type to the *Annales de l'École Normale Supérieure*, in spite of the "mistake" revealed by Lévy and Gevrey. This "mistake" is ultimately nothing but an audacious metaphor to Bachelier's 1900 thesis *The Theory of Speculation.* Obviously, Lévy never knew anything about that.

M.T. : Did Bachelier learn about Lévy's intervention?

B.B. : Yes, he was very upset. He circulated a letter accusing Lévy of having blocked his career and of not knowing his work[27].

M.T. : Do we have Lévy's text?

B.B. : I never saw the Lévy–Gevrey letter. I don't know whether it still exists. On the other hand, what we do have of Lévy are two or three sentences in his books, in that of 1948 on Brownian motion [89][28] and in his 1970 book of memoirs [90]. In the latter, Lévy says he is sorry that he ignored Bachelier's work because of an error in the construction of Brownian motion, but he does

received his doctorate in 1912 (Picard, Poincaré, and Hadamard were on the committee). Paul Lévy contributed not only to probability theory, but also to functional analysis. He was professor at the École Polytechnique from 1920 until his retirement in 1959.

27 Several copies of this letter were found by Ms. Nocton, the head of library at the *Institut Henri Poincaré* in Paris. The article Courtault et. al. [44] contains a number of excerpts from this letter.

28 Here are the footnotes in [89] (second edition) about Bachelier, which mention:
-page 15 footnote (1): the priority of Bachelier over Wiener about Brownian motion.
-page 72 footnote (4): the priority of Bachelier over Kolmogorov about the relation between Brownian motion and the heat equation.
-page 193 footnote (4): the priority of Bachelier over Lévy about the law of the maximum, the joint law of the maximum and Brownian motion, and the joint law of the maximum, the minimum and Brownian motion.

not tell us what the error is, and for good reasons[29]. It seems that it is a late value judgement. Hence, a few cryptic notes on Bachelier which in summary state that "I erred, but Bachelier did too". There is also a letter that Lévy wrote to Benoit Mandelbrot[30]. This is what Lévy writes, about Bachelier:

> *I first heard of him a few years after the publication of my Calcul des Probabilités, that is, in 1928, give or take a year. He was a candidate for a professorship at the University of Dijon. Gevrey, who was teaching there, came to ask my opinion of a work Bachelier published in 1913 ... Gevrey was scandalized by this error. I agreed with him and confirmed it in a letter which he read to his colleagues in Dijon. Bachelier was blackballed. He found out the part I had played and asked for an explanation, which I gave him and which did not convince him of his error. I shall say no more of the immediate consequences of this incident.*
> *I had forgotten it when in 1931, reading Kolmogorov's fundamental paper, I came to "der Bacheliers Fall"[31]. I looked up Bachelier's works, and saw that this error, which is repeated everywhere, does not prevent him from obtaining results that would have been correct if only, instead of v = constant, he had written $v = c\tau^{-1/2}$, and that, prior to Ein-*

[29] Lévy [90] writes (p. 97):

> *The linear Brownian motion function $X(t)$ is often called the function of Wiener. Indeed, it is N. Wiener who, in a celebrated 1923 article, gave the first rigorous definition of $X(t)$. But it would not be right not to remember that there were forerunners, in particular the French Louis Bachelier and the important physicist Albert Einstein. If the work of Bachelier, which appeared in 1900, has not attracted attention, it is because, on one hand, not everything was interesting (this is even more true for his large book "Calcul des Probabilités," published in 1912), and because on the other hand, his definition was at first incorrect. He did not get a coherent body of results about the function $X(t)$. In particular, in relation to the probability law of the maximum of $X(t)$ in an interval $(0, T)$ and also in relation to the fact that the probability density $u(t, x)$ of $X(t)$ is a solution of the heat equation. This latter result was rediscovered in 1905 by Einstein who, evidently, did not know about Bachelier's priority. I myself did not think it useful to continue reading his [Bachelier's] paper, astonished as I was by his initial mistake. It is Kolmogorov who quoted Bachelier in his 1931 article ... and I recognized then the injustice of my initial conclusion.*
>
> [Translation by M.T.].

[30] Letter dated January 25, 1964 from Paul Lévy to Benoit Mandelbrot, in which he recounts the Gevrey incident. Mandelbrot includes excerpts of this letter in a very interesting biographical sketch of Bachelier in [93], pages 392-394. According to Mandelbrot (private communication), the original copy of this letter may be lost.

[31] Der Fall Bacheliers (Bachelier's case).

stein and prior to Wiener, he happens to have seen some important properties of the so-called Wiener or Wiener–Lévy function, namely, the diffusion equation and the distribution of $\max_{0\le\tau\le t} X(t)$.[32]

In this matter with Gevrey, Lévy did not bother to understand what Bachelier wanted to say, namely that once and for all, Brownian motion existed since the time of his thesis where the normalizations were included and the convergences established. The irony of the story is that, while Lévy would publish his beautiful works on Brownian motion beginning in 1938, the same mathematicians (starting with Hadamard) would much mock this $\pm v_0/\sqrt{dt}$ which represents for Lévy as for Bachelier a different kind of speed that"varies constantly in a random way".

M.T. : The British economist John Maynard Keynes seems to have quoted Bachelier.

B.B. : He did so in 1921 in his book on probability [74], quoting Bachelier's texts [12,15] but only in the context of statistical frequency and Laplace's rule of succession[33]. Bachelier's work on finance is not mentioned.

[32] Another excerpt from this letter will be quoted below.

[33] Keynes [73] had reviewed Bachelier's text *Calcul des Probabilités* [12] in 1912. He writes:

> *M. Bachelier's volume is large, and makes large claims. His 500 quarto pages are to be followed by further volumes, in which he will treat of the history and of the philosophy of probability. His work, in the words of the preface, is written with the object, not only of expounding the whole of ascertained knowledge on the calculus of probabilities, but also of setting forth new methods and new results which represent from some points of view une transformation complète de ce calcul. On what he has accomplished it is not very easy to pass judgment. The author is evidently of much ability and perseverance, and of great mathematical ingenuity; and a good many of his results are undoubtedly novel. Yet, on the whole, I am inclined to doubt their value, and, in some important cases, their validity. His artificial hypotheses certainly make these results out of touch to a quite extraordinary degree with most important problems, and they can be capable of few applications. I do not make this judgment with complete confidence, for the book shows qualities of no negligible order. Those who wish to sample his methods may be recommended to read chapter ix, on what he terms Probabilités connexes, as a fair specimen of his original work.*

Keynes notes at the beginning of his review:

> *There never has been a systematic treatise on the mathematical theory of probability published in England, and it is now nearly fifty years since the last substantial volume to deal with this subject from any point of view (Venn's Logic of Chance, 1st edit., 1866) was brought forth here. But a year seldom passes abroad without new books about probability, and the year 1912 has been specially fertile.*

M.T. : Did Bachelier teach in a *lycée*?

B.B. : No, he did not have the necessary diplomas. You had to pass the "aggregation", the competitive examination for *lycée* teachers. He taught only at the university.

M.T. : I've also heard it said that Bachelier made errors while teaching.

B.B. : Yes, it's a rumor that's circulating but I do not know on what it is based. A brilliant candidate, Georges Cerf, obtained the Dijon chair. But after one year, Cerf left for the University of Strasbourg, which was, after Paris, the most famous university in France[34]. Since Cerf had graduated from the École Normale Supérieure (he was *normalien*) and was a specialist on partial differential equations, Gevrey's choice was obvious. Bachelier had no chance.

M.T. : What then happened to Bachelier?

B.B. : Fortunately, Bachelier was saved. He had been lecturer at Besançon and when a position became available in 1927, he obtained it. At Besançon there was a very innovative mathematician who is unfortunately no longer well known, Jules Haag. Haag was at Besançon because he headed the school of chronometry (Besançon is close to Switzerland). In probability, Haag has

He then reviews four books, Poincaré [102], Bachelier [12], Carvallo [39] and Markov [95]. This is what he writes about (the second edition) of Poincaré's text:

> *Poincaré's Calcul des Probabilités originally appeared in 1896 as a reprint of lectures. This new edition includes the whole of the earlier edition, but is now rearranged in chapters according to the subjects treated, in place of the former awkward arrangement into lectures of equal length...*
> *The mathematics remain brilliant and the philosophy superficial – a combination, especially in the parts dealing with geometrical probability, which makes it often suggestive and often provoking. On the whole there is not a great deal in the book which cannot be found, substantially, elsewhere. Poincaré had to lecture on probability, and this is what without giving any very profound attention to the subject, he found to say. This new edition must have been almost the last material to leave his hands before his lamented death. The immense field of Henri Poincaré's achievements had made him one of the greatest mathematicians in Europe, and it must always be a matter of regret to statisticians that modern statistical methods, with their almost equal dependence on mathematics and on philosophy and logic, had not found their way to France in time to receive illumination from his brilliant and speculative intellect. This book has no reference to any of the researches, either German or English, which seek by the union of probability and statistics to forge a new weapon of scientific investigation.*

[34] Baire had been very sick and was often replaced by lecturers. Cerf had taught previously many times in Dijon, in particular from 1919 to 1922 (Bachelier did so later, from 1922 to 1925). René Lagrange got the position in Dijon in 1927 after Cerf was appointed in Strasbourg.

introduced among other things the notion of an exchangeable sequence [63], independently of Finetti. He did some very interesting studies on stochastic algorithms applied to the adjustments that must be done when shooting big guns [62]. The fact remains that he welcomed Bachelier. So the story that Bachelier taught poorly or that he made errors in his teaching, may not be fair. If that story were true, Haag would not have recommended him at Besançon.

M.T. : Where does it come from?

B.B. : I don't know. I know that it's something that had been said about him, but there is contradictory testimony, and in particular at Besançon, where he remained for almost fifteen years teaching analysis. It was probably not a very advanced course, but he must have given it in a very conscientious manner. He undoubtedly found teaching difficult. He was not capable of writing a calculation to the end without notes. In France, we do not like people who copy their notes onto the blackboard.

M.T. : Is this still the case?

B.B. : Yes, but a bit less today because students are less docile than in the past. A course for which there are no prepared notes rapidly becomes a vague and empty discourse with occasional incomprehensible flashes. Borel and Hadamard, contemporaries of Bachelier, brilliant representatives of the French mathematical elite, had reputations in the 20s and 30s of never ending a calculation nor a proof. Students always appreciate a calculation that is well done without notes, but they do not tolerate calculations that come up short. The attitude to lecturing on mathematical subjects at French universities has therefore evolved. There are innumerable anecdotes on the subject. One of the best that I know occurred in the 30s at the time when Einstein decided to leave Berlin. All the great countries offered him a position in their most prestigious universities. In France, on the recommendation of Langevin (the author in 1908 of the stochastic differential equation of Brownian motion [80]), the government decided to create a new chair for Einstein at the Collège de France, the most prominent institution of learning in the country. To Langevin, who was a professor at the Collège de France, and who invited him to accept, Einstein replied that they were doing him a great honor, but his scientific culture was so reduced that his lectures would be a laughing stock. Any ordinary student would know what he knew[35], and he felt like a gypsy who cannot read music and is asked to become first violinist in a symphonic orchestra. Einstein preferred Princeton where he didn't have to teach (with

[35] He writes: *Ich bin eben kein Könner und kein Wisser sondern nur ein Sucher* (In fact, I am neither a man of action nor a man of knowledge but only a seeker).

or without notes)[36]. The letter to Langevin is found in Einstein's correspondence.

M.T. : Did Kolmogorov[37] read Bachelier?

B.B. : Yes. It was Bachelier's article [7] and its extension to the multidimensional case [10] that prompted Kolmogorov toward the end of the 20s to develop his theory, the analytical theory of the Markov processes [76,78]. This is what Kolmogorov wrote in 1931 ([78], Volume 2, p. 63)[38]:

> *In probability theory one usually considers only schemes according to which any changes of the states of a system are only possible at certain moments* $t_1, t_2, \ldots, t_n, \ldots$ *which form a discrete series. As far as I know, Bachelier*[39] *was the first to make a systematic study of schemes in which the probability* $P(t_0, x, t, \mathcal{E})$ *varies continuously with time* t. *We will return to the cases studied by Bachelier in §16 and in the Conclusion. Here we note only that Bachelier's constructions are by no means mathematically rigorous.*

M.T. : Thus, at the time, Kolmogorov knew Bachelier's work better than did other mathematicians[40].

[36] Ironically, a few years later, the situation was reversed. Langevin was arrested in October 1940 by the Gestapo and Einstein then wrote to the American Ambassador William C. Bullitt at the Department of State asking him to offer refuge to Langevin in the U.S.A.

[37] Andrei Nikolaevich Kolmogorov (1903-1987) was one of the greatest mathematicians of the twentieth century. He made fundamental contributions to many areas of pure and applied mathematics, such as trigonometric series, set theory, approximation theory, logic, topology, mechanics, ergodic theory, turbulence, population dynamics, mathematical statistics, information theory, the theory of algorithms and, naturally, probability theory. He is particularly well-known for setting the axioms of probability, for the development of limit theorems of independent random variables and for the analytic theory of Markov processes. Kolmogorov was also very interested in the application of mathematics to the social sciences and linguistics and also in the history and pedagogy of mathematics. (See the overview article [117].)

[38] One of the major contributions of Kolmogorov in his 1931 article is to make rigorous the passage from discrete to continuous schemes. He does that by extending to this setting Lindeberg's method [92] for proving the Central Limit Theorem. In this way the "hyperasymptotic" theory of Bachelier becomes rigorous. One can then derive the parabolic differential equations of Kolmogorov from the difference equations which hold when time is discrete.

[39] I. 'Théorie de la spéculation', *Ann. École Norm. Supér.* **17** (1900), 21; II. 'Les probabilités à plusieurs variables', *Ann. École Norm. Supér.* **27** (1910), 339; III. *Calcul des probabilités*, Paris, 1912.

[40] Kolmogorov told Albert Shiryaev that he has been very influenced by Bachelier (private communication from Shiryaev) [M.T.].

B.B. : There are two important sources for Kolmogorov, Bachelier and Hostinský. Bachelier is a known source; Hostinský, much less so. Hostinský was a Czech mathematician who revived the theory of Markov chains. Markov chains as done by Markov, were meant to generalize the classical probability results to situations where there was no independence. But the development of the physical aspect of chains is due in large part to Hostinský in the last years of the 20s. To understand Kolmogorov's article [76] of 1931, where we find Kolmogorov's equation, we must refer to the two sources, Bachelier and Hostinksý. The conditions of the ergodic theorem are found in Hostinský [65,66], and the idea of continuity in probability under the condition stated by Chapman-Kolmogorov is found in Bachelier [7]. Bachelier considers a case that is not quite general, for he supposes homogeneity.

M.T. : What did Hostinský think of Bachelier?

B.B. : Not much. Hostinský wrote to Fréchet[41] that it was not worth reading Bachelier because there were too many mistakes. In fact, the mathematicians of the 30s who read Bachelier felt that his proofs are not rigorous and they are right, because he uses the language of a physicist who shows the way and provides formulas. But again, there is a difference between using that language and making mistakes. Bachelier's arguments and formulas are correct and often display extreme originality and mathematical richness.

M.T. : What did Bachelier do at Besançon?

B.B. : Bachelier published practically nothing. Obviously he must have been preparing his courses. He was at Besançon between 1927 until his retirement in 1937. He began publishing again once he left Besançon. He published three books at his own expense with Gauthier-Villars [21–23] which are revisions of his pre-war works, but most importantly, in 1941, he published an article [24] at the *Comptes Rendus* that was extremely innovative. It's that paper that Paul Lévy read.

M.T. : How did this happen?

B.B. : Lévy began to take an interest in Brownian motion toward the end of the 1930s through the Polish school, in particular through Marcinkiewicz who was in Paris in 1938. He rediscovered all of Bachelier's results which he had never really seen earlier[42]. Lévy had become enthralled with Brownian

[41] Fréchet archives at the *Académie des Sciences*, Institut de France, quai Conti.
[42] Paul Lévy writes in his book of memoirs [90], p. 123:

> *I learned only after the 1939-1945 war that L. Bachelier had published a new book on Brownian motion just before the war. I do not exclude the possibility that there may be in this book some of the results of my [later] paper. Being busy with other work, I have never checked this.*

[Translation by M.T.]

motion. The book on stochastic processes [89] that he undertook to write was not published until 1948. Lévy was Jewish, and therefore forbidden to publish during the war.

M.T. : Where was Lévy during the Second World War?

B.B. : He went to Lyon since he was professor at the École Polytechnique. The École Polytechnique had relocated to Lyon, a "free zone" under Pétain. There were racist laws. But since he was professor at a military school, he was able to continue teaching for a while. After the American landing in North Africa in 1942, the Germans invaded the free zone. The first large raid on Jews in Paris occurred in July 1942. Lévy hid under an assumed name in Grenoble, and then in Mâcon.

M.T. : Bachelier's paper was 1941.

B.B. : It was while Lévy was still at Lyon. Bachelier, who had retired to Brittany with one of his sisters, must have sent him a reprint. An annotated copy exists in the Lévy archives[43]. Lévy wrote in the margin of that copy that he had written to Bachelier and that Bachelier had told him about additional properties that he knew about. One also finds in the margin comments by Lévy about the obvious enthusiasm that Bachelier has for mathematical research (this was 1942 or thereabouts). The results in this paper of Bachelier, annotated by Lévy, are about excursions of Brownian motion and they were beyond Lévy's latest results. Here is also an excerpt of a letter from Lévy to Fréchet[44] dated September 27, 1943:

> *Concerning priority, I recently had a correspondence with Bachelier, who told me that he had published the equation attributed to Chapman in a math journal in 1906. Can you verify whether that is accurate or have your students verify it? He also gave me some indication about Brownian motion on the surface of a sphere, which would have been studied by Perrin, and I have asked Loève to verify it.*

This excerpt shows that until 1942 or 43, Lévy really knew neither Bachelier's articles from the beginning of the century, not even the thesis [99] of Francis Perrin of 1928. Lévy, who was at that time doing detailed studies of Brownian motion, at last recognized the originality of Bachelier's results. He also wrote to him and apologized[45]:

---

[43] Archives Lévy at the interuniversity mathematics library, Universités Paris VI et VII, Paris.

[44] Box 2 of the Fréchet archives at the *Académie des Sciences*, Institut de France, quai Conti, Paris.

[45] Contination of the letter dated January 25, 1964 from Lévy to Mandelbrot [93].

We became reconciled. I had written him that I regretted that an impression, produced by a single initial error, should have kept me from going on with my reading of a work in which there were so many interesting ideas. He replied with a long letter in which he expressed great enthusiasm for research.

Bachelier, who died in 1946 at the age of 76, thus corresponded with Lévy just before his death[46]. That must have been Bachelier's great satisfaction, to be read by someone, and by the best!

Epilogue

Kiyosi Itô, in Japan, was also influenced by Bachelier, more so than by Wiener[47], and in the United States, Bachelier was read by probabilists such as Paul Erdös, Mark Kac, William Feller and Kai Lai Chung[48] in the forties. But it seems that it is Paul Samuelson[49] who introduced Bachelier to economists in the 50s. This is how it happened[50]:

---

[46] Louis Bachelier died on April 28, 1946 in Saint-Servan-sur-Mer, near Saint Malo in Brittany. He is buried in the Bachelier family's plot in Sanvic, Normandy, near Le Havre.

[47] Personal communication from the economist Robert C. Merton. Itô told this to Merton during the 1994 Wiener symposium at MIT.

[48] See Erdös and Kac [52], Chung [40], and Feller [54] who writes (in a footnote, p. 323):

> *Credit for discovering the connection between random walks and diffusion is due principally to L. Bachelier (1870-). His work is frequently of a heuristic nature, but he derived many new results. Kolmogorov's theory of stochastic processes of Markov type is based largely on Bachelier's ideas. See in particular L. Bachelier Calcul des Probabilités, Paris, 1912.*

Doob [48], in his article on Kolmogorov, also writes positively about Bachelier:

> *Bachelier, in papers from 1900 on, derived properties of the Brownian motion process from asymptotic Bernoulli trial properties. His Brownian motion process was necessarily not precisely defined, but his application of the André reflection principle becomes valid for the Brownian motion process as an application of the strong Markov property. His valuable results were repeatedly rediscovered by later researchers.*

[49] Paul Samuelson received the Nobel prize in Economics in 1970.

[50] As told to M.T. by Paul Samuelson on August 14, 2000. See also [116] for a somewhat similar account. The date 1957, indicated in [116], is probably a little late because Savage's postcard must have been sent no later than 1956, the year of Richard Kruizenga's thesis [79] at MIT (Kruizenga, who was Samuelson's student, quotes Bachelier in his thesis).

Around 1955, Leonard Jimmie Savage, who had discovered Bachelier's 1914 publication in the Chicago or Yale library sent half a dozen "blue ditto" postcards to colleagues, asking "does any one of you know him?" Paul Samuelson was one of the recipients. Samuelson, however, had already heard of Bachelier. First from Stanislaw Ulam, between 1937 and 1940, who then belonged like him to the Society of Fellows at Harvard University. Ulam was a gambler by instinct. He was a topologist who later popularized Monte Carlo methods and worked on the atom bomb at Los Alamos. Samuelson also knew of Bachelier from Feller [54]. But prompted by Savage's postcard, Samuelson looked for and found Bachelier's 1900 thesis at the MIT library. Soon after, in ditto manuscripts and informal talks, Samuelson suggested using geometric Brownian motion as a model for stocks[51].

Today, a full century after his thesis, Bachelier is rightly viewed as the father of mathematical finance.

Acknowledgments: I thank Bernard Bru for having received me so graciously at his home, for granting me this interview and for his continuing valuable input. This meeting was suggested by Jean-Pierre Kahane and Christian Gilain. I am equally grateful to Franck Jovanovic for a careful reading and to Corine Astier, Catriona Byrne, Marie-France Bru, Vladas Pipiras, Jean-Francis Ravoux, Rachelle Taqqu, Gérard Vichniac and Nader Yeganefar for their assistance in formatting the original French version of this paper. And I thank Marie Isenburg for doing the English translation.

[51] The lognormal model was used in several contexts in economics. It was fashionable in Paris in the thirties and forties because of the economist Robert Gibrat [61], who used it instead of the Pareto distribution, to model income. The article Armatte [4] provides many references about that. See also Aitchison and Brown [1], Osborne [97] and Cootner [41].

Dates

1700 – 1800

Pierre Simon, marquis de Laplace	1749 – 1827	(78 years)
Robert Brown	1773 – 1858	(85 years)
Adolphe Quetelet	1796 – 1874	(78 years)

1800 – 1850

Antoine Augustin Cournot	1801 – 1877	(76 years)
Joseph Bertrand	1822 – 1900	(78 years)
Henri Lefèvre	1827 – ?	
Émile Dormoy	1829 – 1891	(62 years)
Désiré André	1840 – 1917	(77 years)
John William Strutt Rayleigh (Lord)	1842 – 1919	(77 years)
Joseph Boussinesq	1842 – 1922	(80 years)
Ludwig Eduard Boltzmann	1844 – 1906	(62 years)

1850 – 1875

Henri Poincaré	1854 – 1912	(58 years)
Paul Appell	1855 – 1930	(75 years)
Émile Picard	1856 – 1941	(85 years)
Jacques Hadamard	1865 – 1963	(98 years)
Louis Bachelier	1870 – 1946	(76 years)
Robert de Montessus	1870 – 1937	(67 years)
Jean Batiste Perrin	1870 – 1942	(72 years)
Émile Borel	1871 – 1956	(85 years)
Paul Langevin	1872 – 1946	(74 years)
Alfred Barriol	1873 – 1959	(86 years)
René Baire	1874 – 1932	(58 years)

1875 – 1900

Maurice René Fréchet	1878 – 1973	(95 years)
Albert Einstein	1879 – 1955	(76 years)
Jules Haag	1882 – 1953	(71 years)
John Maynard Keynes	1883 – 1946	(63 years)
Bohuslav Hostinský	1884 – 1951	(67 years)
Maurice Gevrey	1884 – 1957	(73 years)
Paul Lévy	1886 – 1971	(85 years)
George Pólya	1887 – 1985	(98 years)
Georges Cerf	1888 – 1979	(91 years)
Alexander Yakovlevich Khinchine	1894 – 1959	(65 years)
Norbert Wiener	1894 – 1964	(70 years)

1900 – 1925

Francis Perrin	1901 – 1992	(91 years)
Andrei Nikolaevich Kolmogorov	1903 – 1987	(84 years)
William Feller	1906 – 1970	(64 years)
Stanislaw Ulam	1909 – 1984	(75 years)
Paul Erdös	1913 – 1996	(83 years)
Marc Kac	1914 – 1984	(70 years)
Kiyoshi Itô	1915 –	
Paul A. Samuelson	1915 –	
Kai Lai Chung	1917 –	
Benoit B. Mandelbrot	1924 –	

Regnault's 1863 law on the square root of time[52]

After much thought, we realized that it is not possible to find a relation between stock market gains and losses. It is... with respect to time ... that we found a relation...

In decreasing the time periods to 5 days, 3 days, 2 days ... the mean deviations decrease steadily.

Consequently, the deviations are smaller for shorter time intervals and larger for longer time intervals.

Finally, if one tries to find how these different deviations are related to the different times in which they occur, one notices that as the period decreases by half, the deviation decreases not by half but, roughly, in the proportion 1:1.41; for a period which is three times shorter, the deviation decreases in the proportion 1:1.73, for a time period which is four times shorter, the ratio is 1:2.

There exists therefore a mathematical law which regulates the variations and the mean deviation of stock market prices, and this law, which seems never to have been noticed, is given here for the first time:

THE PRICE DEVIATION IS DIRECTLY PROPORTIONAL TO THE SQUARE ROOT OF TIME.[53]

Hence the investor who wants to sell after the deviation doubles, that is with a difference twice as large between the buy and sell price must wait four times longer, if he wants to sell with triple deviations, [he must wait] nine times longer, and so forth. One multiplies the time by the square of the deviations.

One who leaves only one day between [his buying and] selling, would sell with a deviation which is smaller by one half than one who sells every four days, three times smaller than one who sells every nine days, etc..., dividing the deviations by the square root of time.

Quite a large number of transactions is required, however, in order to make these ratios clearly apparent, and they become strictly correct when the number of transactions is exceedingly great.

Let us understand the reason for this remakable law:

The security varies but is always looking for its real price or an absolute price, which one can represent as the center of a circle whose radius represents the deviation, which may be anywhere on the surface. Time is equal to the surface and the points on the circumference represent extreme deviations. As

[52] Regnault [111], pages 49-52 (text provided by Franck Jovanovic). Translated by M.T.

[53] Capitalized in the original text.

it varies, the security moves either away from or closer to the center, and the basic notions of geometry teach us that the radii or deviations are proportional to the square root of the area, that is of time.

Why is it that the reciprocal law holds when dealing with either gravity or the oscillations of a pendulum, where [in one case] the space traveled or [in the second case] the deviation of the oscillations is proportional to the square of time? It is only because these falling bodies go from the circumference to the center, whereas the stock price in its greatest deviations, is pushed away from the center towards the circumference.

How astonishing and admirable are the ways of Providence, what thoughts come to our mind when observing the marvelous order which presides over the most minute details of the most hidden events! What! The changes in stock market prices are subject to fixed mathematical laws! Events produced by the passing fancy of men, the most unpredictable shocks of the political world, of clever financial schemes, the outcome of a vast number of unrelated events, all this combines and randomness becomes a word without meaning! And now worldly princes, learn and be humble, you who in your pride, dream to hold in your hands the destiny of nations, kings of finance who have at your disposal the wealth and credit of governments, you are but frail and docile instruments in the hands of the One who brings all causes and effects together in harmony and who, as the Bible says, has measured, weighed and parcelled out everything in perfect order.

Man bustles but God leads.

Regnault writes further:

The price of the "Rente," while fluctuating capriciously, remains influenced in final instance by constant causes. The most important one, clearly defined and whose existence is without doubt, is the interest rate. This cause, so feeble in appearance, finally dominates all others. The accidental causes [will] have totally disappeared and, however powerful their effects, however strange and irregular they appear, they always end up after a while cancelling almost completely, revealing the influence of constant and regular causes, however weak the effect [of these regular causes] is[54].

The causes for a drop [in price] are fewer than those for a rise [in price] but, while they are few in number, they make this up by their strength; so that by multiplying number by strength one would obtain a constant value.[55].

The price variations obey two distinct laws. The first is that the deviations are proportional to the square root of time ... The second is that the value [of the stock] whatever its deviation, is constantly attracted towards its average price as the square of its distance [to that price][56].

[54] [111], page 154.
[55] [111], p. 161.
[56] [111], p. 187.

Report on Bachelier's thesis (March 29, 1900)[57]

Le sujet choisi par M. Bachelier s'éloigne un peu de ceux qui sont habituellement traités par nos candidats; sa thèse est intitulée Théorie de la Spéculation et a pour object l'application du Calcul des Probabilités aux Opérations de Bourse. On pourrait craindre d'abord que l'auteur ne se soit fait illusion sur la portée du Calcul des Probabilités, comme on l'a fait trop souvent. Il n'en est rien heureusement; dans son introduction et plus loin dans le paragraphe intitulé "La probabilité dans les Opérations de Bourse", il s'efforce de fixer les limites dans lesquelles on peut avoir légitimement recours à ce genre de Calcul; il n'éxagère donc pas la portée de ses résultats et je ne crois pas qu'il soit dupe de ses formules.

Qu'a-t-on donc légitimement le droit d'affirmer en pareille matière? Il est clair d'abord que les cours relatifs aux diverses sortes d'opérations doivent obéir à certaines lois; ainsi on pourrait imaginer des combinaisons de cours telles que l'on puisse jouer à coup sûr; l'auteur en cite des exemples; il est évident que de pareilles combinaisons ne se produisent jamais, ou que si elles se produisaient elles ne sauraient se maintenir. L'acheteur croit la hausse probable, sans quoi il n'achèterait pas, mais s'il achète, c'est que quelqu'un lui vend; et ce vendeur croit évidemment la baisse probable; d'où il résulte que le marché pris dans son ensemble considère nulle l'espérance mathématique de toute opération et de toute combinaison d'opérations.

Quelles sont les conséquences mathématiques d'un pareil principe? Si l'on suppose que les écarts ne sont pas très grands, on peut admettre que la probabilité d'un écart donné par rapport au cours coté ne dépend pas de la valeur absolue de ce cours; dans ces conditions le principe de l'espérance mathématique suffit pour déterminer la loi des probabilités; on retombe sur la célèbre loi des erreurs de Gauss.

Comme cette loi a été l'objet de démonstrations nombreuses qui pour la plupart sont de simples paralogismes, il convient d'être circonspect et d'examiner cette démonstration de près; ou du moins il est nécessaire d'énoncer d'une manière précise les hypothèses que l'on fait. Ici l'hypothèse que l'on a à faire c'est, comme je viens de le dire, que la probabilité d'un écart donné à partir du cours actuel est indépendante de la valeur absolue de ce cours. L'hypothèse peut être admise, pourvu que les écarts ne soient pas trop grands. L'auteur l'énonce nettement, sans y insister peut-être autant qu'il conviendrait. Il suffit pourtant qu'il l'ait énoncée explicitement pour que ses raisonnements soient corrects.

La manière dont M. Bachelier tire la loi de Gauss est fort originale et d'autant plus intéressante que son raisonnement pourrait s'étendre avec quelques changements à la théorie même des erreurs. Il le développe dans un chapitre dont le titre peut d'abord sembler étrange, car il l'intitule "Rayonnement de

[57] *Registre des thèses de la Faculté des Sciences de Paris*, at the *Archives nationales*, 11 rue des Quatre-Fils, 75003 Paris, classification AJ/16/5537.

la Probabilité." C'est en effet à une comparaison avec la théorie analytique de la propagation de la chaleur que l'auteur a eu recours. Un peu de réflexion montre que l'analogie est réelle et la comparaison légitime. Les raisonnements de Fourier sont applicables presque sans changement à ce problème si différent de celui pour lequel ils ont été créés.

On peut regretter que M. Bachelier n'ait pas développé davantage cette partie de sa thèse. Il aurait pu entrer dans le détail de l'Analyse de Fourier. Il en a dit assez cependant pour justifier la loi de Gauss et faire entrevoir les cas où elle cesserait d'être légitime.

La loi de Gauss étant établie, on peut en déduire assez aisément certaines conséquences susceptibles d'une vérification expérimentale. Telle est par exemple la relation entre la valeur d'une prime et l'écart avec le ferme. On ne doit pas s'attendre à une vérification très exacte. Le principe de l'espérance mathématique s'impose en ce sens que, s'il était violé, il y aurait toujours des gens qui auraient intérêt à jouer de façon à le rétablir et qu'ils finiraient par s'en apercevoir. Mais ils ne s'en apercevront que si l'écart est considérable. La vérification ne peut donc être que grossière. L'auteur de la thèse donne des statistiques où elle se fait d'une façon très satisfaisante.

M. Bachelier examine ensuite un problème qui au premier abord semble devoir donner lieu à des calculs très compliqués. Quelle est la probabilité pour que tel cours soit atteint avant telle date? En écrivant l'équation du problème, on est conduit à une intégrale multiple où on voit autant de signes $\int$ superposés qu'il y a de jours avant la date fixée. Cette équation semble d'abord inabordable. L'auteur la résout par un raisonnement court, simple et élégant; il en fait d'ailleurs remarquer l'analogie avec le raisonnement connu de M. André au sujet du problème du dépouillement d'un scrutin. Mais cette analogie n'est pas assez étroite pour diminuer en quoi que ce soit l'originalité de cet ingénieux artifice. Pour d'autres problèmes analogues, l'auteur s'en sert également avec succès.

En résumé, nous sommes d'avis qu'il y a lieu d'autoriser M. Bachelier à faire imprimer sa thèse et à la soutenir.

Signed: *Appell, Poincaré, J. Boussinesq*

Here is the thesis defense report:

Dans la soutenance de sa premiere thèse, M. Bachelier a fait preuve d'intelligence mathématique et de pénétration. Il a ajouté des résultat intéressants à ceux que contient la thèse imprimée, notamment une application de la méthode des images.

Dans la 2ième thèse, il a montré qu'il possédait à fond les travaux de M. Boussinesq sur le mouvement d'une sphère dans un fluide indéfini.

La Faculté lui a conféré le grade de Docteur avec mention honorable.

Signed: *Le président P. Appell*

Remarks on the bibliography

Louis Bachelier's books are [5,12,15,21–23]. His articles are [6–11,13,14,16–20,24]. The English translation of his thesis [5] can be found in [41]. The best available biography of Louis Bachelier is by Courtault et. al. [44]; we have made use of it here. (Jean-Michel Courtault and Youri Kabanov organized an exhibit on Bachelier at the University of Besançon.) See also the biographical sketch in Mandelbrot [93]. The complicated relations between Émile Borel and Paul Lévy are detailed in Bru [38]. Jules Regnault's book is analyzed in a thesis by Franck Jovanovic, Université de Paris 1 (see also [71]). The Paris financial market of the second empire is described in Pierre Dupont-Ferrier's book [50]. A study on Bachelier's mathematical works that is quite complete and very interesting is now being done by Laurent Carraro of l'École des Mines of Saint-Etienne. Finally, we mention Paul Cootner's introduction [41], the articles of Christian Walter [121,122] on the financial aspects of Bachelier's work, and Jean-Pierre Kahane's article [72] on the mathematical origins of Brownian motion.

References

1. J. C. Aitchison and J. A. C. Brown. *The lognormal distribution, with special reference to its uses in economics.* Cambridge University Press, 1957.
2. S. S. Alexander. Price movements in speculative markets: trends or random walks. *Industrial Management Review*, 2:7–26, 1961. Reprinted in *The Random Character of Stock Market Prices*, P. Cootner editor, MIT Press, 1964, pages 199-218.
3. D. André. Solution directe du problème résolu par M. Bertrand. *Comptes Rendus de l'Académie des Sciences (Paris)*, 105:436–437, 1887. It concerns the famous voting problem stated and solved (using difference equations) by Mr. Bertrand.
4. M. Armatte. Robert Gibrat et la loi de l'effet proportionnel. *Mathématiques, Informatique et Sciences Humaines*, 129:5–35, 1995.
5. L. Bachelier. Théorie de la spéculation. *Annales Scientifiques de l'École Normale Supérieure*, III-17:21–86, 1900. Thesis for the Doctorate in Mathematical Sciences (defended March 29, 1900). Reprinted by Éditions Jacques Gabay, Paris, 1995. English translation in *The random character of stock market prices*, Ed. P. Cootner, pp. 17-78, Cambridge, MIT Press, 1964.
6. L. Bachelier. Théorie mathématique du jeu. *Annales Scientifiques de l'École Normale Supérieure*, 18:143–210, 1901. Reprinted by Éditions Jacques Gabay, Paris, 1992.
7. L. Bachelier. Théorie des probabilités continues. *Journal de Mathématiques Pures et Appliquées*, 2:259–327, 1906. 6ème série.
8. L. Bachelier. Étude sur les probabilités des causes. *Journal de Mathématiques Pures et Appliquées*, 4:395–425, 1908. 6ème série.
9. L. Bachelier. Le problème général des probabilités dans les épreuves répétées. *Comptes-rendus des Séances de l'Académie des Sciences*, 146:1085–1088, 1908. Séance du 25 Mai 1908.

10. L. Bachelier. Les probabilités à plusieurs variables. *Annales Scientifiques de l'École Normale Supérieure*, 27:339–360, 1910. 3ème série.
11. L. Bachelier. Mouvement d'un point ou d'un système matériel soumis à l'action de forces dépendent du hasard. *Comptes-rendus des Séances de l'Académie des Sciences*, 151:852–855, 1910. Séance du 14 Novembre 1910, présentée par H. Poincaré.
12. L. Bachelier. *Calcul des Probabilités*, volume 1. Gauthier-Villars, Paris, 1912. Reprinted by Éditions Jacques Gabay, Paris, 1992. There was no second volume, possibly because of the war.
13. L. Bachelier. Les probabilités cinématiques et dynamiques. *Annales Scientifiques de l'École Normale Supérieure*, 30:77–119, 1913.
14. L. Bachelier. Les probabilités semi-uniformes. *Comptes-rendus des Séances de l'Académie des Sciences*, 156:203–205, 1913. Séance du 20 Janvier 1913, présentée par Appell.
15. L. Bachelier. *Le Jeu, la Chance et le Hasard*. Bibliothèque de Philosophie scientifique. E. Flammarion, Paris, 1914. 320 pp. Reprinted by Éditions Jacques Gabay, Paris, 1993.
16. L. Bachelier. Le périodicité du hasard. *L'Enseignement Mathématique*, 17:5–11, 1915.
17. L. Bachelier. Sur la théorie des corrélations. *Bulletin de la Société Mathématique de France*, 48:42–44, 1920. Comptes-rendus des Séances de la Société Mathématique de France. Séance du 7 Juillet 1920.
18. L. Bachelier. Sur les décimales du nombre π. *Bulletin de la Société Mathématique de France*, 48:44–46, 1920. Comptes-rendus des Séances de la Société Mathématique de France. Séance du 7 Juillet 1920.
19. L. Bachelier. Le problème général de la statistique discontinue. *Comptes-rendus des Séances de l'Académie des Sciences*, 176:1693–1695, 1923. Séance du 11 Juin 1923, présentée par Maurice d'Ocagne.
20. L. Bachelier. Quelques curiosités paradoxales du calcul des probabilités. *Revue de Métaphysique et de Morale*, 32:311–320, 1925.
21. L. Bachelier. *Les lois des grands nombres du calcul des probabilités*. Gauthier-Villars, Paris, 1937.
22. L. Bachelier. *La spéculation et le calcul des probabilités*. Gauthier-Villars, Paris, 1938. 49 pp.
23. L. Bachelier. *Les nouvelles méthodes du calcul des probabilités*. Gauthier-Villars, Paris, 1939. 69 pp.
24. L. Bachelier. Probabilités des oscillations maxima. *Comptes Rendus de l'Académie des Sciences (Paris)*, 212:836–838, 1941. Séance du 19 Mai 1941. Erratum au volume 213 (1941), p. 220.
25. A. Barriol. *Théorie et pratique des opérations financières*. O. Doin, Paris, 1908. 375 pages. A 4th corrected edition appeared in 1931.
26. P. L. Bernstein. *Capital ideas: the improbable origins of modern Wall Street*. Free Press, New York, 1992.
27. J. Bertrand. *Calcul des probabilités*. Gauthier-Villars, Paris, 1888. Second edition 1907. Reprinted, New York: Chelsea, 1972.
28. L. Boltzmann. *Vorlesungen über Gastheorie*. J.A. Barth, Leipzig, 1896. Ludwig Boltzmann (1844-1906). Published in two volumes, 1896 and 1898. Appeared in French in 1902-1905, *Leçons sur la Théorie des Gaz*, Gauthier-Villars, Paris. Published in English by Dover, New York as *Lectures on Gas Theory*, 490p.

29. E. Borel. *Leçons sur la théorie des fonctions.* Gauthier-Villars, Paris, 1898.
30. E. Borel. Remarques sur certaines questions de probabilités. *Bulletin de la Société Mathématique de France*, 33:123–128, 1905. Reprinted in *Oeuvres d'Émile Borel*, Paris: CNRS, volume 4, pp. 985-990.
31. E. Borel. Sur les principes de la théorie cinétique des gaz. *Annales Scientifiques de l'École Normale Supérieure*, III-23:9–32, 1906. Reprinted in *Oeuvres d'Émile Borel*, Paris: CNRS, volume 3, pp. 1669-1692.
32. E. Borel. *Éléments de la Théorie de Probabilités.* Hermann, 1909. Second edition 1910, third edition 1924. New revised edition published in 1950 by Albin Michel, Paris in the series "Bibliothèque d'éducation par la science". English translation by J. E. Freund, Englewood Cliffs: Prentice Hall, 1965.
33. E. Borel. *Notice sur les travaux scientifiques de M. Émile Borel.* Gauthier-Villars, Paris, 1912.
34. E. Borel. *Oeuvres d'Émile Borel.* Editions du Centre National de la Recherche Scientifique (CNRS), Paris, 1972. 4 volumes. Émile Borel lived from 1871 to 1956.
35. J. Boussinesq. *Théorie analytique de la chaleur mise en harmonie avec la thermodynamique et avec la théorie mécanique de la lumière.* Gauthier-Villars, 1901. Cours de physique mathématique de la Faculté des sciences. Vol 1. Vol 2 appears in 1903. Joseph Boussinesq lived from 1842 until 1929.
36. R. Brown. A brief account of microscopical observations made in the months of June, July, and August, 1827, on the particles contained in the pollen of plants; and on the general existence of active molecules in organic and inorganic bodies. *Philosophical Magazine (2nd series)*, 4:161–173, 1828.
37. B. Bru. Doeblin's life and work from his correspondence. In Harry Cohn, editor, *Doeblin and Modern Probability*, volume 149 of *Contemporary Mathematics*, pages 1–64. American Mathematical Society, Providence, R.I., 1993.
38. B. Bru. Borel, Lévy, Neyman, Pearson et les autres. *Matapli*, 60:51–60, 1999.
39. E. Carvallo. *Le Calcul des Probabilités et ses Applications.* Gauthier-Villars, Paris, 1912.
40. K.L. Chung. On the maximum partial sums of sequences of independent random variables. *Proceedings of the National Academy of Sciences of the USA*, 33:133–136, 1947.
41. P. Cootner, editor. *The Random Character of Stock Market Prices.* MIT Press, Cambridge, MA, 1964.
42. A. A. Cournot. *Recherches sur les principes mathématiques de la théorie des richesses.* L. Hachette, Paris, 1838. Also in *Oeuvres complètes*, volume VIII, J. Vrin, Paris, 1980. Published in English as *Mathematical Principles of the Theory of Wealth*, James & Gordon, San Diego, 1995, 187 pages. Antoine Augustin Cournot lived from 1801 to 1877.
43. A. A. Cournot. *Exposition de la théorie des chances et des probabilités.* L. Hachette, F. Didot, Paris, 1843. Also in *Oeuvres complètes*, volume I, J. Vrin, Paris, 1984. The *Oeuvres complètes* are published by the Librairie Philosophique, J. Vrin, Paris, from 1973 to 1989.
44. J.-M. Courtault, Y. Kabanov, B. Bru, P. Crépel, I. Lebon, and A. Le Marchand. Louis Bachelier on the centenary of "Théorie de la Spéculation". *Mathematical Finance*, 10(3):341–353, 2000.
45. B. de Finetti. Sulla possibilità di valori eccezionali per una legge di incrementi aleatori. *Rendiconti della Reale Academia Nazionale dei Lincei*, 10:325–329, 1929.

46. B. de Finetti. Sulle funzioni a incremento aleatorio. *Rendiconti della Reale Academia Nazionale dei Lincei*, 10:163–168, 1929.
47. R. de Montessus. *Leçons élémentaires sur le Calcul des Probabilités.* Gauthier-Villars, Paris, 1908. 191 pages. The author is Robert de Montessus de Ballore, born in 1870 and died in 1937. The book can be found in microfilm in the Mathematics Collection, Brown University Library, reel # 7280, item # 6.
48. J. L. Doob. Kolmogorov's early work on convergence theory and foundations. *The Annals of Probability*, 17:815–821, 1989.
49. E. Dormoy. Théorie mathématique des jeux de hasard. *Journal des Actuaires Français*, 2:38–57, 1873. Émile Dormoy (1829-1891) is an important French actuary.
50. P. Dupont-Ferrier. *Le marché financier de Paris sous le Second Empire.* Presses Universitaires de France, Paris, 1925.
51. A. Einstein. Über die von der molekularkinetischen Theorie der Wärme gefordete Bewegung von in ruhenden Flüssigkeiten suspendierten Teilchen. *Annalen der Physik*, 17:549–560, 1905. Reprinted in A. Einstein, *Investigations on the theory of the Brownian movement*, edited with notes by R. Fürth, translated by A. D. Cowper, London: Methuen, 1926. This English translation appears also in Dover: New York, 1956. Albert Einstein lived from 1879 to 1955.
52. P. Erdös and M. Kac. On certain limit theorems of the theory of probability. *Bulletin of the American Mathematical Society*, 52:292–302, 1946.
53. W. Feller. Zur Theorie der stochastischen Prozesse (Existenz- und Eindeutigkeitssätze). *Mathematische Annalen*, 113:113–160, 1936.
54. W. Feller. *An Introduction to Probability Theory and its Applications*, volume 1. Wiley, New York, 2nd edition, 1957. The first edition appeared in 1950.
55. B. J. Ford. Brownian movement in clarkia pollen: a reprise of the first observations. *The Microscope*, 40:235–241, 1992.
56. M. Fréchet. *La vie et l'oeuvre d'Émile Borel.* Monographie de L'Enseignement Mathématique, Genève, 1965.
57. M. Gevrey. Equations aux dérivées partielles du type parabolique. *Journal de Mathématiques Pures et Appliquées*, série 6, vol. 9:305–471, 1913. See also vol. 10, (1914) 105-148.
58. M. Gevrey. Sur la nature analytique des solutions des équations aux dérivées partielles. *Annales Scientifiques de l'École Normale Supérieure*, série 3, vol. 35:39–108, 1918.
59. M. Gevrey. *Oeuvres de Maurice Gevrey.* Editions du Centre National de la Recherche Scientifique, Paris, 1970. Collected works, 573 pages.
60. M. Gherardt. *Le gain mathématique à la Bourse.* Charles Amat, Paris, 1910.
61. R. Gibrat. *Les Inégalités Économiques.* Sirey, Paris, 1931.
62. J. Haag. Applications au tir. In E. Borel, editor, *Traité du calcul des probabilités et de ses applications*, volume 4, fascicule 1, Paris, 1926. Gauthier-Villars.
63. J. Haag. Sur un problème général de probabilités et ses diverses applications. In *Proceedings of the International Congress of Mathematicians, Toronto 1924*, pages 659–676, Toronto, Canada, 1928. Toronto University Press.
64. J. Hadamard. Sur la solution fondamentale des équations aux dérivées partielles du type parabolique. *Comptes Rendus de l'Académie des Sciences (Paris)*, 152:1148–1149, 1911.
65. B. Hostinský. Sur les probabilités relatives aux transformations répétées. *Comptes Rendus de l'Académie des Sciences (Paris)*, 186:59–61, 1928.

66. B. Hostinský. Sur les probabilités des phénomènes liés en chaîne de Markoff. *Comptes Rendus de l'Académie des Sciences (Paris)*, 189:78–80, 1929.
67. B. Hostinský. Sur la théorie générale des phénomènes de diffusion. *Comptes Rendus du Premier Congrès des Mathématiciens des Pays Slaves, Warszawa 1929*, pages 341–347, 1930.
68. B. Hostinský. *Méthodes générales du calcul des probabilités.* Paris. Gauthier-Villars, 1931. Mémorial des Sciences mathématiques, fascicule 52.
69. F. Jovanovic. Instruments et théorie économiques dans la construction de la "Science de la Bourse" d'Henri Lefèvre. Cahiers de la MSE (Maison des Sciences Économiques, Université de Paris 1) No. 2000.65, 2000.
70. F. Jovanovic. L'origine de la théorie financière: une réévaluation de l'apport de Louis Bachelier. *Revue d'Économie Politique*, 110(3):395–418, 2000.
71. F. Jovanovic and Ph. Le Gall. Does God practice a random walk? A 19th century forerunner in financial theory and econometrics, Jules Regnault. Preprint, 2000.
72. J.-P. Kahane. Le mouvement brownien: un essai sur les origines de la théorie mathématique. In *Matériaux pour l'histoire des mathématiques au XX*[e] *siécle. Actes du colloque à la mémoire de Jean Dieudonné (Nice, 1996)*, volume 3 of *Séminaires et Congrès*, pages 123–155. Société Mathématique de France, 1998.
73. J. M. Keynes. Review of Louis Bachelier's "Calcul des probabilités". *Journal of the Royal Statistical Society*, December 1912. Republished in Volume 11, pp. 567-573 of *The Collected Writings of John Maynard Keynes*, 1971-, London: Macmillan, St. Martin's Press: New York.
74. J. M. Keynes. *A Treatise on Probability.* Macmillan, London, 1921. Republished in Volume 8 of *The Collected Writings of John Maynard Keynes*, 1971-, London: Macmillan, St. Martin's Press: New York.
75. A. Ya. Khinchine. *Asymptotische Gesetze der Wahrscheinlichkeitsrechnung.* Springer, Berlin, 1933. Series "Ergebnisse der Mathematik und ihrer Grenzgebiete". Reissued by Chelsea Pub. Co., New York, 1948.
76. A. N. Kolmogorov. Über die analytischen Methoden in der Wahrscheinlichkeitsrechnung. *Mathematische Annalen*, 104:415–458, 1931.
77. A. N. Kolmogorov. Sulla forma generale di un processo stocastico omogeneo. *Rendiconti della Reale Academia Nazionale dei Lincei*, 15:805–808, 1932. Reprinted in *Selected Works of A.N. Kolmogorov*, Vol. 2 pp. 121-127, Kluwer Academic Publishers, Dordrecht 1992. The second part of the article is called "Ancora sulla forma generale di un processo omogeneo" Vol. 15 of *Rendiconti*, pp. 866-869.
78. A. N. Kolmogorov. *Selected works.* Kluwer Academic Publishers, Dordrecht, 1991. 3 volumes. Andrei Nikolaevich Kolmogorov lived from 1903 to 1987.
79. R. Kruizenga. *Put and call options: a theoretical and market analysis.* PhD thesis, MIT, 1956.
80. P. Langevin. Sur la théorie du mouvement brownien. *Comptes Rendus de l'Académie des Sciences de Paris*, 146:530–533, 1908.
81. P.-S. Laplace. *Théorie Analytique des Probabilités.* Ve Courcier, Paris, 3rd edition, 1820. The book is 560 pages long. It is republished as Volume 7 in the *Oeuvres complètes*, published by Gauthier-Villars, Paris, in 1886. Pierre Simon, marquis de Laplace lived from 1749 until 1827.
82. A. G. Laurent. Comments on "Brownian motion in the stock market". *Operations Research*, 7:806–807, 1959. Comments on an M. F. M. Osborne article, which appears in the same volume, pp. 145-173.

83. H. Lefèvre. *Théorie élémentaire des opérations de bourse.* Chez l'auteur, Bureau du journal des placements financiers, 12 rue Laffite, Paris, 1870.
84. H. Lefèvre. Physiologie et mécanique sociales. *Journal des Actuaires Français*, 2:211–250, 1873. See also pp. 351-388 and volume 3 (1874) 93-118.
85. H. Lefèvre. *Principes de la Science de la Bourse.* Institut Polytechnique, Paris, 1874. Approuvés par la Compagnie des Agents de Change.
86. P. Lévy. *Calcul des probabilités.* Gauthier-Villars, Paris, 1925.
87. P. Lévy. Sur certains processus stochastiques homogènes. *Compositio Mathematica*, 7:283–339, 1939.
88. P. Lévy. Le mouvement brownien plan. *American Journal of Mathematics*, 62:487–550, 1940.
89. P. Lévy. *Processus stochastiques et mouvement brownien.* Gauthier-Villars, Paris, 1st edition, 1948. A second edition appeared in 1965. Reprinted by Éditions Jacques Gabay, Paris, 1992.
90. P. Lévy. *Quelques aspects de la pensée d'un mathématicien.* Albert Blanchard, 9 rue de Médicis, Paris 6e, 1970.
91. P. Lévy. *Oeuvres de Paul Lévy.* Gauthier-Villars, Paris, 1973-. Six volumes. Paul Lévy lived from 1886 to 1971.
92. J. W. Lindeberg. Eine neue Herleitung des Exponentialgesetzes in der Wahrscheinlichkeitsrechnung. *Mathematische Zeitschrift*, 15:211–225, 1922.
93. B. B. Mandelbrot. *The Fractal Geometry of Nature.* W.H. Freeman and Co., New York, 1982. Appears in French as *Les objets fractals : forme, hasard et dimension*, Flammarion, Paris, 1995, 4th edition.
94. J. Marcinkiewicz. Sur une propriété du mouvement brownien. *Acta Litterarum Scientiarum*, pages 77–87, 1939.
95. A. A. Markoff. *Wahrscheinlichkeitsrechnung.* Teubner, Leipzig, 1912. Translated from the second Russain edition.
96. M. F. M. Osborne. Brownian motion in the stock market. *Operations Research*, 7:145–173, 1959. Reprinted in *The Random Character of Stock Market Prices*, P. Cootner editor, MIT Press, 1964, pages 100-128. A reply to a letter of A. G. Laurent, titled *Comments on "Brownian motion in the stock market"* appears in the same volume, pp. 807-811.
97. M. F. M. Osborne. Reply to "Comments on 'Brownian motion in the stock market' ". *Operations Research*, 7:807–811, 1959. Reply to a letter by A. G. Laurent which appears on pp. 806-807.
98. M. F. M. Osborne. *The Stockmarket and Finance from a Physicist's Viewpoint.* Crossgar Press, Minneapolis, MN, 1977.
99. F. Perrin. Étude mathématique du mouvement brownien de rotation. *Annales Scientifiques de l'École Normale Supérieure*, III-45, 1928. Thèse, Paris.
100. J. Perrin. *Les Atomes.* Felix Alcan, Paris, 1912. Nouvelle collection scientifique, directeur Émile Borel.
101. L. Pochet. Géométrie des jeux de Bourse. *Journal des Actuaires Français*, 2:153–160, 1873.
102. H. Poincaré. *Calcul des probabilités.* Gauthier-Villars, Paris, 1896. Leçons professées pendant le second semestre 1893-1894, rédigées par A. Quiquet, ancien élève de l'École Normale Supérieure, Paris. Seconde édition revue et augmentée par l'auteur, Gauthier-Villars, Paris, 1912. Nouveau tirage en 1923. The second edition is reprinted by Éditions Jacques Gabay, Paris, 1987. Henri Poincaré lived from 1854 to 1912.

103. G. Pólya. Über eine Aufgabe der Wahrscheinlichkeitsrechung betreffend die Irrfahrt im Strassennetz. *Mathematische Annalen*, 84:149–160, 1921.
104. G. Pólya. Sur quelques points de la théorie des probabilités. *Annales de l'Institut Henri Poincaré*, 1:117–161, 1930.
105. A. Quetelet. *Lettres à S.A.R. le duc régnant de Saxe-Coburg et Gotha, sur la théorie des probabilités, appliquée aux sciences morales et politiques.* M. Hayez, Bruxelles, 1846. Appears in English as *Letters addressed to H.R.H. the Grand Duke of Saxe Coburg and Gotha, on the theory of probability*, New York: Arno Press, 1981, 309 p. Alphonse Quetelet lived from 1796 to 1874.
106. J. W. S. Rayleigh. *Theory of Sound.* Macmillan, London, 1877. Second edition revised and enlarged, ibid., 1894. Reedited by Dover, New York, 1945. Jonn William Strutt Rayleigh lived from 1842 to 1919.
107. J. W. S. Rayleigh. On the resultant of a large number of vibrations of the same pitch and of arbitrary phases. *Philosophical Magazine*, 10(5):73–78, 1880.
108. J. W. S. Rayleigh. Dynamical problems in illustration of the theory of gases. *Philosophical Magazine*, 32(5):424–445, 1891.
109. J. W. S. Rayleigh. On James Bernoulli's theorem. *Philosophical Magazine*, 42(5):246–251, 1899.
110. J. W. S. Rayleigh. On the problem of random vibrations, and of random flights in one, two and three dimensions. *Philosophical Magazine*, 37(6):321–347, 1919.
111. J. Regnault. *Calcul des chances et philosophie de la Bourse.* Mallet-Bachelier et Castel, Paris, 1863. 219 pages.
112. C. Reid. *Neyman.* Springer Verlag, New York, 1982.
113. P. A. Samuelson. Rational theory of warrant pricing. In P. Cootner, editor, *The Random Character of Stock Market Prices*, pages 506–532. MIT Press, Cambridge, MA, 1964. With an appendix by Henry P. McKean, Jr. *A free boundary problem for the heat equation arising from a problem in Mathematical Economics.*
114. P. A. Samuelson. *The Collected Scientific Papers of Paul A. Samuelson.* MIT Press, Cambridge, MA, 1966. Five volumes, starting in 1966. Volumes 1 and 2 are edited by Joseph E. Stiglitz (1966), Volume 3 by Robert C. Merton (1972), Volume 4 by Hiroaki Nagatani and Kate Crowley (1977), and Volume 5 by Kate Crowley (1986).
115. P. A. Samuelson. Mathematics of speculative price. In R.H. Day and S.M. Robinson, editors, *Mathematical Topics in Economic Theory and Computation.* SIAM, Philadelphia, 1972. Reprinted in Volume 4 of *The Collected Scientific Papers of Paul A. Samuelson*, article 240.
116. P. A. Samuelson. Paul Cootner's reconciliation of economic law with chance. In W. F. Sharpe and C.M. Cootner, editors, *Financial Economics: Essays in Honor of Paul Cootner*, pages 101–117. Prentice Hall, Englewood Cliffs, N.J., 1982. Reprinted in Volume 5 of *The Collected Scientific Papers of Paul A. Samuelson*, article 328.
117. A. N. Shiryaev. Kolmogorov – life and creative activities. *The Annals of Probability*, 17:866–944, 1989.
118. M. Smoluchowski. *Oeuvres de M. Smoluchowski.* Impr. de l'Université jaguellonne, Cracovie, Pologne, 1924. Publiées sous les auspices de l'Académie polonaise des sciences et des lettres par les soins de Ladislas Natanson et Jean Stock.
119. M. S. Taqqu. Bachelier and his times: a conversation with Bernard Bru. *Finance and Stochastics*, 5(1):3–32, 2001.

120. M. S. Taqqu. Bachelier et son époque: une conversation avec Bernard Bru. *Journal de la Société Française de Statistique*, 142, 2001. To appear.
121. C. Walter. Une histoire du concept d'efficience sur les marchés financiers. *Annales d'Histoire Économique et Sociale*, pages 873–905, 1996.
122. C. Walter. The efficient market hypothesis: birth, rise, zenith, crisis and impact on investment management industry. Preprint, 1999.

Modern Finance Theory Within One Lifetime

Paul Samuelson

MIT, E52-383
Department of Economics
50 Memorial Drive
Cambridge, MA 02142, USA

Abstract. We meet to celebrate the century birthday of Louis Bachelier's Paris Ph.D. thesis, *Théorie de la Spéculation.* One hundred is a good round number. However, the text of my sermon today is about the genesis of the present modern theory of finance in one academic lifetime – a rare phenomenon in the annals of any science. Although Bachelier's basic breakthrough came in 1900, its scientific beginning must be placed at about 1950. Science is public knowledge. The treasures that were in Gauss's private notebooks were as if they never existed – until he released them in lecture or publication, or until they were later quasi-independently discovered by scholars other than Gauss. When a tree falls in a forest empty of observers or listeners, it is much as if no tree ever fell there at all.

Discovering Bachelier

Discovery of Bachelier's work is a well-known, rather romantic story. It is worth repeating. It begins with blue ditto hectographed post cards sent out by the late Jimmie Savage to several 1950 theorists asking: Any of you know of a French guy named Bachelier who seems to have written a little 1914 book on speculation? I for one had known the name Bachelier. Back in the late 1930's when I and the brilliant Polish-American topological mathematician Stan Ulam were buddies in Harvard's crack Society of Fellows, Stan had mentioned that name. Occasionally Ulam was asked to fill in at Harvard in giving one of its few courses on probability. Later Ulam gained fame at the war-time atomic bomb program in the Los Alamos Laboratory. It was Ulam who repopularized the Monte Carlo method to solve intractable mathematical problems; and still later it was he who got the bright idea that made Teller's hydrogen fusion bomb workable. Besides, I had vague remembrance of a footnote reference to Bachelier in Feller's 1950 class on *Probability Theory and its Applications, Volume I.* Here are Feller's somewhat patronizing words:

> Credit for discovering the connections between random walks and diffusion is due principally to L. Bachelier (1870–1946). His work is frequently of a heuristic nature, but he derived many new results. Kolmogorov's theory of stochastic processes [first named such, I believe by Bachelier] of the Markov type is based largely on Bachelier's ideas. See in particular L. Bachelier, *Calcul des probabilités*, Paris, 1912.

I may add that it is primarily in Feller's less elementary Volume II, not issued until after the early 1950s, that the Itô-like mathematics gets discussed.

As I expected, the 1914 popular Bachelier exposition was not in the limited MIT library. But a greater treasure was there: the 1900 Paris thesis and the 1912 item. In the superb Harvard Widener Library, I later did find Savage's 1914 reference, along with a post-World War I additional book. After turning the pages of the 1900 masterpiece, I recognized its brilliance: all that was later in Einstein's 1905 brownian motion breakthrough was already in Bachelier and more. When my one-time pupil and colleague Paul Cootner planned to bring out a 1960s anthology of finance memoirs, I urged him to commission an English translation of Bachelier: nuances of original work can get lost on its foreign language readers.

To finish off the Bachelier discovery, his use of the Absolute Gaussian distribution irreducibly resulted in both positive and negative wealth outcomes, something not compatible with the limited-liability feature of common stocks. For that reason I pragmatically replaced his Absolute Gaussians by Log-Normal probabilities: a stock might double or halve at commensurable odds; such a law of proportionate effect, familiar in the theory of skew curves (Gibrat and otherwise), for various additional reasons became the work-horse for Black-Scholes and other option formulas. The astronomer M.F.M. Osborne independently resorted to the same "Geometric Brownian Motion," basing his argument on his own doubtfully persuasive variant of the psychological Weber-Fechner law.

In On the Creation

When I entered the study of economics in 1932 at Chicago, finance theory was not then in the economics curriculum; nor was it in the business school syllabus. Unknown to most of us, Holbrook Working was spending three decades at the Stanford Food Institute working through the statistics of spot commodity prices and futures prices on organized speculative exchanges. I salute him as one of the early heroes of the random-walk story. My task today is to recall and honor these great pioneers of finance, giants who walked the earth in the Neanderthal forests.

To do this I am proposing to fabricate counter-factual science history. If Alfred Nobel can create Nobel Prizes, I can conjure up fictional Samuelson-Nobel Prizes for Finance. Holbrook Working, around 1950, is an early recipient of my Samuelson-Nobel Prize. He also elaborated the theory of hedging and of rational intertemporal storage. From tables of random numbers he constructed diagrams of putative price changes $P_{t+1} - P_t$. When he slyly interspersed their diagrams within undated diagrams of real-market price changes, floor traders could hardly tell the difference. It was like experiments in which laypeople are presented with paintings by (a) modern artists, (b) kindergarten children and (c) inhabitants of mental institutions. In these experiments art critics scored higher in telling them apart than Board of Trader locals did in separating Working's artifacts and the real things.

Let me move past Holbrook Working. Alfred Cowles, III, an heir to a mining and aluminum fortune, contracted tuberculosis and that kept him tied to the salubrious mountain climate of Colorado Springs, Colorado. To satisfy his curiosity and hopefully improve his portfolio performance, Cowles founded the non-profit Cowles Commission to study econometrics. Cowles made some of the first probability studies of forecasting methods, showing that none did better than the best of several ought to do by chance alone. His theoretical tests on runs hinted at random-walk structure for stock prices.

Equally important, Cowles compiled a precursor index to the S&P 500. Every day we see these indexes being used in current appraisals of whether our bull market is or is not grossly overvalued. Cowles therefore also gets early elected to the Samuelson-Nobel Prize.

Prizes must go also to Frank Ramsey, Bruno de Finetti, L.J. Savage, John von Neumann and Jacob Marschak for re-establishing (after Daniel Bernoulli and Laplace) maximizing $\text{Exp}\{U(W)\}$ as a decision criterion.

Nay-sayers can be as important in progress of science as aye-sayers. Awards of my Samuelson-Nobel Prizes should therefore go also to Maurice Allais, whose artful tests have demonstrated that many real-world people will want to violate the Axioms of maximized Expected Utility. Similarly the theoretical generalizations of Mark Machina's non-linear functionals plus the behavioristic experiments rationalized by Amos Tversky and Daniel Kahneman (1974) certainly deserve sharing a top prize in finance. Kenneth Arrow and Gerald Debreu each contributed to Arrow's important concept of complete markets.

Tom Kuhn's 1962 theory of scientific revolutions stressed the key role of new data that refute old notions and launch new ones. The LSE statistician Maurice Kendall, who prided himself on his ignorance of economics, by brute empiricism gave the Royal Statistical Society a 1953 round-up of the serial auto-correlation structure of price-changes in spot commodity markets, futures prices, individual common stock price changes, as well as indexed portfolios of stock price changes. To his surprise and delight, virtually all $r_{\Delta p_t, \Delta p_{t+k}}$ coefficients meandered in the neighborhood of zero, as if by chance alone. He proclaimed this to demonstrate an *absence* of economic law: *white* noise, a tale told by an idiot or by the devil who draws samples of price changes randomly after mixing thoroughly the tickets in a giant urn or hat. (Positive serial correlations that favor runs I later called *blue* noise. Negative serial correlations depicting regression toward the mean, I called *red* noise.)

By report the economist subset in the Royal Statistical Society London audience did not take well this raid from a disciplinary outsider. But with economists' characteristic flexibility and shiftiness, I told them to rally and work the other side of the street: perfect speculation by numerous independent interpreters of arriving new news would be, we came to realize, a sign of higher economic law; as I put it in 1965, an indication of the martingale market theory. "No easy pickings" was the logo on the T-shirt of the prizewinning Kendall and his efficient-market followers.

Mean and variance analysis had informally entered pre-1945 finance theory, as in the heuristic works of Helen Makower & Jacob Marschak and Evsey Domar & Richard Musgrave. But it was Harry Markowitz (1953–59), James Tobin (1958) and William Sharpe (1964), who perfected the important theory of optimal portfolio efficiency by means of quadratic programming. Understandably this (mean, variance) analysis has permeated actual investor practice and richly merits Samuelson-Nobel Prizes, as do also the writings of Modigliani-Miller and Mandelbrot-Fama.

Prizes should not go only to individuals – to Titian and Rembrandt. They should go also to places and institutions: to the Wharton School finance workshops pioneered by Irwin Friend and Stephen Ross; the Berkeley, Stanford and UCLA workshops of Barr Rosenberg, Nils Hakansson and Hayne Leland; the innumerable Chicago experts in finance; the busy Cambridge, Massachusetts workshops of John Lintner, Robert Schlaifer, John Pratt, Richard Zeckhauser, Howard Raiffa, Stewart Myers, John C. Cox, Jonathan Ingersoll, Richard Kruizenga, Paul Cootner and Andrew Lo, to mention only a few. To reward fundamental breakthroughs half a hundred prizes in the 1950–2000 period would not really be enough, particularly in the recent times of Robert Shiller, John Campbell and Jeremy Siegel.

The Holy Grail

I have saved for my all too brief ending the story of the competition to reach the North Pole first. I mean the Black-Scholes-Merton formula for equilibrium option pricing. As the physicist Freeman Dyson documented, Kuhn erred in omitting, as a major cause of scientific revolution breakthroughs, new toolmaking. Behind Galileo and Newton lay the inventor of the telescope. Behind Darwin and Crick-Watson lay the invention of the microscope. Behind Black-Scholes-Merton lay Norbert Wiener, A.N. Kolmogorov and Kyosi Itô. I cannot explicate the point better than by quoting the works of the infallible poet:

> Nature and Nature's law lay hid in night;
> God said, Let Itô be! and all was light.

The stochastic calculus, by being able to model instantaneously rebalancing price changes, put rigor into the brilliant conjecture of Fischer Black and Myron Scholes of an instantaneous variance-free hedge. Suddenly the imperfection of mean-variance analysis evaporated away; suddenly the evaluation of a stock's derivative securities, so that neither buyer nor seller stands to gain, becomes clear. The mathematical seed that Bachelier planted, which Wiener blessed, became through the harvestings of Itô, by Fischer Black, Myron Scholes and Robert Merton the Dyson tool-breakthrough which sparked a revolutionary change in finance science.

Each month in journals all over the world, each day and hour in new markets everywhere, we see at work this skeleton key to the miracles of scientific advance. The saga is only bettered by the opposition to the new paradigm along the way. When an older Milton Friedman pooh-poohed it all as "not even economics at all," this only documented Max Planck's dictum that Science Progresses Funeral by Funeral.

And all this in *one* academic lifetime. My academic lifetime. A wise King Alphonse once said: "If I'd been in on the Creation, I could have done a better job of it." Well, I was in on the creation of finance theory. But I could not have done that better job; if I could have, I would have. And I didn't. Science is public knowledge eked out by all our heroes, two steps forward and one step back. A non-random walk, you will agree.

Future Possibilities in Finance Theory and Finance Practice

Robert C. Merton

Havard Business School
Morgan Hall 397, Soldiers Field
Boston, MA 02163, USA

1 Introduction[1]

The origins of much of the mathematics in modern finance can be traced to Louis Bachelier's 1900 dissertation on the theory of speculation, framed as an option-pricing problem. This work marks the twin births of both the continuous-time mathematics of stochastic processes and the continuous-time economics of derivative-security pricing. In solving his option-pricing problem, Bachelier provides two different derivations of the classic partial differential equation for the probability density of what later was called a Wiener process or Brownian motion process. In one derivation, he writes down a version of what is now commonly called the Chapman–Kolmogorov convolution probability integral in one of the earliest examples of that integral in print. In the other, he uses a limit argument applied to a discrete-time binomial process to derive the continuous-time transition probabilities. Bachelier also develops the method of images (or reflection) to solve for a probability function of a diffusion process with an absorbing barrier. All this in his thesis five years before the publication of Einstein's mathematical theory of Brownian motion.

However, for most of the century, the mathematical and finance branches from Bachelier's work evolved at different paces and independently of one another. On the mathematical side, Kiyoshi Itô was greatly influenced by Bachelier's work in his development in the 1940s and early 1950s of his stochastic calculus, later to become an essential tool in finance.[2] Indeed, at the centennial celebration of Norbert Wiener's birthday at M.I.T. in 1994, Itô told me that Bachelier's thesis was far more influential on his work than Wiener's. Apparently, much the same story holds for the great probabilist A.N. Kolmogorov.

On the financial side, Bachelier's important work was largely lost to financial economists for more than a half century. As we have heard, Paul Samuelson via L.J. Savage brought Bachelier's thesis to the attention of the economics community, including arranging for its translation into English.[3]

[1] Presented in College de France at the First World Congress of the Bachelier Finance Society, Paris, June 28, 2000. Various parts of this paper draw heavily on Merton (1990, 1993b, 1994, 1995b, 1999).

[2] See Itô (1951, 1987).

[3] The English translation by James Boness appears in Cootner (1964).

Furthermore, Samuelson's theory of rational warrant pricing, published in 1965, was centrally motivated by that work.[4] Henry McKean bridged both the mathematical and financial branches as a scientific collaborator with Itô on the mathematics of stochastic processes and with Samuelson on the finance application to warrant pricing.[5] It was not, however, until the end of the 1960s and early 1970s that these two branches of research growing from Bachelier's dissertation were actually reunited. Initially, Itô's mathematics found its way into finance with the development of the continuous-time theory of optimal lifetime consumption and portfolio selection.[6] This theory used diffusion processes to model asset price movements and applied the Itô calculus to analyze the dynamics of continuously traded portfolios. The connection between Itô's work and option pricing was made when that same continuous-trading portfolio modeling tool was used to derive dynamic portfolio strategies that replicate the payoffs to an option, from which the famous Black-Scholes option pricing theory was born.[7]

This unmistakable lineage from Bachelier's dissertation to the Black-Scholes model in both its mathematics and its finance underscores the influence of his work on the development and growth of the listed options market, nearly three-quarters of a century later. Subsequently, that same influence greatly impacted the development, refinement and broad-based practical implementation of contracting technology. Derivative securities such as futures, options, swaps and other financial contractual agreements provide a prime example. Innovations in financial-contracting technology have improved efficiency by expanding opportunities for risk sharing, lowering transactions costs and reducing information and agency costs. The numbers reported for the global use of derivative securities are staggering, $70 trillion worldwide and there are a number of individual banking institutions with multi-trillion dollar off-balance-sheet derivative positions. These reported amounts are notional or face values (and often involve double counting), and thus, they do not measure the market value of the contracts themselves, which is much smaller. Nevertheless, it can surely be said that derivatives are ubiquitous throughout the world financial system and that they are used widely by non-financial firms and sovereigns as well as institutions in virtually every part of their financing and risk-managing activities. The significance of Bachelier's contribution will continue to grow as improved technology, together with growing breadth and experience in the applications of derivatives, expands the scale and scope of

[4] See Samuelson (1965, 1972) and Samuelson and Merton (1969).
For an extensive review of Samuelson's seminal contributions to the theory of option pricing as well as finance more broadly, see Merton (1983b).

[5] See Itô and McKean (1964) and McKean (1965).

[6] See Merton (1969, 1971, 1982, 1992).

[7] Black and Scholes (1973) and Merton (1973). The term the "Black-Scholes model" appeared initially in Merton (1970, 1972). For some of the history of its discovery, see Bernstein (1992, Ch. 11), Black (1989), Merton (1998), Merton and Scholes (1995), and Scholes (1998).

their use and both consumers and producers of derivatives move down the learning curve.[8]

Although the paper will address the practice of finance and the science of finance, it will not focus on the latest option-pricing models, nor is it my aim to introduce state-of-the-art computational tools, which might help implementation of these models. Instead I try my hand at providing a frame of reference for these models and tools by describing something of the interaction between those parts of the science of finance which have direct lineage from Bachelier's thesis and their influence on the practice of finance, and possibilities for future trends in each.

1.1 A Functional Perspective for Forecasting Institutional Change

There are two essentially different frames of reference for trying to analyze and understand innovations in the financial system. One perspective takes as given the *existing* institutional structure of financial-service providers, whether governmental or private-sector, and seeks what can be done to make those institutions perform their particular financial services more efficiently and profitably.

An alternative to this traditional institutional perspective – and the one I favor – is the functional perspective, which takes as given the economic functions served by the financial system and seeks what is the *best* institutional structure to perform those functions.[9] The basic functions of a financial system are essentially the same in all economies, which makes them far more stable, across time and across geopolitical borders, than the identity and structure of the institutions performing them. Thus, a functional perspective offers a more robust frame of reference than an institutional one, especially in a rapidly changing financial environment. It is difficult to use institutions as the conceptual "anchor" for forecasting financial trends when institutional structures are themselves changing significantly, as has been the case for more than two decades and as appears likely to continue well into the future.

Finance theory, which is not institution based, is thus an apt tool for applying the functional perspective to forecast new trends. Indeed, during the last quarter century, finance theory has been a particularly good predictor of future changes in finance practice. That is, when theory seems to suggest that an institution, an instrument, or a service "should be there" and it is not, practice has tended to evolve so that it is. Placed in a normative context, current theory has been a fruitful idea source for subsequent innovations in finance practice.

The Black-Scholes option pricing theory is, of course, the most celebrated instance. However, it is surely not a singular case. The elementary state-

[8] See Jin, Kogan, Lim, Taylor, and Lo (1997) for a live web site with extensive references documenting the wide range of applications of derivatives.

[9] For an in-depth description and application of the functional perspective, see Crane et al (1995) and Merton (1993a, 1995).

contingent securities developed as a theoretical construct by Kenneth Arrow (1953) to explain the function of securities in risk-bearing, were nowhere to be found in the real world until the broad development of the options and derivative-security markets. As we all know, it is now routine for financial engineers to use digital options and other Arrow-like derivative instruments in analyzing and creating new financial products.[10] More broadly, Arrow's notion of "market completeness," long treated as a purely theoretical concept, is now seen as a (nearly) achievable long-run goal for real-world financial markets.[11] Finance theory thus plays useful dual roles: as a *positive* model for predicting the future direction of financial innovation, changes in financial markets and intermediaries, and regulatory design, and as a *normative* model for identifying new product and service opportunities. Although framed in the positive context of "What *will* the trends be?," the discussion to follow could apply equally in the normative context of "What *should* the trends be?"

Just as the science of finance has helped shape the practice of finance, so practice in turn has helped shape the evolving theory. Financial innovation has generated a great variety of new institutions to serve financial functions, presumably more efficiently. Since theory is not institution based, those real-world innovations provide financial scientists a rich opportunity to understand the selective processes mapping institutions to functions. This strong interplay between research and practice is surely exemplified by this Congress where by my count there are roughly equal numbers of academics and practitioners presenting papers, including many individuals who qualify as both.

The view of the future of financial practices as elsewhere in the economic sphere is clouded with significant uncertainties. With this in mind, I will nevertheless try to apply finance theory, specifically the functional perspective, to talk about the possibilities for future trends in both financial products and services[12] – giving examples from each of the four broad classes of customers for financial services – households, endowment institutions, non-financial firms, and governments.

---

[10] For the theory of synthesis and production of Arrow securities from options and dynamic trading strategies, see Banz and Miller (1978), Breeden and Litzenberger (1978), Hakansson (1976) and Ross (1976b).

[11] See Melnikov (1999, forthcoming) and Merton (1993a, 1995a) for the development of the financial innovation spiral as one model for the dynamic interplay between financial institutions and markets driving the evolution of financial instruments toward market completeness as the asymptotic long-run attractor.

[12] See Clasessens, Glaessner, and Klingebiel (2000) for an extensive survey on the impact of electronic and financial technologies on institutions and practices in the world financial system.

1.2 Financial Services for Households in the Future[13]

As a result of major technological innovation and wide-spread deregulation, the household sector of users in the more fully developed financial systems have experienced a major secular trend of *disaggregation*...some call it *disintermediation*...of financial services. Households today are called upon to make a wide range of important and detailed financial decisions that they did not have to in the past. For example, in the United States, there is a strong trend away from defined-benefit corporate pension plans that require no management decisions by the employee toward defined-contribution plans that do. There are more than 7,000 mutual funds and a vast array of other investment products. Along with insurance products and liquidity assets, the household faces a daunting task to assemble these various components into a coherent effective lifetime financial plan.

Some see this trend continuing with existing products such as mutual funds being transported into technologically less-developed financial systems. Perhaps so, especially in the more immediate future, with the widespread growth of relatively inexpensive Internet access to financial "advice engines." However, the creation of all these alternatives combined with the deregulation that made them possible has consequences: Deep and wide-ranging disaggregation has left households with the responsibility for making important and technically complex micro financial decisions involving risk – such as detailed asset allocation and estimates of the optimal level of life-cycle saving for retirement – decisions that they had *not* had to make *in the past*, are *not* trained to make *in the present*, and are *unlikely* to execute efficiently *in the future*, even with attempts at education.

The availability of financial advice over the Internet at low cost may help to address some of the information-asymmetry problems for households with respect to commodity-like products for which the quality of performance promised is easily verified. However, the Internet does not solve the "principal-agent" problem with respect to more fundamental financial advice dispensed by an agent. That is why I believe that the future trend will shift toward more integrated financial products and services, which are easier to understand, more tailored toward individual profiles, and permit much more effective risk selection and control.

The integrated financial services in the impending future, unlike the disaggregated financial services of the recent past, will focus on the *customer* instead of the *product* as the prime unit of attention. That is, the service begins by helping the *customer* design a financial plan to determine his optimal life-cycle needs and *then* finds the *products* necessary to implement that integrated plan in a cost-efficient fashion. The past generation has seen explosive growth in asset management. Since 1974, mutual fund assets in the United States alone have increased 125-fold from \$48 billion to around \$6 trillion. In this time, the financial-service industry has made great strides

[13] This section is an expanded version of Merton (1999).

in developing and improving portfolio-allocation and performance measurement. *However, the central objective function employed, even in sophisticated practice, is still the same basic mean-variance efficient-frontier criterion developed by Markowitz (1952), Tobin (1958), and Sharpe (1964) in the 1950s and 1960s.* This criterion, based on a static one-period model of maximizing the expected utility of end-of-period wealth, is simply not rich enough to capture the myriad of risk dimensions in a real-world lifetime financial plan.

The practical application of this *status quo* model is almost always limited to just the financial assets of the individual. Thus, in the models available to consumers today, there is no formal recognition of either the size or risk characteristics of human capital, which is the largest single asset for most people during much of their lifetime. In addition to taking into account the magnitude of human capital, advice models in the impending future should also capture the important element of its individual risk characteristics: a stock broker, an automobile engineer, a baseball player, a surgeon, or a professor have very different risk profiles. The human capital of a stockbroker will surely be highly correlated with stock market returns. The human capital of the professor much less so. Without holding any equities among his financial assets, a stockbroker has a significant investment exposure to stock returns. Thus, between a stockbroker and a professor with the same total wealth and risk tolerance, the stockbroker should allocate a smaller part of his financial portfolio to equities. As we see, effective models of asset allocation cannot just focus on the expected levels of compensation, but must also consider its volatility and its correlation with other assets' returns.[14]

There are a number of other risks that are important to households besides the uncertainties about the future values of financial assets and about the returns to human capital.[15] In addition to general inflation uncertainty, there is uncertainty about relative prices of individual categories of consumption goods such as local residential housing. There is even uncertainty about the menu of possible consumption goods that will be available in the future. There is uncertainty about one's medical care needs and the age of death. There is uncertainty about one's own tastes in the future, including the importance attached to bequests to transfer wealth to family and other heirs.

One particularly important intertemporal risk faced by households which is not captured in the traditional end-of-period wealth models of choice is uncertainty about the future investment opportunity set.[16] That is, the unpredictable changes in the menu of expected returns and volatilities of returns

[14] For an extension of the mean-variance one-period model to include human capital, see Mayers (1972). For intertemporal dynamic models of optimal consumption and portfolio selection that take into account uncertain human capital, see Bodie, Merton, and Samuelson (1992), Merton (1971, sections 7, 8, 1977b), and Williams (1978, 1979).

[15] For an analytical development, see for examples Breeden (1979), Cox and Huang (1989), and Merton (1970, 1973b, 1977b, 1982).

[16] See Breeden (1979) and Merton (1970, 1973b).

available on investments in the future. To illustrate the point, consider the following choice question: Which would you rather have: $5 million or $10 million? The answer is obvious for all, take the $10 million, *given all other variables are held fixed.* However now consider that choice framed with further elaboration: Which would you rather have: $5 million in an environment in which the only investment available for the rest of time pays a risk-free real interest rate of 10 percent or $10 million in an environment in which the only investment available for the rest of time pays a risk-free real interest rate of 1 percent?

In a one-period model such as the Markowitz-Tobin one, the $10 million selection is still superior because one is presumed to consume all one's wealth at the end of the period and the future rate of return on investments are irrelevant. Note however that the $5 million selection can provide a $500,000 inflation-protected annual cash flow in perpetuity while the $10 million can only provide a $100,000 annual real cash flow in perpetuity. Thus, for anyone with a long enough future consumption horizon (approximately 10 years or longer in this case), the $5 million with a 10 percent interest is the better choice in terms of consumption standard of living.

Move from this simplified example to the general case when future investment rates on both risk-free and risky assets are uncertain. We see that for the household to maintain a stable consumption stream, it is necessary to plan its portfolio to hedge against unanticipated changes in interest rates. Thus, the household's portfolio is such that in future states of the world in which real interest rates are lower than expected, it has more wealth than expected and in states in which real interest rates are higher than expected, it accepts a lower than expected wealth because it doesn't reduce its standard of living. The "natural" financial security to implement such hedging behavior is a long-term inflation-protected bond.

In addition to taking into account the various dimensions of risk, the household products and services of the future will be much more comprehensive and integrative. They will marry risk control and protection with optimal saving plans for lifetime consumption smoothing and bequests. To arrive at the necessary integrated lifetime consumption and asset-allocation decisions, more advanced financial models are required than have been used in past practice. The underlying analysis will have to combine the traditional efficient risk-return tradeoff for the tangible-wealth portfolio, accounting for human-capital risks and returns, hedging the risks of future reinvestment rates and relative consumption goods prices, incorporating mortality and other traditional insurance risks as well as income and estate tax risks.

Exemplifying my theme of "good practice evolving toward good theory" as a guide to predicting future financial innovation, the basic models for implementation can be found in the rich body of published academic research on optimal lifetime consumption and portfolio selection and intertemporal capital asset pricing that has developed since the classic Markowitz–Tobin–Sharpe work.[17]

In the new environment of these integrated retail products, success for financial-service providers will require much more than simply developing these decision models and performing an advisory role. They should also expect to undertake a principal intermediation role as either issuer or guarantor to create financial instruments that eliminate the "short-fall" or "basis" risk for households. One important category for such intermediation is hedging "targeted" expenditures, ones which are almost surely going to be made and the magnitude of which are not likely to depend on changes in the household's overall standard of living. A prime example is tuition, room and board for a child's college education. In the current investment product environment, the household must take the "basis" risk between the amount saved to provide for that education and the subsequent investment performance from those savings and the uncertain inflation rate for college tuition and housing. Basic finance theory suggests that a more efficient approach would be for an intermediary to issue to the household a contract for four years tuition, room and board delivered at a specified future date in return for a fixed price (which can be financed over time, if necessary, just like a car or house is). The intermediary would then bear the basis risk instead of the household.

To serve the households in the future efficiently, providers will find it advantageous to integrate the various risk-management products. To implement this integration will require bundling of some products that cut across traditional provider institutions and the unbundling of others. For example, by bundling long-term care insurance with retirement annuities, there is a prospect for an efficiency gain by reducing traditional selection bias problems for the mortality component. An unbundling of the accumulation for retirement in a pension fund from the life insurance feature of survivor benefits from that fund can provide a more efficient meeting of these two financial needs in different parts of the household life cycle. A third example would be comprehensive value insurance for the household residence, covering value risk from market price changes, fire and natural disasters.

Each of these integrated risk products combines traditional insurance risks with market risks. Effective implementation will require not only greater regulatory flexibility among banks, securities firms and insurance companies, but also that the rigid intellectual barriers between research in the fields of finance and actuarial science become more permeable and flexible as well. This permeability is already underway reflecting changes in real-world practice where insurance functions are extending well beyond traditional actuarial lines to include wide-ranging guarantees of financial performance by both private-sector institutions and governments. Examples are guaranteed income contracts, deposit insurance, pension-benefit guarantees, and guarantees of loan and other

[17] See Merton (1992) for an extensive bibliography.

contractual obligations.[18] The mathematical tools developed to evaluate risks of "nature" (mortality, weather, and fire) are not adequate to analyze those financial guarantees. Instead, the prototype insurance instrument for financial risks is the put option. As we have seen, the mathematical tools for option pricing are found in the finance literature evolving from Bachelier's thesis. Just as insurance is "moving" into the domain of finance, so finance is moving into the realm of insurance. Although only just at its beginnings, there is an effort for a major institutional shift to move much of the catastrophic risk insurance exposures outside insurance companies (or governments) and instead have them borne directly in the capital markets.[19] Thus, cutting-edge research and practice in the future in either field will require a mathematical and substantive knowledge base that spans both fields.

The inadequacies of the current static model have been masked by the compound claim that classic one-period diversification across stocks handles the static risk of investing and that "time diversification" handles the intertemporal dynamic aspects of that risk. The false but oft-claimed belief that investing in stocks become less risky as the investment horizon becomes longer seemingly offers a practical short cut to addressing the multi-year investment and consumption problems of households in the real world. Unfortunately, that view of equities' risk is just plain flawed.[20] A decade-long bull market may have kept the errors of this approach to multi-period investing from becoming apparent. But as we all know too well, that cannot continue indefinitely.

1.3 Production of Integrated Financial Products in the Impending Future

Production of the new brand of integrated, customized financial instruments will be made economically feasible by applying already existing financial pricing and hedging technology that permits the construction of custom products at "assembly-line" levels of cost.

Paradoxically, making the products more user-friendly and simpler to understand for customers will create considerably more complexity for their producers. The good news for the producers is this greater complexity will also

[18] For discussion, see Cummins (1988), Kraus and Ross (1982), Merton (1977a), Merton and Bodie (1992), and Mody (1996).

[19] For discussion and analysis, see Cummins and Geman (1995), Harrington, Mann, and Niehaus (1995), and Hayes, Cole, and Meiselman (1993).

[20] Bodie (1995) makes this point quite dramatically. He shows that the premium for insuring against a shortfall in performance of stocks versus bonds is actually an increasing function of the time horizon over which the insurance is in force instead of a decreasing one, which would be expected with declining risk. A similar fallacy involving the virtues of investing to maximize the geometric mean return as the "dominating" strategy for investors with long horizons was addressed earlier by Samuelson (1971, 1972, 1979).

make reverse engineering and "product knockoffs" by second-movers more difficult and thereby, protect margins and create franchise values for innovating firms. Hence, financial-engineering creativity and the technological and transactional bases to implement that creativity, reliably and cost-effectively, are likely to become a central competitive element in the industry.

A key element for the success of these highly integrated, user-friendly products in the household sector will be to find effective organizational structures for ensuring product performance: that is, that the contingent payments *promised* by the products are actually paid by the issuing institution. The need for assurances on contract performance is likely to stimulate further development of the financial-guarantee business for financial institutions. It is encouraging to note that currently, credit risk analysis and credit-derivative contracting technologies are among the fastest growing areas of development in financial services. In general, the greater complexity in products combined with the greater need for contract performance will require more elaborate and highly quantitative risk-management systems within financial-service firms and a parallel need for more sophisticated approaches to external oversight.[21]

All of these will significantly change the role of the mutual fund from a direct retail customer product to an intermediate or "building block" product embedded in the more integrated products used to implement the consumer's financial plan. The "fund of funds" is an early, crude example. The position and function of the fund in the future will be much like that of individual traded firms today, with portfolio managers, like today's CEOs, selling their stories of superior performance to professional fund analysts, who then make recommendations to retail "assemblers." As we know, commercial marketing is very different from retail marketing, and some fund institutions may have difficulty making the transition. How and what institutional forms will perform the retail assembly and distribution functions is not clear. It does seem, however, that a fully vertically integrated fund complex of the usual kind that limits its front-end assembly operation to using only its *own* funds and products will be at a very distinct disadvantage, because it will not have the breadth of first-quality "building blocks" to assemble the best integrated products.

1.4 Financial Services for Endowment Institutions in the Future

Much the same story to the one on products and advice for households applies to serving endowment institutions. There are, however, significant enough substantive differences between the two to warrant separate attention here.[22]

The standard approach to the management of endowment today is to treat it as if it were the only asset of the institution. As a result, investment advice

[21] See Merton and Bodie (1992) for a discussion of the difference between customer-held and investor-held liabilities in terms of impact on a firm's business and the various approaches to managing default risk.

[22] This section is a revised version of a part of Merton (1993b).

and products for endowments are focused on achieving a mean-variance efficient portfolio with the appropriate level of risk and a prudent withdrawal or dividend rate. Thus, except for choosing the particular point on the risk-return frontier, the investment advice varies little across institutions. Of course, endowment is almost never the only asset of an institution. Specifically, institutions such as universities have a variety of other assets, both tangible and intangible, which are important sources of cash flow. In addition to tuition, there are gifts, bequests, publishing and other business income, and public and private-sector grants for research. Taking explicit account of those assets and their risk and return characteristics can cause the characteristics of the optimal endowment portfolio to change substantially. Although to be concrete the following discussion uses the context of a university, the same principles of analysis would apply to managing the endowments of museums, foundations, and religious organizations.

A procedure for selecting the investments for the endowment portfolio that takes account of non-endowment assets includes the following steps:

1. Estimate the market value that each of the cash flow sources would have if it were a traded asset. Also determine the investment risk characteristics that each of those assets would have as a traded asset.
2. Compute the *total wealth* or net worth of the university by adding the capitalized values of all the cash flow sources to the value of the endowment.
3. Determine the optimal portfolio allocation among traded assets, using the university's total wealth as a base. That is, treat both endowment and cash flow-source assets as if they could be traded.
4. Using the risk characteristics determined in step 1, estimate the "implicit" investment in each traded-asset category that the university has as the result of owning the non-endowment (cash flow-source) assets. Subtract those implicit investment amounts from the optimal portfolio allocations computed in step 3, to determine the optimal "explicit" investment in each traded asset, which is the actual optimal investment allocation for the endowment portfolio.

As a simple illustration, consider a university with \$400 million in endowment assets and a single non-endowment cash flow source. Suppose that the only traded assets are stocks and cash. Suppose further that the university estimates in step 1 that the capitalized value of the cash flow source is \$200 million, with risk characteristics equivalent to holding \$100 million in stock and \$100 million in cash. Thus, the total wealth of the university in step 2 is $(400 + 200 =)$ \$600 million. Suppose that from standard portfolio-selection techniques, the optimal fractional allocation in step 3 is .6 in stocks and .4 in cash, or \$360 million and \$240 million, respectively. From the hypothesized risk characteristics in step 1, the university already has an (implicit) investment of \$100 million in stocks from its non-endowment cash flow source.

Therefore, we have in step 4 that the optimal amount for the endowment portfolio to invest in stocks is \$260 million, the difference between the \$360 million optimal total investment in stocks and the \$100 million implicit part. Similarly, the optimal amount of endowment invested in cash equals (240 − 100 =) \$140 million.

The effect on the composition of the optimal endowment portfolio induced by differences in the size of non-endowment assets can be decomposed into two parts: the wealth effect and the substitution effect. To illustrate the wealth effect, consider two universities with identical preference functions and the same size endowments, but one has non-endowment assets and the other does not. If, as is perhaps reasonable to suppose, the preference function common to each exhibits decreasing absolute risk aversion, then the university with the non-endowment assets (and hence larger net worth) will prefer to have a larger total investment in risky assets. So a university with a \$400 million endowment as its only asset would be expected to choose a dollar exposure to stocks that is smaller than the \$360 million chosen in our simple example by a university with the same size endowment and a non-endowment asset valued at \$200 million. Such behavior is consistent with the belief that wealthier universities can "afford" to take larger risks with their investments. Thus, if the average risk of the non-endowment assets is the same as the risk of the endowment-only university's portfolio, then universities with such assets will optimally invest more of its endowment in risky assets.

The substitution effect on the endowment portfolio is caused by the substitution of non-endowment asset holdings for endowment asset holdings. To illustrate, consider again our simple example of a university with a \$400 million endowment and a \$200 million non-endowment asset. However, suppose that the risk characteristics of the asset are changed so that it is equivalent to holding \$200 million in stocks and no cash. Now, in step 4, the optimal amount for the endowment portfolio to invest in stocks is \$160 million, the difference between the \$360 million optimal total investment in stocks and the \$200 million implicit part represented by the non-endowment asset. The optimal amount of endowment invested in cash rises to (240 − 0 =) \$240 million. If instead the risk characteristics of the asset had changed in the other direction to an equivalent holding of \$0 in stocks and \$200 million in cash, the optimal composition of the endowment portfolio would be (360 − 0 =) \$360 million in stocks and (240 − 200 =) \$40 million in cash.

Note that the changes in risk characteristics do not change the optimal deployment of *total* net worth (\$360 million in stocks and \$240 million in cash). However, the non-endowment assets are not carried in the endowment portfolio. Hence, different risk characteristics for those assets do change the amount of substitution they provide for stocks and cash in the endowment portfolio. Thus, the composition of the endowment portfolio will be affected in both the scale and fractional allocations among assets.

With the basic concept of the substitution effect established, we now apply it in some examples to illustrate its implications for endowment investment policy. Consider a university that on a regular basis receives donations from alums. Clearly, the cash flows from future contributions are an asset of the university, albeit an intangible one. Suppose that the actual amount of gift giving is known to be quite sensitive to the performance of the general stock market. That is, when the market does well, gifts are high; when it does poorly, gifts are low. Through this gift-giving process, the university thus has a "shadow" investment in the stock market. Hence, all else the same, it should hold a smaller portion of its endowment in stocks than would another university with smaller amounts of such market-sensitive gift giving.

The same principle applies to more specific asset classes. If an important part of gifts to a school that specializes in science and engineering comes from entrepreneur alums, then the school de facto has a large investment in venture capital and high-tech companies, and it should therefore invest less of its endowment funds in those areas. Indeed, if a donor is expected to give a large block of a particular stock, then the optimal explicit holding of that stock in the endowment can be negative. Of course, an actual short position may not be truly optimal if such short sales offend the donor. That the school should optimally invest less of its endowment in the science and technology areas where its faculty and students have special expertise may seem a bit paradoxical. But the paradox is resolved by the principle of diversification once the endowment is recognized as representing only a part of the assets of the university.

The same analysis and conclusion apply if alum wealth concentrations are in a different class of assets, such as real estate instead of shares of stock. Moreover, much the same story also applies if we were to change the example by substituting government and corporate grants for private donations and gift giving as the sources of cash flows. That is, the magnitudes of such grant support for engineering and applied science may well be positively correlated with the financial performance of companies in high-tech industries. If so, then the prospect of future cash flows to the university from the grants creates a shadow investment in those companies.

The focus of our analysis is on optimal asset allocation for the endowment portfolio. However, the nature and size of a university's non-endowment assets significantly influence optimal policy for spending endowment. For a given overall expenditure rate as a fraction of the university's total net worth, the optimal spending rate out of endowment will vary, depending on the fraction of net worth represented by non-endowment assets, the expected growth rate of cash flows generated by those assets, and capitalization rates. Hence, neglecting those other assets will generally bias the optimal expenditure policy for endowment.

In addition to taking account of non-endowment assets, our perspective on asset allocation differs from the norm because it takes account of the uncertainty surrounding the costs of the various activities such as education, research, and knowledge storage that define the purpose of the university. The breakdown of activities can of course be considerably more refined. For instance, one activity could be the education of a full-tuition-paying undergraduate, and a second could be the education of an undergraduate who receives financial aid. The unit (net) cost of the former is the unit cost of providing the education less the tuition received, and the unit cost of the latter is the cost plus the financial aid given. An important function of endowment investments is to hedge against unanticipated changes in the costs of university activities.

Consider, for example, the decision as to how much (if any) of the university's endowment to invest in local residential real estate. From a standard mean-variance efficiency analysis, it is unlikely that any material portion of the endowment should be invested in this asset class. However, consider the cost structure faced by the university for providing teaching and research. Perhaps the single largest component is faculty salaries. Universities of the same type and quality compete for faculty from the same talent pools. To be competitive, they must offer a similar standard of living not just similar salaries. Probably the largest part of the differences among universities in the cost of providing this same standard of living is local housing costs. The university that invests in local residential housing hedges itself against this future cost uncertainty by acquiring an asset whose value is higher than expected when the differential cost of faculty salaries is higher than expected. This same asset may also provide a hedge against unanticipated higher costs of off-campus housing for students that would in turn require more financial aid if the university is to compete for the best students. The prescribed targeted investment in very specific real estate assets to hedge against an unanticipated rise in a particular university's costs of faculty salaries and student aid should not be confused with the often-stated (but empirically questionable) assertion that investments in real estate generally are a good hedge against general inflation. Inflation-indexed government bonds, such as Treasury Inflation-Protected Securities are the efficient instruments for that objective.

Similar arguments could be used to justify targeted investment of endowment in various commodities such as oil and natural gas to hedge against unanticipated changes in energy costs. Uncertainty about those costs is especially significant for universities located in extreme climates and for universities with major laboratories and medical facilities that consume large quantities of energy.

In the same fashion, the hedging analysis of whether selling tuition forward is risk reducing for the university cannot be made without understanding the interactive risk structures among both assets and liabilities.

In sum, one cannot properly evaluate the financial decisions of the institution without considering the risks and returns for the total wealth of the institution. Advice and products of the future will adopt this comprehensive perspective.

1.5 Financial Services for Non-Financial Firms in the Future

The optimal management of corporate pension assets follows closely the endowment-model prescription. Indeed, taking into account risks on both sides of the balance sheet is fundamental to providing effective financial services to non-financial firms in general. *Enterprise risk management* is one term for such a unified approach. The movement from tactical to strategic application of currency, interest rate, commodity, and equities hedging is already underway. The next major step is to integrate operational, market, credit and traditional insurance risk management. To implement such integration requires connecting the decisions on operations, on the use of contractual agreements to hedge targeted exposures, and on the choice of capital structure.[23]

A particularly promising area for further development is the management of factor risks, particularly labor. Firms can be leveraged with their explicit and implicit labor contracts in parallel fashion to more traditional financial leverage with debt. Both temporary-employee firms and consulting firms serve the function of "labor intermediaries" that allow more efficient management of the risks for both those who supply labor and those who demand it. Their rapid growth, both in the United States and abroad, is probably a good measure of the significance of these factor risks to enterprises. The point again, *integrated risk management* for firms.

1.6 Government and Financial Services in the Future

A consequence of all this prospective technological change will be the need for greater analytical understanding of valuation and risk management by users, producers, and regulators of financial services. Furthermore, improvements in these products and services will not be effectively realized without concurrent changes in the financial "infrastructure"-the institutional interfaces between intermediaries and financial markets, regulatory practices, organization of trading, clearing, settlement, other back-office facilities, and management-information systems. To perform its functions as both user and overseer of the financial system, government in the future will need to both understand and make use of new financial technology.

Government also serves a critical role as financial intermediary. We already see a major effort underway nearly world wide with respect to restructuring the intermediary roles played by government and the private sector in providing pensions benefits in the retirement segment of the life cycle. Even if the responsibility for retirement benefits shifts largely to the private sector, government must still assess the risks it is underwriting.[24] These can be explicit guarantees as in the case of corporate pension insurance (in the United States

[23] For a compact but comprehensive non-technical overview of modern-day integrated risk management for the firm, see Meulbroek (2000).

[24] See Bodie (1996, 2000), Bodie and Merton (1993), Marcus (1987), Merton (1983a), Smetters (1997), Sosin (1980), and Turvey (1992).

through the Pension Guarantee Insurance Corporation) and implicit ones in its role as the "guarantor of last resort" for a systemic shortfall in benefits that affects an entire generation of retirees. Government is almost surely the only viable provider in adequate size of long-dated, default-free inflation-indexed debt which can be used by private-sector financial intermediaries as the prime hedging asset for issuing life retirement annuity products that are protected against inflation. With all the current discussion in the United States about retiring large portions of the government debt, policy should ensure that an adequate that an adequate supply of such debt is available for this intermediation purpose.

Application of new financial technology is critical to the future provision of risk-accounting standards, designing monetary and fiscal policies, implementing stabilization programs, and financial-system regulation. Many experts on monetary policy[25] have expressed serious concerns about how financial innovation has been eroding the ability of central banks to conduct monetary policy through traditional channels. Much the same concern has been expressed about the effect of financial innovation on some fiscal and regulatory policies as well. Such concerns are manifestly valid to the extent that the effectiveness of these traditional channels rests on large frictions of transaction costs, institutional rigidities, and institutionally defined regulations. Indeed, policymakers who continue to depend on such channel frictions are effectively speculating against the long-run trend of declining transaction costs and growing flexibility in institutional design. However, financial innovation and improved technology also opens new opportunities for government policymakers to perform their financial functions more effectively.[26]

To illustrate how governments of the future might use modern financial technology to pursue their policies more effectively, we present three examples: 1) "automatic" open-market operations for stabilization of interest rates or currencies;[27] 2) providing international diversification to the domestic population under capital controls;[28] 3) measuring and controlling country risks.[29]

In the spring and summer of 1990, the German government issued a sizeable private placement of ten-year maturity Schuldschein bonds with put-option provisions. They were just like standard government bonds, except they had the feature that the holders can put them back to the government for a fixed price.

---

[25] Cf. Friedman (2000).

[26] The Federal Reserve has for some time used implied volatilities derived from prices of traded options on government bonds as an estimate of the market's current assessment of future interest rate uncertainty. See Nasar (1992). Other areas to consider applying option theory are in evaluating physical capital project alternatives (Trigeorgis, 1993) and education and training policy alternatives for human capital development (S.J. Merton, 1992).

[27] For expanded discussion of this example, see Merton (1995b, 1997).

[28] This example is taken from Merton (1990).

[29] See also Merton (1999).

By issuing those bonds, the German government in effect introduced a pre-programmed dynamic stabilization policy. How is that? Suppose that it had issued a standard ten-year bond instead. Suppose further that afterwards interest rates start to rise, and therefore, bond prices fall. Normal ten-year bonds would fall in price in line with interest rate rises. But what happens to the bonds with the put option? The put bonds will not decline as much as the normal ten-year ones. Furthermore, the rate of decline in the put bonds becomes less and less, until they cease to decline at all. At that point, they will actually begin to behave just like a short-term money instrument. In terms of 'hedge ratios' or exposures relative to a normal ten-year bond, what is happening?

To answer, consider a single-factor interest-rate model with dynamics described by a diffusion process. If $B(t)$ denotes the price of a standard ten-year bond, then we can express the price of the puttable bonds as $F(B, t)$, where F is derived from a replicating trading strategy using contingent-claims analysis. From that analysis, the puttable bonds are economically equivalent in exposure to a portfolio of $\partial F/\partial B$ units of the standard ten-year bond and $[F - B\partial F/\partial B]$ invested in the shortest-maturity Treasury bill. It is straightforward to show that $0 \leq \partial F/\partial B \leq 1$ and that F is convex which implies $\partial^2 F/\partial B^2 > 0$. It follows that as the price B falls, the equivalent number of units of B represented by the puttable bonds, $\partial F/\partial B$, also falls.

In effect, because of the puts, the hedge ratio or equivalent number of ten-year bonds for each put bond gets smaller and smaller as the price of the ten-year bond falls. It is thus as if government were repurchasing normal bonds as in a regular open-market operation. In economic effect, the government is taking the interest-rate risk back from holders as if it was purchasing bonds even though it had not actually done so. If instead interest rates were to fall and bond prices rise, then the puts would become more out-of-the-money, the equivalent number of ten-year bonds per put bond rises, and the outstanding bond exposure held by investors would increase, which is effectively the same as the government issuing more bonds. Note that the decrease or increase in the equivalent bond exposure outstanding takes place immediately as interest rates change, without requiring that the bonds actually be put back to the government. So, in effect by issuing those put bonds, the German government put into place an automatic stabilizer to the extent that 'stabilization' means to 'lean' against market movements and buy bonds when bond price goes down and sell bonds when they go up. That is, the put bonds function as the equivalent of a dynamic, 'open-market,' trading operation without any need for actual transactions.

In comparison to traditional open-market activity, the put-option-bond automatic stabilizer will work well even over weekends, over non-trading days, and over crashes, especially in an environment with trading going on around the world because the central bank does not have to be on the scene to do the open-market operations. It automatically 'kicks in' as soon as events occur because it is built into the structure of the securities.

Turning from stabilization policy using open-market operations, we next examine the use of modern financial technology to reduce an unintended risk cost imposed on the domestic population from implementing capital controls. Numerous empirical studies of stock market returns have documented the gains in diversification from investing internationally. By diversifying across the world stock markets, there is significant improvement in the efficient frontier of risk versus expected return. As we know, the last decade has seen widespread implementation of such international diversification among investors in the large developed countries with the major stock markets. However, international diversification has not yet evolved in many smaller countries where indeed it may be relatively more important.

A major barrier to foreign stock market investment by citizens of some of these countries is capital controls, imposed by their governments to prevent flight of domestic capital. A common rationale for such restrictions is that they reduce the risk that the local economy will have inadequate domestic investment to promote growth. Another potential barrier is that the transaction cost paid by foreign investors to buy shares directly in these domestic stock markets can be so large that it offsets any diversification benefits that would otherwise accrue. The cost in lost welfare from less-efficient diversification affects both large-country and small-country citizens. However, the per capita magnitude of the cost is much larger for the latter, since the potential gains from international diversification are greatest for citizens of the smaller countries with domestic economies that are by necessity less well diversified. An additional cost may be that domestic physical investment is driven to become more diversified than would otherwise be efficient according to the principle of comparative advantage.

Of course, one (and perhaps the best) solution is to eliminate capital flow restrictions and open capital markets. However, with the capital controls taken as a given, a constrained solution exists that separates the capital-flow effects of investment from its risk-sharing aspects. Suppose that small-country domestic investors who already own the domestic equity (perhaps through domestic mutual funds or financial intermediaries) were to enter into "swap" agreements with large foreign investors. In the swap, the total return per dollar on the small country's stock market is exchanged annually for the total return per dollar on a market-value weighted-average of the world stock markets. This exchange of returns could be in a common currency, dollars, as described or adjusted to different currencies along similar lines to currency-swap agreements. The magnitudes of the dollar exchanges are determined by the "notional" or principal amount of the swap to which per dollar return differences apply. As is the usual case with swaps, there is no initial payment by either party to the other for entering the agreement.

Without pursuing further the details of implementation, we see that the swap agreement effectively transfers the risk of the small-country stock market to foreign investors and provides the domestic investors with the risk-return pattern of a well-diversified world portfolio. Since there are no initial payments between parties, there are no initial capital flows in or out of the country. Subsequent payments which may be either inflows or outflows involve only the *difference* between the *returns* on the two stock market indices, and no "principal" amounts flow. For example, on a notional or principal amount of $1 billion, if, ex post, the world stock market earns 10 percent and the small-country market earns 12 percent, there is only a flow of $(.12-.10)\times\$1$ billion or $20 million out of the country. Furthermore, the small-country investors make net payments out precisely when they can "best" afford it: namely, when their local market has outperformed the world markets. In those years in which the domestic market underperforms the world stock markets, the swap generates net cash flows into the country to its domestic investors. Hence, in our hypothetical example, if the small-country market earns 8 percent and the world stock market earns 11 percent, then domestic investors receive $(.11-.08)\times\$1$ billion = $30 million, a net cash inflow for the small country. Moreover, with this swap arrangement, trading and ownership of actual shares remain with domestic investors.

Foreign investors also benefit from the swap by avoiding the costs of trading in individual securities in the local markets and by not having the problems of corporate control issues that arise when foreigners acquire large ownership positions in domestic companies. Unlike standard cash investments in equities or debt, the default or expropriation exposure of foreign investors is limited to the difference in returns instead of the total gross return plus principal (in our example, $20 million versus $1.12 billion).

The potential exposure of foreign investors to manipulation by local investors is probably less for the swap than for direct transactions in individual stocks. It is more difficult to manipulate a broad market index than the price of a single stock. Even if settlement intervals for swaps are standardized at six months or one year, the calendar settlement dates will differ for each swap, depending upon the date of its initiation. Hence, with some swaps being settled every day, manipulators would have to keep the prices of shares permanently low to succeed. Furthermore, with the settlement terms of swaps based on the per-period rate of return, an artificially low price (and low rate of return) for settlement this year will induce an artificially high rate of return for settlement next year. Thus, gains from manipulation in the first period are given back in the second, unless the price can be kept low over the entire life of the swap. Since typical swap contract maturities range from two to ten years (with semi-annual or annual settlements), this would be difficult to achieve.

Note that this derivative-security innovation is *not* designed to circumvent the stated objective of the capital-control regulation, to prevent domestic capital flight. Instead, it is designed to eliminate (or at least reduce) the unintended and undesirable "side effects" of this policy on efficient risk bearing and diversification. Whether or not this proposed solution using a swap turns out to be an effective real-world solution is not the central point of the exercise here. Rather, it is to demonstrate how a modern financial technological innovation by government could help reduce the social cost of using "blunt" traditional policy tools that affect a number of countries around the world.

A similar but considerably broader prospective application of modern financial technology by government is the measurement and management of country risk. That is, we ask "How do we explain different countries' relative performance and the variations in performance across regions and what to do about it?"

That question prompts another one: How much of what we observe *ex post* as differences in performance is a consequence of *ex ante* different risk profiles versus different management and government policy decisions? For instance, Taiwan is heavily into electronics but produces no automobiles. More generally, few countries, if any, are well diversified when measured against the world market portfolio, the theoretically best-diversified portfolio if all assets including human capital, were traded or could be hedged and there were no transaction costs.

A non-traditional approach to address the performance issue and its implication for evaluating policy is to apply the technology of a well-studied problem in risk and performance measurement for investment management and financial firms. This is the problem of configuring all the decomposition and reintegration of risk-factor exposures that must be determined within a financial institution before the aggregate risk measures such as value-at-risk (*VAR*) can be applied. I believe that this technology, if properly adapted, can be used to measure country risk exposures.

In practice, measuring the differences in country exposures is not a simple task since many asset classes are not traded at all. But this is structurally the same problem faced in the risk measurement of non-traded assets and liabilities in financial institutions. In short, it is like the challenges of extending the VAR and stress-testing concepts to include the domain of non-traded assets and liabilities. But as with the application to financial institutions, I see this as a tough engineering problem, not one of new science... we know how to approach it in principle and what we need to model, but actually doing it is the challenge.

As with conventional private-sector applications, the country risk exposures give us important information about the dynamics of future changes that cannot be inferred from the standard "country" accounting statements, either the country balance sheet or the country income or flow-of-funds statements. That is, information not extractable from an accurate listing of the value of assets including foreign reserves, or from the trade flows or the capital flows, even if they are all mark-to-market numbers.

As we discover with more conventional applications of risk management systems, once we can measure the risk exposures we have, it is difficult to resist exploring whether we could improve economic efficiency and risk sharing by changing those exposures. Again, take the example of Taiwan. Suppose it decided to try to align its risk exposures more with the world portfolio. In the past, that might lead to an industrial policy to develop an automobile industry... a truly inefficient solution!

However, as another application of the ubiquitous swap contract, it is now feasible to separate the risk exposure decisions from the investment decisions. Instead of physically building a new industry, we can imagine Taiwan implementing its risk policy by entering into swap contracts in which it is a payer of the returns on a world electronics portfolio and a receiver of the returns on a world automobile portfolio.

Would such a swap be feasible? It is certainly structurally attractive. On the ability to pay, Taiwan is a net payer when electronics outperforms autos and a net receiver when electronics underperforms. There is no moral hazard or major asymmetric information problem for the country's counterparts, because payments are not based on country-specific performance in an industry, but are instead based on its global performance.

For the same reason, it avoids the political economic issue that the country's government can be accused of "giving away" its best assets as sometimes happens when foreigners buy the shares of its industries, because the country gets to keep its "alpha," Expropriation risk is also minimized, both because there is no principal exposure and because (returning to the first structural point) the likely ability to pay is aligned with the liability. Finally, while the useful implementation of such a swap obviously requires a large-size market, there are natural counterparts: other countries seeking alignment of their risk exposures.

These points seem to mitigate the usual incentive and information asymmetry problems for transactions with sovereigns. The technical problems of building a set of surrogate portfolios to use as benchmarks for risk measurement and contract specification are well understood. Initially at least, using mixtures of traded indices as the underlying asset for swap purposes would make the liquidity much better and the settlement mechanics easier. Contract credit risk is important but here too we know a lot about designing solutions, whether by a combination of mark-to-market collateral, purchase of private-sector performance guarantees, or efforts involving government and quasi-government institutional guarantees.

While the benefits of country-risk management systems and the associated markets would be expected to accrue to all, those in smaller countries with developing financial systems have the greater potential to benefit. With more concentrated investment opportunities, they should gain disproportionately from developing global access for capital and, perhaps more importantly, from more efficient allocation of risk.

Moreover, if they design their financial system using the most-up-to-date financial technology, these countries can "leap-frog" existing systems in terms of efficiency. In so doing, they can dramatically reduce the cost of investment capital and thereby materially increase national wealth.

With the developed countries, Japan and EMU Europe in particular, and the emerging ones both working on major changes in their financial systems, this may be an especially opportune time to explore country risk management. It is certainly an opportune time to be a finance professional, financial engineer, or financial architect.

As we take stock on this centennial of Bachelier's thesis, the influence of that work on 20th Century science and practice of finance is unmistakable. Its impact has been global and truly extraordinary. I feel secure with the forecast that when reviewed again at its bicentennial, Bachelier's influence and impact on both 21st Century finance science and finance practice will be even more extraordinary.

References

Arrow, K.J. (1953), "Le Rôle des Valeurs Boursières pour la Répartition la Meilleure des Risques," Econometrie, Colloques Internationaux du Centre National de la Recherche Scientifique, Vol. XI, Paris, 41–7.

Bachelier, L. (1900), "Théorie de la Spéculation," Annales de l'Ecole Normale Supérieure, 3, Paris: Gauthier-Villars. English translation in Cootner, P.H. (1964).

Banz, R.W. and M.H. Miller (1978), "Prices for State-Contingent Claims: Some Estimates and Applications," Journal of Business, 51 (October): 653–72.

Bernstein, P.L. (1992), Capital Ideas: The Improbable Origins of Modern Wall Street, New York: Free Press.

Black, F. (1989), "How We Came Up With the Option Formula," Journal of Portfolio Management, 15 (Winter): 4–8.

Black F. and M. Scholes (1973), "The Pricing of Options and Corporate Liabilities," Journal of Political Economy, 81 (May-June): 637–54.

Bodie, Z. (1995), "On the Risk of Stocks in the Long Run," Financial Analysts Journal, (May-June): 18–22.

Bodie, Z. (1996), "What the Pension Benefit Guaranty Corporation Can Learn from the Federal Savings-and-Loan Insurance Corporation," Journal of Financial Services Research, 10 (January): 83–100.

Bodie, Z. (2000), "Financial Engineering and Social Security Reform," in Risk Aspects of Investment-Based Social Security Reform, eds. J. Campbell and M. Feldstein, Chicago: University of Chicago Press.

Bodie, Z. and R.C. Merton (1993), "Pension Benefit Guarantees in the United States: A Functional Analysis," in R. Schmitt, ed., The Future of Pensions in the United States, Pension Research Council, Philadelphia: University of Pennsylvania Press: 195–234.

Bodie, Z., R.C. Merton and W. Samuelson (1992), "Labor Supply Flexibility and Portfolio Choice in a Life-Cycle Model," Journal of Economic Dynamics and Control, 16: 427–449.

Breeden, D.T. (1979), "An Intertemporal Asset Pricing Model with Stochastic Consumption and Investment Opportunities," Journal of Financial Economics, 7 (September): 265–96.

Breeden, D.T. and R. Litzenberger (1978), "Prices of State-Contingent Claims Implicit in Option Prices," Journal of Business, 51 (October): 621–51.

Claessens, S., T. Glaessner and D. Klingebiel (2000), "Electronic Finance: Reshaping the Financial Landscape Around the World," Financial Sector Discussion Paper No. 4 (September) The World Bank.

Cootner, P.H., ed. (1964), The Random Character of Stock Market Prices, Cambridge, MA: MIT Press.

Cox, J.C. and C. Huang (1989), "Optimum Consumption and Portfolio Policies When Asset Prices Follow a Diffusion Process," Journal of Economic Theory, 49 (October): 33–83.

Crane, D., et al (1995), The Global Financial System: A Functional Perspective, Boston: Harvard Business School Press.

Cummins, J.D. (1988), "Risk-Based Premiums for Insurance Guarantee Funds," Journal of Finance, 43 (September): 823–89.

Cummins, J.D. and H. Geman (1995), "Pricing Catastrophe Insurance Futures and Call Spreads: An Arbitrage Approach," Journal of Fixed Income, (March): 46–57.

Friedman, B.M. (2000), "Decoupling at the Margin: The Threat to Monetary Policy from the Electronic Revolution in Banking," National Bureau of Economic Research Working Paper No. W7955, Cambridge, MA: October.

Hakansson, N.H. (1976), "The Purchasing Power Fund: A New Kind of Financial Intermediary," Financial Analysts Journal, 32 (November-December): 49–59.

Harrington, S., S. Mann, and G. Niehaus (1995), "Insurer Capital Structure Decisions and the Viability of Insurance Derivatives," Journal of Risk and Insurance, 62 (September): 483–508.

Hayes, J., J. Cole, and D. Meiselman (1993), "Health Insurance Derivatives: The Newest Application of Modern Financial Risk Management," Business Economics, 28 (April): 36–40.

Itô, K. (1951),"On Stochastic Differential Equations," Memoirs of American Mathematical Society, 4: 1–51.

Itô, K. (1987), Kiyoshi Itô Selected Papers, New York: Springer-Verlag.

Itô, K. and H.P. McKean, Jr. (1964), Diffusion Processes and Their Sample Paths, New York: Academic Press.

Kraus, A. and S.A. Ross (1982), "The Determination of Fair Profits for the Property-Liability Insurance Firm," Journal of Finance, 37 (September): 1015–28.

Jin, L., L. Kogan, T. Lim, J. Taylor, and A. Lo (1997), "The Derivatives Sourcebook: A Bibliography of Applications of the Black-Scholes/Merton Option-Pricing Model," Sloan School of Management, Massachusetts Institute of Technology Working Paper. Web site: http://forbin.mit.edu/dsp/

Marcus, A.J. (1987), "Corporate Pension Policy and the Value of PBGC Insurance," in Z. Bodie, J.B. Shoven, and D.A. Wise, eds., Issues in Pension Economics, Chicago, IL: University of Chicago Press.

Markowitz, H. (1952), "Portfolio Selection," Journal of Finance,7 (March): 77–91.

Mayers, D. (1972), "Non-Marketable Assets and Capital Market Equilibrium Under Uncertainty," in M. Jensen, ed., Studies in the Theory of Capital Markets, New York: Praeger.

McKean, Jr., H.P. (1965), "Appendix: A Free Boundary Problem for the Heat Equation Arising From a Problem in Mathematical Economics," Industrial Management Review, 6 (Spring): 32–9.

Melnikov, A.V. (1999), "On Innovation and Risk Aspects of Financial System Evolution," Questions of Risk Analysis, 1,1:22–27 (in Russian).

Melnikov, A.V. (forthcoming), "Financial System: Innovations and Pricing Risks," Journal of Korean Mathematical Society.

Merton, R.C. (1969), "Lifetime Portfolio Selection Under Uncertainty: The Continuous-Time Case," Review of Economics and Statistics, 51 (August): 247–57.

Merton, R.C. (1970), "A Dynamic General Equilibrium Model of the Asset Market and its Application to the Pricing of the Capital Structure of the Firm," Working Paper No. 497-70, Sloan School of Management, Massachusetts Institute of Technology, Cambridge, MA. Reprinted in Merton (1992, Ch. 11).

Merton, R.C. (1971), "Optimum Consumption and Portfolio Rules in a Continuous-Time Model," Journal of Economic Theory, 3 (December): 373–413.

Merton, R.C. (1972), "Appendix: Continuous-Time Speculative Processes," in R.H. Day and S.M. Robinson, eds., Mathematical Topics in Economic Theory and Computation, Philadelphia, PA: Society for Industrial and Applied Mathematics. Reprinted in SIAM Review, 15 (January 1973): 34–8.

Merton, R.C. (1973a), "Theory of Rational Option Pricing," Bell Journal of Economics and Management Science, 4 (Spring): 141–83.

Merton, R.C. (1973b), "An Intertemporal Capital Asset Pricing Model," Econometrica, 41 (September): 867–87.

Merton, R.C. (1977a), "An Analytic Derivation of the Cost of Deposit Insurance and Loan Guarantees: An Application of Modern Option Pricing Theory," Journal of Banking and Finance, 1 (June): 3–11.

Merton, R.C. (1977b), "A Reexamination of the Capital Asset Pricing Model," in Studies in Risk and Return, J. Bicksler and I. Friend, eds., Cambridge, MA: Ballinger, Vol. I & II: 141–159.

Merton, R.C. (1982), "On the Microeconomic Theory of Investment Under Uncertainty," in Handbook of Mathematical Economics, Volume II, K. Arrow and M. Intriligator, eds., Amsterdam: North-Holland Publishing Company.

Merton, R.C. (1983a), "On Consumption-Indexed Public Pension Plans," in Financial Aspects of the U.S. Pension System, Z. Bodie and J. Shoven, eds., Chicago: University of Chicago Press.

Merton, R.C. (1983b), "Financial Economics," in E.C. Brown and R.M. Solow, eds., Paul Samuelson and Modern Economic Theory, New York: McGraw-Hill.

Merton, R.C. (1990), "The Financial System and Economic Performance," Journal of Financial Services Research, 4, (December): 263–300.

Merton, R.C. (1992), Continuous-Time Finance, Revised Edition, Cambridge, MA: Basil Blackwell.

Merton, R.C. (1993a), "Operation and Regulation in Financial Intermediation: A Functional Perspective," in P. Englund, ed., Operation and Regulation of Financial Markets, Stockholm: The Economic Council.

Merton, R.C. (1993b), "Optimal Investment Strategies for University Endowment Funds," in C. Clotfelter and M. Rothschild, eds., Studies of Supply and Demand in Higher Education, Chicago: University of Chicago Press.

Merton, R.C. (1994), "Influence of Mathematical Models in Finance on Practice: Past, Present and Future," Philosophical Transactions of the Royal Society of London, Vol. 347, (June): 451–463.

Merton, R.C. (1995a), "A Functional Perspective of Financial Intermediation," Financial Management, Vol. 24, No. 2, (Summer): 23–41.

Merton, R.C. (1995b), "Financial Innovation and the Management and Regulation of Financial Institutions," Journal of Banking and Finance, 19 (June): 461–481.

Merton, R.C. (1997), "On the Role of the Wiener Process in Finance Theory and Practice: The Case of Replicating Portfolios," in D. Jerison, I.M. Singer, and D.W. Stroock, eds., The Legacy of Norbert Wiener: A Centennial Symposium, PSPM Series, Vol. 60, Providence, RI: American Mathematical Society.

Merton, R.C. (1998), "Applications of Option-Pricing Theory: Twenty-Five Years Later," Les Prix Nobel 1997, Stockholm: Nobel Foundation; reprinted in American Economic Review, (June): 323–349.

Merton, R.C. (1999), "Commentary: Finance Theory and Future Trends: The Shift to Integration," Risk, (July): 48–50.

Merton, R.C. and Z. Bodie (1992), "On the Management of Financial Guarantees," Financial Management, 21 (Winter): 87–109.

Merton, R.C. and Z. Bodie (1993), "Deposit Insurance Reform: A Functional Approach," in A. Meltzer and C. Plosser, eds., Carnegie-Rochester Conference Series on Public Policy, Vol. 38, (June): 1–34.

Merton, R.C. and M.S. Scholes (1995), "Fischer Black," Journal of Finance, 50 (December): 1359–1370.

Merton, S.J. (1992), "Options Pricing in the Real World of Uncertainties: Educating Towards a Flexible Labor Force," B.A. Thesis, Department of Economics, Harvard University, March 18, 1992.

Meulbroek, L.K. (2000), "Total Strategies for Company-Wide Risk Control," Financial Times, 9 May, Mastering Risk series: 1–4.

Mody, A. (1996), "Valuing and Accounting for Loan Guarantees," World Bank Research Observer, 11, No. 1:119–42.

Nasar, S. (1992), "For Fed, a New Set of Tea Leaves," New York Times, 5 July, sec. D:1.

Neal, R.S. (1996), "Credit Derivatives: New Financial Instruments for Controlling Credit Risk," Federal Reserve Bank of Kansas City Economic Review, 81 (Second Quarter): 14–27.

Ross, S. A. (1976a), "Arbitrage Theory of Capital Asset Pricing," Journal of Economic Theory, 13 (December): 341–60.

Ross, S. A. (1976b), "Options and Efficiency," Quarterly Journal of Economics, 90 (February): 75–89.

Samuelson, P.A. (1965), "Rational Theory of Warrant Pricing," Industrial Management Review, 6 (Spring): 13–31.

Samuelson, P.A. (1971), "The 'Fallacy' of Maximizing the Geometric Mean in Long Sequences of Investing or Gambling," Proceedings of the National Academy of Sciences, 68: 2493–2496.

Samuelson, P.A. (1972), "Mathematics of Speculative Price," in R.H. Day and S.M. Robinson, eds., Mathematical Topics in Economic Theory and Computation, Philadelphia, PA: Society for Industrial and Applied Mathematics. Reprinted in SIAM Review, 15 (January 1973): 1–42.

Samuelson, P.A. (1979), "Why We Should Not Make Mean Log of Wealth Big Though Years to Act Are Long," Journal of Banking and Finance, 3: 305–307.

Samuelson, P.A. and R.C. Merton (1969), "A Complete Model of Warrant Pricing that Maximizes Utility," Industrial Management Review, 10 (Winter): 17–46.

Scholes, M.S. (1998), "Derivatives in a Dynamic Environment," Les Prix Nobel 1997, Stockholm: Nobel Foundation.

Sharpe, W. F. (1964), "Capital Asset Prices: A Theory of Market Equilibrium Under Conditions of Risk," Journal of Finance, 19 (September): 425–42.

Smetters, K. (1997), "Investing the Social Security Trust Fund in Equities: An Option Pricing Approach," Technical Paper Series, Macroeconomic Analysis and Tax Analysis Divisions, Washington, DC.

Sosin, H. (1980), "On the Valuation of Federal Loan Guarantees to Corporations," Journal of Finance, 35 (December): 1209–21.

Tobin, J. (1958), "Liquidity Preference as Behavior Towards Risk," Review of Economic Studies, 25 (February): 68–85.

Trigeorgis, L. (1993), "Real Options and Interactions with Financial Flexibility," Financial Management, 22 (Autumn): 202–24.

Turvey, C. (1992), "Evaluating Premiums for Farm Income Insurance Policy," Canadian Journal of Agricultural Economics, 40, No. 2:183–98.

Williams, J. (1978), "Risk, Human Capital, and the Investor's Portfolio," Journal of Business, 51: 65–89.

Williams, J. (1979), "Uncertainty and the Accumulation of Human Capital Over the Life Cycle," Journal of Business, 52: 521–48.

Brownian Motion and the General Diffusion: Scale & Clock

Henry P. McKean

CIMS, 251 Mercer Street, New York, NY 10012 USA

1 Introduction

People in financial mathematics make their living (in part) by Itô's lemma: $df(b) = f'(b)db + \frac{1}{2}f''(b)dt$ in which f is a smooth function and $b(t) : t \geq 0$ is the standard Brownian motion $\mathbf{BM}(1)$ starting at $b(0) = 0$. It lies at the back of Itô's local representation $dx = \sigma(x)db + m(x)dt$ of the general 1-dimensional diffusion $\mathfrak{x}$ with paths $x(t) : t \geq 0$ and infinitesimal operator $\mathfrak{G} = \frac{1}{2}\sigma^2(x)d^2/dx^2 + m(x)d/dx$, as suggested by the simplest example $x(t) = x(0) + \sigma b(t) + mt$ with constant σ and m. I will speak about a less familiar reduction of $\mathfrak{x}$ to standard Brownian motion, by change of scale and clock, and other aspects of this main idea. It is appropriate to do so on the centenary of Bachelier's thesis in which is to be found the very first use of Brownian paths per se. P. Lévy, K. Itô, and P. Malliavin, to name only the most illustrious, have gone much more deeply into the path function picture, but Bachelier had the idea first, way before its general recognition as the best way of thinking about motion subject to chance. I do not give financial applications, but see for instance Geman & Yor [1993] and Donati-Martin, Matsumoto, & Yor [2000].

2 Scale

The "natural" scale for the general diffusion $\mathfrak{x}$ is the invention of Feller [1954]. Let x^* be an increasing solution of $\mathfrak{G}x^* \equiv 0$, determined up to an affine substitution $x^* \to Ax^* + B$. Then $x^*[x(t)] = x^*(t)$ is a martingale, and, with $a < x < b$ and $T =$ the exit time min $[t \geq 0 : x(t) = a \text{ or } b]$, you have

$$P_x[x(T) = b] = E_x\left[\frac{x^*(T) - a^*}{b^* - a^*}\right] = E_x\left[\frac{x^*(0) - a^*}{b^* - a^*}\right] = \frac{x^* - a^*}{b^* - a^*},$$

with a self-evident notation, and also, by Itô's lemma

$$dx^*(t) = x^{*\prime}\sigma(x)db(t) + \mathfrak{G}x^*(x)dt = \sigma^*[x^*(t)]db(t)$$

with $\sigma^*(x^*) = x^{*\prime}\sigma(x)$, so that, by change of scale, the drift is removed, $\mathfrak{G}$ is reduced to $\frac{1}{2}\sigma^2(x)d^2/dx^2$, and you have $P_x[x(T) = b] = \frac{x-a}{b-2}$ plain. I keep this reduced format in the next article, but first I make a brief additional advertisement for Feller's scale in which so many aspects of 1-dimensional diffusion are capable of transparent expression. Take $\mathfrak{G} = \frac{1}{2}\sigma^{*2}(x^*)d^2/dx^{*2}$ before reduction to natural scale, then if $x^*(\infty) < \infty$ and $\int_0^\infty \sigma^{*-2}(x^*)dx^* < \infty$ (and only then) the particle is capable both of running out to ∞ and coming back, and it must be told what to do out there: whether to die, or to "reflect", or whatever. The uneasy feeling that ∞ is "different" is only an illusion due to looking at things in the wrong scale.

3 Clocks

Now the reduced diffusion obeys $dx(t) = \sigma[x(t)]db(t)$ and $x(t) = x(0)+\sigma b(t)$ if σ is constant, or, what is the same, $x(t) = x(0)+b(\sigma^2 t)$ in view of the Brownian scaling $\sigma b(t/\sigma^2) \equiv b(t)$[1]. This simple remark can be adapted to non-constant σ in the form $x(t) = x(0) + b(T)$ with the "clock" $T(t) = \int_0^t \sigma^2[x(t)']dt'$, the rule being[2] $db(T) = \sqrt{dT/dt}db^*(t) = \sigma[x(t)]db^*(t)$ with a new Brownian motion $b^*(t) : t \geq 0$ depending on the original in a somewhat involved way. I clarify the content as follows: the several values of the inverse clock $T^{-1}(t)$ are stopping times for $\mathfrak{x}$ and $b^*(t) = x[T^{-1}(t)]$ is a standard Brownian motion. Here is a more direct, more helpful version of the recipe: if $b(t) : t \geq 0$ is standard Brownian motion and if $T(t)$ is the clock $\int_0^t \sigma^{-2}[b(t)']dt'$, then the several values of $T^{-1}(t)$ are Brownian stopping times and $x(t) = x(0) + b[T^{-1}(t)]$ is a copy of the desired diffusion: in fact,

$$dx(t) = db(T^{-1}) = \sqrt{dT^{-1}/dt}db^*(t) = \sigma[b(T^{-1})]db^*(t) = \sigma[x(t)]db^*(t)$$

with (as before) a new Brownian motion $b^*(t) : t \geq 0$[3].

4 An Extreme Example

Let $\sigma(x) = 1$ or $+\infty$ according as $x \geq 0$ or $x < 0$. The corresponding (formal) diffusion is standard Brownian motion to the right. To the left, it moves so fast that it spends no time there at all. What does the recipe of Sect. 3 say? The clock in the display is the cumulative time the Brownian motion spends to the right of the origin. It is flat on left-hand excursions, so the substitution $t \to T^{-1}(t)$ into the Brownian path erases these, pushing the right-hand excursions down until they abut, as in Fig. 1. The diffusion $x(t) = b(T^{-1})$ is one and the same as the reflecting Brownian motion; compare Itô-McKean [1965:81].

[1] The symbol $\equiv$ signifies "identical in law", *i.e.* with the same statistics.

[2] See McKean [1969] for such matters.

[3] See Itô-McKean [1965:167] for the full treatment and Volkonskii [1958 and 1961] for priority.

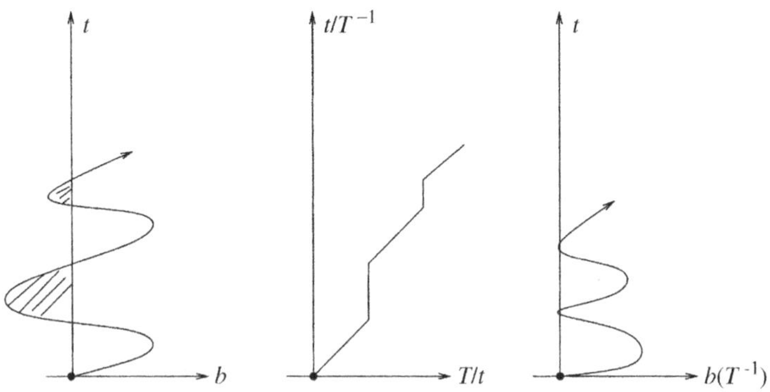

Fig. 1.

5 Higher Dimensions

The same thing works (with complications) in dimensions $d \geq 2$. Let $\mathfrak{x}$ be a diffusion in R^d and, for any fixed domain D and any starting point $x \in D$, introduce the hitting probabilities $h(x, dy) = P_x[x(T) \in dy]$, T being the exit time of $\mathfrak{x}$ from D[4]. These play the role of scale in dimension 1. Think of them as a road map showing where you can get to. The speed at which you go tells the rest: in fact, any two diffusions with the same road map can be reduced one to the other, by change of clock much as in Sect. 3[5], but with this complication: In dimension 1, any two road maps differ by change of scale, but in higher dimensions $d \geq 2$, there are infinitely many classes of road maps, inequivalent under change of coordinates. A nice example is provided by the diffusion performed by the eigenvalues $\lambda_+ > \lambda_-$ of the 2×2 symmetric matrix

$$\begin{bmatrix} \sqrt{2}b_{11}(t) & b_{12}(t) \\ b_{12}(t) & \sqrt{2}b_{22}(t) \end{bmatrix}$$

in which you see three independent copies of $\mathbf{BM}(1)$. Here

$$x_1(t) \equiv \frac{1}{2}(\lambda_+ + \lambda_-) = \frac{b_{11} + b_{22}}{\sqrt{2}} \text{ and } x_2(t) \equiv \frac{1}{2}(\lambda_+ - \lambda_-) = \sqrt{\left(\frac{b_{11} - b_{22}}{\sqrt{2}}\right)^2 + b_{12}^2}$$

are independent copies of $\mathbf{BM}(1)$ and $\mathbf{BES}(2)$. Reduction to $\mathbf{BM}(2)$ via a (smooth) change of coordinates $x \to y(x)$ and a (nice) change of clock requires 1) $\det(\partial y/\partial x) \neq 0$, 2) $\Delta y + (1/x_2)\partial y/\partial x_2 = 0$, and 3) $(\nabla y_1) \cdot (\nabla y_2) = 0$

[4] For the standard Brownian motion, $h(x, \bullet)$ is classical harmonic measure of ∂D as viewed from x inside.

[5] See McKean-Tanaka [1961] and Blumenthal-Getoor-McKean [1962].

with $|\nabla y_1| = |\nabla y_2|$. But 1) and 3) lead quickly to $\Delta y = 0$ and now you are in trouble: 2) requires y to be independent of x_2!

6 P. Lévy on the Conformal Invariance of BM(2)[6]

Let $\mathfrak{z}$ be **BM**(2) in complex format, as in $z(t) = b_1(t) + \sqrt{-1}b_2(t)$, and let $f(z)$ be analytic in a domain $D \subset \mathbb{C}$. Then, up to the exit time of $\mathfrak{z}$ from D, you have $df(z) = f'(z)dz + \frac{1}{2}f''(z)(dz)^2$ in which

$$(dz)^2 = (db_1)^2 + 2\sqrt{-1}db_1db_2 - (db_2)^2 = dt + 2\sqrt{-1} \times 0 - dt = 0 ,$$

by Itô's lemma. Thus, with $f[z(0)] = 0$, say, you have

$$f[z(t)] = \int_0^t f'(z)dz = \int_0^t |f'(z)|e^{\sqrt{-1}\theta}dz = \int_0^t |f'(z)|dz^*$$

with $\theta = \arg f'(z)$ and a new (complex) Brownian motion $\mathfrak{z}^*$, the point being that the isotropic quality of the Gaussian differential dz cannot be spoiled by an independent phase shift, whence $z^*(t) = \int_0^t e^{\sqrt{-1}\theta}dz$ is, itself, a Brownian motion. But now you can go one step further and write $f[z(t)] = \int_0^t |f'(z)|dz^* = z^{**}(T)$ with yet another Brownian motion $\mathfrak{z}^{**}$ and the clock $T(t) = \int_0^t |f'(z)|^2dt'$, which is to say that $z^{**}(t) = f[z(T^{-1})]$ is a standard Brownian motion. In short, application of a (local) analytic map to the Brownian locus produces a new locus with the same (Brownian) statistics: only the clock is changed. Here are four pretty examples.

Example 6.1. Let $D = \mathbb{C}$ and $f(z) = e^z$. Then, with $z(0) = 0$, say, $\exp[z(t)] = z^*(T)$ with $T(t) = \int_0^t |e^z|^2dt'$ and $z^*(0) = 1$. Now $\mathfrak{z}$ is unbounded, so $T(\infty) = \infty$ and, to the right, the whole Brownian locus $z^*[0,\infty)$ is seen. To the left, $\exp(z) \neq 0$, whence **BM**(2) does not hit any point $z \neq z(0)$ named in advance.

Example 6.2. Let $D = \mathbb{C} - 0$ and $f(z) = 1/z$. Then, with $z(0) = 1$, say, $1/z(t) = z^*(T)$ with $T(t) = \int_0^t |z|^{-4}dt'$ and $z^*(0) = 1$. Because $\mathfrak{z}$ is unbounded, $z^*(T)$ comes as close to the origin as you want, forcing $T(\infty) = \infty$ by Ex. 6.1, with the result that the locus $z[0,\infty)$ is (statistically) invariant under the inversion $z \to 1/z$; in particular, $\mathfrak{z}$ enters every disc centered at the origin, and it is a small step to the full statement that it hits every plane disc infinitely often as $t \uparrow \infty$.

Example 6.3. Let $D = \mathbb{C} - 0$ as before and $f(z) = \ln z$. The function is not single-valued but never mind: it can be made so by passing to its Riemann surface depicted in Fig. 2. Now if $z(0) = 1$, say, then $z(t) \neq 0$ for any $t \geq 0$, and the branch of $\ln[z(t)]$ starting at $\ln[z(0)] = 0$ looks like a new Brownian

[6] P. Lévy [1948:270].

motion $z^*(T)$ with clock $T(t) = \int_0^t |z|^{-2}dt$. $T(\infty) = \infty$ being obvious, it follows from Ex. 6.2 that $\mathfrak{z}$ returns to the vicinity of $z(0) = 1$, with winding number $0, \pm 1, \pm 2, \pm 3$, what you will, infinitely often as $t \uparrow \infty$; see Sect. 7 next below for more information obtained by the machinery of Sects. 2 and 3.

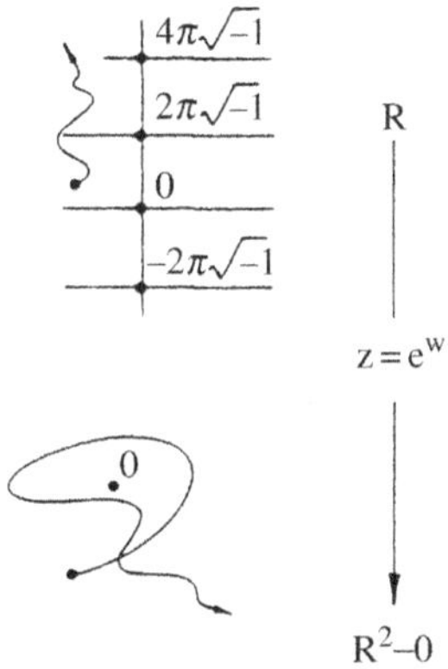

Fig. 2.

Example 6.4. Let D be the twice-punctured plane $\mathbb{C} - 0 - 1$ and let the open upper half-plane H be interpreted as its "universal cover". The points of H are deformation classes of paths $z(t) : 0 \le t \le T$ in the "base" D, starting at $z(0) = 2$, say, and recording, together with the end point $z(T)$ the winding of the path about the punctures 0 and 1. Thus, over every base point $z(T)$ below is stacked an infinite number of covering points, one such to each winding history, and there is an (analytic) "projection" k^2, which is indifferent to winding, mapping the whole stack of covering points to $z(T)$.[7] Fig. 3 depicts H divided up into "sheets" according to the action of the stability group $PSL(2, \mathbb{Z})$ of k^2. This is window-dressing and does not matter. What is important is that if you want to know how the Brownian motion downstairs winds about the punctures 0 and 1, then the thing to do is to lift it up to the cover via any branch of the function inverse to k^2 and see what happens there. But this is easy: The lifted motion is just another Brownian motion run with a complicated clock $T(t)$. Now Brownian motion in a half-plane $(x_2 > 0)$ hits the bordering line $(x_1 = 0)$. It follows that $T(\infty) < \infty$ is nothing but this passage time, whence the original Brownian motion in the punctured plane below returns infinitely often to the vicinity of its starting point in a progressively more and more complicated state of winding that never gets undone. Lyons-McKean [1984] and McKean-Sullivan [1984] proved a more surprising fact: that if you

[7] k^2 is "Jacobi's modulus" alias the "elliptic modular function".

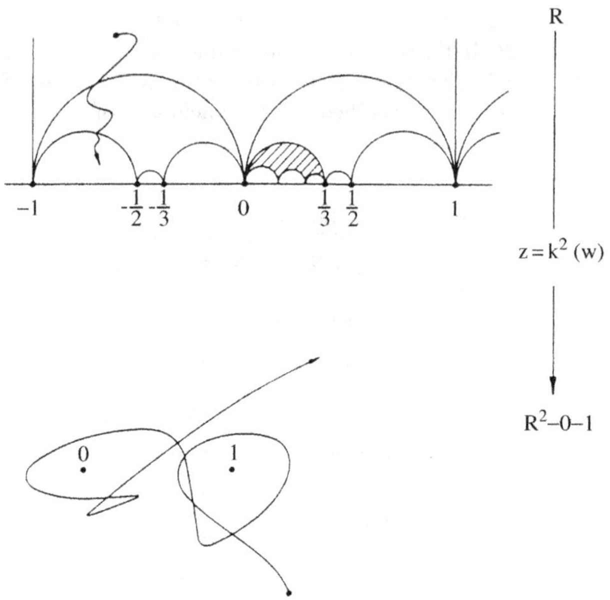

Fig. 3.

keep track only of the winding numbers about the two punctures, ignoring the full history, then the same thing happens – the path gets inextricably tangled up. Now the winding is recorded on an intermediate cover which looks like a parking garage of infinitely many floors: each floor is a copy of $\mathbb{C}$ provided with an infinite array of helical stairways, like the "stair" in Fig. 2, located at the points $2\pi\sqrt{-1}\mathbb{Z}$, leading, indifferently, to the floors above and below. The lifted path appears as a Brownian motion run with a clock $T(t)$, only now $T(\infty) = \infty$, so you see the full locus of the Brownian motion in the cover, and this is so vast that the traveller gets lost far far away and never comes back.

7 More on How BM(2) Winds About a Point

Let $\mathfrak{x}$ be standard 2-dimensional Brownian motion starting at $x(0) = (1,0)$, say, written in polar coordinates $0 < r < \infty$ and $0 \leq \theta < 2\pi$. You have

$$\mathfrak{G} = \frac{1}{2}\Delta = \frac{1}{2}\left(\frac{\partial^2}{\partial r^2} + \frac{1}{r}\frac{\partial}{\partial r}\right) + \frac{1}{r^2} \times \frac{1}{2}\frac{\partial^2}{\partial \theta^2}$$

from which it is plain that the total algebraic angle swept out by $x(t') : 0 \leq t' \leq t$ is of the form $\theta(t) = a(T)$ in which $a(t) : t \geq 0$ is a copy of $\mathbf{BM}(1)$ and $T(t)$ is the clock $\int_0^t r^{-2} dt'$ run with an independent Bessel process $\mathbf{BES}(2)$

with paths $r(t) : t \geq 0$ and infinitesimal operator $\frac{1}{2}(r^2/dr^2 + r^{-1}d/dr)$. The investigation of $\theta(t)$ for $x(0) = 1$, say, and $t \uparrow \infty$ provides a nice illustration of scale and clock, amplifying Ex. 6.3. Note first that $dr = db + (2r)^{-1}dt'$ with a copy of $\mathbf{BM}(1)$. The drift is removed by the natural scale $r^* = \ln r$: in fact, $dr^*(t) = \mathrm{e}^{-r^*}db(t)$, whence $r^*(t) = \ln r(t) = b[R^{-1}(t)]$ with the clock $R(t) = \int_0^t \mathrm{e}^{2b}dt'$ with yet another copy of $\mathbf{BM}(1)$ starting at $b(0) = 0$. Now

$$T(t) = \int_0^t r^{-2}dt' = \int_0^t \mathrm{e}^{-2b}(R^{-1})dt' = \int_0^{R^{-1}(t)} \mathrm{e}^{-2b(t')}dR(t') = R^{-1}(t) ,$$

so $\theta(t) = a[R^{-1}(t)]$, and, for each separate large time t, $R(t)$ is identical in law to

$$\int_0^t \mathrm{e}^{2\sqrt{t}b(t'/t)}dt' = t\int_0^1 \mathrm{e}^{2\sqrt{t}b(t')}dt' \cong \mathrm{e}^{2\sqrt{t}m}$$

with $m = \max[b(t') : t' \leq 1]$. Thus, $R^{-1}(t)$ is well-approximated (in law) by $(\ln t/2m)^2$. But

$$P_0\left(\frac{1}{m^2} < \xi\right) = P_0\left(\frac{1}{\sqrt{\xi}} < m\right) = \sqrt{\frac{2}{\pi}}\int_{1/\sqrt{\xi}}^{\infty} \mathrm{e}^{-x^2/2}dx = \int_0^{\xi} \mathrm{e}^{-1/2\eta}\frac{d\eta}{\sqrt{2\pi\eta^3}} ,$$

which is to say that $1/m^2$ is a one-sided stable variable with exponent $1/2$. The upshot is that $\theta(t) \cong a[(\ln t/2m)^2]$ is a copy of $\frac{1}{2}\ln t \times a(1/m^2)$ or very nearly so, $a(1/m^2)$ being the composition of $\mathbf{BM}(1)$ and the independent stable variable $1/m^2$. As such, it is distributed à la Cauchy:

$$\lim_{t\uparrow\infty} P_0\left[\frac{2\theta(t)}{\ln t} \leq \xi\right] = \frac{1}{\pi}\int_{-\infty}^{\xi}\frac{d\xi}{1+\eta^2} .$$

The proof is rough but the statement is correct. The logarithmic scaling could have been forseen: most of the winding takes place while $|x(t)| < 1$ and measure $[t' \leq t : |x(t')| < 1]$ is more or less identical in law to $\frac{1}{2}\ln t \times$ an exponential holding time[8].

8 Hyperbolic Brownian Motion

Let the open upper half-plane $x_2 > 0$ be equipped with Poincaré's hyperbolic line element $x_2^{-1}\sqrt{(dx_1)^2 + (dx_2)^2}$. This makes the Gaussian curvature everywhere equal to -1. The natural Laplacian invariant under the hyperbolic rigid motions is $\Delta = x_2^2(\partial^2/\partial x_1^2 + \partial^2/\partial x_2^2)$. $\Delta/2$ is the infinitesimal operator of the hyperbolic Brownian motion $\mathfrak{x}$ which may be seen under two different aspects.

[8] See, for example, Itô-McKean [1965:277].

Aspect 1 is a self-evident version of Sect. 3: $x(t) = b[T^{-1}(t)]$ with a standard 2-dimensional Brownian motion $b(t) : t \geq 0$, starting at $x(0)$, and the clock $T(t) = \int_0^t b_2^{-2} dt'$. Let m be the passage time $\min[t : b_2(t) = 0]$. $T(m) = \infty$ in view of the overestimate $|b_2(m-t)| \leq 2\sqrt{t \ln(1/t)}$ for $t \downarrow 0$, so $T^{-1}(\infty) = m$, which is to say that $\mathfrak{x}$ tends to a definite point $x(\infty) = [b_1(m), 0]$ of the bordering line; moreover, if $x_2(0) = h$, then $b_1(m)$ is the composition of 1-dimensional Brownian motion and the one-sided stable variable m of exponent 1/2 with density $h(2\pi t)^{-3/2} \exp(-h^2/2t)$, whence $x_1(\infty)$ is Cauchy distributed as in

$$P[x_1(\infty) \leq \xi] = \frac{h}{\pi} \int_{-\infty}^{\xi} \frac{d\eta}{\eta^2 + h^2} \; .$$

Aspect 2 imitates Sect. 7. The vertical motion x_2 is itself a diffusion: in fact, $dx_2 = x_2 db$ with a copy of $\mathbf{BM}(1)$, by which you recognize x_2 as the "geometric" Brownian motion $x_2(t) = \exp[b(t) - t/2]$. The horizontal motion $\mathfrak{x}_1$ is now seen under the aspect $x_1(t) = a[T(t)]$ with an independent copy of $\mathbf{BM}(1)$ and the clock $T(t) = \int_0^t x_2^2 dt'$; obviously, $T(\infty) < \infty$ so that $\mathfrak{x}$ stabilizes as $t \uparrow \infty$, as you know already.

Bibliography

Blumenthal, R., R. Getoor, and H. P. McKean: Markov processes with identical hitting probabilities. *Ill. J. Math.* **6** (1962) 402–420.

Donati-Martin, C., H. Matsumoto, and M. Yor: The law of geometric Brownian motion and its integral, revisited; application to conditional moments. This volume, p. 221–243.

Feller, N.: The general diffusion operator and positivity-preserving semi-groups in one dimension. *Ann. Math.* **60** (1954) 417–436.

Geman, H., and M. Yor: Asian options, Bessel processes, and perpetuities, *Math. Finance* **3**, No. 4 (1993) 349–375.

Itô K., and H. P. McKean: *Diffusion Processes and their Sample Paths.* Springer-Verlag, Berlin, Heidelberg, New York, 1965.

Lévy, P.: *Processus Stochastiques et Mouvement Brownien.* Gauthier-Villars, Paris, 1948.

Lyons, T., and H. P. McKean: Winding of the plane Brownian motion. *Adv. Math.* **51** (1984) 212–225.

McKean, H. P.: *Stochastic Integrals.* Academic Press, New York and London, 1969.

McKean, H. P., and D. Sullivan: Brownian motion and harmonic functions on the class surface of the thrice-punctured sphere. *Adv. Math.* **51** (1984) 203–211.

McKean, H. P., and H. Tanaka: Additive functionals of the Brownian path. *Mem. Coll. Sci. Kyôto* **33** (1961) 479–506.

Volkonskii, V.A.: Random substitution of time in strong Markov processes. *Teor. Veroyatnost.* **3** (1958) 332–350.

Volkonskii, V.A.: Construction of inhomogeneous Markov processes by means of a random time substitution. *Teor. Veroyatnost.* **6** (1961) 47–56.

Rare Events, Large Deviations

S.R.S. Varadhan

Courant Institute of Mathematical Sciences
New York University
251, Mercer Street,
New York, NY 10012
USA

1 Introduction

There are many situations where we want to estimate the probability of a rare event with some precision. For example if we toss a fair coin n times the probability of getting n heads in a row is clearly small, and for large n is very small. We know its exact value as 2^{-n} and in logarithmic scale we can write it as $e^{-n\log 2}$. While in this case the exact probability is easy to evaluate, there are many situations in which a direct exact calculation is impossible and we need to develop indirect methods that will provide us with estimates.

2 An Example

The first example of such a kind in fact came from the insurance industry in Scandinavia [6]. Suppose that the number of claims against the company in a single period is Poisson distributed with some parameter λ and the amounts of different claims are independently distributed with a common distribution F; we want to estimate the probability that the total amount from all the claims does not exceed $k\lambda$ for some k.

The distribution of the total claim X is a compound Poisson and its distribution has a Laplace transform

$$\mathbf{E}[e^{\theta X}] = \exp\Big[\lambda\int(e^{\theta\, x} - 1)dF(x)\Big] = \exp\Big[\lambda[h(\theta) - h(0)]\Big]$$

where

$$h(\theta) = \int e^{\theta\, x} dF(x)$$

By Tchebychev's inequality, for $k > \int x\, dF(x)$ and $\theta > 0$

$$P\Big[X \geq k\lambda\Big] \leq e^{\lambda[h(\theta)-1-k\,\theta]}$$

One can optimize over $\theta > 0$, to get

$$P\Big[X \geq k\lambda\Big] \leq e^{-\lambda\psi(k)}$$

where

$$\psi(k) = -\inf_{\theta>0}[h(\theta) - 1 - k\,\theta] = \sup_{\theta>0}[k\,\theta - h(\theta) + 1]$$

Since $k > m = \int x dF(x) = h'(0)$ it is easy to check that $\psi(k) > 0$ for all $k > m$.

It was proved subsequently by Cramér [1] that the above upper bound is in fact the best possible by showing that

$$\lim_{\lambda\to\infty} \frac{1}{\lambda} \log P\Big[X \geq k\lambda\Big] = -\psi(k)$$

or all $k > m$. There is of course a similar formula on the down side, perhaps not of the same importance to the insurance company, but nevertheless:

$$\lim_{\lambda\to\infty} \frac{1}{\lambda} \log P\Big[X \leq k\lambda\Big] = -\psi(k)$$

where now

$$\psi(k) = -\inf_{\theta<0}[h(\theta) - 1 - k\,\theta] = \sup_{\theta<0}[k\,\theta - h(\theta) + 1]$$

It is elementary to check that we do not change anything if we just define

$$\psi(k) = -\inf_{\theta}[h(\theta) - 1 - k\,\theta] = \sup_{\theta}[k\,\theta - h(\theta) + 1]$$

Then $\psi(k) > 0$ for $k \neq m$, and the interpretation of $\psi(\cdot)$ is that

$$P\Big[\frac{X}{\lambda} \sim k\Big] \simeq \exp[-\lambda\psi(k)]$$

3 Large Deviations

This is in fact the principle of large deviations where we have a family P_λ of probability measures on some space $\mathcal{X}$ and all the probability is getting concentrated near a point $x_0 \in \mathcal{X}$. The probabilities P_λ decay fast at points away from x_0 with a local rate $\psi(x)$. The rate function has the property, $\psi(x_0) = 0$ and $\psi(x) > 0$ for $x \neq x_0$. This vague description is made more precise by asking that the relation

$$\lim_{\lambda\to\infty} \frac{1}{\lambda} \log P_\lambda(A) = -\inf_{x\in A} \psi(x)$$

hold for nice sets. The correct formulation turns out to be

$$\limsup_{\lambda\to\infty} \frac{1}{\lambda} \log P_\lambda(A) \leq -\inf_{x\in C} \psi(x) \quad \text{for } C \quad \text{closed}$$

$$\liminf_{\lambda\to\infty} \frac{1}{\lambda} \log P_\lambda(G) \geq - \inf_{x\in G} \psi(x) \quad \text{for } G \quad \text{open}$$

Roughly then

$$P_\lambda[A] \simeq \exp[-\lambda \inf_{x\in A} \psi(x)]$$

Over the last forty years, diverse contexts where the principle holds and the rate function $\psi(x)$ can be explicitly calculated have been identified. See for instance, [3], [5], [2], [11], or [7]. I will select some familiar and some not so familiar situations and describe the results.

4 Ventcel–Freidlin Theory

Here we start with an Itô equation with a small noise.

$$dx(t) = b(t, x(t))dt + \epsilon\sigma(t, x(t))d\beta(t)$$

with $x(0) = x_0$. As $\epsilon \to 0$ we expect the solution to converge to the solution of the ODE without noise

$$dx(t) = b(t, x(t))dt$$

with the same initial condition $x(0) = x_0$. We fix a curve $f(t)$ in $[0, T]$ and ask how small the probability

$$P_\epsilon\Big[x(\cdot) \sim f(\cdot) \quad \text{in} \quad [0, T]\Big]$$

is. The answer is

$$P_\epsilon\Big[x(\cdot) \sim f(\cdot) \quad \text{in} \quad [0, T]\Big] \simeq \exp[-\frac{1}{\epsilon^2}\psi(f)]$$

where

$$\psi(f) = \frac{1}{2}\int_0^T \frac{(f'(t) - b(t, f(t)))^2}{\sigma^2(t, f(t))} dt$$

Such estimates have interesting applications. For instance if $f_0(t)$ is the solution of the ODE and $a \neq f_0(T)$, one can calculate

$$\inf_{f : f(T)=a} \psi(f) = I(a)$$

and that would give us an estimate on the probability

$$P_\epsilon\Big[x(T) \sim a\Big] \simeq \exp[-\frac{1}{\epsilon^2} I(a)]$$

and moreover the minimizing trajectory $f_a(t)$ will tell us the most likely history of the trajectory that by chance made it to a.

5 Donsker–Varadhan–Gärtner Theory

Here we look at the large time behavior of a continuous time Markov process on a finite state space. The infinitesimal transition rates are $a(x,y)$ and we are interested in estimating

$$P\Big[\frac{1}{T}\int_0^T V(x(s))ds \sim a\Big] \simeq \exp[-T\,\psi(a)]$$

It is more convenient to estimate the large deviations of the empirical distribution

$$p_T(x) = \frac{1}{T}\int_0^T \mathbf{1}_{\{x\}}(x(t))dt$$

of the proportion of time spent at the different sites. If $\mu(x)$ is the invariant probability we expect that $p_T \sim \mu$ with probability close to 1. It is natural to estimate

$$P\Big[p_T(\cdot) \sim \alpha(\cdot)\Big] \simeq \exp[-T\,I(\alpha)]$$

where α is a probability measure different from μ. The following formula for $I(\alpha)$ was derived in [4] and [8].

$$I(\alpha) = \sup_{u>0}\Big[-\sum_x \frac{\alpha(x)}{u(x)}\sum_y a(x,y)u(y)\Big]$$

If the detailed balance condition $a(x,y)\mu(x) = a(y,x)\mu(y)$ holds, then

$$\begin{aligned} I(\alpha) &= -\sum_{x,y} a(x,y)\mu(x)\sqrt{\frac{\alpha(y)}{\mu(y)}}\sqrt{\frac{\alpha(x)}{\mu(x)}} \\ &= \frac{1}{2}\sum_{x,y} a(x,y)\mu(x)\Bigg[\sqrt{\frac{\alpha(y)}{\mu(y)}} - \sqrt{\frac{\alpha(x)}{\mu(x)}}\Bigg]^2 \end{aligned}$$

We go from I to ψ in the usual manner by

$$\psi(a) = \inf_{\alpha:\sum \alpha(x)V(x)=a} I(\alpha)$$

The variational formula

$$\lambda(V) = \sup_{u:\sum_x u^2(x)\mu(x)=1}\Big\{\sum_x V(x)u^2(x) + \sum a(x,y)\mu(x)u(y)u(x)\Big\}$$

has a natural interpretation in this context.

6 Link to Control Theory

There is a way of viewing these as control problems that is often useful. Let us look at question of perturbing the rates $a(x,y)$ into $b(x,y)$ such that the perturbed processs has α as the invariant probability instead of μ. Let us denote by $\mathcal{B}(\alpha)$ the class of such rates. The relative entropy of the new modified process with respect to the old one can be computed. It is linear in time proportional to

$$C(b) = \sum_{x \neq y} \alpha(x)[b(x,y) \log \frac{b(x,y)}{a(x,y)} - b(x,y) + a(x,y)]$$

The rate function can now be recovered as

$$I(\alpha) = \inf_{b \in \mathcal{B}_\alpha} C(b)$$

This converts the problem of determining the rate function to a problem in control theory where the cost function is the entropy $C(b)$.

7 Hydrodynamic Scaling Models: Symmetric Case

We now consider a class of examples where the model is prescribed at one scale and the questions are raised at a much larger scale. The simplest such model with interaction is the simple exclusion process. This has particles on some of the sites on $\mathbf{Z}$, but with at most one particle per site. Particles wait for an exponential holding time and then try to jump to new site picking one at random with a probability $p(z)$ for a step of size z. If the particle is at x and z is chosen and $x+z$ is empty the particle jumps there. If $x+z$ already has a particle then the particle has to forget about the jump and wait patiently for another exponential time. The generator of the process can be written down

$$(\mathcal{A}f)(\eta) = \sum_{x,y} \eta(x)(1-\eta(y))p(y-x)[f(\eta^{x,y}) - f(\eta)]$$

For now we will be interested in the case when $p(z) = p(-z)$. Moreover $p(\cdot)$ will have finite range. We can look at the problem just as well on a periodic lattice of size N. If N is large, rather than keep the micro-information about the exact sites that are occupied, we may just worry about the density profile $\rho_0(q)$ for q in the unit interval modulo 1 or the circle $\mathcal{S}$ of perimeter 1.

$$\rho_0(q)dq = \lim_{N\to\infty} \frac{1}{N} \sum_x \delta_{\frac{x}{N}} \eta_0(x)$$

Since the particles have no preference for moving in one direction over the other, they will diffuse and so any appreciable shift in density that requires a migration over $O(N)$ sites will take time $O(N^2)$. One should therefore speed up time by a factor of N^2 and look at the evoution of the approximate density

$$\frac{1}{N}\sum_x \delta_{\frac{x}{N}} \eta_{N^2 t}(x).$$

The 'law of large numbers' will say that the quantity above has a nonrandom limit $\rho(t,q)dq$ and $\rho(\cdot,\cdot)$ will be a solution of

$$\rho_t = \frac{D}{2}\rho_{qq}$$

where the diffusion coefficient D is just $\sum_z z^2 p(z)$. This is deceptively simple, because if we change the condition from symmetry of p to the weaker condition $\sum_z zp(z) = 0$, the equation becomes

$$\rho_t = \frac{1}{2}[D(\rho)\rho_q]_q$$

becoming nonlinear. One can only say that $D(\rho) \geq \sum z^2 p(z)$ with strict inequality holding in general. Let us return to the symmetric case. The particles have identities and one can ask what happens to the measure

$$R_N = \frac{1}{N}\sum_i \delta_{\frac{1}{N}x_i(N^2\cdot)}$$

viewed as a random measure on the space of paths from some $[0,T]$ into $\mathcal{S}$. This is determined by another function $S(\rho)$ called the self-diffusion coefficient that describes how a tagged particle diffuses in a uniform environment of constant density ρ. It is known that R_N has a limit in probability which is a diffusion process with generator

$$\mathcal{L}_t = \frac{1}{2}\frac{d}{dq}S(\rho(t,q))\frac{d}{dq} + (S(\rho(t,q)) - D)\frac{\rho_q(t,q)}{2\rho(t,q)}\frac{d}{dq}$$

where $\rho(t,q)$ is the solution of the heat equation

$$\rho_t = \frac{D}{2}\rho_{qq}$$

with initial distribution ρ_0. For an arbitrary measure Q it is possible to find the large deviation rate function $I_{\rho_0}(Q)$ and it is interesting that it depends only on the macroscopic information ρ_0. See [10].

8 Hydrodynamic Scaling Models: Asymmetric Case

Here we look at the case $p(1) = 1$ and $p(z) = 0$ otherwise. The particles just want to move to the right all the time. The natural time scale to look at now is Nt and if we consider

$$\frac{1}{N}\sum_x \delta_{\frac{x}{N}} \eta_{Nt}(x)$$

the limit now satisfies Burgers equations (with zero viscosity)

$$\rho_t + [\rho(1-\rho)]_q = 0$$

This equation may not always have smooth solutions. While smooth solutions, when they exist, are unique, nonsmooth solutions are not unique. They have to be picked out of all weak solutions by what are called entropy conditions. If h is a convex function, the entropy condition says that, with g determined by $g'(\rho) = (2\rho - 1)h'(\rho)$, $\rho(\cdot,\cdot)$ satisfies

$$[h(\rho(t,q)]_t + [g(\rho(t,q)]_q = \mu \leq 0$$

in the distributional sense, with μ being a non positive measure.

The large deviations principle has been studied in this context, see [9], and has an interesting answer. $I(\rho) = +\infty$ unless $\rho(\cdot,\cdot)$ is a weak solution of the Burgers equation with initial condition ρ_0. But the entropy condition may be violated. We use the special entropy $h(\rho) = \rho \log \rho + (1-\rho)\log(1-\rho)$ and now

$$[h(\rho(t,q)]_t + [g(\rho(t,q)]_q = \mu$$

can be a signed measure and $I(\rho(\cdot,\cdot)) = \mu^+(\mathcal{S} \otimes [0,T])$.

9 Conclusion

The last two examples are some of the easy ones in this context. A stochastic evolution model is specified at the microscopic level, answers are expected for questions raised at the macroscopic level. The analysis was made easier because the microstructure is the lattice which though large is quite regular. I am sure that there are similar problems in economics and other social sciences that are much harder because the underlying microstructure itself is irregular and perhaps even random.

Acknowledgements. This article was supported by NSF Grant DMS-9803140

References

1. Cramér, H.: Sur un nouveau théorème-limites de la théorie des probabilités, Actualités Scientifiques et Industrielles 736 (1938), 5-23. Colloque consacré à la théorie des probabilités, Vol 3, Hermann, Paris
2. Dembo, Amir; Zeitouni, Ofer: Large deviations – techniques and applications. Second edition. Applications of Mathematics, 38. Springer-Verlag, New York, 1998.
3. Deuschel, Jean-Dominique; Stroock, Daniel W.: Large deviations. Pure and Applied Mathematics, 137. Academic Press, Inc., Boston, MA, 1989.
4. Donsker, M. D.; Varadhan, S. R. S.: Asymptotic evaluation of certain Markov process expectations for large time. I. II. Comm. Pure Appl. Math. 28 (1975), 1–47; ibid. 28 (1975), 279–301.

5. Dupuis, Paul; Ellis, Richard S.: A weak convergence approach to the theory of large deviations. Wiley Series in Probability and Statistics: Probability and Statistics. A Wiley-Interscience Publication. John Wiley & Sons, Inc., New York, 1997.
6. Esscher, F.: On the probability function in the collective theory of risk, Skandinavisk Aktuarietidskrift, 15 (1932) 175-195
7. Freidlin, M. I.; Wentzell, A. D.: Random perturbations of dynamical systems. Translated from the 1979 Russian original by Joseph Szücs. Second edition. Grundlehren der Mathematischen Wissenschaften, 260. Springer-Verlag, New York, 1998.
8. Gärtner, Jürgen: On large deviations from an invariant measure. (Russian) Teor. Verojatnost. i Primenen. 22 (1977), no. 1, 27–42.
9. Jensen, Leif: Large Deviations of the asymmetric simple exclusion process. Ph.D. thesis, New York University, (2000).
10. Quastel, J.; Rezakhanlou, F.; Varadhan, S. R. S.: Large deviations for the symmetric simple exclusion process in dimensions $d \geq 3$. (English. English summary) Probab. Theory Related Fields 113 (1999), no. 1, 1–84.
11. Shwartz, Adam; Weiss, Alan: Large deviations for performance analysis. Queues, communications, and computing. With an appendix by Robert J. Vanderbei. Stochastic Modeling Series. Chapman & Hall, London, 1995

Conquering the Greeks in Monte Carlo: Efficient Calculation of the Market Sensitivities and Hedge-Ratios of Financial Assets by Direct Numerical Simulation

Marco Avellaneda and Roberta Gamba*

Courant Institute of Mathematical Sciences, New York University,
251 Mercer Street, New York, NY, 10012

Abstract. The calculation of price-sensitivities of contingent claims is formulated in the framework of Monte Carlo simulation. Rather than perturbing the parameters that drive the economic state-variables of the model, we perturb the vector of probabilities of simulated paths in a neighborhood of the uniform distribution. The resulting hedge-ratios (sensitivities with respect to input prices) are characterized in terms of statistical moments of simulated cashflows. The computed sensitivities dislay excellent agreement with analytic closed-form solutions whenever the latter are available, *e.g.* with the Greeks of the Black-Scholes model, and with approximate analytic solutions for Basket Options in multi-asset models. The advantage of the new sensitivities is that they are "universal" (non-parametric) and simple to compute: they do not require performing multiple MC simulations, discrete-differentiation, or re-calibration of the simulation.

1 Introduction

The goal of this note is to clarify recent proposals made for pricing and hedging derivative securities using Monte Carlo simulations with weighted paths (Avellaneda, Buff, Friedman, Grandchamp, Kruk and Newman (1999)). A typical Monte Carlo (MC) simulation assigns equal probability to each simulation path. In the weighted MC framework, paths can be assigned different probabilities. This added flexibility allows us to better fit model prices of benchmark instruments to the actual prices observed in the market. It also gives rise to a new method for computing price-sensitivities. The question of interest here to analyze the sensitivities and the hedges that we compute by this new method and to compare them with standard approaches.

We begin by briefly reviewing the Weighted Monte Carlo approach. Fitting prices by assigning different probabilities to the simulated future market scenarios leads to the problem of how to choose the probabilities. Since the number of paths is typically much greater than the number of prices to be fitted, there exist many probability vectors which are consistent with a given

* This research was partially funded by the National Science Foundation (DMS-9973226) and by Lamb Analytics Corp.

set of observed prices. Naturally, we would like to select the pricing probabilities by a procedure which gives rise to a unique solution and makes the pricing scheme stable with respect to perturbations in the input prices.

To be specific, let N represent the number of paths and let M represent the number of benchmark instruments under consideration. Denote by $C_1, \ldots, C_M$ the spot market prices of these instruments. For each of the benchmark instruments, we consider the vector $\mathbf{g}^{(j)} = (g_1^{(j)}, \ldots, g_N^{(j)})$ where the entries represent the present value of the cash-flows of the instrument if the ith path occurs $(i = 1, \ldots, N)$. If the path i has probability p_i, the M prices satisfy the equations

$$C_j = \sum_{i=1}^{N} g_i^{(j)} p_i \,, \quad j = 1, \ldots M \,. \tag{1}$$

In general, we have $M \ll N$ so this system of equation admits many solutions. A possible selection mechanism consists of choosing the probabilities p_i so as to minimize the quantity

$$\sum_{i=1}^{N} \Psi(p_i) \tag{2}$$

where $\Psi(x)$ is a convex function. The selection of a set of probabilities p_i which satisfy (1) and minimize (2) can be made by solving a Lagrange multipliers problem of the type

$$\min_{\lambda,\mu} \left\{ \sum_{i=1}^{N} \Psi(p_i) + \sum_{j=1}^{M} \lambda_j \left[\sum_{i=1}^{N} g_i^{(j)} p_i - C_j \right] + \mu \left[\sum_{i=1}^{N} p_i - 1 \right] \right\} .$$

The formal solution of the optimization problem is given by

$$p_i = (\Psi')^{-1} \left[- \sum_{j=1}^{M} \lambda_j g_i^{(j)} - \mu \right]$$

or, equivalently,

$$p_i = \Psi^{*\prime} \left[- \sum_{j=1}^{M} \lambda_j g_i^{(j)} - \mu \right] . \tag{3}$$

Here, $\Psi^*(x)$ is the convex conjugate of $\Psi(x)$, in the sense of Legendre. The latter equation can be interpreted as defining an M-parameter family of distributions (corresponding to the probability distributions obtained by varying the vector of Lagrange multipliers $(\lambda_1, \ldots \lambda_M)$). The parameter μ is determined uniquely from the fact that the sum of the p_i's is one.

There exist three special cases of the Ψ-function that are noteworthy. First,

$$\Psi(x) = x \ln(x)$$

which corresponds to the Kullback-Leibler relative entropy under constraints. Secondly, the function

$$\Psi(x) = \sqrt{x}\,,$$

which corresponds to the Skorohod distance. The third case corresponds to the quadratic function

$$\Psi(x) = \left(x - \frac{1}{N}\right)^2 .$$

Unfortunately, the quadratic penalization function may lead to negative probabilities – unlike the Kullback and Skorohod distances. It should therefore be used with caution.

We shall focus mostly on the relative entropy distance. The parametric family of probabilities which arises in this case is the well-known *Boltzman* distribution

$$p_i(\lambda) = \frac{\exp\left(\sum_{j=1}^M \lambda_j g_i^{(j)}\right)}{\sum_{i=1}^N \exp\left(\sum_{j=1}^M \lambda_j g_i^{(j)}\right)} = \frac{1}{Z(\lambda)} \exp\left(\sum_{j=1}^M \lambda_j g_i^{(j)}\right) \tag{4}$$

where

$$Z(\lambda) = \sum_{i=1}^N \exp\left(\sum_{j=1}^M \lambda_j g_i^{(j)}\right) \tag{5}$$

is the partition function. A useful identity for finding the Lambdas numerically is

$$C_j = \frac{\partial \ln Z(\lambda)}{\partial \lambda_j}\,, \tag{6}$$

which follows from substituting the expression for the p_i's in the pricing identity (1). This means that the values of the Lagrange multipliers that produce the correct probabilities in the entropic case are obtained by minimizing the *objective function*

$$\ln(Z(\lambda)) - \sum_{j=1}^M \lambda_j C_j\,.$$

The minimization can be implemented numerically with a gradient-based optimization algorithm such as BFGS, or with the classical Newton algorithm that utilizes both the gradient and the Hessian of $\ln(Z(\lambda))$. The gradient-based method uses $1+M$ function evaluations, each of which has complexity $O(N)$. The full Newton approach requires instead $1+M+\frac{M(M+1)}{2}$ function evaluations. So far, we have only tested the gradient-based method and obtained satisfactory results with up to 50–100 instruments. This may have to do with the fact that BFGS is a highly efficient algorithm which combines linesearches with quasi-Newton steps using a pseudo-Hessian. The quasi-Newton method seems well-fitted to the objective functions that arise in the entropy optimization problem, which, despite being convex, are quite "flat" in most regions of space. In deciding whether to choose the quasi-Newton or the "pure" Newton approach, the user must therefore weigh the benefits of using an explicit Hessian against having a more costly evaluation step in the search.[1]

2 Computing Sensitivities with Respect to the Input Prices

We have come to the main subject of this paper, which is the calculation of price-sensitivities. One of the main conclusions of the section will be to show that the sensitivities are closely related to regression coefficients and that this result is, in some sense, independent of the penalization function Ψ.

Let $h=(h_1,\dots h_N)$ represent the payoff vector of a portfolio of contingent claims. The model value of the portfolio is

$$E(h)=\sum_{i=1}^{N} p_i h_i = \sum_{i=1}^{N} \Psi_i^{*\prime} h_i \tag{7}$$

where we set $\Psi_i^{*\prime}=\Psi^{*\prime}[-\lambda\cdot g_i-\mu]$.

The problem of computing the sensitivities of such a portfolio can be cast formally as the calculation of the partial derivatives

$$\frac{\partial E(h)}{\partial C_j}\,, \quad j=1,\dots M\,.$$

[1] The case for computing with the full Newton algorithm has been made to me by Paul Fackler (private communication, April 2000). Also, Raphael Douady (private communication 1999) proposed a modified Newton algorithm that used the Hessian matrix to compute the minimum entropy solution.

If we consider the case of probability measures obtained using a penalty function $\Psi(x)$, as described above, we find that

$$
\begin{aligned}
\frac{\partial E(h)}{\partial \lambda_j} &= \sum_{i=1}^{N} \frac{\partial p}{\partial \lambda_j} h_i \\
&= -\sum_{i=1}^{N} \left[\Psi_i^{*\prime\prime} g_i^{(j)} + \Psi_i^{*\prime\prime} \frac{\partial \mu}{\partial \lambda_j} \right] h_i \\
&= -\sum_{i=1}^{N} \Psi_i^{*\prime\prime} g_i^{(j)} h_i - \frac{\partial \mu}{\partial \lambda_j} \sum_{i=1}^{N} \Psi_i^{*\prime\prime} h_i
\end{aligned} \tag{8}
$$

where we used the fact that $p_i = \Psi^{*\prime}[-\lambda \cdot g_i - \mu]$ and the abbreviation $\Psi_i^{*\prime\prime} = \Psi^{*\prime\prime}[-\lambda \cdot g_i - \mu]$. To obtain an "explicit" expression for $\frac{\partial \mu}{\partial \lambda_j}$, we differentiate equation (3) with respect to λ_j. The result is

$$
\sum_{i=1}^{N} \left[\Psi_i^{*\prime\prime} g_i^{(j)} + \Psi_i^{*\prime\prime} \frac{\partial \mu}{\partial \lambda_j} \right] = 0 ,
$$

which implies that

$$
\frac{\partial \mu}{\partial \lambda_j} = - \frac{\sum_{i=1}^{N} \Psi_i^{*\prime\prime} g_i^{(j)}}{\sum_{i=1}^{N} \Psi_i^{*\prime\prime}} .
$$

Substituting back into (8), we find that

$$
\begin{aligned}
\frac{\partial E(h)}{\partial \lambda_j} &= -\sum_{I=1}^{N} \Psi_i^{*\prime\prime} g_i^{(j)} + \frac{\sum_{i=1}^{N} \Psi_i^{*\prime\prime} g_i^{(j)}}{\sum_{i=1}^{N} \Psi_i^{*\prime\prime}} \sum_{i=1}^{N} \Psi_i^{*\prime\prime} h_i \\
&= -\left(\sum_{i=1}^{N} \Psi_i^{*\prime\prime} \right) \cdot \left(\frac{\sum_{I=1}^{N} \Psi_i^{*\prime\prime} g_i^{(j)} h_i}{\sum_{I=1}^{N} \Psi_i^{*\prime\prime}} - \frac{\sum_{i=1}^{N} \Psi_i^{*\prime\prime} g_i^{(j)}}{\sum_{i=1}^{N} \Psi_i^{*\prime\prime}} \cdot \frac{\sum_{i=1}^{N} \Psi_i^{*\prime\prime} h_i}{\sum_{i=1}^{N} \Psi_i^{*\prime\prime}} \right) .
\end{aligned} \tag{9}
$$

Notice that the expression in parenthesis can be interpreted as a covariance. More precisely, if we define the new expectation operator (and its associated probability) by the equation

$$
E^*(f) = \frac{\sum_{i=1}^{N} \Psi_i^{*\prime\prime} f_i}{\sum_{i=1}^{N} \Psi_i^{*\prime\prime}} ,
$$

we can write, more concisely,

$$
\frac{\partial E(h)}{\partial \lambda_j} = -\left(\sum_{i=1}^{N} \Psi_i^{*\prime\prime} \right) \cdot Cov^*(g^{(j)}, h) .
$$

Clearly, the same argument can be applied to the case $h = g^{(k)}$ so we also have

$$\frac{\partial E(g^{(k)})}{\partial \lambda_j} = -\left(\sum_{i=1}^{N} \Psi_i^{*\prime\prime}\right) \cdot Cov^*(g^{(j)}, g^{(k)}) .$$

Finally, since we have $C_k = E(g^{(k)})$ we derive an expression for the sensitivities with respect to the *input prices*, namely,

$$\frac{\partial E(h)}{\partial C_k} = \sum_{j=1}^{M} \frac{\partial E(h)}{\partial \lambda_j} \cdot \frac{\partial \lambda_j}{\partial C_k}$$

or, in matrix notation,

$$\nabla_C E(h) = Cov^*(\mathbf{g}, h) \cdot (Cov^*(\mathbf{g}, \mathbf{g}))^{-1} . \tag{10}$$

This implies that

Proposition 1. *The vector of price-sensitivities obtained with the penalization function $\Psi(x)$ is equal to the vector of regression coefficients (under the measure induced by $E^*(\bullet)$) of the payoff of the target portfolio on the linear space generated by the cash-flow vectors of the input instruments $(\mathbf{g}^{(1)}, \ldots \mathbf{g}^{(M)})$. More precisely, we have*

$$\frac{\partial E(h)}{\partial C_k} = \beta_k$$

where

$$\begin{aligned} \boldsymbol{\beta} &= \arg\min_{\beta} E^* \left(h - \sum_{j=1}^{M} \beta_j \mathbf{g}^{(j)} \right)^2 \\ &= \arg\min_{\beta} \frac{E\left[\left(\frac{\Psi^{*\prime\prime}}{\Psi^{*\prime}} \right) \left(h - \sum_{j=1}^{M} \beta_j \mathbf{g}^{(j)} \right)^2 \right]}{E\left[\left(\frac{\Psi^{*\prime\prime}}{\Psi^{*\prime}} \right) \right]} . \end{aligned} \tag{11}$$

It is important to note at this point the special role played by the Kullback-Leibler entropy distance. Since in this case we have $\Psi(x) = x \ln(x)$, and hence

$$\Psi^*(x) = \mathrm{e}^{x-1} = \Psi^{*\prime}(x) = \Psi^{*\prime\prime}(x) ,$$

we have

$$\Psi_i^{*\prime\prime} = \Psi_i^{*\prime} = p_i .$$

Therefore,

Proposition 2. *If the penalization corresponds to the Kullback-Leibler relative entropy distance, the hedge-ratios are equal to the regression coefficients of the cashflow vector of the portfolio onto the cash-flow vectors of the benchmark instruments under the pricing measure.*

This result was derived previously in Avellaneda (1998) and in Avellaneda, Buff, Friedman, Grandchamp, Kruk and Newman (1999).

In situations of practical interest, the prior probability measure may be such that the simulation is already calibrated to the observed prices of benchmark instruments. This would arise, for instance, if the underlying model parameters were fitted using a classical procedure, such as least-squares fitting or another optimization procedure. Under these circumstances, the method of Weighted Monte Carlo can still be used to compute sensitivities and hedge ratios and gives non-trivial results. In fact, assume that the model is calibrated to the prices of benchmark instruments with $p =$ the uniform measure on paths. If this is the case, we have $\Psi_i^{*\prime\prime} = const.$ and $\Psi_i^{*\prime} = \frac{1}{N}$. In particular, we have $E^*(\bullet) = E(\bullet) =$ expectation over the uniform measure, and thus the

Corollary 3. *If the Monte Carlo simulation is calibrated with $\lambda = 0$, the hedge-ratios produced by all Ψ-penalizations are equal. They are equal to the coefficients of the least-squares projection of the vector of cash- flows of the target portfolio on the space generated by the cash-flows of the benchmark instruments.*

3 The Parametric Approach

It is natural to compare the above results with "parametric approach" to sensitivity-analysis, which consists of embedding the risk-neutral measure in an M-parameter family of risk-neutral measures and performing a perturbation analysis. Consider therefore a parametric, but "non-entropic", family of probability vectors $p_i(\theta), i = 1, \ldots, N, \theta \in R^M$.

The pricing equations are now

$$E(h) = \sum_{i=1}^{N} p_i(\theta) h_i$$

and

$$E(g^{(j)}) = \sum_{i=1}^{N} p_i(\theta) g_i^{(j)} ,$$

for $j = 1, \ldots, M$. (For simplicity, we assume that the number of parameters is the same as the number of input instruments). Differentiation of these equations with respect to θ gives

$$\begin{aligned} \nabla_\theta E(h) &= \sum_{i=1}^{N} \nabla_\theta p_i(\theta) h_i \\ &= E\left(\frac{\nabla_\theta p_i(\theta)}{p(\theta)} h\right) \\ &= Cov\left(\frac{\nabla_\theta p(\theta)}{p(\theta)} h\right) \end{aligned}$$

and

$$\begin{aligned} \nabla_\theta E(g^{(j)}) &= \sum_{i=1}^{N} \nabla_\theta p_i(\theta) g_i^{(j)} \\ &= E\left(\frac{\nabla_\theta p(\theta)}{p(\theta)} g^{(j)}\right) \\ &= Cov\left(\frac{\nabla_\theta p(\theta)}{p(\theta)}, g^{(j)}\right) . \end{aligned}$$

(Notice that we used here the identity

$$E\left(\frac{\nabla_\theta p(\theta)}{p(\theta)}\right) = \mathbf{0} ,$$

which holds because $p(\theta)$ is a probability.) We conclude that the "parametric" expression for hedge-ratios takes the form

$$\nabla_C E(h) = Cov\left(\frac{\nabla_\theta p(\theta)}{p(\theta)}, h\right) \cdot \left[Cov\left(\frac{\nabla_\theta p(\theta)}{p(\theta)}, \mathbf{g}\right)\right]^{-1} . \tag{12}$$

Thus, as in the "non-parametric approach" the hedge-sensitivities with respect to prices are expressed in terms of covariances involving the different vectors of cash-flows. The main difference is that the parametric form of the distribution is manifested with the appearance of the gradient of the "likelihood function"

$$\nabla_\theta \ln p(\theta) = \frac{\nabla_\theta p(\theta)}{p(\theta)} .$$

The reader will note that these formulas can be used, in conjunction with the explicit expressions for $p(\theta) = p(\lambda)$ of the non-parametric case to rederive the results of the previous sections. The quantity that corresponds to the likelihood function in the case of a Ψ-perturbation is simply the ratio

$$\frac{\Psi_i^{*\prime\prime}}{\Psi_i^{*\prime}} .$$

Formulas for computing Greek sensitivities for option prices along these lines were first proposed by Broadie and Glasserman (1996).

This "parametric approach" is based on differentiation of the probability measure on path space with respect to the parameter vector θ. With the exception of simple cases Broadie and Glasserman (1996), the dependence between the parameters and the probabilities is not explicit and requires, for instance, the use of Calculus of Variations in path-space, or Malliavin Calculus (see Fournie, Lasry, Lebuchoux, Lions and Touzi (1998)). Unfortunately, a "direct" Malliavin calculus approach – i.e. the calculation of the *exact* derivatives of the expectation with respect to the parameters θ – will be difficult if not impossible to implement in practice due to the complexity of the typical pricing models. This is more so as we require that the probability measure be fitted to the price of many benchmark instruments (requiring the fitting of may parameters in the calibration step). In contrast, the approach that we advocate here requires only performing regressions of different cash-flow vectors – without carrying out any explicit differentiation with respect to the "internal" model parameters.

A theoretical comparison between the "parametric hedge-ratios" which arise by varying the parameter θ and the non-parametric hedge-ratios obtained by regressions is developed in the Appendix. There, we show that the WMC method is consitent with (i.e. converges to) the true sensitivities as the number of benchmarks increases (in a suitable sense). In practice, simulations show that we obtain surprisingly good results with a relatively small set of benchmark instruments.

In the following section, we develop explicit formulas for Delta and Gamma using the technique of Weighted Monte Carlo. The numerical results are left for the last section.

4 Computation of the Classical "Greeks" in MC Simulations

The weighted MC method produces the sensitivities with respect to *input instruments* by computing a least-squares regression of cash-flows. On the other hand, traders might be interested in sensitivities with respect to *spot prices*, such as Deltas and Gammas. The method presented here can be used to calculate these "Greek" sensitivities, provided that the sensitivities of the benchmark prices to the parameters of interest can be computed in closed-form.

To fix ideas, we shall consider a Black-Scholes world in which the input instruments are either options or forwards. We assume that the prices of the benchmark instruments are now given by the standard Black–Scholes expressions

$$C_j = C_j(S)$$

where S is the spot price. The Delta of the contingent claim with payoff h is

$$\begin{aligned}\frac{\partial E(h)}{\partial S} &= \sum_{j=1}^{M} \frac{\partial E(h)}{\partial C_j}\frac{\partial C_j}{\partial S} \\ &= \nabla_C E(h) \cdot \nabla_S \mathbf{C} \\ &= \boldsymbol{\beta} \cdot \nabla_S \mathbf{C} \,. \end{aligned} \tag{13}$$

Thus, the total exposure to a movement in the spot price is computed by converting the exposures to different instruments into "Delta-equivalents", using the Black-Scholes formula to compute the Delta of each benchmark.

The calculation of Gamma is more complicated, but follows the same approach:

$$\begin{aligned}\frac{\partial^2 E(h)}{\partial S^2} &= \frac{\partial}{\partial S}\left(\sum_{j=1}^{M} \beta_j \frac{\partial C_j}{\partial C_k}\right) \\ &= \sum_{j=1}^{M} \frac{\partial \beta_j}{\partial S}\frac{\partial C_j}{\partial S} + \sum_{j=1}^{M} \beta_j \frac{\partial^2 C_j}{\partial S^2} \,. \end{aligned} \tag{14}$$

The key point is the computation of the quantities $\frac{\partial \beta_j}{\partial S}$. For this, we proceed as follows

$$\begin{aligned}\frac{\partial \beta_j}{\partial S} &= \sum_{k=1}^{M} \frac{\partial \beta_j}{\partial \lambda_j}\frac{\partial C_k}{\partial S} \\ &= \sum_{k=1}^{M}\sum_{l=1}^{M} \frac{\partial \beta_j}{\partial \lambda_l}\frac{\partial \lambda_l}{\partial C_k}\frac{\partial C_k}{\partial S} \,. \end{aligned} \tag{15}$$

To continue, we use the correspondence between λ-derivatives and covariances, which holds for the Kullback-Leibler perturbations. Accordingly, let

$$Q = (Cov(\mathbf{g}, \mathbf{g}))^{-1} \,.$$

We have

$$\frac{\partial Q}{\partial \lambda_l} = -Q \cdot \left(\frac{\partial}{\partial \lambda_l} Cov(\mathbf{g}, \mathbf{g}\right) \cdot Q$$

and hence

$$\begin{aligned}\frac{\partial \boldsymbol{\beta}}{\partial \lambda_l} &= \frac{\partial Q}{\partial \lambda_l} \cdot Cov(h, \mathbf{g}) + Q \cdot \left(\frac{\partial}{\partial \lambda_l} Cov(h, \mathbf{g})\right) \\ &= -Q \cdot \left(\frac{\partial}{\partial \lambda_l} Cov(\mathbf{g}, \mathbf{g})\right) \cdot Q \cdot Cov(h, \mathbf{g}) + Q \cdot \left(\frac{\partial}{\partial \lambda_l} Cov(h, \mathbf{g})\right) \end{aligned} \tag{16}$$

The derivatives inside this expression can be computed explicitly using the identity

$$\frac{\partial E(f)}{\partial \lambda_l} = Cov(f, g^{(l)}) \, ,$$

which is valid for all f. The final result, after substituting the expressions for $\frac{\partial \beta_j}{\partial \lambda_l}$ into the expression for Gamma, is

$$\frac{\partial^2 E(h)}{\partial S^2} = \sum_{j=1}^{M} \beta_j \frac{\partial^2 C_j}{\partial S^2} - \sum_{j,k=1}^{M} \frac{\partial C_j}{\partial S} M_{jk} \frac{\partial C_k}{\partial S} \, , \tag{17}$$

where

$$M_{jk} = \sum_{p,q=1}^{M} Q_{jp} N_{pq} Q_{qk} \, ,$$

with

$$\begin{aligned} N_{pq} &= E(\gamma^{(p)} \gamma^{(q)} \eta) \, , \\ \gamma_i^{(p)} &= g_i^{(p)} - C_p \, , \\ \gamma_i^{(q)} &= g_i^{(q)} - C_q \, , \end{aligned}$$

and

$$\eta_i = h_i - \boldsymbol{\beta} \cdot \mathbf{g}_i - (E(h) - \boldsymbol{\beta} \cdot \mathbf{C}) \, .$$

These expressions can be easily evaluated numerically.

5 Numerical Results

We conducted several numerical experiments in the case of MC simulations based on lognormal pricing models. In the first experiment, we considered a single-asset Black-Scholes model with the following inputs

$$\begin{aligned} S &= \$146 \\ \sigma &= 58\% \\ r &= 5\% \\ d &= 0\% \end{aligned}$$

The day count basis was 365 days/yr. The benchmark instruments chosen were:

Type	Expiration (days)	Strike	Mkt. Price
fwd	135	–	–
call	135	143	22.98
fwd	115	–	–
call	115	143	21.27
fwd	105	–	–
call	105	143	20.37
fwd	95	–	–
call	95	143	19.42
fwd	85	–	–
call	85	143	18.42
fwd	75	–	–
call	75	143	17.37
fwd	65	–	–
call	65	143	16.25
fwd	55		–
call	55	143	15.04
fwd	45		–
call	45	143	13.73
fwd	35	–	–
call	35	143	12.26
fwd	25	–	–
call	25	143	10.58
fwd	15		–
call	15	143	8.54
fwd	5		–
call	5	143	5.65
fwd	125	–	–
call	125	143	22.14
fwd	90	–	–
fwd	60	–	–

These prices correspond exactly to the Black-Scholes option prices with the above parameters, so that the model is assumed to be nearly calibrated with $\lambda = 0$. We simulated 7000 paths in each simulation and performed 10 simulations for each pricing event to eliminate the dependence of the seed. Each pricing event was calibrated exactly using the Kullback entropy method. The average magnitude for the entropy distance was on the order of 10^{-3}. For each pricing event, we computed the prices, deltas and gammas of the different options. We carried out two types of simulations: ones which involved minimum-entropy calibration (and thus $\lambda \neq 0$) and others in which we did not calibrate (thus $\lambda = 0$). In the latter cases, we still used formulas (13) and (17) to compute the Deltas and Gammas, respectively.

Example 1. European put, expiration: 90 days, strike \$135.

–	Price	Delta	Gamma
BS	10.4462	−0.323301	0.00853
Weighted MC	10.4519	−0.320309	0.00874
MC ($\lambda = 0$)	10.4395	−0.320268	0.00873

Example 2. European call, expiration 60 days, strike \$190.

–	Price	Delta	Gamma
BS	2.74417	0.166784	0.00728
Weighted MC	2.72814	0.162294	0.00711
MC ($\lambda = 0$)	2.70869	0.161862	0.00708

Example 3. European call, expiration 60 days, strike \$145.

–	Price	Delta	Gamma
BS	14.6858	0.57118	0.011377
Weighted MC	14.6759	0.57204	0.011677
MC ($\lambda = 0$)	14.6322	0.57195	0.011668

Additional experiments on European-style options on a single asset seem to indicate that the WMC procedure and the formulas for the Greeks gives accurat deltas and gammas for a wide range of strikes and maturities. The robustness of the method is quite good in the sense that the sensitivities are computed with high accuracy even for options which are deeply out of the money (e.g. 15% deltas).

We also experimented with pricing Basket Options on equities. We considered the following parameters:

–	Stock 1	Stock 2	Stock 3
Spot Price	86	340	73.25
Vol	60	60	35

and correlation matrix

–	Stock 1	Stock 2	Stock 3
Stock 1	100	43.42	24.76
Stock 2	43.42	100	20.19
Stock 3	24.76	20.19	100

The benchmark securities used in the calibration were

Stock	Type	Expiration	Strike	Price
1	fwd	45	–	–
1	call	45	85	9.36
1	fwd	40	–	–
1	fwd	35	–	–
1	call	35	80	10.98
1	fwd	32	–	–
1	fwd	30	–	–
1	call	30	90	5.54
1	fwd	25	–	–
1	fwd	15	–	–
2	fwd	45	–	–
2	call	45	341	34.73
2	fwd	40	–	–
2	fwd	35	–	–
2	fwd	25	–	–
2	fwd	15	–	–
3	fwd	45	–	–
3	call	45	73	4.66
3	fwd	40	–	–
3	fwd	35	–	–
3	call	35	75	3.18
3	fwd	25	–	–
3	fwd	15	–	–

The interest rate was taken to be 4% and stock dividends were neglected. We considered a few examples and compared the corresponding results to the well-known Black-Scholes approximation for the value and Greeks of basket options, which is commonly used by equity derivatives traders.

The Black-Scholes approximation for the price of the basket is computed by inputting a constant volatility of

$$\sigma^2_{\text{basket}} = \frac{\sum_i n_i^2 S_i^2 \sigma_i^2 + \sum_{i \neq j} n_i n_j S_i S_j \sigma_i \sigma_j}{(\sum N_i S_i)^2} .$$

The "basket delta" for MC was calculated by the formula:

$$\varDelta_{\text{basket}} = \frac{1}{3}\left(\frac{\delta_1}{n_1} + \frac{\delta_2}{n_2} + \frac{\delta_3}{n_3}\right) ,$$

where δ_i represent the deltas with respect to each stock in the multiasset model and n_i represents the amounts of shares of each stock in the basket. In the case of BS, individual Deltas are obtained by multiplying the Basket Delta (given in closed form) by the number or shares of each stock held. With the data used for this example, we have $\sigma_{\text{basket}} = 38.69\%$. We used a day-count convemtion of 255 days/yr., simulations with 5000 paths, and priced each option 10 times. The results represent the averages of the 10 pricing events.

Example 4. Put option on a basket composed of \$100 in each of the three stock at inception. Expiration: 40 days, strike price: \$300.

–	Price	Delta(1)	Delta(2)	Delta(3)	Basket Delta
BS	17.3301	−0.52720	−0.13335	−0.61896	−0.45410
WMC	17.3672	−0.51555	−0.12877	−0.66648	−0.45339
MC($\lambda = 0$)	17.3083	−0.51548	−0.12875	−0.66656	−0.45644

We also computed de Gamma matrix:

Γ	Stock 1	Stock 2	Stock 3
Stock 1	0.0113		
Stock 2	0.0026	0.0008	
Stock 3	0.0156	0.0034	0.0154

Example 5. Call option on a basket composed of \$100 in each of the three stock at inception. Expiration: 30 days, strike price: \$350.

–	Price	Delta(1)	Delta(2)	Delta(3)	Basket Delta
BS	2.79104	0.16853	0.04263	0.19787	0.14494
WMC	2.84153	0.16861	0.04747	0.17639	0.13107
MC($\lambda = 0$)	2.79141	0.16862	0.04748	0.17502	0.13037

A more detailed study of the sensitivities for basket options, including Vega-sensitivities will appear in a separate article.

6 Appendix: Comparison Between the True Sensitivities and the Ones Obtained by Regression

In this section, we assume that the risk-neutral measure is given by a parametric family of distributions $p_i(\theta)$, where θ is a parameter. We also assumed that the model is calibrated to match the prices of M reference instruments. The sensitivities computed by the standard method can be obtained by differentiation with respect to θ and applying the chain rule to obtain sensitivities with respect to prices. The sensitivities obtained by the "entropy method" or, more generally, by Ψ-penalizations, correspond to the "betas" of linear regressions of cash-flows. In this Appendix, we show how the two concepts are mathematically related.

Let $\boldsymbol{\beta}_p$ and $\boldsymbol{\beta}_{np}$ denote the vectors of sensitivities obtained by the parametric and non-parametric methods, respectively. Recall that

$$\boldsymbol{\beta}_p = \left(Cov\left(\frac{\nabla_\theta p(\theta)}{p(\theta)}, \mathbf{g} \right) \right)^{-1} \cdot Cov\left(\frac{\nabla_\theta p(\theta)}{p(\theta)}, h \right)$$

and

$$\boldsymbol{\beta}_{np} = (Cov(\mathbf{g},\mathbf{g}))^{-1} \cdot Cov(\mathbf{g}, h) .$$

Treating $\frac{\nabla_\theta p(\theta)}{p(\theta)}$ as a random vector, we consider a linear regression of this vector on the space generated by the cash-flow vectors $g^{(j)} - E(g^{(j)})$, $j = 1, \ldots M$. Accordingly, we have

$$\frac{1}{p(\theta)} \frac{\partial p(\theta)}{\partial \theta_j} = \sum_{k=1}^{M} a_{jk}(g^{(j)} - E(g^{(j)})) + \epsilon_j$$

where $a_{jk} = a_{jk}(\theta)$ are regression coefficients and ϵ_j are random variables (residuals) with mean zero. (Notice that we use here the fact that $\frac{\nabla_\theta p(\theta)}{p(\theta)}$ has mean zero.) We can rewrite the above equation in matrix-vector notation as

$$\frac{\nabla_\theta p(\theta)}{p(\theta)} = A(\theta) \cdot (\mathbf{g} - E(\mathbf{g})) + \boldsymbol{\epsilon} . \tag{18}$$

In particular, it follows that

$$Cov\left(\frac{\nabla_\theta p(\theta)}{p(\theta)}, \mathbf{g}\right) = A(\theta) \cdot Cov(\mathbf{g},\mathbf{g})$$

Similarly, we have

$$Cov\left(\frac{\nabla_\theta p(\theta)}{p(\theta)}, h\right) = A(\theta) \cdot Cov(\mathbf{g}, h) + Cov(\boldsymbol{\epsilon}, h) .$$

Hence,

$$\begin{aligned}
\boldsymbol{\beta}_p &= \left(Cov\left(\frac{\nabla_\theta p(\theta)}{p(\theta)}, \mathbf{g}\right)\right)^{-1} \cdot Cov\left(\frac{\nabla_\theta p(\theta)}{p(\theta)}, h\right) \\
&= (Cov(\mathbf{g},\mathbf{g}))^{-1} \cdot (A(\theta))^{-1}[A(\theta) \cdot Cov(\mathbf{g}, h) + Cov(\boldsymbol{\epsilon}, h)] \\
&= \boldsymbol{\beta}_{np} + (A(\theta))^{-1} \cdot Cov(\boldsymbol{\epsilon}, h) \\
&= \boldsymbol{\beta}_{np} + (Cov(\mathbf{g},\mathbf{g}))^{-1} \cdot (A(\theta))^{-1} \cdot Cov(\boldsymbol{\epsilon}, h - \boldsymbol{\beta}_{np} \cdot \mathbf{g}) .
\end{aligned}$$

The last expression was obtained using the fact that the residual of $\frac{\nabla_\theta p(\theta)}{p(\theta)}$ is uncorrelated with $\mathbf{g}$. Thus, formally, we have

Proposition 1. *The difference between the parametric and non-parametric hedge-ratios satisfies:*

$$\begin{aligned}
\boldsymbol{\beta}_p - \boldsymbol{\beta}_{np} &= (Cov(\mathbf{g},\mathbf{g}))^{-1} \cdot (A(\theta))^{-1} \cdot Cov(\boldsymbol{\epsilon}, h - \boldsymbol{\beta}_{np} \cdot \mathbf{g}) \\
&= (Cov(\mathbf{g},\mathbf{g}))^{-1} \cdot (A(\theta))^{-1} \cdot Cov\left(\frac{\nabla_\theta p(\theta)}{p(\theta)} - A(\theta) \cdot \mathbf{g}, h - \boldsymbol{\beta}_{np} \cdot \mathbf{g}\right) .
\end{aligned}$$

The difference depends on the covariance of the residuals of $\frac{\nabla_\theta p(\theta)}{p(\theta)}$ and h with respect to projections on the space generated by the cash-flows of the input instruments.

In particular, the difference between the two sets of coefficients will be small if either one or the other residual is small or if the residuals are nearly uncorrelated.

This result provides a qualitative understanding of the difference between parametric and non-parametric "Greeks". Despite the obvious difficulties in obtaining quantitative estimates from the above proposition, we note that if the reference instruments form a "reasonable" spanning set in the L^2 sense, the residuals will be small and the approximation will be good. Finally, the fact that the error is controlled by the residual of $\frac{\nabla_\theta p(\theta)}{p(\theta)}$ indicates that the non-parametric sensitivities can be reasonably good even if the target payoff h is poorly correlated with the input instruments. This last remark is consistent with the fact that the MC algorithm gives excellent numerical approximations to the Greeks even for out of the money options with deltas as small as 15%.

7 References

Avellaneda, M. (1988), Minimum-relative-entropy calibration of assetpricing models, *International Journal of Theoretical and Applied Finance*, Vol. 1. No. 4, 447–472

Avellaneda, M., R. Buff, C. Friedman, N. Grandchamp, L. Kruk and J. Newman (1999), Weighted Monte Carlo: a new approach for calibrating asset-pricing models, International Journal of *Theoretical and Applied Finance*, Vol. 4, No. 1 (2001), 91–119

Broadie Mark and P. Glasserman, (1994), Estimating Security Price Derivatives Using Simulation, *Management Science*, Vol. 42, No. 2, p. 269–285.

Fournié, E.; J.-M. Lasry, J. Lebuchoux, P.-L. Lions, and N. Touzi (1998), Applications of Malliavin calculus to Monte-Carlo methods in Finance, *Finance and Stochastics*

On the Term Structure of Futures and Forward Prices *

Tomas Björk[1]** and Camilla Landén[2]

[1] Department of Finance, Stockholm School of Economics
Box 6501, SE-113 83 Stockholm, SWEDEN
e-mail: tomas.bjork@hhs.se

[2] Department of Mathematics, Royal Institute of Technology
SE-100 44 Stockholm, SWEDEN
e-mail: camilla@math.kth.se

Abstract. We investigate the term structure of forward and futures prices for models where the price processes are allowed to be driven by a general marked point process as well as by a multidimensional Wiener process. Within an infinite dimensional HJM-type model for futures and forwards we study the properties of futures and forward convenience yield rates. For finite dimensional factor models, we develop a theory of affine term structures, which is shown to include almost all previously known models. We also derive two general pricing formulas for futures options. Finally we present an easily applicable sufficient condition for the possibility of fitting a finite dimensional futures price model to an arbitrary initial futures price curve, by introducing a time dependent function in the drift term.

Keywords

term structure, futures price, forward price, options, jump-diffusion model, affine term structure.
JEL classification: E43, G13

1 Introduction

The object of this paper is to study the properties of forward and futures prices (as well as their derivatives) within a reasonably general framework, and in particular we are interested in the case when the underlying asset is non-financial, i.e. when we have a non zero convenience yield.

The literature on forward and futures contracts is a rich one. In [1], [5], [10], [17], and [18] the models have purely Wiener driven price dynamics, while [11] also allows for a point process. In this paper we are in particular inspired by the exposition in [17].

* We are grateful to K. Miltersen for some very helpful comments.
** The financial support of ITM is gratefully acknowledged.

The main contributions of the present paper are as follows.

- In Section 2 we present a general framework for the term structure dynamics of forward and futures prices, allowing for a general marked point process as well as for a multidimensional Wiener process in the price dynamics. The approach is to model the entire term structure a la Heath-Jarrow-Morton, and the main novelty is the introduction of the point process.
- In Section 3 we study how it is possible to model the forward and futures term structure by modeling the term structure of forward and futures convenience yield rates. This approach has earlier been taken in [1] and [17]. We extend the earlier results to the point process case, and we also give new (even for the pure Wiener case) results about the forward convenience yield rates drift condition. Furthermore we provide new results on the relations between forward and futures convenience yield rates and the conditional expectation of the future value of the spot convenience yield.
- In Section 4 we consider finite dimensional factor models and develop a theory of affine term structures for forward and futures prices. This is done very much as in the interest rate case (see [7]) and we show that almost all previously known factor models for forwards and futures belong to the affine class. In particular we show that the natural (from an affine point of view) spot price models are the ones where the local rate of return and the squared volatility are affine in the log of the spot price. We also provide new affine factor models.
- Section 5 is devoted to the pricing of futures options within the general framework of Section 2. Since the futures price process is not the spot price process of a traded asset, the general Geman-El Karoui-Rochet option pricing formula (see [9]) is not applicable. Instead, by introducing two hitherto new types of martingale measures, we manage to provide two different general option pricing formulas and we also discuss the economic interpretation of these formulas.
- Finally, in Section 6 we discuss the problem of fitting a given finite dimensional factor model to a given initial futures term structure. We present a reasonably large class of models for which the fitting can be done by means of a deterministic perturbation of the drift term of the spot price, and it is seen that most existing models in the literature belong to this class.

2 Basics

We consider a financial market living on a stochastic basis (filtered probability space) $(\Omega, \mathcal{F}, \mathbf{F}, Q)$ where $\mathbf{F} = \{\mathcal{F}_t\}_{t\geq 0}$. The basis is assumed to carry a multidimensional Wiener process W as well as a marked point process $\mu(dt, dy)$ on a measurable Lusin mark space $(E, \mathcal{E})$ with predictable compensator $\nu(dt, dy)$. The predictable σ-algebra is denoted by $\mathcal{Q}$, and we make the definition $\tilde{\mathcal{Q}} = \mathcal{Q} \otimes \mathcal{E}$. We assume that $\nu([0,t] \times E) < \infty$ Q-a.s. for all finite t, i.e. μ is a multivariate point process in the terminology of [13]. For simplicity we also assume that μ has an intensity λ, i.e. the compensator has the form

$$\nu(dt, dy) = \lambda(t, dy)dt.$$

The compensated point process $\tilde{\mu}$ is defined by $\tilde{\mu}(dt, dx) = \mu(dt, dy) - \nu(dt, dy)$.

The primitive assets to be considered on the market are forward and futures contracts, written on a given underlying asset, with different delivery dates. We denote the forward price at time t of a forward contract with delivery date T by $G(t,T)$. The futures price at time t with delivery date T is denoted by $F(t,T)$. The induced spot price process S_t is given by a standard arbitrage argument as

$$S_t = F(t,t) = G(t,t). \tag{1}$$

We assume that there is an idealized market (liquid, frictionless, unlimited short selling allowed etc.) for forward and futures contracts for every delivery date T. We do, however, **not** assume that the asset underlying the futures and forward market is traded on an idealized market. The market for the underlying could for example be very thin, there could be transactions costs, prohibitive storage costs or shortselling constraints. Typical examples would be a commodity market or a market for electric energy. We will also have to consider the bond market, and we let $p(t,T)$ denote the price, at time t, of a zero coupon bond maturing at T. The corresponding forward rates are denoted by $f(t,T)$, where as usual

$$f(t,T) = -\frac{\partial}{\partial T} \ln p(t,T). \tag{2}$$

The short rate is denoted by $r(t)$, and defined by $r(t) = f(t,t)$. The money account is defined as usual by $B(t) = \exp \int_0^t r(s)ds$. We assume that the market for bonds, futures and forwards is arbitrage free in the sense that the probability measure Q is a martingale measure (for the numeraire B) for the economy. For the rest of the paper we will, either by implication or by assumption, consider dynamics of the following type.

Forward price dynamics

$$\begin{aligned} dG(t,T) &= G(t,T)\alpha_G(t,T)dt + G(t,T)\sigma_G(t,T)dW_t \\ &+ G(t-,T)\int_E \delta_G(t,y,T)\mu(dt,dy), \end{aligned} \tag{3}$$

Futures price dynamics

$$\begin{aligned} dF(t,T) &= F(t,T)\alpha_F(t,T)dt + F(t,T)\sigma_F(t,T)dW_t \\ &+ F(t-,T)\int_E \delta_F(t,y,T)\mu(dt,dy), \end{aligned} \tag{4}$$

Spot price dynamics

$$dS(t) = S(t)\alpha_S(t)dt + S(t)\sigma_S(t)dW_t + S(t-)\int_E \delta_S(t,y)\mu(dt,dy), \tag{5}$$

Short rate dynamics

$$dr(t) = \alpha_r(t)dt + \sigma_r(t)dW_t + \int_E \delta_r(t,y)\mu(dt,dy), \tag{6}$$

Bond price dynamics

$$dp(t,T) = p(t,T)\alpha_p(t,T)dt + p(t,T)\sigma_p(t,T)dW_t + p(t-,T)\int_E \delta_p(t,y,T)\mu(dt,dy), \tag{7}$$

Forward rate dynamics

$$df(t,T) = \alpha_f(t,T)dt + \sigma_f(t,T)dW_t + \int_E \delta_f(t,y,T)\mu(dt,dy). \tag{8}$$

In the above formulas the coefficient processes are assumed to meet standard conditions required to guarantee that the various processes are well defined.

We recall the following basic results (see e.g. [2]).

Proposition 2.1. *Let Q^T denote the T-forward martingale measure. Then the following hold.*

- *For a fixed T, the futures price process $F(t,T)$ is a Q-martingale, and in particular we have*

$$F(t,T) = E^Q\left[S(T)\middle|\mathcal{F}_t\right] \tag{9}$$

- *The forward price process $G(t,T)$ is a martingale under Q^T, and in particular we have*

$$G(t,T) = E^{Q^T}\left[S(T)\middle|\mathcal{F}_t\right]. \tag{10}$$

For future use, we recall the following relation between the volatilities of the forward rates and the bond prices.

Proposition 2.2. *With notation as in (7)-(8) we have*

$$\sigma_p(t,T) = -\int_t^T \sigma_f(t,s)ds, \tag{11}$$

$$\delta_p(t,y,T) = e^{-D_f(t,y,s)} - 1, \tag{12}$$

where

$$D_f(t,T,y) = \int_t^T \delta_f(t,y,s)ds \tag{13}$$

Proof. See [4].

Since the modeling above is done directly under a martingale measure, there will be "drift conditions", relating the drift terms to the volatilities of the various processes above.

Proposition 2.3. *Under the martingale measure Q, the following relations hold.*

$$\alpha_G(t,T) = \sigma_G(t,T)\int_t^T \sigma_f^\star(t,s)ds - \int_E \delta_G(t,y,T)e^{-D_f(t,y,T)}\lambda(t,dy), \quad (14)$$

$$\alpha_F(t,T) = -\int_E \delta_F(t,y,T)\lambda(t,dy), \quad (15)$$

$$\alpha_p(t,T) = r_t - \int_E \left\{e^{-D_f(t,y,T)} - 1\right\}\lambda(t,dy), \quad (16)$$

$$\alpha_f(t,T) = \sigma_f(t,T)\int_t^T \sigma_f^\star(t,s)ds - \int_E \delta_f(t,y,T)e^{-D_f(t,y,T)}\lambda(t,dy). \quad (17)$$

Proof. For (16)-(17) we refer to [4]. In order to derive (15), we rewrite the F-dynamics on semimartingale form by compensating the point process. We thus obtain

$$\begin{aligned} dF(t,T) = {} & F(t,T)\left\{\alpha_F(t,T) + \int_E \delta_F(t,y,T)\lambda(t,dy)\right\}dt \\ & + F(t,T)\sigma_F(t,T)dW_t \\ & + F(t-,T)\int_E \delta_F(t,y,T)\tilde{\mu}(dt,dy) \end{aligned}$$

This formula gives the futures price process as a sum of a predictable finite variation process and two martingales. Since the futures price process is a Q-martingale, the dt-term must vanish, and we are finished.

To derive the drift condition for the forward price process, we now change measure from Q to the T-forward measure Q^T. From general theory (see [9]) we know that the likelihood process for this measure transformation is given by

$$L_t^T = \frac{p(t,T)}{B(t)p(0,T)},$$

where

$$L_t^T = \frac{dQ^T}{dQ}, \quad \text{on } \mathcal{F}_t.$$

Using (7), (12), (16) and the Itô formula, we thus have the L^T-dynamics

$$dL_t^T = L_t^T\sigma_p(t,T)dW_t + L_{t-}^T\int_E \delta_p(t,y,T)\tilde{\mu}(dt,dy).$$

From the Girsanov Theorem (see [13]) it now follows that we can write

$$dW_t = \sigma_p^\star(t,T)dt + dW_t^T, \tag{18}$$

where W^T is a Q^T-Wiener process. It also follows from the Girsanov Theorem that, under Q^T, the point process μ has an intensity λ^T given by

$$\lambda^T(t,dy) = (\delta_p(t,y,T)+1)\,\lambda(t,dy). \tag{19}$$

We can thus write the Q^T-semimartingale dynamics for $G(t,T)$ as

$$\begin{aligned} dG(t,T) = G(t,T)\Big\{&\alpha_G(t,T) + \sigma_G(t,T)\sigma_p^\star(t,T) \\ &+ \int_E \delta_G(t,y,T)\,(\delta_p(t,y,T)+1)\,\lambda(t,dy)\Big\}\,dt \\ &+ G(t,T)\sigma_G(t,T)dW_t^T \\ &+ G(t-,T)\int_E \delta_G(t,y,T)\left\{\mu(dt,dy) - \lambda^T(t,dy)dt\right\}. \end{aligned}$$

Since $G(t,T)$ is a Q^T-martingale the dt-term thus has to vanish and, using Proposition 2.2, we obtain (14).

3 Modeling the Forward and Futures Convenience Yield

In this section we will study how it is possible to model the term structure of forwards and futures by modeling the spot price and the term structure of the forward and futures convenience yields. This approach goes back to [1], [5] and [17].

3.1 Basic Definitions

Definition 3.1. With notations as above we define the following objects.

- The term structure of futures convenience yields $\varphi(t,T)$, for $0 \le t \le T$, is defined by the relation

$$F(t,T) = S(t)e^{\int_t^T [f(t,s)-\varphi(t,s)]ds}. \tag{20}$$

- The term structure of forward convenience yields $\gamma(t,T)$, for $0 \le t \le T$, is defined by the relation

$$G(t,T) = S(t)e^{\int_t^T [f(t,s)-\gamma(t,s)]ds}. \tag{21}$$

- The spot convenience yield $c(t)$ is defined by the relation

$$\alpha_S(t) = r(t) - c(t) - \int_E \delta_S(t,y)\lambda(t,dy) \tag{22}$$

In order to connect with elementary theory we note that, for the case when $S(t)$ is the price of an underlying asset traded on an idealized market, we have the relation

$$G(t,T) = S(t)p(t,T)^{-1} = S(t)e^{\int_t^T f(t,s)ds}.$$

The forward convenience yields thus measures the deviation from this idealized situation. We also note that, by the definition above, the spot price has the Q-dynamics

$$dS_t = S(t)(r_t - c_t)dt + S(t)\sigma_S(t)dW + \int_E \delta_S(t,y)\tilde{\mu}(dt,dy). \tag{23}$$

Thus, as usual under the martingale measure, the local mean rate of return of the spot price equals the short rate minus the spot convenience yield.

We will start by investigating some elementary properties of the various yields, and then we go on to discuss how to model the yield dynamics.

3.2 Elementary Properties

It is easy to see that if the spot convenience yield c_t is deterministic, then we have, for all $0 \leq t \leq T$,

$$c_T = \gamma(t,T).$$

In the general situation with stochastic c_t, this result raises the question if, $\gamma(t,T)$ and/or $\varphi(t,T)$ can be viewed as predictors (at time t) of the spot yield at time T. The following result provides an answer to this question. In order to shorten notation, we use Cov_t^Q to denote the Q-covariance, conditioned on $\mathcal{F}_t$.

Proposition 3.2.

- *The forward convenience yield satisfies*

$$\gamma(t,T) = E^Q\left[c_T|\,\mathcal{F}_t\right] + \frac{Cov_t^Q\left[e^{-\int_t^T r(s)ds}S_T, c_T\right]}{E^Q\left[e^{-\int_t^T r(s)ds}S_T\middle|\,\mathcal{F}_t\right]} \tag{24}$$

- *The futures convenience yield satisfies*

$$\begin{aligned}\varphi(t,T) = E^Q\left[c_T|\,\mathcal{F}_t\right] + \frac{Cov_t^Q\left[S_T, c_T\right]}{E^Q\left[S_T|\,\mathcal{F}_t\right]} \\ + f(t,T) - E^Q\left[r_T|\,\mathcal{F}_t\right] - \frac{Cov_t^Q\left[S_T, r_T\right]}{E^Q\left[S_T|\,\mathcal{F}_t\right]}\end{aligned} \tag{25}$$

- *In particular we have*

$$c_t = \varphi(t,t) = \gamma(t,t). \tag{26}$$

Proof. We start by noting that (26) follows immediately from (24) and (25). From the definition of γ we have

$$-\frac{\partial}{\partial T}\ln G(t,T) = \gamma(t,T) - f(t,T). \tag{27}$$

From general theory we also know that

$$\begin{aligned} G(t,T) &= \frac{1}{p(t,T)} E^Q\left[e^{-\int_t^T r(s)ds} S(T) \middle| \mathcal{F}_t \right] \\ &= e^{\int_t^T f(t,s)ds} E^Q\left[e^{-\int_t^T r(s)ds} S(T) \middle| \mathcal{F}_t \right] \end{aligned}$$

Using subscript to denote partial derivatives, we obtain

$$G_T(t,T) = f(t,T)G(t,T) + e^{\int_t^T f(t,s)ds} \frac{\partial}{\partial T} E^Q\left[e^{-\int_t^T r(s)ds} S(T) \middle| \mathcal{F}_t \right]$$

In order to compute the last term, let us (for a fixed t) define the process Z_T by $Z_T = e^{-\int_t^T r(s)ds} S(T)$. An application of the Itô formula, together with (23), gives us

$$\begin{aligned} dZ_T &= -r_T Z_T dT + e^{-\int_t^T r(s)ds} dS_T \\ &= -r_T Z_T dT + e^{-\int_t^T r(s)ds} S_T (r_T - c_T) dT + dM \\ &= -e^{-\int_t^T r(s)ds} S_T c_T dT + dM, \end{aligned}$$

where dM denotes a generic Q-martingale increment. From this we obtain

$$\begin{aligned} \frac{\partial}{\partial T} E^Q\left[e^{-\int_t^T r(s)ds} S(T) \middle| \mathcal{F}_t \right] &= \lim_{h\to 0} \frac{1}{h} E^Q\left[Z_{T+h} - Z_T \middle| \mathcal{F}_t \right] \\ &= -\lim_{h\to 0} \frac{1}{h} E^Q\left[\int_T^{T+h} e^{-\int_t^u r(s)ds} S_u c_u du \middle| \mathcal{F}_t \right] \\ &= -E^Q\left[e^{-\int_t^T r(s)ds} S_T c_T \middle| \mathcal{F}_t \right]. \end{aligned}$$

We thus obtain

$$-\frac{\partial}{\partial T}\ln G(t,T) = -\frac{G_T(t,T)}{G(t,T)} = -f(t,T) + \frac{E^Q\left[e^{-\int_t^T r(s)ds} c_T S_T \middle| \mathcal{F}_t \right]}{E^Q\left[e^{-\int_t^T r(s)ds} S_T \middle| \mathcal{F}_t \right]}.$$

Comparing this expression with (27) gives us

$$\gamma(t,T) = \frac{E^Q\left[e^{-\int_t^T r(s)ds} c_T S_T \middle| \mathcal{F}_t \right]}{E^Q\left[e^{-\int_t^T r(s)ds} S_T \middle| \mathcal{F}_t \right]}.$$

Using the formula $E[XY] = E[X] \cdot E[Y] + Cov(X,Y)$, this proves (24).

For the futures convenience yield we have, from (20),

$$-\frac{\partial}{\partial T}\ln F(t,T) = \varphi(t,T) - f(t,T).$$

From general theory we also have

$$F(t,T) = E^Q\left[S_T | \mathcal{F}_t\right].$$

From this we obtain (arguing as above)

$$F_T(t,T) = E^Q\left[S_T(r_T - c_T)| \mathcal{F}_t\right].$$

Thus we have

$$-\frac{\partial}{\partial T}\ln F(t,T) = \frac{E^Q\left[S_T(c_T - r_T)| \mathcal{F}_t\right]}{E^Q\left[S_T| \mathcal{F}_t\right]},$$

so

$$\varphi(t,T) - f(t,T) = \frac{E^Q\left[S_T(c_T - r_T)| \mathcal{F}_t\right]}{E^Q\left[S_T| \mathcal{F}_t\right]},$$

which proves (25).

We also have some easy consequences from this result.

Corollary 3.3.

- *If the spot convenience yield c is deterministic, then*

$$\gamma(t,T) = c_T, \quad \forall 0 \le t \le T. \tag{28}$$

- *If the short rate r is deterministic, then*

$$\varphi(t,T) = \gamma(t,T) = E^Q\left[c_T | \mathcal{F}_t\right] + \frac{Cov_t^Q\left[S_T, c_T\right]}{E^Q\left[S_T | \mathcal{F}_t\right]}. \tag{29}$$

Proof. The relation (28) follows immediately from (24). The formula (29) follows also from (28) together with the fact that $F = G$ (and thus $\varphi = \gamma$) when interest rates are deterministic.

3.3 Drift Conditions for the Yields

By specifying the dynamics of the spot price and the futures (forward) convenience yield, the dynamics of futures (forward) prices are completely specified. Since we are modeling under a martingale measure we will have the drift conditions for futures and forward price dynamics given by Proposition 2.3, and these conditions will obviously imply drift conditions on the yield dynamics. Before starting on this investigation we need a small technical lemma.

Lemma 3.4. *Assume that, for each T the process $X(t,T)$ has dynamics, for $t \leq T$, given by*

$$dX(t,T) = \alpha_X(t,T)dt + \sigma_X(t,T)dW_t + \int_E \delta_X(t,y,T)\mu(dt,dy), \tag{30}$$

where the coefficient processes are assumed to meet standard conditions required to guarantee that the X process is well defined. Assume furthermore that the coefficients are regular enough to allow for an application of the stochastic Fubini Theorem. If, for every T, the process $Z(t,T)$ is defined by

$$Z(t,T) = \int_t^T X(t,s)ds,$$

then the stochastic differential of Z is given by

$$dZ(t,T) = \{A_X(t,T) - X(t,t)\}\, dt + S_X(t,T)dW_t + \int_E D_X(t,y,T)\mu(dt,dy),$$

where

$$A_X(t,T) = \int_t^T \alpha_X(t,s)ds, \tag{31}$$

$$S_X(t,T) = \int_t^T \sigma_X(t,s)ds, \tag{32}$$

$$D_X(t,y,T) = \int_t^T \delta_X(t,y,s)ds. \tag{33}$$

Proof. Fubini.

We may now state and prove the martingale measure drift conditions for the futures and forward convenience yield dynamics.

Proposition 3.5. *Assume that the dynamics of the futures and forward convenience yields, under the martingale measure Q, are given by*

$$d\varphi(t,T) = \alpha_\varphi(t,T)dt + \sigma_\varphi(t,T)dW_t + \int_E \delta_\varphi(t,y,T)\mu(dt,dy),$$

$$d\gamma(t,T) = \alpha_\gamma(t,T)dt + \sigma_\gamma(t,T)dW_t + \int_E \delta_\gamma(t,y,T)\mu(dt,dy).$$

Then the futures and forward yield drift terms are given by

$$\begin{aligned}\alpha_\varphi(t,T) &= \sigma_f(t,T)\int_t^T \sigma_f^\star(t,s)ds - \int_E \delta_f(t,y,T)e^{-D_f(t,y,T)}\lambda(t,dy) \\ &+ \{\sigma_f(t,T)-\sigma_\varphi(t,T)\}\left\{\sigma_S^\star(t) + \int_t^T \left[\sigma_f^\star(t,s)-\sigma_\varphi^\star(t,s)\right]ds\right\} \\ &+ \int_E \{\delta_f(t,y,T)-\delta_\varphi(t,y,T)\}\left\{e^{D_f(t,y,T)-D_\varphi(t,y,T)}\right\} \\ &\times (\delta_S(t,y)+1)\,\lambda(t,dy).\end{aligned} \tag{34}$$

$$\begin{aligned}\alpha_\gamma(t,T) &= \sigma_\gamma(t,T)\int_t^T \sigma_\gamma^\star(t,s)ds - \sigma_\gamma(t,T)\sigma_S^\star(t) \\ &- \int_E \delta_\gamma(t,y,T)e^{-D_\gamma(t,y,T)}\,(\delta_S(t,y)+1)\,\lambda(t,dy).\end{aligned} \tag{35}$$

Proof. We have by definition

$$F(t,T) = S_t e^{Z(t,T)},$$

with Z defined as

$$Z(t,T) = \int_t^T [f(t,s)-\varphi(t,s)]\,ds.$$

From Lemma 3.4, together with the facts that $f(t,t)=r_t$ and $\varphi(t,t)=c_t$ we obtain (suppressing (t,T))

$$\begin{aligned}dZ &= \{c_t - r_t + A_f - A_\varphi\}\,dt + \{S_f - S_\varphi\}\,dW_t \\ &+ \int_E (D_f - D_\varphi)(t,y,T)\mu(dt,dy).\end{aligned}$$

Defining, for each T, the process $X(t,T)$ by $X(t,T)=e^{Z(t,T)}$, Itô's formula now gives us (suppressing T)

$$\begin{aligned}dX_t &= X_t\left\{c_t - r_t + A_f - A_\varphi + \frac{1}{2}\|S_f - S_\varphi\|^2\right\}dt + X_t\{S_f - S_\varphi\}\,dW_t \\ &+ X_{t-}\int_E \left\{e^{D_f - D_\varphi} - 1\right\}\mu(dt,dy).\end{aligned}$$

From this expression, the spot price dynamics (5) and the relation (22), an application of the Itô formula to the expression $F(t,T) = S_t X(t,T)$ gives us

$$\begin{aligned}
dF_t &= F_t \left\{ c_t - r_t + A_f - A_\varphi + \frac{1}{2}\|S_f - S_\varphi\|^2 \right\} dt \\
&+ F_t \left\{ r_t - c_t - \int_E \delta_S(t,y)\lambda(t,dy) \right\} dt \\
&+ F_t \left(S_f - S_\varphi \right) dW_t \\
&+ F_t \sigma_S dW_t \\
&+ F_t \left(S_f - S_\varphi \right) \sigma_S^\star dt \\
&+ F_{t-} \int_E \left\{ e^{D_f - D_\varphi} \left(1 + \delta_S \right) - 1 \right\} \mu(dt, dy).
\end{aligned}$$

Compensating the point process part and collecting terms give us the semi-martingale dynamics

$$\begin{aligned}
dF_t &= F_t \left\{ A_f - A_\varphi + \frac{1}{2}\|S_f - S_\varphi\|^2 + \left(S_f - S_\varphi \right) \sigma_S^\star \right. \\
&\left. + \int_E \left(e^{D_f - D_\varphi} - 1 \right) \left(1 + \delta_S \right) \lambda(t,dy) \right\} dt \\
&+ F_t \left\{ S_f - S_\varphi + \sigma_S \right\} dW_t \\
&+ F_{t-} \int_E \left\{ e^{D_f - D_\varphi} \left(1 + \delta_S \right) - 1 \right\} \tilde{\mu}(dt, dy),
\end{aligned}$$

Since $F(t,T)$ is a Q-martingale for each fixed T, the dt-term must vanish, so the following identity must hold for all t and T with $0 \le t \le T$.

$$\begin{aligned}
0 &= A_f(t,T) - A_\varphi(t,T) + \frac{1}{2}\|S_f(t,T) - S_\varphi(t,T)\|^2 \\
&+ \left(S_f(t,T) - S_\varphi(t,T) \right) \sigma_S^\star(t) \\
&+ \int_E \left(e^{D_f(t,y,T) - D_\varphi(t,y,T)} - 1 \right) \left(1 + \delta_S(t,y) \right) \lambda(t,dy).
\end{aligned}$$

Differentiating this identity w.r.t. T, using (17) and rearranging, we obtain the drift condition for α_φ.

In order to derive the forward yield drift condition we obtain, in the same way as for dF,

$$\begin{aligned}
dG_t &= G_t \left\{ A_f - A_\gamma + \frac{1}{2}\|S_f - S_\gamma\|^2 + \left(S_f - S_\gamma \right) \sigma_S^\star - \int_E \delta_S \lambda(t,dy) \right\} dt \\
&+ G_t \left\{ S_f - S_\gamma + \sigma_S \right\} dW_t \\
&+ G_{t-} \int_E \left\{ e^{D_f - D_\gamma} \left(1 + \delta_S \right) - 1 \right\} \mu(dt, dy),
\end{aligned}$$

Applying Proposition 2.3 to the G-dynamics thus derived and taking the T-derivative, gives us the result for α_γ.

4 Affine Term Structures for Forwards and Futures

Modeling the entire term structure of forward or futures prices results in an infinite dimensional state variable. Therefore it is sometimes more convenient to model a finite dimensional state process Z, and to assume that forward and futures prices are given as functions of this state process. Just as in interest rate theory (see [7], [8]), the term structures defined by functions which are exponentially affine in the state variables are computationally very tractable, and below we give necessary and sufficient conditions in terms of the dynamics of the state process Z for the forward and futures term structures to be affine.

Assumption 4.1. *The m-dimensional Markov process Z is assumed to have a stochastic differential given by*

$$dZ_t = \alpha_Z(t, Z_t)dt + \sigma_Z(t, Z_t)dW_t + \int_E \delta_Z(t, Z_{t-}, y)\mu(dt, dy)$$

under the martingale measure Q. Furthermore we assume that the compensator ν of μ can be written as $\nu(\omega; dt, dy) = \lambda(t, Z_{t-}(\omega), dy)dt$.

4.1 Futures

Assumption 4.2. *We assume that the futures prices can be written on the following form*

$$F(t, T) = H_F(t, Z_t, T), \tag{36}$$

where $H_F : R^3 \to R$ is a smooth function. In particular we assume that the spot price S is given by

$$S(t) = H_F(t, Z_t, t) = h(t, Z_t).$$

Lemma 4.3. *If futures prices are given by (36), then H_F satisfies the following partial differential equation*

$$\begin{cases} \dfrac{\partial H_F}{\partial t}(t, z, T) + \mathcal{A}H_F(t, z, T) = 0, \\ \qquad\qquad H_F(T, z, T) = h(T, z), \end{cases} \tag{37}$$

where $\mathcal{A}$ is given by

$$\begin{aligned} \mathcal{A}H(t, z, T) = &\sum_{i=1}^m \alpha_Z^i(t, z)\frac{\partial H}{\partial z_i} + \frac{1}{2}\sum_{i,j=1}^m C_{ij}(t, z)\frac{\partial^2 H}{\partial z_i}\partial z_j \\ &+ \int_E [H(t, z + \delta_Z(t, z, y), T) - H(t, z, T)]\lambda(t, z, dy). \end{aligned} \tag{38}$$

In the expression (38) the matrix C is defined by

$$C = \sigma_Z \sigma_Z^*, \tag{39}$$

where $*$ denotes transpose and all the partial derivatives of H should be evaluated at (t, z, T).

Proof. Dynkin's formula on

$$H_F(t, Z_t, T) = E^Q[h(T, Z_T)|\mathcal{F}_t].$$

Definition 4.4. The term structure of futures prices is said to be **affine** if the function H_F from (36) is of the following form

$$\ln H_F(t, z, T) = A_F(t, T) + B_F^*(t, T)z, \tag{40}$$

where A_F and B_F are deterministic functions.

Proposition 4.5. *Suppose that Assumption 4.2 is in force and that the functions α_Z, σ_Z, δ_Z, λ and h are of the following form*

$$\begin{aligned}
\alpha_Z(t, z) &= a_1(t) + a_2(t)z, \\
\sigma_Z\sigma_Z^*(t, z) &= k_0(t) + \sum_{i=1}^{m} k_i(t)z_i, \\
\delta_Z(t, z, y) &= \delta_Z(t, y), \\
\lambda(t, z, y) &= l_1(t, y) + l_2^*(t, y)z, \\
\ln h(t, z) &= c(t) + d^*(t)z.
\end{aligned} \tag{41}$$

Then the term structure of futures prices is affine, that is H_F from (36) can be written on the form (40) where A_F and B_F solve the following system of ordinary differential equations.

$$\begin{cases} \dfrac{\partial A_F(t,T)}{\partial t} + a_1^*(t)B_F(t,T) + \dfrac{1}{2}B_F^*(t,T)k_0(t)B_F(t,T) \\ \qquad + \displaystyle\int_E \left(e^{B_F^*(t,T)\delta_Z(t,y)} - 1\right) l_1(t, dy) = 0 \\ \qquad A_F(T,T) = c(T), \end{cases} \tag{42}$$

and

$$\begin{cases} \dfrac{\partial B_F(t,T)}{\partial t} + a_2^*(t)B_F(t,T) + \dfrac{1}{2}\beta_F^*(t,T)K(t)B_F(t,T) \\ \qquad + \displaystyle\int_E \left(e^{B_F^*(t,T)\delta_Z(t,y)} - 1\right) l_2(t, dy) = 0 \\ \qquad B_F(T,T) = d(T), \end{cases} \tag{43}$$

where

$$K(t) = \begin{bmatrix} k_1(t) \\ k_2(t) \\ \vdots \\ k_m(t) \end{bmatrix}, \tag{44}$$

and $\beta_F(t,T)$ *is the* $m^2 \times m$*-matrix*

$$\beta_F(t,T) = \begin{bmatrix} B_F(t,T) & 0 & \dots & 0 \\ 0 & B_F(t,T) & & \\ \vdots & & \ddots & \\ 0 & & & B_F(t,T) \end{bmatrix}.$$

Proof. This follows from the fact that $\exp\{A_F(t,T) + B_F(t,T)\}$, where A_F and B_F solve (42) and (43), respectively, solves the PDE (37), which uniquely characterizes the futures prices in this setting.

Remark 4.6. As in [7] it can be shown that under non degeneracy conditions and a possible reordering of indices a σ_Z of the form in (41) can be written as

$$\sigma_Z(t,Z_t) = \Sigma(t) \begin{bmatrix} \sqrt{u_1(t,Z_t)} & 0 & \dots & 0 \\ 0 & \sqrt{u_2(t,Z_t)} & \dots & 0 \\ & & \ddots & \\ 0 & \dots & 0 & \sqrt{u_m(t,Z_t)} \end{bmatrix}$$

where the matrix $\Sigma(t)$ depends only on t and

$$u_i(t,z) = k_0^i(t) + \sum_{j=1}^{m} k_j^i(t) z_j.$$

In most models which have appeared in the literature, the spot price is one factor. Note that if we want an affine term structure we can not use the spot price itself as a factor, but we must use the logarithm of the spot price. Also note that if the logarithm of the spot price is a factor, then the boundary conditions of A_F and B_F will be uniquely determined. We summarize these observations in the following corollary of Proposition 4.5.

Corollary 4.7. *Assume that* $Z^0(t) = \ln S(t)$ *and that* α_Z, $\sigma_Z\sigma_Z^*$, δ_Z *and* λ *are of the form stated in Proposition 4.5. Then*

$$h(t,z) = e^{z^0},$$

and the term structure of futures prices is affine. That is, H_F *from (36) can be written on the form (40) where* A_F *and* B_F *solve the system (42)-(43) of ordinary differential equations with the boundary conditions replaced by*

$$A_F(T,T) = 0,$$

and

$$B_F(T,T) = e_0,$$

where $e_0 = (1, 0, \dots, 0)^*$.

The stochastic differential of Z^0 *has the required form if and only if the stochastic differential of the spot price* S *can be written as*

$$dS_t = \left[\beta_1(t) + \beta_2(t)\ln S_t + \sum_{i=1}^{m} \beta_3^i(t) Z_t^i\right] S_t dt \\ +\sigma_S(t, Z_t) S_t dW_t \\ +S_{t-}\int_E \delta_S(t, y)\mu(dt, dy),$$

where

$$\sigma_S(t, Z_t)\sigma_S^*(t, Z_t) = \Gamma_1(t) + \Gamma_2(t)\ln S_t + \sum_{i=1}^{m} \Gamma_3(t) Z_t^i.$$

4.2 Examples

A number of factor models have been proposed in the literature, and the corresponding futures prices have been computed on a case by case basis. We now give a list of the most well known factor models and it follows immediately from Corollary 4.7 that all these models will give rise to an affine term structure of futures prices, which thus easily can be computed. We also provide some new examples of affine factor models. All the models are given directly under an equivalent martingale measure Q, and unless indicated otherwise, the coefficients of the models are assumed to be constant.

The Schwartz Spot Price Model. The following spot price model was studied in [18].

$$dS_t = \kappa(\alpha_S - \ln S_t) S_t dt + \sigma_S S_t dW_t., \tag{45}$$

Here W is a one-dimensional Wiener process. Since this model is a special case of the model below we defer computing the futures prices for this model to the next section.

The General Affine Wiener Driven Spot Price Model. Without loosing the affine term structure, we can extend the one-factor model by Schwartz to include a volatility of Cox-Ingersoll-Ross type. The model then looks as follows

$$dS_t = [a_1(t) + a_2(t)\ln S_t] S_t dt + S_t\sqrt{k_0(t) + k_1(t)\ln S_t} dW_t. \tag{46}$$

The functions A_F and B_F for this model satisfy the following ordinary differential equations ($Z_t = \ln S_t$ is used)

$$\frac{\partial A_F(t,T)}{\partial t} + \left(a_1(t) - \frac{1}{2}k_0(t)\right) B_F(t,T) + \frac{1}{2}k_0(t) B_F^2(t,T) = 0, \\ A_F(T,T) = 0,$$

and

$$\frac{\partial B_F(t,T)}{\partial t} + \left(a_2(t) - \frac{1}{2}k_1(t)\right) B_F(t,T) + \frac{1}{2}k_1(t)B_F^2(t,T) = 0,$$
$$B_F(T,T) = 1.$$

The solutions are given by

$$B_F(t,T) = \frac{1}{X(t,T)},$$

where

$$X(t,T) = e^{-\int_t^T [a_2(s)-\frac{1}{2}k_1(s)]ds} - \frac{1}{2}\int_t^T e^{-\int_t^u [a_2(s)-\frac{1}{2}k_1(s)]ds} k_1(u)du,$$

and

$$A_F(t,T) = \int_t^T \left[\left(a_1(s) - \frac{1}{2}k_0(s)\right) B_F(s,T) + \frac{1}{2}k_0(s)B_F^2(s,T)\right] ds.$$

As a special case, inserting

$$\begin{aligned} a_1(t) &= \kappa\alpha_S, \\ a_2(t) &= -\kappa, \\ k_0(t) &= \sigma^2, \\ k_1(t) &= 0, \end{aligned}$$

we find the futures prices for the Schwartz one-factor model (see [18]).

The Gibson–Schwartz Two-Factor Model. The following two-factor model uses the spot price and the spot convenience yield as factors. It is based on the model in [10] and appears in [18].

$$\begin{aligned} dS_t &= (r - c_t)S_t dt + S_t\sigma_S dW_t, \\ dc_t &= \kappa(\alpha_c - c_t)dt + \sigma_c dW_t. \end{aligned} \tag{47}$$

Here W is a two-dimensional Wiener process and

$$\sigma_S\sigma_c^* = \rho\|\sigma_S\| \cdot \|\sigma_c\|.$$

Again, this model is a special case of the next and therefore we defer computing the futures prices for this model to the next paragraph.

Note that for the one and two-factor models presented so far, forward and futures prices agree, since interest rates are assumed to be deterministic.

The Schwartz Three-Factor Model. Including the short rate r as a third factor makes forward and futures prices different. The following model can be found in [18].

$$
\begin{aligned}
dS_t &= (r_t - c_t)S_t dt + S_t \sigma_S dW_t, \\
dc_t &= \kappa_c(\alpha_c - c_t)dt + \sigma_c dW_t, \\
dr_t &= \kappa_r(\alpha_r - r_t)dt + \sigma_r dW_t,
\end{aligned} \tag{48}
$$

where W is a three-dimensional Wiener process, and

$$
\begin{aligned}
\sigma_S \sigma_c^* &= \rho_{Sc} \|\sigma_S\| \cdot \|\sigma_c\|, \\
\sigma_c \sigma_r^* &= \rho_{cr} \|\sigma_c\| \cdot \|\sigma_r\|, \\
\sigma_S \sigma_r^* &= \rho_{Sr} \|\sigma_S\| \cdot \|\sigma_r\|,
\end{aligned}
$$

Let $Z_t = [\ln S_t, c_t, r_t]^*$. The functions A_F and $B_F = [B_S, B_c, B_r]^*$ for this model satisfy the following ordinary differential equations

$$
\begin{aligned}
\frac{\partial A_F(t,T)}{\partial t} - \frac{1}{2}\|\sigma_S\|^2 B_S(t,T) + \kappa_c \alpha_c B_c(t,T) + \kappa_r \alpha_r B_r(t,T)& \\
+ \frac{1}{2} B_F(t,T)^* k_0 B_F(t,T) &= 0, \\
A_F(T,T) &= 0,
\end{aligned}
$$

where k_0 is the covariance matrix

$$
k_0 = \begin{bmatrix} \|\sigma_S\|^2 & \rho_{Sc}\|\sigma_S\| \cdot \|\sigma_c\| & \rho_{Sr}\|\sigma_S\| \cdot \|\sigma_r\| \\ \rho_{Sc}\|\sigma_S\| \cdot \|\sigma_c\| & \|\sigma_c\|^2 & \rho_{cr}\|\sigma_c\| \cdot \|\sigma_r\| \\ \rho_{Sr}\|\sigma_S\| \cdot \|\sigma_r\| & \rho_{cr}\|\sigma_c\| \cdot \|\sigma_r\| & \|\sigma_r\|^2 \end{bmatrix},
$$

and

$$
\begin{aligned}
\frac{\partial B_F(t,T)}{\partial t} + a_2^* B_F(t,T) &= 0, \\
B_F(T,T) &= [1, 0, 0]^*,
\end{aligned}
$$

where the matrix a_2 is given by

$$
a_2 = \begin{bmatrix} 0 & -1 & 1 \\ 0 & -\kappa_c & 0 \\ 0 & 0 & -\kappa_r \end{bmatrix}.
$$

The solutions are given by

$$
\begin{aligned}
B_S(t,T) &\equiv 1, \\
B_c(t,T) &= \frac{1}{\kappa_c}\left(e^{-\kappa_c(T-t)} - 1\right), \\
B_r(t,T) &= \frac{1}{\kappa_r}\left(1 - e^{-\kappa_r(T-t)}\right),
\end{aligned}
$$

and

$$
\begin{aligned}
A_F(t,T) = & \frac{\kappa_c \alpha_c + \rho_{Sc}\|\sigma_S\| \cdot \|\sigma_c\|}{\kappa_c^2}(1 - e^{-\kappa_c(T-t)} - \kappa_c(T-t)) \\
& - \frac{\kappa_r \alpha_r + \rho_{Sr}\|\sigma_S\| \cdot \|\sigma_r\|}{\kappa_r^2}(1 - e^{-\kappa_r(T-t)} - \kappa_r(T-t)) \\
& + \frac{\|\sigma_c\|^2}{4\kappa_c^3}(2\kappa_c(T-t) - 3 + 4e^{-\kappa_c(T-t)} - e^{-2\kappa_c(T-t)}) \\
& + \frac{\|\sigma_r\|^2}{4\kappa_r^3}(2\kappa_r(T-t) - 3 + 4e^{-\kappa_r(T-t)} - e^{-2\kappa_r(T-t)}) \\
& + \frac{\rho_{cr}\|\sigma_c\| \cdot \|\sigma_r\|}{\kappa_c + \kappa_r}\left\{\frac{1 - e^{-\kappa_c(T-t)} - e^{-\kappa_r(T-t)} + e^{-(\kappa_c+\kappa_r)(T-t)}}{\kappa_c\kappa_r}\right. \\
& \left. + \frac{1}{\kappa_c^2}(1-e^{-\kappa_c(T-t)} - \kappa_c(T-t)) + \frac{1}{\kappa_r^2}(1-e^{-\kappa_r(T-t)} - \kappa_r(T-t))\right\}.
\end{aligned}
$$

Remark 4.8. Using the explicit expression for bond prices for this model (see [19]) we see that the futures prices can be written in the following manner

$$
F(t,T) = \frac{S_t}{p(t,T)} e^{B_c(t,T)c_t} e^{\bar{A}(t,T)},
$$

where

$$
\bar{A}(t,T) = A_F(t,T) + \frac{\|\sigma_r\|^2}{4\kappa_r^3}(2\kappa_r(T-t) - 3 + 4e^{-\kappa_r(T-t)} - e^{-2\kappa_r(T-t)})
$$

This result will hold even if α_r is allowed to be time-dependent.

The Hilliard–Reis Three-Factor Model. This model was suggested in [11], and coincides with the Schwartz three-factor model except for the facts that the spot price process includes jumps and that the drift of the short rate is time dependent. The function $\alpha_r(t)$ below is chosen so that the initial bond prices produced by the model agree with the observed bond prices. The model is defined by

$$
\begin{aligned}
dS_t &= (r_t - c_t)S_t dt + S_t \sigma_S dW_t + S_{t-} \textstyle\int_R y\tilde{\mu}(dt, dy), \\
dc_t &= \kappa_c(\alpha_c - c_t)dt + \sigma_c dW_t, \qquad (49) \\
dr_t &= \kappa_r(\alpha_r(t) - r)dt + \sigma_r dW_t,
\end{aligned}
$$

where W and σ_S, σ_c and σ_r are as in Section 4.2, and as before $\tilde{\mu}$ is given by $\tilde{\mu}(dt, dy) = \mu(dt, dy) - \nu(dt, dy)$. Here the marked point process μ has mark space $(R, \mathcal{B})$, where $\mathcal{B}$ is the Borel algebra, and the compensator ν is given by

$$
\nu(dt, dy) = \lambda dt \frac{1}{\sqrt{2\pi\eta^2}} \cdot \frac{1}{y+1} \exp\left\{-\frac{[\ln(y+1) - \xi]^2}{2\eta^2}\right\} dy,
$$

for $-1 < y < \infty$. This means that the spot price process will jump according to a Poisson process with intensity λ and that if δ denotes the relative jump size, then $1+\delta$ has a log-normal distribution: $\ln(1+\delta) \sim N(\xi, \eta^2)$.

Since the spot price for this model equals the spot price for the Schwartz three-factor model (with a time dependent α_r, see Remark 4.8) plus a Q-martingale, the futures prices will be the same for these models (as was pointed out in [11]).

A Three-Factor Model with Positive Short Rate If we want to be sure that the short rate is positive, we could replace the short rate process assumed in the models above by a Cox-Ingersoll-Ross type process. With a wise choice of the parameters κ_r, α_r and σ_r the following model will have a positive short rate (see [6]). This model has (to our knowledge) not been studied previously.

$$\begin{aligned} dS_t &= (r_t - c_t)S_t dt + S_t \sigma_S dW_t, \\ dc_t &= \kappa_c(\alpha_c - c_t)dt + \sigma_c dW_t, \\ dr_t &= \kappa_r(\alpha_r - r_t)dt + \sqrt{r_t}\sigma_r dW_t. \end{aligned} \tag{50}$$

Again W is assumed to be a three-dimensional Wiener process, and

$$\begin{aligned} \sigma_S \sigma_c^* &= \rho_{Sc}\|\sigma_S\| \cdot \|\sigma_c\|, \\ \sigma_c \sigma_r^* &= 0, \\ \sigma_S \sigma_r^* &= 0, \end{aligned}$$

Let $Z_t = [\ln S_t, c_t, r_t]^*$. The functions A_F and $B_F = [B_S, B_c, B_r]^*$ for this model satisfy the following ordinary differential equations

$$\begin{aligned} \frac{\partial A_F(t,T)}{\partial t} - \frac{1}{2}\|\sigma_S\|^2 B_S(t,T) + \kappa_c\alpha_c B_c(t,T) + \kappa_r\alpha_r B_r(t,T) \\ + \frac{1}{2}B_F(t,T)^* k_0 B_F(t,T) = 0, \\ A_F(T,T) = 0, \end{aligned}$$

where k_0 is the matrix

$$k_0 = \begin{bmatrix} \|\sigma_S\|^2 & \rho_{Sc}\|\sigma_S\| \cdot \|\sigma_c\| & 0 \\ \rho_{Sc}\|\sigma_S\| \cdot \|\sigma_c\| & \|\sigma_c\|^2 & 0 \\ 0 & 0 & 0 \end{bmatrix},$$

and

$$\begin{aligned} \frac{\partial B_F(t,T)}{\partial t} + a_2^* B_F(t,T) + \frac{1}{2}B_F^*(t,T)k_r B_F(t,T) = 0, \\ B_F(T,T) = [1,0,0]^*, \end{aligned}$$

where the matrix a_2 is given by

$$a_2 = \begin{bmatrix} 0 & -1 & 1 \\ 0 & -\kappa_c & 0 \\ 0 & 0 & -\kappa_r \end{bmatrix},$$

and the matrix k_r by

$$k_r = \begin{bmatrix} 0 & 0 & 0 \\ 0 & 0 & 0 \\ 0 & 0 & \|\sigma_r\|^2 \end{bmatrix}.$$

The solutions are given by

$$B_S(t,T) \equiv 1,$$
$$B_c(t,T) = \frac{1}{\kappa_c}\left(e^{-\kappa_c(T-t)} - 1\right),$$
$$B_r(t,T) = -\frac{2}{\|\sigma_r\|^2} \cdot \frac{(2r_2 + \|\sigma_r\|^2)r_1 e^{-r_1(T-t)} - (2r_1 + \|\sigma_r\|^2)r_2 e^{-r_2(T-t)}}{(2r_2 + \|\sigma_r\|^2)e^{-r_1(T-t)} - (2r_1 + \|\sigma_r\|^2)e^{-r_2(T-t)}},$$

where

$$r_1 = \frac{\kappa_r}{2} + \sqrt{\frac{\kappa_r^2}{4} + \frac{\|\sigma_r\|^2}{2}}$$
$$r_2 = \frac{\kappa_r}{2} - \sqrt{\frac{\kappa_r^2}{4} + \frac{\|\sigma_r\|^2}{2}}$$

Once B_F has been determined A_F can obtained via numerical integration.

4.3 Forwards

Assumption 4.9. *We assume that the zero-coupon bond prices are of the form*

$$p(t,T) = H_P(t, Z_t, T), \tag{51}$$

where $H_P : R^3 \to R$ is a smooth function. Furthermore, we assume that the forward prices can be written on the following form

$$G(t,T) = H_G(t, Z_t, T), \tag{52}$$

where $H_G : R^3 \to R$ is a smooth function. In particular we assume that the spot price S is given by

$$S(t) = H_G(t, Z_t, t) = h(t, Z_t).$$

Lemma 4.10. *If zero-coupon bond prices are given by (51) and forward prices are given by (52), then H_G satisfies the following partial differential equation*

$$\begin{cases} \dfrac{\partial H_G}{\partial t}(t,z,T) + \mathcal{A}^T H_G(t,z,T) = 0, \\ \qquad\qquad\qquad\qquad H_G(T,z,T) = h(T,z), \end{cases} \tag{53}$$

where $\mathcal{A}^T$ is given by

$$\mathcal{A}^T H(t,z,T) = \sum_{i=1}^{m} (\alpha_Z^T)^i(t,z)\frac{\partial H}{\partial z_i} + \frac{1}{2}\sum_{i,j=1}^{m} C_{ij}(t,z)\frac{\partial^2 H}{\partial z_i}\partial z_j$$
$$+ \int_E [H(t, z + \delta_Z(t,z,y), T) - H(t,z,T)]\lambda^T(t,z,dy). \tag{54}$$

Here

$$\alpha_Z^T(t,z) = \alpha_Z(t,z) + \sigma_Z(t,z)\sigma_P^*(t,z,T),$$
$$\lambda^T(t,z,dy) = [\delta_P(t,z,y,T) + 1]\lambda(t,z,dy),$$
$$\sigma_P(t,z,T) = \frac{\nabla_z H_P(t,z,T)}{H_P(t,z,T)}\sigma_Z(t,z), \tag{55}$$
$$\delta_P(t,z,y,T) = \frac{H_P(t, z + \delta_Z(t,z,y), T) - H_P(t,z,T)}{H_P(t,z,T)}.$$

The matrix C was defined in (39). In the expression (54) all the partial derivatives of H should be evaluated at (t,z,T).

Proof. Itô's formula applied to $p(t,T) = H_P(t, Z_t, T)$ gives the expressions for σ_P and δ_P. The dynamics of Z under the T-forward measure Q^T can be found using (18) and (19) and they are given by

$$dZ_t = \alpha_Z^T(t, Z_t)dt + \sigma_Z(t, Z_t)dW_t^T + \int_E \delta_Z(t, Z_{t-}, y)\mu(dt, dy)$$

where the intensity of μ is $\lambda^T(t, Z_{t-}(\omega), dy)$. The result now follows from an application of Dynkin's formula to

$$H_G(t, Z_t, T) = E^T[h(T, Z_T)|\mathcal{F}_t].$$

Definition 4.11. The term structure of interest rates is said to be **affine** if the function H_P from (51) is of the following form

$$\ln H_P(t,z,T) = A_P(t,T) + B_P^*(t,T)z, \tag{56}$$

where A_P and B_P are deterministic functions. Analogously, the term structure of forward prices is said to be **affine** if the function H_G from (52) is of the form

$$\ln H_G(t,z,T) = A_G(t,T) + B_G^*(t,T)z, \tag{57}$$

where A_G and B_G are deterministic functions.

Proposition 4.12. *Suppose that Assumption 4.9 is in force. Furthermore, suppose that the term structure of interest rates is affine, that is the function H_P from (51) can be written on the form (56) and that the functions α_Z, σ_Z, δ_Z, λ and h are of the form given in (41). Then the term structure of forward prices is affine, that is H_F from (52) can be written on the form (57) where A_G and B_G solve the following system of ordinary differential equations.*

$$\begin{cases} \dfrac{\partial A_G(t,T)}{\partial t} + B_G^*(t,T)[a_1(t) + k_0(t)B_P(t,T)] \\ \qquad + \dfrac{1}{2}B_G^*(t,T)s_0(t)B_G(t,T) \\ \qquad + \displaystyle\int_E \left(e^{B_G^*(t,T)\delta_Z(t,y)} - 1\right) e^{B_P^*(t,T)\delta_Z(t,y)} l_1(t,dy) = 0, \\ A_G(T,T) = c(T), \end{cases} \tag{58}$$

and

$$\begin{cases} \dfrac{\partial B_G(t,T)}{\partial t} + a_2^*(t)B_G(t,T) + \beta_G^*(t,T)K(t)B_P(t,T) \\ \qquad + \dfrac{1}{2}\beta_G^*(t,T)K(t)B_G(t,T) \\ \qquad + \displaystyle\int_E \left(e^{B_G^*(t,T)\delta_Z(t,y)} - 1\right) e^{B_P^*(t,T)\delta_Z(t,y)} l_2(t,dy) = 0 \\ B_G(T,T) = d(T), \end{cases} \tag{59}$$

where K is given by (44) and

$$\beta_G(t,T) = \begin{bmatrix} B_G(t,T) & 0 & \dots & 0 \\ 0 & B_G(t,T) & & \\ \vdots & & \ddots & \\ 0 & & & B_G(t,T) \end{bmatrix}.$$

Proof. This follows from the fact that $\exp\{A_G(t,T) + B_G(t,T)\}$, where A_G and B_G solve (58) and (59), respectively, solves the PDE (53), which uniquely characterizes the forward prices in this setting.

5 Options on Futures Prices

5.1 General Formula

In this section we will consider pricing options on futures. To obtain pricing formulas we will use the change of numeraire technique developed in [9]. As before we let Q denote the martingale measure. Apart from Q the measures

Q^T,Q^{F_T}, Q^F and Q^{T_F} will appear. They are defined as follows, starting with the measure Q^F

$$dQ^F = L_t^F dQ, \quad \text{on } \mathcal{F}_t$$

where

$$L_t^F = \frac{F(t, T_1)}{F(0, T_1)}.$$

Using Itô's formula the dynamics of L^F are obtained as

$$dL_t^F = L_t^F \sigma_F(t, T_1) dW_t + L_{t-}^F \int_E \delta_F(t, y, T_1) \tilde{\mu}(dt, dy). \tag{60}$$

Now, the measure Q^{T_F} is defined by

$$dQ^{T_F} = R^{T_F} dQ^F, \quad \text{on } \mathcal{F}_T$$

where

$$R^{T_F} = \frac{\exp\left\{-\int_0^T r_s ds\right\}}{E^F\left[\exp\left\{-\int_0^T r_s ds\right\}\right]}.$$

Here the super index F indicates that the expectation should be taken under Q^F.

As we have seen before the T-forward measure Q^T is defined by

$$dQ^T = L_t^T dQ, \quad \text{on } \mathcal{F}_t,$$

where

$$L_t^T = \frac{p(t, T)}{B(t) p(0, T)}.$$

Using Itô's formula the dynamics of L^T are obtained as

$$dL_t^T = L_t^T \sigma_p(t, T) dW_t + L_{t-}^T \int_E \delta_p(t, y, T) \tilde{\mu}(dt, dy). \tag{61}$$

Finally, the measure Q^{F_T} is defined by

$$dQ^{F_T} = L_t^{F_T} dQ^T, \quad \text{on } \mathcal{F}_t,$$

where

$$\begin{aligned} L_t^{F_T} = \exp\Bigg\{-\int_0^t \Bigg[& \sigma_F(u, T_1) \sigma_p^*(u, T) \\ & + \int_E \delta_F(u, y, T_1) \delta_p(u, y, T) \lambda(u, dy)\Bigg] du\Bigg\} \frac{F(t, T_1)}{F(0, T_1)}. \end{aligned}$$

The Q^T-dynamics of $F(t, T_1)$ can be found using the Girsanov Theorem. Itô's formula then gives us the dynamics of L^{F_T} under Q^T

$$dL_t^{F_T} = L_t^{F_T} \sigma_F(t, T_1) dW_t^T + L_{t-}^{F_T} \int_E \delta_F(t, y, T_1) \tilde{\mu}(dt, dy). \tag{62}$$

The point process μ has the intensity λ^T under Q^T, where $\lambda^T(t, dy)$ is given by $\lambda^T(t, dy) = (1 + \delta_p(t, y, T))\lambda(t, dy)$.

Lemma 5.1. *The price, at date zero, of a European call option with exercise date T and exercise price K written on the futures price with delivery date T_1 can be computed from either one of the following two formulas*

$$\begin{aligned} C_F = F(0, T_1)E^F \left[e^{-\int_0^T r(s)ds}\right] Q^{T_F}(F(T, T_1) \geq K) \\ - Kp(0, T)Q^T(F(T, T_1) \geq K), \end{aligned} \tag{63}$$

where the super index F indicates that the expectation should be taken under Q^F, or

$$\begin{aligned} C_F = p(0, T)F(0, T_1)E^{F_T} \left[\exp\left\{\int_0^T \kappa_s ds\right\} I\{F(T, T_1) \geq K\}\right] \\ - Kp(0, T)Q^T(F(T, T_1) \geq K), \end{aligned} \tag{64}$$

where the super index F_T indicates that the expectation should be taken under Q^{F_T}, and κ is given by

$$\kappa(t) = \sigma_F(t, T_1)\sigma_p^*(t, T) + \int_E \delta_F(t, y, T_1)\delta_p(t, y, T)\lambda(t, dy). \tag{65}$$

Proof. Write the option as

$$X = \max\{F(T, T_1) - K, 0\} = [F(T, T_1) - K] \cdot I\{F(T, T_1) \geq K\},$$

where I is the indicator function, i.e.

$$I\{F(T, T_1) \geq K\} = \begin{cases} 1 \text{ if } F(T, T_1) \geq K, \\ 0 \text{ if } F(T, T_1) < K. \end{cases}$$

We then have that the price of the option, at date zero, is given by

$$\begin{aligned} C_F &= E^Q[B(T)^{-1}[F(T, T_1) - K] \cdot I\{F(T, T_1) \geq K\}] \\ &= E^Q\left[\exp\left\{-\int_0^T r_s ds\right\} F(T, T_1) \cdot I\{F(T, T_1) \geq K\}\right] \\ &\quad -KE^Q\left[\exp\left\{-\int_0^T r_s ds\right\} \cdot I\{F(T, T_1) \geq K\}\right] \end{aligned}$$

To prove the first formula we change to the measure Q^F in first term to obtain

$$\begin{aligned} &E^Q\left[\exp\left\{-\int_0^T r_s ds\right\} F(T, T_1) \cdot I\{F(T, T_1) \geq K\}\right] \\ &= F(0, T_1)E^F\left[\exp\left\{-\int_0^T r_s ds\right\} \cdot I\{F(T, T_1) \geq K\}\right], \end{aligned}$$

and then we change to the measure Q^{T_F}. Note that it is the T-forward measure as seen from Q^F.

For the second term we use the T-forward measure directly.

To prove the second formula we use the T-forward measure Q^T in first term to obtain

$$
\begin{aligned}
&E^Q\left[\exp\left\{-\int_0^T r_s ds\right\} F(T,T_1)\cdot I\{F(T,T_1)\geq K\}\right] \\
&= p(0,T)E^T\left[F(T,T_1)I\{F(T,T_1)\geq K\}\right] \\
&= p(0,T)F(0,T_1)E^T\left[e^{-\int_0^T \kappa_s ds}\frac{F(T,T_1)}{F(0,T_1)}\cdot e^{\int_0^T \kappa_s ds} I\{F(T,T_1)\geq K\}\right]
\end{aligned}
$$

and then we change to the measure Q^{F_T}.

For the second term, again, we use the T-forward measure directly.

We may now ask ourselves whether it is possible to interpret the above measures as martingale measures for some numeraire asset. This question is answered by the following lemma, the proof of which is easy and therefore omitted.

Lemma 5.2.
I: *Under Q^F the process $V^F(t)$ defined by $V^F(t) = B(t)F(t,T_1)$ acts as a numeraire asset. Its Q-dynamics are given by*

$$
\begin{aligned}
dV^F(t) &= r(t)V^F(t)dt + V^F(t)\sigma_F(t,T_1)dW(t) \\
&\quad + V^F(t-)\int_E \delta_F(t,y,T_1)\tilde{\mu}(dt,dy)
\end{aligned}
$$

The process is the value process of a self-financing portfolio consisting of $B(t)$ T_1-futures options and $F(t,T_1)$ units of the money account (i.e. $F(t,T_1)/B(t)$ dollars invested in the money account).

II: *Under Q^{F_T} the process $V^{F_T}(t) = e^{-\int_0^t \kappa_s ds}p(t,T)F(t,T_1)$ acts as a numeraire asset. Here κ is the process given by*

$$
\kappa(t) = \sigma_F(t,T_1)\sigma_p^*(t,T) + \int_E \delta_F(t,y,T_1)\delta_p(t,y,T)\lambda(t,dy).
$$

The Q-dynamics of V^{F_T} are given by

$$
\begin{aligned}
\frac{dV^{F_T}(t)}{V^{F_T}(t-)} &= r(t)dt + \left[\sigma_p(t,T) + \sigma_F(t,T_1)\right]dW(t) \\
&\quad + \int_E \left\{\delta_F(t,y,T_1)\left[1+\delta_p(t,y,T)\right] + \delta_p(t,y,T)\right\}\tilde{\mu}(dt,dy).
\end{aligned}
$$

The process V^{F_T} is the value process of a self-financing portfolio consisting of $\exp\left\{-\int_0^t \kappa_s ds\right\} \cdot F(t-,T_1)$ *bonds with maturity* T, $\exp\{-\int_0^t \kappa_s ds\}p(t-,T)$ *units of T_1-futures options, and $V(t-)$ units of an asset with price process $\Pi(t) \equiv 0$ and cumulative dividend process D defined by*

$$dD(t) = \int_E \delta_p(t,y,T)\delta_F(t,y,T_1)][\mu(dt,dy) - \lambda(t,dy)dt].$$

5.2 Special Cases

Deterministic Volatilities Consider the following model under the martingale measure Q

$$\begin{aligned} df(t,T) &= \alpha_f(t,T)dt + \sigma_f(t,T)dW_t + \int_E \delta_f(t,y,T)\mu(dt,dy), \\ dF(t,T) &= F(t,T)\alpha_F(t,T)dt + F(t,T)\sigma_F(t,T)dW_t \\ &\quad + F(t-,T)\int_E \delta_F(t,y,T)\mu(dt,dy). \end{aligned}$$

Here σ_f and σ_F are assumed to be deterministic functions of the time parameters. Also δ_f and δ_F are assumed to be deterministic functions, now of the time parameters and the mark space variable y. Finally, α_f and α_F are given by the drift conditions (17) and (15), respectively.

Remark 5.3. Instead of modeling the futures prices directly we could model the futures convenience yield and the spot price dynamics. Suppose we do this as

$$d\varphi(t,T) = \alpha_\varphi(t,T)dt + \sigma_\varphi(t,T)dW_t + \int_E \delta_\varphi(t,y,T)\mu(dt,dy),$$

$$dS_t = \alpha_S(t)S_t dt + S(t)\sigma_S(t)dW_t + S_{t-}\int_E \delta_S(t,y)\mu(dt,dy),$$

where σ_φ and σ_S are assumed to be deterministic functions of the time parameters, and δ_φ and δ_S are assumed to be deterministic functions of the time parameters and the mark space variable y, and, finally, α_φ is given by the drift condition (34), whereas α_S is given by (22). Then it is easily seen from (20) that the futures prices resulting from these specifications will have deterministic volatilities, i.e. they will be of the form considered above.

Using the second pricing formula, (64), we see that the price, at date zero, of a European call option with exercise date T and exercise price K written on the futures price with delivery date T_1 is given by

$$\begin{aligned} C_F &= p(0,T)F(0,T_1)e^{\int_0^T \kappa_s ds}Q^{F_T}(F(T,T_1) \geq K) \\ &\quad - Kp(0,T)Q^T(F(T,T_1) \geq K), \end{aligned} \tag{66}$$

since the process κ defined in (65) is deterministic for this model (recall formula (11)). The dynamics of $F(t,T_1)$ under Q^T and Q^{F_T} are easily found using the dynamics of L^T and L^{F_T} given in (61) and (62), respectively, together with the Girsanov Theorem. Under Q^T they are given by

$$dF(t,T_1) = \left[\sigma_F(t,T_1)\sigma_p^*(t,T) - \int_E \delta_F(t,y,T_1)\lambda(t,dy)\right] F(t,T_1)dt$$

$$+ F(t,T_1)\sigma_F(t,T_1)dW_t^T + F(t-,T_1)\int_E \delta_F(t,y,T_1)\mu(dt,dy).$$

The point process μ has intensity λ^T under Q^T, where

$$\lambda^T(t,dy) = (1+\delta_p(t,y,T))\lambda(t,dy).$$

Under Q^{F_T} the dynamics of $F(t,T_1)$ are

$$\frac{dF(t,T_1)}{F(t-,T_1)} = \left[\sigma_F(t,T_1)\sigma_F^*(t,T_1) + \sigma_F(t,T_1)\sigma_p^*(t,T) - \int_E \delta_F(t,y,T_1)\lambda(t,dy)\right] dt$$

$$+ \sigma_F(t,T_1)dW_t^{F_T} + \int_E \delta_F(t,y,T_1)\mu(dt,dy)$$

Under Q^{F_T} the point process μ has intensity λ^{F_T}, where

$$\lambda^{F_T}(t,dy) = (1+\delta_F(t,y,T_1))(1+\delta_p(t,y,T))\lambda(t,dy).$$

The Hilliard–Reis Three-Factor Model. Consider again the Hilliard–Reis model defined in (49). The bond price volatilities for this model are given by (see [12])

$$\sigma_p(t,T) = \frac{\sigma_r}{\kappa_r}(e^{-\kappa_r(T-t)} - 1).$$

Using the explicit expression for the futures prices, the futures price dynamics can be shown to be (see [11])

$$dF(t,T) = -F(t,T)\bar{\delta}\lambda dt + F(t,T)\sigma_F(t,T)dW_t + F(t-,T)\int_R y\mu(dt,dy). \tag{67}$$

Here, as before, the marked point process μ has mark space $(R,\mathcal{B})$, where $\mathcal{B}$ is the Borel alebra, and a compensator

$$\nu(dt,dy) = \lambda dt \frac{1}{\sqrt{2\pi\eta^2}} \cdot \frac{1}{y+1} \exp\left\{-\frac{[\ln(y+1)-\xi]^2}{2\eta^2}\right\} dy$$

for $-1 < y < \infty$. Recall that this means that the spot price process will jump according to a Poisson process with intensity λ and that if δ denotes the relative jump size, then $1+\delta$ has a log-normal distribution: $\ln(1+\delta) \sim N(\xi, \eta^2)$. Furthermore, $\bar{\delta} = E^Q[\delta]$ in (67) denotes the expected relative jump size under Q, and we have that

$$\sigma_F(t,T)\sigma_F^*(t,T) = \sigma_S^2 + \sigma_c^2 B_c^2(t,T) + \sigma_r^2 B_r^2(t,T) + 2\|\sigma_S\| \cdot \|\sigma_c\| \rho_{Sc} B_c(t,T)$$
$$+ 2\|\sigma_S\| \cdot \|\sigma_r\| \rho_{Sr} B_r(t,T) + 2\|\sigma_c\| \cdot \|\sigma_r\| \rho_{cr} B_c(t,T) B_r(t,T),$$
$$\sigma_F^*(t,T_1)\sigma_p(t,T) = -\|\sigma_S\| \cdot \|\sigma_r\| \rho_{Sr} B_r(t,T) - \|\sigma_c\| \cdot \|\sigma_r\| \rho_{cr} B_r(t,T) B_c(t,T_1)$$
$$- \|\sigma_r\|^2 B_r(t,T) B_r(t,T_1).$$

Specializing the formulas for the dynamics of $F(t,T_1)$ under Q^T and Q^{F_T} to this case, we see that conditional on that there has been n jumps $F(T,T_1)$ can be written as

$$F(T,T_1) = F(0,T_1) \prod_{i=1}^{n} (1+\delta_{T_i}) \times$$
$$\times \exp\left\{ \int_0^T \left(\sigma_F(t,T_1)\sigma_p^*(t,T) - \bar{\delta}\lambda - \frac{1}{2}\sigma_F(t,T_1)\sigma_F^*(t,T_1) \right) dt + \int_0^T \sigma_F(t,T_1) dW_t^T \right\},$$

where δ_{T_i}, $i = 1, \ldots, n$ are i.i.d. random variables and $\ln(1+\delta_{T_i}) \in N(\xi, \eta^2)$ under Q^T. The variables δ_{T_i} are also independent of W^T. Alternatively we can write $F(T,T_1)$ as

$$F(T,T_1) = F(0,T_1) \prod_{i=1}^{n} (1+\Delta_{T_i}) \times$$
$$\times \exp\left\{ \int_0^T \left(\sigma_F(t,T_1)\sigma_p^*(t,T) - \bar{\delta}\lambda + \frac{1}{2}\sigma_F(t,T_1)\sigma_F^*(t,T_1) \right) dt + \int_0^T \sigma_F(t,T_1) dW_t^{F_T} \right\},$$

where Δ_{T_i}, $i = 1, \ldots, n$ are i.i.d. with $\ln(1+\Delta_{T_i}) \in N(\xi+\eta^2, \eta^2)$ under Q^{F_T}. The variables Δ_{T_i} are also independent of W^{F_T}. Given this we can express the two probabilities in (66) as follows

$$Q^{F_T}(F(T,T_1) \geq K) = \sum_{n=0}^{\infty} \left[\frac{e^{-\lambda' T}(\lambda' T)^n}{n!} \right] N(d_{1n}),$$

$$Q^T(F(T,T_1) \geq K) = \sum_{n=0}^{\infty} \left[\frac{e^{-\lambda T}(\lambda T)^n}{n!} \right] N(d_{2n}),$$

where

$$\begin{aligned} d_{1n} &= \frac{1}{\sqrt{\zeta^2 + n\eta^2}} \left(\ln \left(\frac{F(0,T_1)Y(0,T,T_1)}{K} \right) + n\xi - \bar{\delta}\lambda T + \frac{1}{2}\zeta^2 + n\eta^2 \right), \\ d_{2n} &= d_1 - \sqrt{\zeta^2 + n\eta^2}, \end{aligned}$$

$$Y(t,T,T_1) = e^{\int_t^T \sigma_F(s,T_1)\sigma_p^*(s,T)ds},$$

$$\zeta^2 = \int_0^T \sigma_F(t,T_1)\sigma_F(t,T_1)^* dt,$$

$$\lambda' = \lambda e^{\xi + \eta^2/2}.$$

The price, at date zero, of a European call option with exercise date T and exercise price K written on the futures price with delivery date T_1 is therefore given by the following formula after some rewriting where we use the fact that $\exp\{\xi + \eta^2/2\} = \bar{\delta} + 1$

$$\begin{aligned} C_F = p(0,T) \sum_{n=0}^{\infty} \left[\frac{e^{-\lambda T}(\lambda T)^n}{n!} \right] \Big[& F(0,T_1)Y(0,T,T_1)e^{-\lambda\bar{\delta}T + n\ln(1+\bar{\delta})} N[d_{1n}] \\ & - KN[d_{2n}] \Big]. \end{aligned}$$

Here N denotes the cumulative distribution function of a normally distributed random variable with expectation zero and variance one, and d_{1n}, d_{2n} and $Y(t,T,T_1)$ have been defined above. This reproduces the results in [11].

The Gaussian Case. Consider the following model, which is a special case of the model of the previous section, under the martingale measure Q.

$$\begin{aligned} df(t,T) &= \alpha_f(t,T)dt + \sigma_f(t,T)dW_t, \\ dF(t,T) &= F(t,T)\alpha_F(t,T)dt + F(t,T)\sigma_F(t,T)dW_t. \end{aligned}$$

Here σ_f and σ_F are assumed to be deterministic functions of the time parameters, whereas α_f and α_F are given by the drift conditions (17) and (15), respectively.

Specializing the formulas for the dynamics of $F(t,T_1)$ to this case, we see that $F(t,T_1)$ follows a geometric Brownian motion both under Q^T and under Q^{F_T}. The probabilities can therefore be computed and we find that the price, at date zero, of a European call option with exercise date T and exercise price K written on the futures price with delivery date T_1 is given by

$$\begin{aligned} C_F = p(0,T)F(0,T_1) \exp\left\{ \int_0^T \kappa_u du \right\} N[d_1] \\ - Kp(0,T)N[d_2], \end{aligned}$$

where N denotes the cumulative distribution function of a normally distributed random variable with expectation zero and variance one. Furthermore, d_1 and d_2 are given by

$$d_1 = \frac{1}{\sqrt{\int_0^T \|\sigma_F(u,T_1)\|^2 du}} \left(\ln\left(\frac{F(0,T_1)}{K}\right) + \int_0^T \left\{ \kappa_s ds + \frac{1}{2}\|\sigma_F(s,T_1)\|^2 \right\} ds \right),$$

$$d_2 = d_1 - \sqrt{\int_0^T \|\sigma_F(u,T_1)\|^2 du}.$$

The process κ was defined in (65). This reproduces the result in [17] (in the referenced work the futures convenience yield and the spot price dynamics are modeled instead of the futures price dynamics, but as was pointed out in Remark 5.3, deterministic volatilities for the convenience yield and the spot price imply deterministic futures price volatilities, and thus the above pricing formula is applicable).

Quadratic Interest Rates Consider a model specified by the following equations

$$dZ_1(t) = [\alpha_1 - \beta_1 Z_1(t)]dt + \rho_1 dW_1(t),$$

$$dZ_2(t) = [\alpha_2 - \beta_2 Z_2(t)]dt + \rho_2 dW_2(t),$$

$$dF(t,T) = \sigma_1 F(t,T)dW_1(t) + \sigma_2 F(t,T)dW_2(t),$$

and

$$r(t) = \frac{1}{2}(Z_1^2(t) + Z_2^2(t)).$$

The specifications are all made under a martingale measure Q, and W_1 and W_2 are two independent standard Wiener processes. We assume that α_i, β_i, ρ_i and σ_i, $i = 1, 2$ are all constants.

For this model the T-maturity zero-coupon bond price can be shown to be of the form

$$p(t,T) = \exp\left\{ -\sum_{i=1}^{2} \left[\frac{1}{2} B_i(t,T) Z_i^2(t) + b_i(t,T) Z_i(t) \right] - c(t,T) \right\},$$

where B_i, b_i, $i = 1, 2$ and c solve the following ordinary differential equations

$$\frac{\partial B_i}{\partial t} = 2\beta_i B_i + \rho_i^2 B_i^2 - 1,\ B_i(T,T) = 0, \tag{68}$$

$$\frac{\partial b_i}{\partial t} - (\beta_i + \rho_i^2 B_i) b_i + \alpha_i B_i = 0,\ b_i(T,T) = 0, \tag{69}$$

$$\frac{\partial c}{\partial t} + \sum_{i=1}^{2} \left(\alpha_i b_i + \frac{1}{2}\rho_i^2 [B_i - b_i^2] \right) = 0,\ \ c(T,T) = 0. \tag{70}$$

(see [14] for details).

We will now attempt to compute the price, at date zero, of a European call option with exercise date T and exercise price K written on the futures price with delivery date T_1, using the first pricing formula (63). The dynamics of Z_1, Z_2 and $F(t,T_1)$ under Q^F are easily found using the dynamics of L^F given in (60) together with the Girsanov Theorem. They are

$$dZ_1(t) = [\alpha_1 + \rho_1\sigma_1 - \beta_1 Z_1(t)]dt + \rho_1 dW_1^F(t),$$

$$dZ_2(t) = [\alpha_2 + \rho_2\sigma_2 - \beta_2 Z_2(t)]dt + \rho_2 dW_2^F(t),$$

$$dF(t,T) = (\sigma_1^2 + \sigma_2^2)F(t,T_1)dt + \sigma_1 F(t,T)dW_1^F(t) + \sigma_2 F(t,T)dW_2^F(t),$$

where W_1^F and W_2^F are two independent Q^F-Wiener processes. From this we see that the state variables Z_1 and Z_2 are still Gaussian. Using Q^F as a martingale measure we can therefore compute bond prices in exactly the same way as before, except for that α_i is replaced by $\tilde{\alpha}_i = \alpha_i + \sigma_i\rho_i$, $i = 1, 2$. These modified bond prices, which we will denote by $\tilde{p}(t,T)$, give us the first expectation in (63). From the above dynamics we see that the futures price $F(T,T_1)$ is log-normally distributed, and hence the second probability in (63) is given by

$$Q^T(F(T,T_1) \geq K) = N\left[\frac{1}{\|\sigma_F\|\sqrt{T}}\left[\ln\left(\frac{F(0,T_1)}{K}\right) + \frac{1}{2}\|\sigma_F\|^2 T\right]\right],$$

where $\sigma_F = (\sigma_1, \sigma_2)$ and N denotes the cumulative distribution function of a normally distributed random variable with mean zero and variance one.

The Radon-Nikodym derivative of Q^{T_F} with respect to Q^F can now be written as

$$R^{T_F} = \frac{\exp\left\{-\int_0^T r_s ds\right\}}{\tilde{p}(0,T)}.$$

This means that the likelihood process is given by

$$L_t^{T_F} = \frac{\tilde{p}(t,T)}{B(t)\tilde{p}(0,T)}$$

(thus, Q^{T_F} is the T-forward measure "as seen from Q^F"). Let $\tilde{B}_i$, $\tilde{b}_i$, $i = 1, 2$ and $\tilde{c}$ denote the functions you obtain solving the equations (68), (69) and (70), respectively, with α_i replaced by $\tilde{\alpha}_i = \alpha_i + \sigma_i\rho_i$, $i = 1, 2$. Then we have that

$$\tilde{p}(t,T) = \exp\left\{-\sum_{i=1}^{2}\left[\frac{1}{2}\tilde{B}_i(t,T)Z_i^2(t) + \tilde{b}_i(t,T)Z_i(t)\right] - \tilde{c}(t,T)\right\}.$$

Using Itô's formula we then see that the dynamics of L^{T_F} are

$$\begin{aligned} dL_t^T &= -L_t^T(\tilde{B}_1(t,T)Z_1(t) + \tilde{b}_1(t,T))\rho_1 dW_1^F(t) \\ &\quad - L_t^T(\tilde{B}_2(t,T)Z_2(t) + \tilde{b}_2(t,T))\rho_2 dW_2^F(t). \end{aligned}$$

The Girsanov Theorem then gives us the following dynamics under Q^{T_F}.

$$\begin{aligned}
dZ_1(t) &= [\alpha_1 + \rho_1\sigma_1 - \rho_1^2\tilde{b}_1(t,T) - (\beta_1 + \rho_1^2\tilde{B}_1(t,T))Z_1(t)]dt \\
&\quad + \rho_1 dW_1^{T_F}(t), \\
dZ_2(t) &= [\alpha_2 + \rho_2\sigma_2 - \rho_2^2\tilde{b}_2(t,T) - (\beta_2 + \rho_2^2\tilde{B}_2(t,T))Z_2(t)]dt \\
&\quad + \rho_2 dW_2^{T_F}(t), \\
dF(t,T) &= [\sigma_1^2 + \sigma_2^2 - (\tilde{B}_1(t,T)Z_1(t) + \tilde{b}_1(t,T))\rho_1\sigma_1 \\
&\quad - (\tilde{B}_2(t,T)Z_2(t) + \tilde{b}_2(t,T))\rho_2\sigma_2]F(t,T)dt \\
&\quad + \sigma_1 F(t,T)dW_1^{T_F}(t) + \sigma_2 F(t,T)dW_2^{T_F}(t),
\end{aligned}$$

where $W_1^{T_F}$ and $W_2^{T_F}$ are two independent Q^{T_F}-Wiener processes. Now let $Y(t) = (Z_1(t), Z_2(t), \ln F(t,T_1))^*$. If we apply Itô's formula to the third component of this process we see that the process satisfies the following stochastic differential equation

$$dY_t = [\alpha_Y(t) - \beta_Y(t)Y_t]dt + \sigma_Y(t)dW_t,$$

where

$$\alpha_Y(t) = \begin{bmatrix} \alpha_1 + \rho_1\sigma_1 - \rho_1^2\tilde{b}_1(t,T) \\ \alpha_2 + \rho_2\sigma_2 - \rho_2^2\tilde{b}_2(t,T) \\ \frac{1}{2}\|\sigma_F\|^2 - \rho_1\sigma_1\tilde{b}_1(t,T) - \rho_2\sigma_2\tilde{b}_2(t,T) \end{bmatrix},$$

$$\beta_Y(t) = \begin{bmatrix} \beta_1 + \rho_1^2\tilde{B}_1(t,T) & 0 & 0 \\ 0 & \beta_2 + \rho_2^2\tilde{B}_2(t,T) & 0 \\ \rho_1\sigma_1\tilde{B}_1(t,T) & \rho_2\sigma_2\tilde{B}_2(t,T) & 0 \end{bmatrix},$$

and

$$\sigma_Y(t) = \begin{bmatrix} \rho_1 & 0 \\ 0 & \rho_2 \\ \sigma_1 & \sigma_2 \end{bmatrix}$$

Suppose that Y_0 is deterministic. We then see that Y has a three dimensional normal distribution, where the mean $m(t) = E^{T_F}[Y_t]$ and the covariance matrix $V = E[\|(Y_t - m(t))\|^2]$ are obtained as the solutions to the following linear equations (see for instance [15]).

$$\begin{aligned}
\dot{m}(t) &= -\beta_Y(t)m(t) + \alpha_Y(t), \\
\dot{V}(t) &= -\beta_Y(t)V(t) - V(t)\beta_Y^*(t) + \sigma_Y(t)\sigma_Y^*(t).
\end{aligned}$$

The probability in the pricing formula (63) is therefore given by

$$Q^{T_F}\left(F(T,T_1) \geq K\right) = N\left[\frac{1}{\sqrt{V_{33}(T)}}\left[\ln\left(\frac{F(0,T_1)}{K}\right) + m_3(T)\right]\right].$$

6 Inverting the Term Structure

Consider a given finite dimensional factor model for the futures term structure, say of the form

$$\begin{aligned} F(t,T) &= H_F(t,Z_t,T) \\ dZ_t &= \alpha_Z(t,Z_t)dt + \sigma_Z(t,Z_t)dW_t + \int_E \delta_Z(t,Z_{t-},y)\mu(dt,dy), \end{aligned}$$

under the martingale measure Q. For a given initial value z_0 of Z_0 the model will produce the theoretical initial term structure $F(0,T) = H_F(0,z_0,T)$. Assuming that we have a liquid futures market for all delivery dates, the market will provide us with an observed initial futures term structure $F^\dagger(0,T)$, and we would of course like our theoretical initial term structure to coincide with the observed one. We would thus like to choose the parameters in the Z-dynamics such that

$$F(0,T) = F^\dagger(0,T), \quad \forall T \geq 0,$$

and since this is an infinite dimensional system of equations, we will need an infinite dimensional parameter vector. The entire project is thus completely parallel to that of inverting the yield curve in interest rate theory.

6.1 Conditionally Affine Models

The general problem of when and how it is possible to fit an arbitrarily given initial term structure, for a parameterized family of futures price models, is a very hard one and there seems to be no strong general results (see however [16]). Here we will instead present a nontrivial particular class of models for which the initial term structure in fact can be inverted, and it turns out that most existing factor models belong to this class. The class is characterized by the facts that the spot price S is one of the factors, and we furthermore impose a particular structure on the factor dynamics. See [3] for a similar approach in the context in interest rate theory.

Definition 6.1. We say that a factor model for futures prices is a **conditionally affine model** if the following conditions hold.

- The factor vector can be decomposed as (S_t, Z_t) where $Z \in R^d$.

– The factor dynamics are of the form

$$dS_t = S_t \{f(Z_t) - \kappa \ln S_t\} dt + S_t g(Z_t) dW_t + S_{t-} \int_E \delta_S(t, Z_{t-}, y) \mu(dt, dy), \tag{71}$$

$$dZ_t = a(Z_t)dt + b(Z_t)dW_t + \int_E \delta_z(t, Z_{t-}, y)\mu(dt, dy). \tag{72}$$

Here μ is a point process, W is an m-dimensional Wiener process and κ is a (typically positive) real number. The functions a,b, f and g are nonlinear functions of appropriate dimensions (Z is viewed as a column vector process). The jump volatility functions $\delta_S(t, z, y)$ and $\delta_z(t, z, y)$ are assumed to be determininstic functions of the variables t, z and y with $\delta_S > -1$.

Given a conditionally affine model (S, Z) as above, as well as a deterministic R-valued function $\varphi(t)$, the corresponding **perturbed model** (S^φ, Z) is defined by the relations

$$dS_t^\varphi = S_t^\varphi \{\varphi(t) + f(Z_t) - \kappa \ln S_t^\varphi\} dt + S_t^\varphi g(Z_t) dW_t + S_{t-}^\varphi \int_E \delta_S(t, Z_{t-}, y)\mu(dt, dy),$$

$$dZ_t = a(Z_t)dt + b(Z_t)dW_t + \int_E \delta_z(t, Z_{t-}, y)\mu(dt, dy).$$

The perturbed model is assumed to have the same initial data as the original model. The futures prices generated by the original model and the perturbed model are denoted by $F(t, T)$ and $F^\varphi(t, T)$ respectively.

Note that the dynamics of the Z process do not involve S at all. The deeper significance of the S-dynamics is that with this particular form, and for a given Z-trajectory, we can write $S_t = e^{\xi_t}$ where ξ satisfies a **linear** SDE. The above dynamics are thus (conditional on the Z-trajectory) the natural extension of the standard Black-Scholes stock price dynamics (including jumps). We also note that Z is not affected by the choice of φ and that the original model corresponds to $\varphi = 0$.

Given an initial (observed) term structure, $\{F^\dagger(0, T);\ T \geq 0\}$ the problem is to see if it is possible to choose φ in such a way that $F^\varphi(0, T) = F^\dagger(0, T)$ for all $T \geq 0$. The reader will note the similarity between this perturbation approach and the way in which Hull-White extend the Vasicek short rate model in order to invert the yield curve. We have the following strong result.

Proposition 6.2. *Let $F^\dagger(0, T)$ be any smooth (i.e. C^1) initial term structure such that $F^\dagger(0, 0) = F^0(0, 0)$. Define the function H by*

$$H(T) = \ln\left(\frac{F^\dagger(0, T)}{F^0(0, T)}\right) \tag{73}$$

Then the following hold:

- *The perturbed model can be fitted to* $F^\dagger$. *In fact, by choosing* φ *as*

$$\varphi(t) = H'(t) + \kappa H(t), \tag{74}$$

we have

$$F^\varphi(0,T) = F^\dagger(0,T), \quad \forall T \geq 0. \tag{75}$$

- *With* $S_0^0 = S_0^\varphi$ *and with the above choice of* φ *we have, for all* $0 \leq t \leq T$,

$$F^\varphi(t,T) = \frac{F^\dagger(0,T)}{F^0(0,T)} F^0(t,T). \tag{76}$$

- *Assume that the short rate r is deterministic. For any contingent T-claim of the form* $\Phi(S_T, Z_T)$ *we denote the corresponding arbitrage free pricing functions by* $P^0(t,s,z)$ *and* $P^\varphi(t,s,z)$ *for the original model and the perturbed model respectively. Then the following relation hold.*

$$P^\varphi(t,s,z) = P^0\left(t, se^{\int_t^T e^{-\kappa(T-u)}\varphi(u)du}, z\right). \tag{77}$$

- *Assume in particular that* φ *is chosen such that the initial term structure is completely fitted as above. Then we have*

$$P^\varphi(t,s,z) = P^0\left(t, s \cdot e^{\left\{H(T) - e^{-\kappa(T-t)}H(t)\right\}}, z\right). \tag{78}$$

Remark 6.3. The point of (77)-(78) is that if we have computed derivatives pricing formulas in the original model, then these formulas can be used in the perturbed model, by simply modifying the value of the observed spot price.

Proof. Defining ξ^φ by $\xi^\varphi = \ln S^\varphi$ it is easy to see that

$$\begin{aligned} d\xi_t^\varphi &= \{-\kappa\xi_t^\varphi + \varphi(t) + f_\xi(Z_t)\}\, dt + g(Z_t)dW_t \\ &+ \int_E \delta_\xi(t, Z_{t-}, y)\mu(dt, dy), \end{aligned}$$

where

$$\begin{aligned} f_\xi(z) &= f(z) - \frac{1}{2}g^2(z), \\ \delta_\xi(t,z,y) &= \ln\{1 + \delta_S(t,z,y)\}. \end{aligned}$$

Using the Itô formula one verifies readily that

$$\begin{aligned} \xi_T^\varphi = e^{-\kappa T}\xi_0^\varphi &+ \int_0^T e^{-\kappa(T-u)}\varphi(u)du + \int_0^T e^{-\kappa(T-u)} f_\xi(Z_u)du \\ &+ \int_0^T e^{-\kappa(T-u)} g(Z_u)dW_u \\ &+ \int_0^T \int_E e^{-\kappa(T-u)}\delta_\xi(u, Z_{u-}, y)\mu(du, dy), \end{aligned}$$

and, since $\xi_0^\varphi = \xi_0^0 = \ln S_0$, we have

$$\xi_T^\varphi = \xi_T^0 + \int_0^T e^{-\kappa(T-u)}\varphi(u)du,$$

and hence,

$$S_T^\varphi = S_T^0 \cdot e^{\int_0^T e^{-\kappa(T-u)}\varphi(u)du}. \tag{79}$$

From this we obtain, for all t and T with $0 \le t \le T$

$$\begin{aligned} F^\varphi(t,T) = E^Q\left[S_T^\varphi \middle| \mathcal{F}_t\right] &= e^{\int_0^T e^{-\kappa(T-u)}\varphi(u)du} E^Q\left[S_T^0 \middle| \mathcal{F}_t\right] \\ &= e^{\int_0^T e^{-\kappa(T-u)}\varphi(u)du} F^0(t,T). \end{aligned} \tag{80}$$

Thus, in order to fit the given initial term structure $F^\dagger$ we have to find φ such that

$$F^\dagger(0,T) = e^{\int_0^T e^{-\kappa(T-u)}\varphi(u)du} F^0(0,T). \tag{81}$$

This equation can, with H defined as above, be written as

$$\int_0^T e^{\kappa u}\varphi(u)du = e^{\kappa T} H(T),$$

and, taking the T-derivative, we obtain (74). The relation (76) follows immediately from (80)-(81).

If $\Phi(S_T, Z_T)$ is a contingent T-claim, then the corresponding arbitrage free price process $\Pi^\varphi(t,\Phi)$ for the perturbed model is given by

$$\Pi^\varphi(t,\Phi) = e^{-\int_t^T r_u du} E^Q\left[\Phi(S_T^\varphi, Z_T) \middle| \mathcal{F}_t\right].$$

From the Markovian setup we have in fact

$$E^Q\left[\Phi(S_T^\varphi, Z_T) \middle| \mathcal{F}_t\right] = P_0^\varphi\left(t, S_t^\varphi, Z_t\right),$$

where the real valued function $P_0^\varphi : R_+ \times R_+ \times R^d \to R$ is defined by

$$P_0^\varphi(t,s,z) = E^Q\left[\Phi(S_T^\varphi, Z_T) \middle| S_t^\varphi = s, Z_t = z\right].$$

The pricing function P^φ is thus given by $P^\varphi(t,s,z) = e^{-\int_t^T r(u)du} P_0^\varphi(t,s,z)$. Now, given that $S_t^\varphi = s$, we obtain as above,

$$\begin{aligned} S_T^\varphi = s^{e^{-\kappa(T-t)}} &\cdot \exp\left\{\int_t^T e^{-\kappa(T-u)}\left\{\varphi(u) + f_\xi(Z_u)\right\}du\right\} \\ &\times \exp\left\{\int_t^T e^{-\kappa(T-u)} g(Z_u)dW_u\right\} \\ &\times \exp\left\{\int_t^T \int_E e^{-\kappa(T-u)}\delta_\xi(u, Z_{u-}, y)\mu(du,dy)\right\} \end{aligned}$$

From direct inspection of this formula we see that the distribution of (S_T^φ, Z_T) given $S_t^\varphi = s$ and $Z_t = z$ is identical with the distribution of (S_T^0, Z_T) given $S_t^0 = s \cdot e^{\int_t^T e^{-\kappa(T-u)}\varphi(u)du}$ and $Z_t = z$. This proves (77)-(78).

6.2 An Example

To exemplify the theory above we now give a brief sketch of how to fit the Extended Schwartz Three Factor Model in Section 4.2 to an initial future price curve $F^\dagger(0,T)$. The perturbed model is given by

$$dS_t = (r_t + \varphi(t) - c_t)S_t dt + S_t \sigma_S dW_t,$$
$$dc_t = \kappa_c(\alpha_c - c_t)dt + \sigma_c dW_t,$$
$$dr_t = \kappa_r(\alpha_r - r_t)dt + \sigma_r dW_t,$$

and φ is obtained from Proposition 6.2 as

$$\varphi(t) = \frac{\partial}{\partial t}\ln F^\dagger(0,t) + f(0,t) - \frac{\partial}{\partial t}B_c(0,t)c_0 - \frac{\partial}{\partial t}\bar{A}(0,t)$$

Here $f(0,t)$ denotes the forward rates in the Vasiček short rate model, whereas B_c and $\bar{A}$ are given in Remark 4.8.

It is worth noticing that with this perturbation, the process c_t no longer has the interpretation of being the spot convenience yield. The spot convenience yield c^φ in the perturbed model is instead given by $c_t^\varphi = c_t - \varphi(t)$. Using S, c^φ and r as state variables we easily obtain the alternative dynamics

$$dS_t = (r_t - c_t^\varphi)S_t dt + S_t \sigma_S dW_t,$$
$$dc_t^\varphi = \kappa_c\left[\alpha_c - \varphi(t) - \kappa_c^{-1}\varphi'(t) - c_t^\varphi\right]dt + \sigma_c dW_t,$$
$$dr_t = \kappa_r(\alpha_r - r_t)dt + \sigma_r dW_t.$$

References

1. Amin, K. I., Ng, V., and Pirrong, S. Valuing energy derivatives. In *Managing Energy Price Risk*. Risk Publications, 1995, pp. 57–70.
2. Björk, T. *Arbitrage Theory in Continuous Time.* Oxford University Press, 1998.
3. Björk, T., and Hyll, M. *On the ínversion of yield curve.* Working paper, Stockholm School of Economics, 2000.
4. Björk, T., Kabanov, Y., and Runggaldier, W. Bond market structure in the presence of a marked point process. *Mathematical Finance* 7, 2 (1995), 211–239.
5. Cortazar, G., and Schwartz, E. The valuation of commodity contingent claims. *Journal of Derivatives* (1994), 27–39.

6. Cox, J., Ingersoll, J., and Ross, S. A theory of the term structure of interest rates. *Econometrica 53* (1985), 385–408.
7. Duffie, D., and Kan, R. A yield factor model of interest rates. *Mathematical Finance 6*, 4 (1996), 379–406.
8. Duffie, D., Pan, J., and Singleton, K. Transform analysis and asset pricing for affine jump diffusions. Forthcoming in *Econometrica* (2000).
9. Geman, H., El Karoui, N., and Rochet, J.-C. Changes of numéraire, changes of probability measure and option pricing. *Journal of Applied Probability 32* (1995), 443–458.
10. Gibson, R., and Schwartz, E. Stochastic convenience yield and the pricing of oil contingent claims. *Journal of Finance 45*, 3 (1990), 959–976.
11. Hilliard, J., and Reis, J. Valuation of commodity futures and options under stochastic convenience yields, interest rates, and jump diffusion in the spot. *JFQA 33*, 1 (1998), 61–86.
12. Hull, J., and White, A. Pricing interest-rate-derivative securities. *Review of Financial Studies 3* (1990), 573–592.
13. Jacod, J., and Shiryaev, A. *Limit Theorems for Stochastic Processes.* Springer verlag, 1987.
14. Jamshidian, F. Bond, futures and option evaluation in the quadratic interest rate model. *Applied Mathematical Finance 3* (1996), 93–115.
15. Karatzas, I., and Shreve, S. *Brownian motion and stochastic calculus.* Springer, 1991.
16. Landén, C. Spot price realizations of futures price term structures. Tech. rep., Department of Mathematics, KTH, Stockholm, 2000.
17. Miltersen, K., and Schwartz, E. Pricing of options on commodity futures with stochastic term structures of convenience yields and interest rates. *JFQA 33* (1998), 33–59.
18. Schwartz, E. The stochastic behaviour of commodity prices. *Journal of Finance 52*, 3 (1997), 923–973.
19. Vasiček, O. An equilibrium characterization of the term structure. *Journal of Financial Economics 5*, 3 (1977), 177–188.

Displaced and Mixture Diffusions for Analytically-Tractable Smile Models

Damiano Brigo and Fabio Mercurio

Product and Business Development Group
Banca IMI, SanPaolo IMI Group
Corso Matteotti, 6
20121 Milano, Italy

Abstract. We propose two different classes of analytical models for the dynamics of an asset price that respectively lead to skews and smiles in the term structure of implied volatilities. Both classes are based on explicit SDEs that admit unique strong solutions whose marginal densities are also provided. We then consider some particular examples in each class and explicitly calculate the European option prices implied by these models.

Keywords

Local-Volatility Models, Stock-Price Dynamics, Risk-Neutral Density, Analytically-Tractable Models, Explicit Option Pricing, Lognormal-Mixture Dynamics, Volatility Skew, Volatility Smile, Fokker-Planck Equation.

1 Introduction

The Black and Scholes [7] model cannot consistently price all European options that are quoted in one specific market, since the assumption of a constant volatility (for pricing derivatives on the same underlying asset) fails to hold true in practice. Indeed, in real markets, the implied volatility curves typically have skewed or smiley shapes. The term "skew" is used to indicate those structures where low-strikes implied volatilities are higher than high-strikes implied volatilities. The term "smile" is used instead to denote those structures with a minimum value around the underlying forward price.

If, for every fixed maturity, the implied volatilities were equal for different strikes but different along the time-to-maturity dimension, we could use the following simple extension of the Black-Scholes model to exactly retrieve the market option prices:

$$dS_t = \mu S_t dt + \sigma_t S_t dW_t,$$

where σ_t is a time-dependent (deterministic) volatility function.

However, more complex volatility structures are observed in real financial markets. A possible way to tackle this issue is by introducing a more articulated form of volatility coefficient in the stock-price dynamics, leading to the so called "local-volatility models". This is the approach we follow in this paper. We in fact propose two different classes of stock-price models by specifying the stock-price dynamics under the risk-neutral measure. The volatility σ_t we introduce is in both cases a function of time t and of the stock-price S_t at the same time. The examples we consider in the two classes lead respectively to skews and smiles in the term structure of implied volatilities.

Many researchers have tried to address the problem of a good, possibly exact, fitting of market option data. We now briefly review the major approaches that have been proposed.

A first approach is based on assuming an *alternative explicit dynamics* for the stock-price process that immediately leads to volatility smiles or skews. In general this approach does not provide sufficient flexibility to properly calibrate the whole volatility surface. An example is the general CEV process being analyzed by Cox [17] and Cox and Ross [18]. A general class of processes is due to Carr et al. [16]. The models in the first class we propose, based on a shifting technique and leading in particular to the "shifted-lognormal" and "shifted-CEV" dynamics, also fall into this *alternative explicit dynamics* category. The shifting technique, however, does not completely solve the flexibility issue, though adding flexibility with respect to the previously-known examples.

A second approach is based on the assumption of a *continuum of traded strikes* and goes back to Breeden and Litzenberger [8]. Successive developments are due to Dupire [22], Derman and Kani [20], [21] and Dempster and Richard [19] who derive an explicit expression for the Black-Scholes volatility as a function of strike and maturity. This approach has the major drawback that one needs to smoothly interpolate option prices between consecutive strikes in order to be able to differentiate them twice with respect to the strike itself. Explicit expressions for the risk-neutral stock-price dynamics are also derived by Avellaneda et al. [3] by minimizing the relative entropy to a prior distribution, and by Brown and Randall [14] by assuming a quite flexible analytical function describing the local volatility surface.

Another approach, pioneered by Rubinstein [37], consists of finding the risk-neutral probabilities in a binomial/trinomial model for the stock price that lead to a best fitting of market option prices with respect to some smoothness criterion. We refer to this approach as to the *lattice* approach. Further examples are in Jackwerth and Rubinstein [27] and Andersen and Brotherton-Ratcliffe [2] who use instead finite-difference grids. A different lattice approach is due to Britten-Jones and Neuberger [13].

A last approach is given by what we may refer to as *incomplete-market* approach. It includes stochastic-volatility models, such as those of Hull and White [26], Heston [25] and Tompkins [40], [41], and jump-diffusion models, such as that of Prigent, Renault and Scaillet [33].

In general the problem of finding a risk-neutral distribution that consistently prices all quoted options is largely undetermined. A possible solution is given by assuming a particular *parametric risk-neutral distribution* depending on several, possibly time-dependent, parameters and then use such parameters for the volatility calibration. An example of this approach is the work by Shimko [38]. But the question remains of finding an asset-price dynamics consistent with the chosen parametric form of the risk-neutral density. The second class we propose, the "general-mixture dynamics" leading in particular to the "lognormal-mixture dynamics", addresses this question by finding a dynamics leading to a parametric risk-neutral distribution that is flexible enough for practical purposes. The resulting process combines therefore the *parametric risk-neutral distribution* approach with the *alternative dynamics approach*, providing explicit dynamics leading to flexible parametric risk-neutral densities.

The major example we propose in our second class is indeed based on the assumption of a marginal density that is given by a mixture of lognormals. The introduction of such a risk-neutral distribution leads to i) *analytical formulas* for European options, so that the calibration to market data and the computation of Greeks can be extremely rapid, ii) *explicit asset-price dynamics*, so that exotic claims can be priced through Monte Carlo simulations and iii) *recombining lattices*, so that instruments with early-exercise features can be valued via backward calculation in the tree.[1]

The paper is structured as follows. Section 2 proposes a class of analytical asset-price models based on the shifting technique and leading to volatility skews. The cases of a shifted CEV process and a shifted geometric Brownian motion are treated as particular examples. Section 3 proposes a different class of analytical asset-price models based on a mixture-dynamics and leading instead to volatility smiles. The example of a mixture of lognormal distributions is then considered. Section 4 concludes the paper. Some technical results are written in appendices.

2 Displacing a Known Diffusion

Let us denote by S the asset-price process and assume that interest rates are constant through time and equal to $r > 0$ for all maturities. We assume that the asset price under the risk-neutral measure has the following dynamics

$$dS_t = \mu S_t dt + \nu(t, S_t) dW_t, \tag{2.1}$$

with S_0 given and deterministic and where W is a standard Brownian motion, ν is a smooth function of t and S_t and μ, the risk-neutral drift rate associated

[1] Alternative methods of extracting a risk-neutral distribution from option prices are in Malz [29] and Pirkner et al. [31]. An alternative model with explicit formulas for European option has been proposed by Li [28].

to the process S, is a real number. For example, if the asset is a stock paying a continuous dividend yield q, then $\mu = r - q$. If the asset is instead an exchange rate, then $\mu = r - r_f$, where r_f is the (assumed constant) risk-free rate for the foreign currency.

We want to determine the diffusion coefficient ν in such a way that S is given by an affine transformation of a diffusion process X whose marginal density is known. The marginal density of S will be thus obtained by shifting the known density of X. In formulas, we want to find the deterministic functions (of time) a and b such that

$$S_t = a_t + b_t X_t,$$

where the process X satisfies

$$dX_t = \varphi(t, X_t)dt + \psi(t, X_t)dW_t,$$

with φ and ψ regular functions such that the marginal density of X is known.

By Ito's lemma, we can write

$$dS_t = [a_t' + b_t' X_t + b_t \varphi(t, X_t)]dt + b_t \psi(t, X_t)dW_t, \tag{2.2}$$

with $'$ denoting the time derivative. Imposing the classical risk-neutral drift condition, which is evident in the dynamics (2.1), we thus obtain that the functions a and b must satisfy

$$a_t' + b_t' X_t + b_t \varphi(t, X_t) = \mu S_t, \tag{2.3}$$

for each t.

To make such condition more explicit, we will consider the case where X is a general candidate CEV process. As we shall see in the sequel, this choice is motivated by the model analytical tractability.

Let us then assume that X follows the general candidate CEV process[2]

$$dX_t = \gamma_t X_t dt + \eta_t X_t^\rho dW_t, \tag{2.4}$$

where γ_t and η_t are deterministic functions of time and ρ is any real constant in $[1/2, 1]$. We assume $\gamma_t \geq 0$. Notice that in some cases it is possible to take also $0 < \rho < 1/2$, but in this case it may be necessary to specify an additional boundary condition in order to ensure uniqueness of the solution (see for example Andersen and Andreasen's [1] extension of the LIBOR market model).

Under the model (2.4), the dynamics (2.2) becomes

$$dS_t = \left[a_t' - a_t\left(\frac{b_t'}{b_t} + \gamma_t\right) + \left(\frac{b_t'}{b_t} + \gamma_t\right) S_t\right] dt + b_t \eta_t \left(\frac{S_t - a_t}{b_t}\right)^\rho dW_t, \tag{2.5}$$

[2] See also Cox [17] and Cox and Ross [18].

so that condition (2.3) is equivalent to the following system

$$\begin{cases} a_t' - a_t\left(\dfrac{b_t'}{b_t} + \gamma_t\right) = 0 \\ \dfrac{b_t'}{b_t} + \gamma_t = \mu \end{cases}$$

whose solution is

$$\begin{cases} a_t = a_0 e^{\mu t} \\ b_t = b_0 e^{\int_0^t (\mu - \gamma_u) du} \end{cases}$$

for any real constants a_0 and b_0. From (2.5), it is then easy to see that we can set $b_t = 1$ for each t, with no loss of generality, so that the required candidate asset-price dynamics is

$$dS_t = \mu S_t dt + \eta_t (S_t - a_t)^\rho dW_t. \tag{2.6}$$

Indeed, the effect of a more general b can be absorbed into η.

Remark 2.1. We started from a basic candidate model consisting of the general time-varying-coefficients formulation of the CEV model, which will be particularly useful in the time-homogeneous case due its analytical tractability. However, the procedure can be used to shift any other distribution coming from a pre-selected dynamics. We could for example shift the hyperbolic diffusion model of Bibby and Sørensen [6], or the dynamics suggested by Platen [32], although the benefits of such a procedure are less clear when analytical tractability of the basic model is missing.

In this section, by directly considering the asset-price dynamics under the risk-neutral measure, we have implicitly assumed existence and uniqueness of such a measure. In the following examples, in fact, it can be easily proved that the risk-neutral measure exists unique when starting from real-world dynamics that differ from the given risk-neutral ones in the (constant) drift rate only.

2.1 The Shifted CEV Process with Deterministic Coefficients

We now assume that η_t is constant and equal to η for each t. In fact, restricting our attention to the case with constant coefficients allows for the derivation of explicit marginal densities and analytical formulas. We also assume that $\rho \neq 1$, meaning that we are excluding the lognormal case. Notice that it might seem necessary to assume $\mu \neq 0$ in order to ensure existence of the coefficients below. However, if $\mu = 0$ such formulas (and in particular k) still hold by substituting the relevant quantities with their limits for $\mu \to 0$ (see for example Andersen and Andreasen [1] for the case $\mu = 0$, also for the possibility of a time-varying η handled through a time-change). We then have the following.

Proposition 2.2. *If we set $\alpha := a_0$, the SDE (2.6) admits a unique strong solution that is given by*

$$S_t = P_t + \alpha e^{\mu t}, \quad t \geq 0, \tag{2.7}$$

where P is the CEV process

$$dP_t = \mu P_t dt + \eta P_t^{\rho} dW_t. \tag{2.8}$$

Moreover, the continuous part of the density function of S_T conditional on S_t, $t < T$, is

$$p_{S_T|S_t}(x) = 2(1-\rho)k^{1/(2-2\rho)}(uw^{1-4\rho})^{1/(4-4\rho)}e^{-u-w}I_{1/(2-2\rho)}(2\sqrt{uw}), \tag{2.9}$$

where

$$\begin{aligned} k &= \frac{\mu}{\eta^2(1-\rho)[e^{2\mu(1-\rho)(T-t)}-1]} \\ u &= k(S_t - \alpha e^{\mu t})^{2(1-\rho)}e^{2\mu(1-\rho)(T-t)} \\ w &= k(x - \alpha e^{\mu T})^{2(1-\rho)} \end{aligned} \tag{2.10}$$

and I_q denotes the modified Bessel function of the first kind of order q. Denoting by $g(y,z) = \frac{e^{-z}z^{y-1}}{\Gamma(y)}$ the gamma density function and by $G(y,x) = \int_x^{+\infty} g(y,z)dz$ the complementary gamma distribution, the probability that $S_T = \alpha e^{\mu T}$ conditional on S_t is given by $G\Big(\frac{1}{2(1-\rho)}, u\Big)$.

Proof. We just have to notice that $b_t = 1$ for each t implies that $\gamma_t = \mu$ for each t, so that (2.8) holds with $X = P$ and hence (2.7) follows. The density (2.9) is then simply obtained by shifting the density of the CEV process (2.8), which for instance can be found in Schroder [39], by the relevant quantity at each time and by writing $P_t = S_t - \alpha e^{\mu t}$.

It is well known that by modeling the asset-price dynamics with the process (2.8) we can derive explicit formulas for European options on the asset. The same analytical feature applies to the process (2.7) as well. Indeed, let us consider a European option with maturity T, strike K and written on the asset. The arbitrage-free option price under the model (2.7) is given in the following.

Proposition 2.3. *Under the assumption that $\alpha e^{\mu T} < K$, the European call option value at any time $t < T$ is*

$$\begin{aligned} C_t = {} & (S_t - \alpha e^{\mu t})e^{(\mu-r)(T-t)} \sum_{n=0}^{+\infty} g(n+1,u)\, G\Big(\xi, k(K-\alpha e^{\mu T})^{2(1-\rho)}\Big) \\ & - (K - \alpha e^{\mu T})e^{-r(T-t)} \sum_{n=0}^{+\infty} g\,(\xi, u)\; G\Big(n+1, k(K-\alpha e^{\mu T})^{2(1-\rho)}\Big). \end{aligned} \tag{2.11}$$

where k and u are defined as in (2.10) and $\xi := n + 1 + \frac{1}{2(1-\rho)}$.

Proof. Denoting by E_t the time-t conditional expectation under the risk-neutral measure, we have

$$C_t = e^{-r(T-t)} E_t \left\{ (S_T - K)^+ \right\} = e^{-r(T-t)} E_t \left\{ [P_T - (K - \alpha e^{\mu T})]^+ \right\}.$$

Since P is a CEV process with the proper risk-neutral drift, the last discounted expectation is simply given by the CEV option-price formula[3] with the strike being equal to $K - \alpha e^{\mu T}$. Formula (2.11) is then obtained by remembering that $P_t = S_t - \alpha e^{\mu t}$.

Remark 2.4. The assumption $\alpha e^{\mu T} < K$ is rather natural since, otherwise, the payoff would lose its optionality. Indeed, if $\alpha e^{\mu T} \geq K$ we obtain (at time 0):

$$E_0[P_T - (K - \alpha e^{\mu T})]^+ = E_0[P_T - (K - \alpha e^{\mu T})] = e^{\mu T} S_0 - K .$$

In such a case the shift is so large as to render the payoff linear again, which is of course not desirable.

The option price (2.11) leads to skews in the implied volatility structure. This is intuitive since already the basic CEV process, corresponding to $\alpha = 0$, shows such property. However, we now have three parameters, α, η and ρ, that can be employed for a better fitting of the market volatility curves. An example of the volatility structure that is implied by the model (2.7) is shown in Fig. 1.

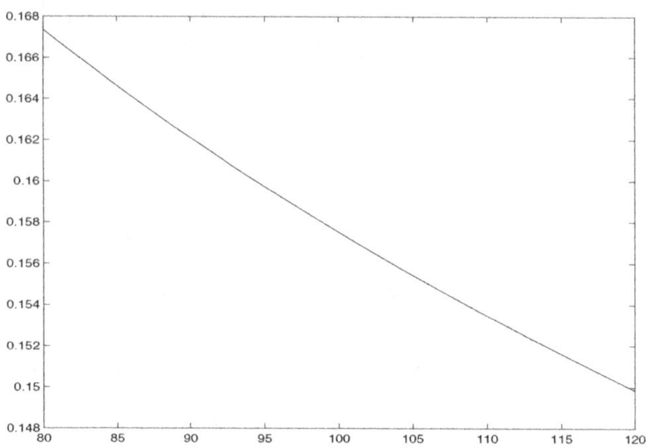

Fig. 1. Volatility structure implied by the option prices (2.11) at time $t = 0$, where we set $\mu = r = 0.035$, $T = 1$, $\alpha = -10$, $\eta = 1.5$, $\rho = 0.5$ and $S_0 = 100$

Besides pricing European options analytically, the model (2.7) leads to fast numerical procedures for pricing exotic derivatives. In fact, the known

[3] As seen, for instance, in Beckers [5].

transition density (2.9) renders the implementation of Monte Carlo procedures easier and more efficient when pricing a large variety of path-dependent claims. Moreover, binomial or trinomial trees for S can be constructed by first building the tree for P and then by properly displacing the corresponding nodes.

2.2 The Shifted-Lognormal Case

The second case we consider is that of a basic process X following, under the risk-neutral measure, the geometric Brownian motion[4]

$$dX_t = \mu X_t dt + \beta_t X_t dW_t, \tag{2.12}$$

so that the asset price $S_t = X_t + \alpha e^{\mu t}$ evolves under such measure according to

$$dS_t = \mu S_t dt + \beta_t (S_t - \alpha e^{\mu t}) dW_t. \tag{2.13}$$

We then have the following.

Proposition 2.5. *The asset price S can be explicitly written as*

$$S_t = \alpha e^{\mu t} + (S_0 - \alpha) e^{\int_0^t (\mu - \frac{1}{2}\beta_u^2) du + \int_0^t \beta_u dW_u}, \tag{2.14}$$

and the distribution of the asset price S_T, conditional on S_t, $t < T$, is a shifted lognormal distribution with density

$$p_{S_T|S_t}(x) = \frac{1}{(x - \alpha e^{\mu T}) V_t \sqrt{2\pi}} \exp\left\{ -\frac{1}{2} \left(\frac{\ln(x - \alpha e^{\mu T}) - M_t}{V_t} \right)^2 \right\}, \quad x > \alpha e^{\mu T}, \tag{2.15}$$

where

$$\begin{aligned} M_t &= \ln(S_t - \alpha e^{\mu t}) + \int_t^T \left(\mu - \frac{1}{2}\beta_u^2 \right) du \\ V_t &= \sqrt{\int_t^T \beta_u^2 du}. \end{aligned} \tag{2.16}$$

Proof. The explicit form (2.14) is obtained by integration of the SDE (2.12) and by remembering the previous general results. The density (2.15) is then obtained by suitably shifting the lognormal density of X and by writing $X_t = S_t - \alpha e^{\mu t}$.

As before, knowledge of the risk-neutral distribution of the process S allows us to explicitly derive formulas for European options. This is accomplished in the following.

[4] A displaced diffusion model has been first considered by Rubinstein [36].

Proposition 2.6. *Consider a European option with maturity T, strike K and written on the asset. Then, under the assumption that $\alpha e^{\mu T} < K$, the option value at any time $t < T$ is given by*

$$O_t = \omega \Bigg[(S_t - \alpha e^{\mu t}) e^{(\mu - r)(T-t)} \Phi\Bigg(\omega \frac{\ln \frac{S_t - \alpha e^{\mu t}}{K - \alpha e^{\mu T}} + \mu(T-t) + \frac{1}{2} V_t^2}{V_t} \Bigg) - (K - \alpha e^{\mu T}) e^{-r(T-t)} \Phi\Bigg(\omega \frac{\ln \frac{S_t - \alpha e^{\mu t}}{K - \alpha e^{\mu T}} + \mu(T-t) - \frac{1}{2} V_t^2}{V_t} \Bigg) \Bigg], \tag{2.17}$$

where $\omega = 1$ for a call and $\omega = -1$ for a put and with Φ denoting the standard normal cumulative distribution function.

Proof. As before, let us denote by E_t the time-t conditional expectation under the risk-neutral measure. Noting that $\ln(X_T)$ conditional on X_t is normally distributed with mean M_t and variance V_t^2, we have

$$\begin{aligned} O_t &= e^{-r(T-t)} E_t \left\{ [\omega(S_T - K)]^+ \right\} = e^{-rT} E_t \left\{ [\omega X_T - \omega(K - \alpha e^{\mu T})]^+ \right\} \\ &= e^{-r(T-t)} \int_{-\infty}^{+\infty} [\omega e^x - \omega(K - \alpha e^{\mu T})]^+ \frac{1}{V_t \sqrt{2\pi}} e^{-\frac{1}{2}\left(\frac{x - M_t}{V_t}\right)^2} dx \\ &= e^{-r(T-t)} \omega \Bigg[e^{M_t + \frac{1}{2} V_t^2} \Phi\Bigg(\omega \frac{M_t - \ln(K - \alpha e^{\mu T}) + V_t^2}{V_t} \Bigg) \\ &\quad - (K - \alpha e^{\mu T}) \Phi\Bigg(\omega \frac{M_t - \ln(K - \alpha e^{\mu T})}{V_t} \Bigg) \Bigg], \end{aligned}$$

which immediately leads to (2.17).

Notice that the option price (2.17) is nothing but the Black-Scholes price with the asset spot price and the strike being respectively replaced with $S_t - \alpha e^{\mu t}$ and $K - \alpha e^{\mu T}$ and where the implied volatility is $\sqrt{\frac{1}{T-t} \int_t^T \beta_u^2 du}$.

Model (2.13) is a simple analytical extension of the Black-Scholes model allowing for skews in the option implied volatility. An example of the skewed volatility structure that is implied by (2.13) is provided in Fig. 2. Appendix A shows that for strikes close to the forward asset price, the implied volatility structure is indeed monotone in the strike. Similarly to the model (2.7), therefore, model (2.13) can be used to fit market volatility structures that are skewed. Even though we have now one less parameter than in model (2.7), since ρ is set to one, the time dependence of β allows, however, for a better fitting along the time-to-maturity dimension.

As to pricing of exotic derivatives, the decomposition $S_t = X_t + \alpha e^{\mu t}$ allows for a fast Monte Carlo generation of asset-price scenarios, through the generation of paths of the geometric Brownian motion (2.12). Again, binomial or trinomial trees for S can be constructed by starting from the basic ones for X and displacing the corresponding nodes by the relevant (deterministic) quantity at each time step.

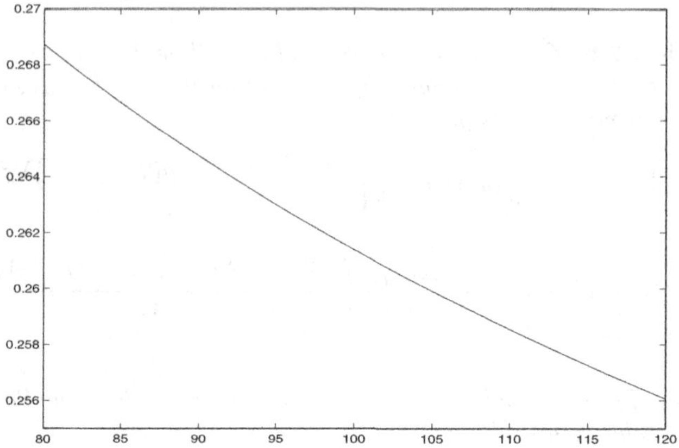

Fig. 2. Volatility structure implied by the option prices (2.17) at time $t = 0$, where we set $\mu = r = 0.035$, $T = 1$, $\alpha = -30$, $\beta_t = 0.2$ for each t and $S_0 = 100$.

3 The General Mixture-Dynamics Class

In this section we propose a new class of analytically-tractable models for the asset-price dynamics that are capable to fit more general volatility structures. The diffusion processes we obtain follow from assuming a particular risk-neutral distribution for the asset price S. Precisely, we assume that the marginal density of S under the risk-neutral measure is the weighted average of known densities of some given diffusion processes.

Let us then consider N diffusion processes with risk-neutral dynamics given by

$$dS_t^i = \mu S_t^i dt + v_i(t, S_t^i) dW_t, \quad i = 1, \dots, N, \tag{3.18}$$

with initial value S_0^i, where W is a standard Brownian motion under the risk-neutral measure Q and $v_i(t, y)$'s are real functions satisfying regularity conditions to ensure existence and uniqueness of the solution to the SDE (3.18). In particular we assume that, for a suitable $L_i > 0$, the following linear-growth condition holds:

$$v_i^2(t, y) \le L_i(1 + y^2) \;\text{ uniformly in } t. \tag{3.19}$$

For each t, we denote by $p_t^i(\cdot)$ the density function of S_t^i, i.e., $p_t^i(y) = d(Q\{S_t^i \le y\})/dy$, where, in particular, $p_0^i(y)$ is the δ-Dirac function centered in S_0^i.

Let us now assume that the dynamics of the asset price S under the risk-neutral measure Q is given by

$$dS_t = \mu S_t dt + \sigma(t, S_t) S_t dW_t, \tag{3.20}$$

where $\sigma(\cdot, \cdot)$ satisfies, for a suitable positive constant L, the linear-growth condition

$$\sigma^2(t, y) y^2 \le L(1 + y^2) \;\text{ uniformly in } t\,. \tag{3.21}$$

Such growth condition ensures existence of a strong solution. We assume, moreover, that σ satisfies some further condition assuring uniqueness of the solution.[5]

The problem we want to address is the derivation of the "local volatility" $\sigma(t, S_t)$ such that the risk-neutral density of S satisfies

$$p_t(y) := \frac{d}{dy} Q\{S_t \leq y\} = \sum_{i=1}^{N} \lambda_i \frac{d}{dy} Q\{S_t^i \leq y\} = \sum_{i=1}^{N} \lambda_i p_t^i(y), \tag{3.22}$$

where each S_0^i is set to S_0, and λ_i's are strictly positive constants such that $\sum_{i=1}^{N} \lambda_i = 1$. Indeed, $p_t(\cdot)$ is a proper risk-neutral density function since, by definition,

$$\int_0^{+\infty} y p_t(y) dy = \sum_{i=1}^{N} \lambda_i \int_0^{+\infty} y p_t^i(y) dy = \sum_{i=1}^{N} \lambda_i S_0 e^{\mu t} = S_0 e^{\mu t}.$$

Remark 3.1. Notice that in the last calculation we were able to recover the proper risk-neutral expectation thanks to our assumption that all the processes (3.18) share the same drift parameter μ. However, the role of the processes S^i is merely instrumental, and there is no need to assume their drift to be of the form μS_t^i if not for simplifying calculations. In particular, what matters in obtaining the right expectation as in the last formula above is the marginal distribution p_i of the S^i's. We could generate alternative dynamics leading to the same marginal densities as in the S_i's by selecting arbitrary diffusion coefficients and then by defining appropriate drifts according to the results in Brigo and Mercurio [9], [10], although computations would generally become involved.

As already noticed by many authors,[6] the above problem is essentially the reverse to that of finding the marginal density function of the solution of an SDE when the coefficients are known. In particular, $\sigma(t, S_t)$ can be found by solving the Fokker-Planck equation

$$\frac{\partial}{\partial t} p_t(y) = -\frac{\partial}{\partial y} (\mu y p_t(y)) + \frac{1}{2} \frac{\partial^2}{\partial y^2} \left(\sigma^2(t, y) y^2 p_t(y)\right), \tag{3.23}$$

given that each density $p_t^i(y)$ satisfies itself the Fokker-Planck equation

$$\frac{\partial}{\partial t} p_t^i(y) = -\frac{\partial}{\partial y} (\mu y p_t^i(y)) + \frac{1}{2} \frac{\partial^2}{\partial y^2} \left(v_i^2(t, y) p_t^i(y)\right). \tag{3.24}$$

Applying the definition (3.22) and the linearity of the derivative operator, (3.23) can be written as

$$\sum_{i=1}^{N} \lambda_i \frac{\partial}{\partial t} p_t^i(y) = \sum_{i=1}^{N} \lambda_i \left[-\frac{\partial}{\partial y} (\mu y p_t^i(y)) \right] + \sum_{i=1}^{N} \lambda_i \left[\frac{1}{2} \frac{\partial^2}{\partial y^2} \left(\sigma^2(t, y) y^2 p_t^i(y)\right) \right],$$

5 The classical example is the local Lipschitz condition. In this section, however, we need not write it explicitly.

6 See for instance Dupire [23].

which, by substituting from (3.24), becomes

$$\sum_{i=1}^{N} \lambda_i \left[\frac{1}{2} \frac{\partial^2}{\partial y^2} \left(v_i^2(t,y) p_t^i(y) \right) \right] = \sum_{i=1}^{N} \lambda_i \left[\frac{1}{2} \frac{\partial^2}{\partial y^2} \left(\sigma^2(t,y) y^2 p_t^i(y) \right) \right].$$

Using again the linearity of the second order derivative operator,

$$\frac{\partial^2}{\partial y^2} \left[\sum_{i=1}^{N} \lambda_i v_i^2(t,y) p_t^i(y) \right] = \frac{\partial^2}{\partial y^2} \left[\sigma^2(t,y) y^2 \sum_{i=1}^{N} \lambda_i p_t^i(y) \right].$$

If we look at this last equation as to a second order differential equation for $\sigma(t,\cdot)$, we find easily its general solution

$$\sigma^2(t,y) y^2 \sum_{i=1}^{N} \lambda_i p_t^i(y) = \sum_{i=1}^{N} \lambda_i v_i^2(t,y) p_t^i(y) + A_t y + B_t, \tag{3.25}$$

with A and B suitable real functions of time. The regularity conditions (3.19) and (3.21) imply that the LHS of the equation has zero limit for $y \to \infty$.[7] As a consequence, the RHS must have a zero limit as well. This holds if and only if $A_t = B_t = 0$, for each t. We therefore obtain that the expression for $\sigma(t,y)$ that is consistent with the marginal density (3.22) and with the regularity constraint (3.21) is, for $(t,y) > (0,0)$,

$$\sigma(t,y) = \sqrt{\frac{\sum_{i=1}^{N} \lambda_i v_i^2(t,y) p_t^i(y)}{\sum_{i=1}^{N} \lambda_i y^2 p_t^i(y)}}. \tag{3.26}$$

Notice that by setting

$$\Lambda_i(t,y) := \frac{\lambda_i p_t^i(y)}{\sum_{i=1}^{N} \lambda_i p_t^i(y)} \tag{3.27}$$

for each $i = 1, \dots, N$ and $(t,y) > (0,0)$, we can write

$$\sigma^2(t,y) = \sum_{i=1}^{N} \Lambda_i(t,y) \frac{v_i^2(t,y)}{y^2}, \tag{3.28}$$

so that the square of the volatility σ can be written as a (stochastic) convex combination of the squared volatilities of the basic processes (3.18). In fact, for each (t,y), $\Lambda_i(t,y) \geq 0$ for each i and $\sum_{i=1}^{N} \Lambda_i(t,y) = 1$. Moreover, by (3.19) and setting $L := \max_{i=1,\dots,N} L_i$, the condition (3.21) is fulfilled since

$$\sigma^2(t,y) y^2 = \sum_{i=1}^{N} \Lambda_i(t,y) v_i^2(t,y) \leq \sum_{i=1}^{N} \Lambda_i(t,y) L_i (1+y^2) \leq L(1+y^2).$$

[7] Notice in fact that the linear-growth condition (3.19) implies that each $p_t^i(\cdot)$ has a second moment, hence that $\lim_{y\to\infty} y^2 p_t^i(y) = 0$ for each i and t (we assume existence of such limits).

The function σ may then be extended to the semi-axes $\{(t, 0) : t > 0\}$ and $\{(0, y) : y > 0\}$ according to the specific choice of the basic densities $p_t^i(\cdot)$.

Formula (3.26) leads to the following SDE for the asset price under the risk–neutral measure Q:

$$dS_t = \mu S_t dt + \sqrt{\frac{\sum_{i=1}^N \lambda_i v_i^2(t, S_t) p_t^i(S_t)}{\sum_{i=1}^N \lambda_i S_t^2 p_t^i(S_t)}} S_t dW_t. \tag{3.29}$$

This SDE, however, must be regarded as defining some candidate dynamics that leads to the marginal density (3.22). Indeed, if σ is bounded, then

$$E\left\{\int_0^t \sigma^2(u, S_u)du\right\} < \infty,$$

so that the SDE

$$d\ln(S_t) = \left[\mu - \frac{1}{2}\sigma^2(t, S_t)\right] dt + \sigma(t, S_t)dW_t$$

is well defined, and so is (3.29). But the conditions we have imposed so far are not sufficient to grant the uniqueness of the strong solution, so that a verification must be done on a case-by-case basis.

Let us now give for granted that the SDE (3.29) has a unique strong solution. Then, remembering the definition (3.22), it is straightforward to derive the model option prices in terms of the option prices associated to the basic models (3.18). Indeed, let us consider a European option with maturity T, strike K and written on the asset. Then, if $\omega = 1$ for a call and $\omega = -1$ for a put, the option value $\mathcal{O}$ at the initial time $t = 0$ is given by

$$\begin{aligned} \mathcal{O} &= e^{-rT} E^Q \left\{[\omega(S_T - K)]^+\right\} \\ &= e^{-rT} \int_0^{+\infty} [\omega(y - K)]^+ \sum_{i=1}^N \lambda_i p_T^i(y)dy \\ &= \sum_{i=1}^N \lambda_i e^{-rT} \int_0^{+\infty} [\omega(y - K)]^+ p_T^i(y)dy \\ &= \sum_{i=1}^N \lambda_i \mathcal{O}_i, \end{aligned} \tag{3.30}$$

where E^Q denotes expectation under Q and $\mathcal{O}_i$ denotes the option price associated to (3.18).

Remark 3.2. We can now motivate our assumption that the asset marginal density be given by the mixture of known basic densities. Notice that, starting from a general asset-price dynamics, it may be quite problematic to come up with analytical formulas for European options. Here, instead, the use of analytically-tractable densities p^i immediately leads to explicit option prices for the process S. Moreover, a virtually unlimited number of parameters can be introduced in the asset-price dynamics, and hence used for a better fitting of market data. Finally, we notice that S can viewed as a process whose density at time t coincides with the basic density p_t^i with probability λ_i.

The above derivation shows that a dynamics leading to a marginal density for the asset price that is the convex combination of basic densities induces the same convex combination among the corresponding option prices. Furthermore, due to the linearity of the derivative operator, the same convex combination applies to all option Greeks. In particular this ensures that starting from analytically tractable basic densities one finds a model that preserves the analytical tractability.

3.1 The Mixture-of-Lognormals Case

Let us now consider the particular case where the densities $p_t^i(\cdot)$'s are all lognormal. Precisely, we assume that, for each i,

$$v_i(t, y) = \sigma_i(t)y, \tag{3.31}$$

where all σ_i's are deterministic functions of time that are bounded from above and below by positive constants. Notice that if moreover σ_i's are continuous and we take a finite time–horizon, then boundedness from above is automatic, and the only condition to be required explicitly is boundedness from below by a positive constant. The marginal density of S^i conditional on S_0 is then given by

$$\begin{aligned} p_t^i(y) &= \frac{1}{yV_i(t)\sqrt{2\pi}} \exp\left\{ -\frac{1}{2V_i^2(t)} \left[\ln\frac{y}{S_0} - \mu t + \tfrac{1}{2}V_i^2(t) \right]^2 \right\}, \\ V_i(t) &:= \sqrt{\int_0^t \sigma_i^2(u)du}. \end{aligned} \tag{3.32}$$

The case where the risk-neutral density is a mixture of lognormal densities has been originally studied by Ritchey [34][8] and subsequently used by Melick and Thomas [30], Bhupinder [15] and Guo [24]. However, their works are mainly empirical: They simply assumed such risk-neutral density and then studied the resulting fitting to option data.

In this paper, instead, we develop the model from a theoretical point of view and derive the specific asset-price dynamics that implies the chosen distribution. As a consequence, not only do we propose an asset-price model that

[8] Indeed, Ritchey [34] assumed a mixture of normal densities for the density of the asset log-returns. However, it can be easily shown that this is equivalent to assuming a mixture of lognormal densities for the density of the asset price.

is capable to fit real option data, but also we can construct efficient procedures for the pricing of exotic derivatives that are path-dependent or have early-exercise features.

Proposition 3.3. *Let us assume that each σ_i is also continuous and that there exists an $\varepsilon > 0$ such that $\sigma_i(t) = \sigma_0 > 0$, for each t in $[0, \varepsilon]$ and $i = 1, \ldots, N$. Then, if we set*

$$\sigma(t,y) = \sqrt{\frac{\sum_{i=1}^{N} \lambda_i \sigma_i^2(t) \frac{1}{V_i(t)} \exp\left\{-\frac{1}{2V_i^2(t)}\left[\ln\frac{y}{S_0} - \mu t + \frac{1}{2}V_i^2(t)\right]^2\right\}}{\sum_{i=1}^{N} \lambda_i \frac{1}{V_i(t)} \exp\left\{-\frac{1}{2V_i^2(t)}\left[\ln\frac{y}{S_0} - \mu t + \frac{1}{2}V_i^2(t)\right]^2\right\}}}, \tag{3.33}$$

for $(t, y) > (0, 0)$ and $\sigma(t, y) = \sigma_0$ for $(t, y) = (0, S_0)$, the SDE (3.20) has a unique strong solution whose marginal density is given by the mixture of lognormals

$$p_t(y) = \sum_{i=1}^{N} \lambda_i \frac{1}{y V_i(t)\sqrt{2\pi}} \exp\left\{-\frac{1}{2V_i^2(t)}\left[\ln\frac{y}{S_0} - \mu t + \tfrac{1}{2}V_i^2(t)\right]^2\right\}. \tag{3.34}$$

Moreover, for $(t, y) > (0, 0)$, we can write

$$\sigma^2(t,y) = \sum_{i=1}^{N} \Lambda_i(t,y)\sigma_i^2(t), \tag{3.35}$$

where, for each (t, y) and i, $\Lambda_i(t, y) \geq 0$ and $\sum_{i=1}^{N} \Lambda_i(t, y) = 1$. As a consequence

$$0 < \tilde{\sigma} \leq \sigma(t,y) \leq \hat{\sigma} < +\infty \quad \text{for each } t, y > 0. \tag{3.36}$$

where

$$\tilde{\sigma} := \inf_{t \geq 0}\left\{\min_{i=1,\ldots,N} \sigma_i(t)\right\},$$

$$\hat{\sigma} := \sup_{t \geq 0}\left\{\max_{i=1,\ldots,N} \sigma_i(t)\right\}.$$

Proof. See Appendix B.

The above proposition provides us with the analytical expression for the diffusion coefficient in the SDE (3.20) such that the resulting equation has a unique strong solution whose marginal density is given by (3.34). Moreover, the square of the "local volatility" $\sigma(t, y)$ can be viewed as a weighted average of the squared "basic volatilities" $\sigma_1^2(t), \ldots, \sigma_N^2(t)$, where the weights are all functions of the lognormal marginal densities (3.32). In particular, the "local volatility" $\sigma(t, y)$ always lies in the interval $[\tilde{\sigma}, \hat{\sigma}]$.[9]

[9] This property relates our model to that of Avellaneda et al. [4] who considered a stochastic volatility evolving within a predefined band.

Remark 3.4 (Market completeness). Also in this section, we have directly considered the asset-price dynamics under the risk-neutral measure. Notice, indeed, that we can easily prove the existence and uniqueness of such a measure for the above model. To this end, let us assume that under the real-world measure Q_0, the asset price evolves according to

$$dS_t = \nu S_t dt + \sigma(t, S_t) S_t dW_t^0,$$

where ν is a real constant and W^0 is a Q_0-standard Brownian motion. Then the Radon-Nicodym derivative defining the change of measure from Q_0 to Q is expressed in terms of the "market price of risk"

$$\theta(t, S_t) = \frac{\nu - \mu}{\sigma(t, S_t)},$$

which is bounded due to (3.36). As a consequence, the Novikov condition, ensuring the feasibility of such change of measure, is immediately satisfied, and $dW_t = dW_t^0 + \theta(t, S_t)dt$.

As we have already noticed, the pricing of European options under the asset-price model (3.20) with (3.33) is quite straightforward. Indeed, we have the following.

Proposition 3.5. *Consider a European option with maturity T, strike K and written on the asset. Then, the option value at the initial time $t = 0$ is given by the following convex combination of Black-Scholes prices*

$$\begin{aligned} \mathcal{O} = \omega \sum_{i=1}^{N} \lambda_i \Bigg[S_0 e^{(\mu - r)T} \Phi\left(\omega \frac{\ln \frac{S_0}{K} + \left(\mu + \frac{1}{2}\eta_i^2\right) T}{\eta_i \sqrt{T}} \right) \\ - K e^{-rT} \Phi\left(\omega \frac{\ln \frac{S_0}{K} + \left(\mu - \frac{1}{2}\eta_i^2\right) T}{\eta_i \sqrt{T}} \right) \Bigg], \end{aligned} \tag{3.37}$$

where $\omega = 1$ for a call and $\omega = -1$ for a put and

$$\eta_i := \frac{V_i(T)}{\sqrt{T}} = \sqrt{\frac{\int_0^T \sigma_i^2(t)dt}{T}}. \tag{3.38}$$

Proof. We just have to apply (3.30) and notice that $\mathcal{O}_i = e^{-rT} \int_0^{+\infty} [\omega(y - K)]^+ p_T^i(y)dy$ is nothing but the Black-Scholes call/put price corresponding to the volatility η_i.

The option price (3.37) leads to smiles in the implied volatility structure. An example of the shape that can be reproduced in shown in Fig. 3. Indeed, the volatility implied by the option prices (3.37) has a minimum exactly at a strike equal to the forward asset price $S_0 e^{\mu T}$. This is formally proven in Appendix C.

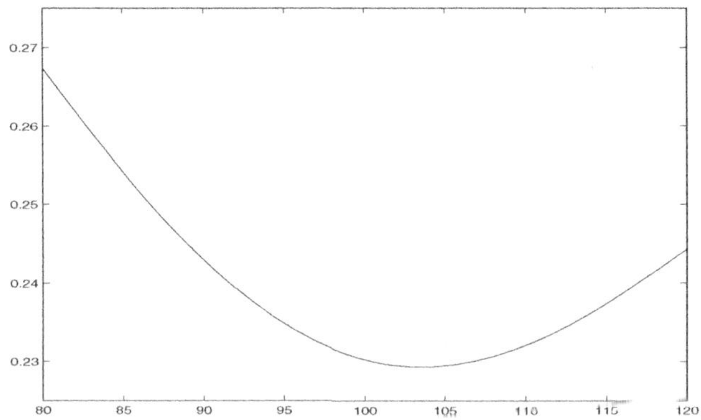

Fig. 3. Volatility structure implied by the option prices (3.37), where we set $\mu = r = 0.035$, $T = 1$, $N = 3$, $(\eta_1, \eta_2, \eta_3) = (0.5, 0.1, 0.2)$, $(\lambda_1, \lambda_2, \lambda_3) = (0.2, 0.3, 0.5)$ and $S_0 = 100$.

Like the models falling into the class considered in the previous section, also these second-class models are quite appealing when pricing exotic derivatives. Notice, indeed, that having explicit dynamics implies that the asset-price paths can be simulated by discretizing the associated SDE with a numerical scheme. Hence we can use Monte Carlo procedures to price path-depending derivatives. Claims with early-exercise features can be priced with grids or lattices that can be constructed given the explicit form of the asset-price dynamics.

4 Conclusions and Suggestions for Further Research

We have considered two different classes of asset-price models allowing respectively for skews and smiles in the implied volatility structure. The first class is based on shifting the marginal distribution of a known diffusion process for which there exist analytical formulas for European options. As a result, the processes in this class have known dynamics and marginal densities, which also lead to explicit European option prices. The cases of a CEV process with constant coefficients and a general geometric Brownian motion are considered as major applications.

The second class is instead based on asset-price processes whose marginal density is given by the mixture of some suitably chosen densities. In particular, if the basic densities are associated to specific (risk-neutral) asset-price dynamics that imply analytical option prices, so does their mixture. Indeed, we have derived the diffusion process followed by the asset price under the risk-neutral measure such that its marginal density is given by the chosen mixture. As a major example, we have considered the case where the process density is a mixture of lognormal densities. The use of a mixture of lognormal densities for fitting the market volatility structure is not new in the financial literature. However, it is now clear how to relate such asset-price density to some specific dynamics. This is very important, because it allows the pricing and hedging of general derivatives either through a Monte Carlo procedure or through a lattice implementation.

The classes proposed in this paper are defined in a quite general way, so that many more examples could be considered and studied. For instance, processes in the first class could act as bricks for building a second-class process. Vice versa, we could take a process in the second class and obtain a first-class process by shifting its dynamics. In particular, the risk-neutral asset-price process

$$dS_t = \mu S_t dt + \sqrt{\frac{\sum_{i=1}^N \lambda_i \sigma_i^2(t) \frac{1}{V_i(t)} \exp\left\{-\frac{1}{2V_i^2(t)}\left[\ln\frac{S_t - \alpha e^{\mu t}}{S_0 - \alpha} - \mu t + \frac{1}{2}V_i^2(t)\right]^2\right\}}{\sum_{i=1}^N \lambda_i \frac{1}{V_i(t)} \exp\left\{-\frac{1}{2V_i^2(t)}\left[\ln\frac{S_t - \alpha e^{\mu t}}{S_0 - \alpha} - \mu t + \frac{1}{2}V_i^2(t)\right]^2\right\}}} \cdot (S_t - \alpha e^{\mu t}) dW_t,$$

with $\alpha \neq 0$, has a marginal density that is given by shifting a mixture of lognormal densities by the quantity $\alpha e^{\mu t}$ at each time t. The corresponding option prices lead to an implied volatility structure whose minimum point is shifted from the asset forward price. We are thus able to obtain more flexible structures to better fit the real market volatility data. An example of this model fitting quality to real market data has been investigated by Brigo and Mercurio [11] on the Italian stock index option market.

A final remark concerns the type of financial market being considered in this paper. Our asset, in fact, can be viewed as an equity stock or index or as an exchange rate. However, our treatment perfectly works also for forward LIBOR rates, as soon as we think of them as special assets. In a such a case, we just have to replace the risk-neutral measure with the corresponding forward measure and remember that, under its canonical measure, a forward rate has null drift, i.e., $\mu = 0$. See Brigo and Mercurio [12] for the details.

Appendix A

We prove here that the ATM-forward volatility that is implied by the model (2.13) has a non-zero derivative with respect to the strike. The Black-Scholes volatility σ that is implied by the call option price (2.17), with $\omega = 1$, is implicitly defined by

$$C^{BS}(S_0, K, T, \mu, \sigma) = C^{BS}(S_0 - \alpha, K - \alpha e^{\mu T}, T, \mu, v), \tag{A.1}$$

where $v := \sqrt{\frac{1}{T}\int_0^T \beta_u^2 du}$, and $C^{BS}(S_0, K, T, \mu, \sigma)$ is the Black-Scholes price when the underlying asset price is S_0, the strike price is K, the maturity is T, the risk-free rate is μ and the volatility is σ. Notice that the equation (A.1) always has a unique solution $\sigma = \sigma(K)$ given that $\alpha < Ke^{-\mu T}$.

Since $\frac{\partial C^{BS}}{\partial \sigma} \neq 0$, by Dini's theorem

$$\frac{d\sigma}{dK}(K) = \frac{\frac{\partial C^{BS}}{\partial K}(S_0 - \alpha, K - \alpha e^{\mu T}, T, \mu, v) - \frac{\partial C^{BS}}{\partial K}(S_0, K, T, \mu, \sigma(K))}{\frac{\partial C^{BS}}{\partial \sigma}(S_0, K, T, \mu, \sigma(K))}$$
$$= -\frac{e^{-\mu T}\left[\Phi\!\left(\frac{\ln\frac{S_0-\alpha}{K-\alpha e^{\mu T}} + \left(\mu - \frac{1}{2}v^2\right)T}{v\sqrt{T}}\right) - \Phi\!\left(\frac{\ln\frac{S_0}{K} + \left(\mu - \frac{1}{2}\sigma^2(K)\right)T}{\sigma(K)\sqrt{T}}\right)\right]}{S_0\sqrt{T}\Phi'\!\left(\frac{\ln\frac{S_0}{K} + \left(\mu + \frac{1}{2}\sigma^2(K)\right)T}{\sigma(K)\sqrt{T}}\right)},$$

where Φ' denote the standard normal density function. In particular, for an ATM-forward strike $\bar{K} = S_0 e^{\mu T}$,

$$\frac{d\sigma}{dK}(\bar{K}) = -\frac{e^{-\mu T}\left[\Phi\!\left(-\frac{1}{2}v\sqrt{T}\right) - \Phi\!\left(-\frac{1}{2}\sigma(\bar{K})\sqrt{T}\right)\right]}{S_0\sqrt{T}\Phi'\!\left(\frac{1}{2}\sigma(\bar{K})\sqrt{T}\right)}, \tag{A.2}$$

while (A.1) becomes

$$2S_0\Phi\!\left(\frac{1}{2}\sigma(\bar{K})\sqrt{T}\right) = 2(S_0 - \alpha)\Phi\!\left(\frac{1}{2}v\sqrt{T}\right) + \alpha,$$

or equivalently

$$2S_0\left[\Phi\!\left(\frac{1}{2}\sigma(\bar{K})\sqrt{T}\right) - \Phi\!\left(\frac{1}{2}v\sqrt{T}\right)\right] = \alpha\left[1 - 2\Phi\!\left(\frac{1}{2}v\sqrt{T}\right)\right].$$

Hence $\alpha > 0$ implies that $\sigma(\bar{K}) < v$,[10] and, by (A.2), that $\frac{d\sigma}{dK}(\bar{K}) > 0$, with the sign of the derivative that is preserved in a neighborhood of $\bar{K}$. Conversely, $\alpha < 0$ implies that $\sigma(\bar{K}) > v$, and that $\frac{d\sigma}{dK}(\bar{K}) < 0$.

Appendix B: Proof of Proposition 3.3

The formula (3.33) for $(t, y) > (0, 0)$ is trivially obtained from (3.26) by using (3.31) and (3.32). Equation (3.35), which immediately derives from (3.28), implies that

$$\min_{i=1,\dots,N} \sigma_i(t) \leq \sigma(t, y) \leq \max_{i=1,\dots,N} \sigma_i(t),$$

for each $(t, y) > (0, 0)$, so that the value $\sigma(0, S_0)$ is set by continuity. The inequalities (3.36) immediately follow from (3.35) and the boundedness of each function σ_i. If we write $S_t = \exp(Z_t)$, where

$$dZ_t = \left[\mu - \frac{1}{2}\sigma^2\left(t, e^{Z_t}\right)\right]dt + \sigma\left(t, e^{Z_t}\right)dW_t, \tag{B.1}$$

[10] Remember we have assumed $\alpha \neq 0$.

the SDE (3.20) then admits a unique strong solution since the SDE (B.1) admits a unique strong solution. In fact, its coefficients are bounded and hence satisfy the usual linear-growth condition. Moreover, setting $u(t,z) := \sigma(t, e^z)$, we have

$$\frac{\partial u^2}{\partial z}(t,z) = (\ln(S_0) + \mu t - z)\frac{\sum_{i=1}^N \sum_{j=1}^N \lambda_i \lambda_j A_i A_j \frac{\sigma_i^2(t)}{V_i(t)V_j(t)}\left(\frac{1}{V_i^2(t)} - \frac{1}{V_j^2(t)}\right)}{\sum_{i=1}^N \sum_{j=1}^N \lambda_i \lambda_j A_i A_j \frac{1}{V_i(t)V_j(t)}},$$

where

$$A_i := \exp\left[-\frac{1}{2V_i^2(t)}\left(z - \ln(S_0) - \mu t + \frac{1}{2}V_i^2(t)\right)^2\right],$$

so that $\frac{\partial u^2}{\partial z}(t,z)$ is well defined and continuous for each $(t,z) \in (0, M] \times (-\infty, +\infty)$, $M > 0$, due to the continuity of each σ_i and V_i, and

$$\lim_{t\to 0} \frac{\partial u^2}{\partial z}(t,z) = 0,$$

since u is constant for $t \in [0,\epsilon]$. Therefore, $\frac{\partial u^2}{\partial z}(t,z)$ is bounded on each compact set $[0,M] \times [-M,M]$, and so is $\frac{\partial u}{\partial z}(t,z) = \frac{1}{2u(t,z)}\frac{\partial u^2}{\partial z}(t,z)$ since σ is bounded from below. Hence, the function u is locally Lipschitz in the sense of Theorem 12.1 in Section V.12 of Rogers and Williams [35]. In view of this theorem, the SDE (B.1) then admits a unique strong solution.

Finally, the density (3.34) follows from the construction procedure developed at the beginning of Section 3.

Appendix C

We prove here that the volatility implied by the option price (3.37) has a derivative with respect to the strike that is zero at the forward asset price $S_0 e^{\mu T}$. The implied volatility σ is implicitly defined by

$$C^{BS}(S_0, K, T, \mu, \sigma) = \sum_{i=1}^N \lambda_i C^{BS}(S_0, K, T, \mu, \eta_i), \tag{C.1}$$

where η_i is defined by (3.38). Notice, indeed, that the equation (C.1) always has a unique solution $\sigma = \sigma(K)$, where only the parameter K is treated as a variable.

Since $\frac{\partial C^{BS}}{\partial \sigma} \neq 0$, by Dini's theorem

$$\begin{aligned}\frac{d\sigma}{dK}(K) &= \frac{\sum_{i=1}^N \lambda_i \frac{\partial C^{BS}}{\partial K}(S_0, K, T, \mu, \eta_i) - \frac{\partial C^{BS}}{\partial K}(S_0, K, T, \mu, \sigma(K))}{\frac{\partial C^{BS}}{\partial \sigma}(S_0, K, T, \mu, \sigma(K))}\\ &= -\frac{e^{-\mu T}\left[\sum_{i=1}^N \lambda_i \Phi\left(\frac{\ln\frac{S_0}{K} + (\mu - \frac{1}{2}\eta_i^2)T}{\eta_i\sqrt{T}}\right) - \Phi\left(\frac{\ln\frac{S_0}{K} + (\mu - \frac{1}{2}\sigma^2(K))T}{\sigma(K)\sqrt{T}}\right)\right]}{S_0\sqrt{T}\Phi'\left(\frac{\ln\frac{S_0}{K} + (\mu + \frac{1}{2}\sigma^2(K))T}{\sigma(K)\sqrt{T}}\right)},\end{aligned}$$

which, for an ATM-forward strike $\bar{K} = S_0 e^{\mu T}$, becomes

$$\frac{d\sigma}{dK}(\bar{K}) = -\frac{e^{-\mu T}\left[\sum_{i=1}^{N}\lambda_i \Phi\Big(-\frac{1}{2}\eta_i\sqrt{T}\Big) - \Phi\Big(-\frac{1}{2}\sigma(\bar{K})\sqrt{T}\Big)\right]}{S_0\sqrt{T}\Phi'\Big(\frac{1}{2}\sigma(\bar{K})\sqrt{T}\Big)}.$$

Then, we have that $\frac{d\sigma}{dK}(\bar{K}) = 0$, since for $K = \bar{K}$ (C.1) reduces to

$$\Phi\Big(-\frac{1}{2}\sigma(\bar{K})\sqrt{T}\Big) = \sum_{i=1}^{N}\lambda_i\Phi\Big(-\frac{1}{2}\eta_i\sqrt{T}\Big). \tag{C.2}$$

The proof that $K = \bar{K}$ is a minimum point is completed by showing that $\frac{d^2\sigma}{dK^2}(\bar{K}) > 0$. Indeed, we have that

$$\begin{aligned}\frac{d^2\sigma}{dK^2}(\bar{K}) &= \frac{\sum_{i=1}^{N}\lambda_i\frac{\partial^2 C^{BS}}{\partial K^2}(S_0,\bar{K},T,\mu,\eta_i) - \frac{\partial^2 C^{BS}}{\partial K^2}(S_0,\bar{K},T,\mu,\sigma(\bar{K}))}{\frac{\partial C^{BS}}{\partial\sigma}(S_0,\bar{K},T,\mu,\sigma(\bar{K}))}\\ &= \frac{e^{-\mu T}\left[\sum_{i=1}^{N}\lambda_i\frac{\Phi'(-\frac{1}{2}\eta_i\sqrt{T})}{\bar{K}\eta_i\sqrt{T}} - \frac{\Phi'(-\frac{1}{2}\sigma(\bar{K})\sqrt{T})}{\bar{K}\sigma(\bar{K})\sqrt{T}}\right]}{S_0\sqrt{T}\Phi'\Big(\frac{1}{2}\sigma(\bar{K})\sqrt{T}\Big)},\end{aligned}$$

so that $\frac{d^2\sigma}{dK^2}(\bar{K}) > 0$ if and only if

$$\sum_{i=1}^{N}\lambda_i\frac{\Phi'\Big(-\frac{1}{2}\eta_i\sqrt{T}\Big)}{\bar{K}\eta_i\sqrt{T}} > \frac{\Phi'\Big(-\frac{1}{2}\sigma(\bar{K})\sqrt{T}\Big)}{\bar{K}\sigma(\bar{K})\sqrt{T}}. \tag{C.3}$$

Now, setting $x_i := -\frac{1}{2}\eta_i\sqrt{T}$ and $\bar{x} := -\frac{1}{2}\sigma(\bar{K})\sqrt{T}$, (C.2) and (C.3) become

$$\begin{aligned}\Phi(\bar{x}) &= \sum_{i=1}^{N}\lambda_i\Phi(x_i)\\ \frac{\Phi'(\bar{x})}{\bar{x}} &> \sum_{i=1}^{N}\lambda_i\frac{\Phi'(x_i)}{x_i}.\end{aligned}$$

To prove the last inequality, we look for a real constant ρ such that

$$\rho\Big[\Phi(\bar{x}) - \Phi(x_i)\Big] \le \frac{\Phi'(\bar{x})}{\bar{x}} - \frac{\Phi'(x_i)}{x_i}, \quad \forall i = 1,\dots,n. \tag{C.4}$$

Indeed, the existence of such ρ implies that, multiplying both members of (C.4) by λ_i and summing over $i = 1,\dots,n$,

$$0 = \rho\Big[\Phi(\bar{x}) - \sum_{i=1}^{N} \lambda_i \Phi(x_i)\Big] \leq \frac{\Phi'(\bar{x})}{\bar{x}} - \sum_{i=1}^{N} \lambda_i \frac{\Phi'(x_i)}{x_i}, \tag{C.5}$$

with the inequality being actually strict, as we shall prove in the sequel.

Assuming, without lack of generality, that $x_1 < x_2 < \cdots < x_s \leq \bar{x} \leq x_{s+1} < \cdots < x_N < 0$,[11] since Φ is increasing, (C.4) is equivalent to the following system

$$\begin{cases} \rho \leq \dfrac{\frac{\Phi'(\bar{x})}{\bar{x}} - \frac{\Phi'(x_i)}{x_i}}{\Phi(\bar{x}) - \Phi(x_i)}, & \forall i = 1, \ldots, s, \\ \rho \geq \dfrac{\frac{\Phi'(\bar{x})}{\bar{x}} - \frac{\Phi'(x_i)}{x_i}}{\Phi(\bar{x}) - \Phi(x_i)}, & \forall i = s+1, \ldots, N, \end{cases} \tag{C.6}$$

where we understand that if $\bar{x} = x_s$ (resp. $\bar{x} = x_{s+1}$), the s-th (resp. $(s+1)$-th) inequality does not appear in the system, since the corresponding inequality in (C.4) is automatically verified. This system has a solution ρ if and only if

$$\min_{i=1,\ldots,s} \frac{\frac{\Phi'(\bar{x})}{\bar{x}} - \frac{\Phi'(x_i)}{x_i}}{\Phi(\bar{x}) - \Phi(x_i)} \geq \max_{j=s+1,\ldots,N} \frac{\frac{\Phi'(\bar{x})}{\bar{x}} - \frac{\Phi'(x_j)}{x_j}}{\Phi(\bar{x}) - \Phi(x_j)}.$$

But this inequality holds true since the function

$$z \longmapsto \frac{\frac{\Phi'(\bar{x})}{\bar{x}} - \frac{\Phi'(z)}{z}}{\Phi(\bar{x}) - \Phi(z)} \tag{C.7}$$

is strictly decreasing for each $\bar{x} < 0$ and z in $(-\infty, 0)$, as it can be easily proven.

Finally, the inequality in (C.5) is actually strict since if the equality held true, all the inequalities in (C.4) would be equalities as well. Hence

$$\rho = \frac{\frac{\Phi'(\bar{x})}{\bar{x}} - \frac{\Phi'(x_1)}{x_1}}{\Phi(\bar{x}) - \Phi(x_1)} = \frac{\frac{\Phi'(\bar{x})}{\bar{x}} - \frac{\Phi'(x_n)}{x_n}}{\Phi(\bar{x}) - \Phi(x_n)},$$

which is false since $x_1 < \bar{x} < x_n$ and the function (C.7) is strictly decreasing in $(-\infty, 0)$.

Acknowledgements. We are grateful to Aleardo Adotti, head of the Product and Business Development Group at Banca IMI for encouraging us in the prosecution of the most speculative side of research in mathematical finance. We are also grateful to Chris Rogers and Giulio Sartorelli for their suggestions and remarks.

[11] Since Φ is increasing, $\Phi(x_1) < \sum_{i=1}^{N} \lambda_i \Phi(x_i) < \Phi(x_N)$. Hence $x_1 < \bar{x} < x_N$, since Φ^{-1} is increasing too.

References

1. L. Andersen and J. Andreasen: Volatility skews and extensions of the Libor market model. Applied Mathematical Finance **7** (2000) 1–32
2. L. Andersen and R. Brotherton-Ratcliffe: The Equity Option Volatility Smile: An Implicit Finite-Difference Approach. Journal of Computational Finance **1** (1997) 5–38
3. M. Avellaneda, C. Friedman, R. Holmes and D. Samperi: Calibrating Volatility Surfaces via Relative-Entropy Minimization. Preprint. Courant Institute of Mathematical Sciences. New York University (1997)
4. M. Avellaneda, A. Levy and A. Paras: Pricing and Hedging Derivative Securities in Markets with Uncertain Volatilities. Applied Mathematical Finance **2** (1995) 73–88
5. S. Beckers: The Constant Elasticity of Variance Model and its Implications for Option Pricing. The Journal of Finance **35** (1980) 661–673
6. B. M. Bibby and M. Sørensen: A hyperbolic diffusion model for stock prices. Finance and Stochastics **1** (1997) 25–42
7. F. Black and M. Scholes: The Pricing of Options and Corporate Liabilities. Journal of Political Economy **81** (1973) 637–659
8. D. T. Breeden and R. H. Litzenberger: Prices of State-Contingent Claims Implicit in Option Prices. Journal of Business **51** (1978) 621–651
9. D. Brigo and F. Mercurio: Discrete Time vs Continuous Time Stock-price Dynamics and Implications for Option Pricing. Internal Report. Banca IMI (1999). Available on the internet at `http://www.damianobrigo.it` and at `http://www.fabiomercurio.it`.
10. D. Brigo and F. Mercurio: Option pricing impact of alternative continuous time dynamics for discretely observed stock prices. Finance and Stochastics **4**(2) (2000) 147–159
11. D. Brigo and F. Mercurio: A mixed-up smile. Risk, September 2000, 123–126
12. D. Brigo and F. Mercurio: *Interest Rate Models: Theory and Practice.* Springer-Verlag, Heidelberg 2001
13. M. Britten-Jones and A. Neuberger: Option Prices, Implied Price Processes and Stochastic Volatility. Preprint. London Business School (1999)
14. G. Brown and C. Randall: If the Skew Fits. Risk, April 1999, 62–67
15. B. Bhupinder: Implied Risk-Neutral Probability Density Functions from Option Prices: A Central Bank Perspective, in J. Knight and S. Satchell (eds.) *Forecasting Volatility in the Financial Markets*, 137–167. Butterworth Heinemann. Oxford 1998
16. P. Carr, M. Tari and T. Zariphopoulou: Closed Form Option Valuation with Smiles. Preprint. NationsBanc Montgomery Securities (1999)
17. J. Cox: Notes on Option Pricing I: Constant Elasticity of Variance Diffusions. Working paper. Stanford University (1975)
18. J. Cox and S. Ross: The Valuation of Options for Alternative Stochastic Processes. Journal of Financial Economics **3** (1976) 145–166
19. M. A. H. Dempster and D. G. Richard: Pricing Exotic American Options Fitting the Volatility Smile. Preprint. Central for Financial Research. University of Cambridge (1999)
20. E. Derman and I. Kani: Riding on a Smile. Risk, February 1994, 32–39

21. E. Derman and I. Kani: Stochastic Implied Trees: Arbitrage Pricing with Stochastic Term and Strike Structure of Volatility. International Journal of Theoretical and Applied Finance **1** (1998) 61–110
22. B. Dupire: Pricing with a Smile. Risk, January 1994, 18–20
23. B. Dupire: Pricing and Hedging with Smiles, in M. A. H. Dempster and S. R. Pliska (eds.) *Mathematics of Derivative Securities.* Cambridge University Press 1997
24. C. Guo: Option Pricing with Heterogeneous Expectations. The Financial Review **33** (1998) 81–92
25. S. Heston: A Closed Form Solution for Options with Stochastic Volatility with Applications to Bond and Currency Options. Review of Financial Studies **6** (1993) 327–343
26. J. Hull and A. White: The Pricing of Options on Assets with Stochastic Volatilities. Journal of Financial and Quantitative Analysis **3** (1987) 281–300
27. J. C. Jackwerth and M. Rubinstein: Recovering Probability Distributions from Option Prices. Journal of Finance **51** (1996) 1611–1631
28. A. Li: A one Factor Volatility Smile Model with Closed Form Solutions for European Options. European Financial Management **5** (1999) 203–222
29. A. M. Malz: Option-Implied Probability Distributions and Currency Excess Returns. Preprint. Federal Reserve Bank of New York (1997)
30. W. R. Melick and C. P. Thomas: Recovering an Asset's Implied PDF from Option Prices: An Application to Crude Oil During the Gulf Crisis. Journal of Financial and Quantitative Analysis **32** (1997) 91–115
31. C. D. Pirkner, A. S. Weigend and H. Zimmermann: Extracting Risk-Neutral Densities from Option Prices Using Mixture Binomial Trees. Preprint. Stern School of Business (1999)
32. E. Platen: A complete market model. Preprint. School of Finance and Economics. University of Technology Sidney (1999)
33. J. L. Prigent, O. Renault and O. Scaillet: An Autoregressive Conditional Binomial Option Pricing Model. Preprint. Université de Cergy (2000)
34. R. J. Ritchey: Call Option Valuation for Discrete Normal Mixtures. Journal of Financial Research **13** (1990) 285–296
35. L. C. G. Rogers and D. Williams: *Diffusions, Markov Processes and Martingales* Vol. 2. John Wiley & Sons. Chichester 1996
36. M. Rubinstein: Displaced Diffusion Option Pricing. Journal of Finance **38** (1983) 213–217
37. M. Rubinstein: Implied Binomial Trees. Journal of Finance **49** (1994) 771–818
38. D. Shimko: Bounds of Probability. Risk, April 1993, 33–37
39. M. Schroder: Computing the Constant Elasticity of Variance Option Pricing Formula. The Journal of Finance **44** (1989) 211–219
40. R. G. Tompkins: Stock Index Futures Markets: Stochastic Volatility Models and Smiles. To appear in: Journal of Futures Markets, 2000
41. R. G. Tompkins: Fixed Income Futures Markets: Stochastic Volatility Models and Smiles. Preprint. Department of Finance. Vienna University of Technology (2000)

The Theory of Good-Deal Pricing in Financial Markets

Aleš Černý and Stewart Hodges

Imperial College Management School
Financial Options Research Centre, University of Warwick

Introduction

The term 'no-good-deal pricing' in this paper encompasses pricing techniques based on the absence of attractive investment opportunities – good deals – in equilibrium. We borrowed the term from [8] who pioneered the calculation of price bands conditional on the absence of high Sharpe Ratios. Alternative methodologies for calculating tighter-than-no-arbitrage price bounds have been suggested by [4], [6], [12]. The theory presented here shows that any of these techniques can be seen as a generalization of no-arbitrage pricing. The common structure is provided by the Extension and Pricing Theorems, already well known from no-arbitrage pricing, see [15]. We derive these theorems in no-good-deal framework and establish general properties of no-good-deal prices. These abstract results are then applied to no-good-deal bounds determined by von Neumann-Morgenstern preferences in a finite state model[1]. One important result is that *no-good-deal bounds generated by an unbounded utility function are always strictly tighter than the no-arbitrage bounds.* The same is *not true for bounded utility functions.* For smooth utility functions we show that one will obtain the no-arbitrage and the representative agent equilibrium as the two opposite ends of a spectrum of no-good-deal equilibrium restrictions indexed by the maximum attainable certainty equivalent gains.

A sizeable part of finance theory is concerned with the valuation of risky income streams. In many cases this valuation is performed against the backdrop of a frictionless market of basis assets. Whenever the payoff of the focus asset can be synthesized from the payoffs of basis assets the value of the focus asset is uniquely determined and this valuation process is preference-free – any other price of the focus asset would lead to an arbitrage opportunity. In reality, however, the perfect replication is an unattainable ideal, partly due to market frictions and partly due to genuine sources of unhedgeable risk presenting themselves, for example, as stochastic volatility. When perfect replication

[1] Each of the no-good-deal restrictions mentioned above is in fact derived from a utility function: for Bernardo and Ledoit it is the Domar-Musgrave utility, for [8] it is the truncated quadratic utility, and for [12] it is the negative exponential utility, see [6].

is not possible, a situation synonymous with an 'incomplete market', the standard Black-Scholes pricing methodology fails because the price of the focus asset is no longer unique.

One way to overcome this difficulty is to single out one price of the focus asset consistent with the price of basis assets. This can be achieved via the representative agent equilibrium, where the 'special' pricing functional is obtained from the marginal utility of the optimized representative agent's consumption, see [20].

A valid objection against the representative agent equilibrium is that it imposes very strong assumptions about the way the equilibrium is generated. Alternative route is to look for preference-free price bounds, in the spirit of [16], which leads to the calculation of super-replication bounds [2]. However, these bounds have a practical shortcoming in that they tend to be rather wide and hence not very informative.

Recently a new approach has emerged whereby it is accepted that the price of a non-redundant contingent claim is not unique, but an attempt is made to render the price bound more informative by restricting equilibrium outcomes beyond no arbitrage. Typically, one tries to hedge the focus asset with a self-financed portfolio of basis assets to maximize a given 'reward for risk' measure and rules out those focus asset prices that lead to a highly desirable hedging strategy. Such a procedure gives a price interval for every contingent claim where the interval is the wider the more attractive investment one allows to exist in equilibrium.

The idea of good deals as an analogy of arbitrage comes naturally at this point. Recall that arbitrage is an opportunity to purchase an unambiguously positive claim, that is a claim that pays strictly positive amount in some states and non-negative amounts in all other states, at no cost. While the absence of arbitrage is surely a necessary condition for the existence of a market equilibrium, it is still a rather weak requirement. Considering a claim with zero price that either earns \$1000 or loses \$1 with equal probability, one feels that, although not an arbitrage, such investment opportunity still should not exist in equilibrium. One can then define 'approximate' arbitrage, or as we say here 'good deal', as an opportunity to buy a *desirable claim* at no cost.

Historically, good deals have been associated with high Sharpe Ratios. The Arbitrage Pricing Theory of [18] is a prime example of ruling out high Sharpe Ratios. Further breakthrough came with the work of [11] who established a duality link equating the maximum Sharpe Ratio available in the market and the minimum volatility of discount factors consistent with all prices. While [11] use this result to construct an empirical lower bound on discount factor volatility, [8] realize that it can be used in the opposite direction, namely to limit the discount factor volatility and thus to infer the no-good-deal prices conditional on the absence of high Sharpe Ratios.

[2] See [17] for a one-period finite state setting and [10] for a continuous time model.

It is well known that outside the elliptic world the absence of high Sharpe Ratios does not generally imply the absence of arbitrage. Other researchers therefore tried to come up with reward for risk measures that would automatically capture all arbitrage opportunities. [4] base the definition of good deals on the gain-loss ratio and [12] uses a generalized Sharpe Ratio derived from the negative exponential utility function. [6] calculates the Hansen-Jagannathan duality link for good deals defined by an arbitrary smooth utility function and proposes a reward for risk measure generated by the CRRA utility class.

In this paper we point out that, regardless of the specific definition of good deals, the nature of the duality restrictions is formalized in the *extension theorem*, already well known from the no-arbitrage theory[3]. The extension theorem states that any incomplete market without good deals can be augmented by adding new securities in such a way that the resulting complete market has no good deals. The important point is that the set of complete market state prices which do not allow good deals is independent both of the basis and the focus assets. The pricing theorem uses the above fact to assert that any no-good-deal price of a focus asset must be supported by a complete market no-good-deal pricing functional. These results are crucial both for establishing the theoretical properties of no-good-deal prices, which will be discussed here, and for practical applications, see for example [6].

The paper is organized as follows: The first section reviews the essentials of no-arbitrage theory and builds the no-good-deal theory in analogous way. The second section derives abstract versions of the Extension Theorem and the Pricing Theorem in no-good-deal framework. Section three applies the theory to desirable claims defined by von Neumann-Morgenstern preferences in finite state space. The results of this section are summarized in the no-good-deal pricing theorem of section four, where we also discuss the similarities and differences between the finite and infinite state space. The fifth section gives a geometric illustration of the theory by taking an example from the literature – desirable claims determined by Sharpe ratio as in [8]. Finally, section six concludes.

1 Arbitrage and Good Deals

In this section we briefly describe the axiomatic theory of no-arbitrage pricing[4] and show how it can be analogously used to define good deals and no-good-deal prices. The model of security market is an abstract one, the application to a multiperiod security market is spelled out in Clark and in the references mentioned in the introduction. One should bear in mind that in this section

[3] Incidentally, it is Ross again (we already mentioned his APT contribution) who has introduced the extension theorem to finance in his 1978 paper on the valuation of risky streams. The extension theorem in no-arbitrage setting has been studied extensively in the realm of mathematical finance, starting with [15].

[4] This section is based on [7].

the origin 0 is not to be taken literally as a position with zero wealth, rather it is the position relative to an initial endowment.

We will have a topological vector space X of all contingent claims. The space of all continuous linear forms on X (strong dual) is denoted X^*. The vector space X will be endowed with a natural ordering $\geq$ which defines the *positive cone*[5] $X_+ \equiv \{x \in X : x \geq 0\}$. The cone of *strictly positive claims* is

$$X_{++} \equiv \{x \in X : x > 0\} = X_+ \backslash \{0\}.$$

Suppose we have a collection of claims with predetermined prices, so called *basis assets.* These claims generate the marketed subspace M and their prices define a price correspondence p on this subspace. The cone of strictly positive claims has the following role:

Definition 1.1. A strictly positive claim with zero or negative price is called *arbitrage.*

Definition 1.2. Let M be a linear subspace of X. A linear functional $p : M \to \mathbb{R}$ is *positive* if $p(m) \geq 0$ for all $m \in M \cap X_+$. We say that p is *strictly positive* if $p(m) > 0$ for all $m \in M \cap X_{++}$.

Standing assumption 1 There is a strictly positive marketed claim.

Clark shows that under this assumption no arbitrage implies that the price correspondence p is in fact a strictly positive linear functional. This result guarantees, among others, unique price for each marketed claim.

Now we move on to define generalized arbitrage opportunities – good deals. Suppose we have a convex set K disjoint from the origin which we interpret as the set of all desirable claims. At the moment we do not specify how the set of desirable claims is obtained or what are its additional properties. The relationship between arbitrage and strictly positive claims is generalized as follows:

Definition 1.3. A desirable claim with zero or negative price is called a *good deal.*

Frictionless trading leads to the following definition:

[5] In no-arbitrage pricing one works with natural (canonical) ordering. Thus, for example, positive cone in $\mathbb{R}^n$ is formed by n-tuples with non-negative coordinates, positive cone in L^p by non-negative random variables etc. Note that a claim $(1, 0, 0) \in \mathbb{R}^3$ is strictly positive in the canonical ordering on $\mathbb{R}^3$ but at the same time it is equal to zero with positive probability, therefore 'strictly positive in X' is not to be confused with 'strictly positive with probability one'. The term 'strictly positive' will only be used when we have in mind the canonical ordering on X.

Definition 1.4. A claim is *virtually desirable* if some positive scalar multiple of it is desirable. The set of all virtually desirable claims is denoted C_{++},

$$C_{++} = \bigcup_{\lambda>0} \lambda K$$

A virtually desirable claim at zero or negative price constitutes a *virtually good deal*. When markets are frictionless the presence of a virtually good deal implies the existence of a good deal, simply by re-scaling the portfolio which gives the virtually good deal. Thus the absence of good deals implies absence of virtually good deals and vice versa.

Proposition 1.5. *There are no good deals if and only if there are no virtually good deals.*

Geometrically the set of all virtually desirable claims is the convex cone with vertex at 0 generated as a convex hull of 0 and the set of desirable claims K.

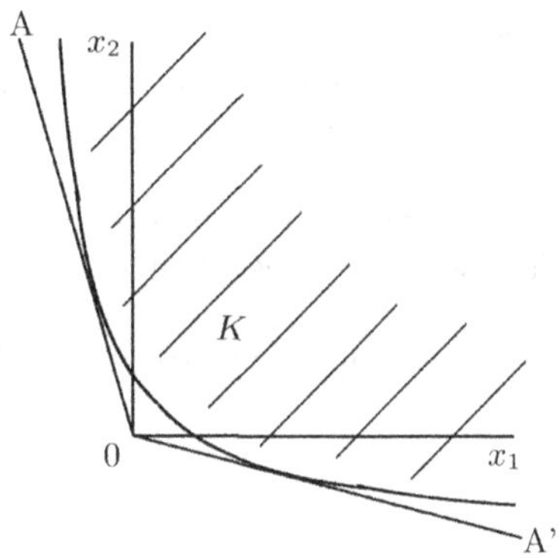

Fig. 1. The cone of virtually good deals C (AOA') generated by the set of good deals K

To benefit fully from the analogy between arbitrage and good deals we have to realize that, similarly to X_+, the cone $C \equiv C_{++} \cup \{0\}$ defines ordering on the space of all contingent claims by putting $x_1 \succeq x_2$ when $x_1 - x_2 \in C$ and $x_1 \succ x_2$ when $x_1 - x_2 \in C_{++}$. Similarly as in Definition 1.2 we can speak of C-positive functionals and C-strictly positive functionals. The key point is that the link between no arbitrage and strictly positive pricing rule carries over to good deals.

Theorem 1.6. *Suppose that there is a (virtually) desirable marketed claim and the price correspondence p on the marketed subspace M gives no good deals. Then* $p : M \to \mathbb{R}$ *is a C-strictly positive linear functional, i.e.* $p(m)$ *is*

unique for all $m \in M$, $p(m_1 + m_2) = p(m_1) + p(m_2)$ for all $m_1, m_2 \in M$ and p assigns strictly positive price to all (virtually) desirable marketed claims.

Proof. The proof follows from the proof of Theorem 1 in [7] when X_+ is replaced with C, or equivalently $>$ with $\succ$.

Since we are guaranteed that p is a linear functional we can define a subspace $M_0(p)$ of all claims with zero price which plays essential role in the extension theorem.

2 Extension Theorem

2.1 The Idea

The extension theorem states that an incomplete market without good deals can be augmented by adding new securities in such a way that the resulting complete market has no good deals. The important point is that the set of complete market state prices which provide no good deals is independent both of the basis and the focus assets present in the market. The pricing theorem uses the above fact to provide a complete characterization of the no-good-deal price region since any no-good-deal price of a focus asset must be supported by a complete market no-good-deal pricing functional.

The pricing function on the marketed subspace defines a yet smaller subspace of marketed claims with zero price, denoted $M_0(p)$.

Definition 2.1. For a given strictly positive pricing functional p on M we say that

$$M_0(p) = \{m \in M : p(m) = 0\}$$

is a *zero investment marketed subspace*[6].

In the absence of good deals this subspace must be disjoint from the set of good deals K,

$$M_0(p) \cap K = \emptyset.$$

As the figure 2 suggests it is quite natural to expect that if $M_0(p)$ is disjoint from K then there is a hyperplane H containing $M_0(p)$ and still disjoint from K. The separating hyperplane H is interpreted as the zero investment subspace of the completed market. The fact that H is disjoint from K guarantees that there are no good deals in the completed market.

[6] The term 'zero investment portfolio' was introduced by [13], alternatively one could use the term 'zero cost marketed subspace'.

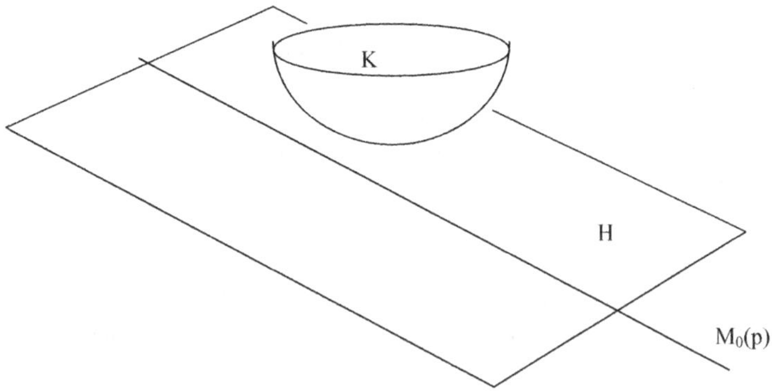

Fig. 2. Illustration to the Extension Theorem

2.2 Technicalities

Mathematicians distinguish among three types of separation of two convex sets. Weak separation means the separating hyperplane may touch both sets. Strict separation signifies that the separating hyperplane does not touch either of the convex bodies but can come arbitrarily close to each of them. Strong separation occurs when there is a uniform gap between the separating hyperplane and both of the convex sets. It is hard to find references to semistrict separation, which is what we need here, because the separating hyperplane will touch $M_0(p)$ but we would like it to be disjoint from K.

By drawing pictures in $\mathbb{R}^2$ one is tempted to conjecture that semistrict separation is always possible in finite dimension. However, this conjecture is false, as a three dimensional example in Lemma A.1 shows. Thus the situation in finite dimension is quite clear: K and $M_0(p)$ can always be weakly separated, and it follows from Lemma A.1 that in general one cannot expect more. When K is closed and bounded the two sets can be strongly separated, and when K is open the two sets can be separated semistrictly, see for example [2].

In infinite dimension not even weak separation is available automatically, for a nice counterexample see [21]. Weak separation is available when K has non-empty interior and semistrict separation is possible when K is open. Strong separation becomes possible when K is compact. These facts motivate the following definition.

Definition 2.2. We say that the set of desirable claims K is *boundedly generated* if there is a closed bounded set $B \subset K$ such that any desirable claim in K can be obtained as a scalar multiple of a desirable claim in B.

For boundedly generated sets of desirable claims we obtain a clear-cut result both for the extension and pricing theorem thanks to weak compactness of bounded sets in standard probability spaces.

Theorem 2.3 (Extension Theorem). *Suppose* $X = L^q(\Omega, \mathcal{F}, P), 1 < q < +\infty$, *the set of good deals* K *is closed and boundedly generated and the zero investment marketed subspace is closed. Then there is a* C*-strictly positive continuous linear extension of the pricing rule* p *on the marketed subspace to the whole market if and only if there is no good deal.*

Proof. See Appendix A.

Definition 2.4. Suppose we fix a set of desirable claims K with the implied cone C of virtually desirable claims. A continuous C-strictly positive functional on X is called a *no-good-deal pricing functional*[7]. The set of all such functionals is denoted C^*_{++}

$$C^*_{++} = \{\varphi \in X^* : \varphi(x) > 0 \text{ for all } x \in K\}.$$

Making use of the Extension Theorem we can completely characterize the no-good-deal price region for several focus assets jointly[8].

Theorem 2.5 (Pricing Theorem). *Suppose* X *is an* L^p *space ,* $1 < p < +\infty$, *and the set of desirable claims is closed and boundedly generated. Let us have a closed marketed subspace* M *in which prices are given by a* C*-strictly positive and continuous linear functional* ϕ*. Let there be further* m *focus assets with payoffs* $y_1, y_2, \dots, y_m$*, no-good-deal price of which we want to find. Then*

a) the no good deal price region P *for these claims is given as*

$$P = \left\{(\varphi(y_1), \dots, \varphi(y_m)) \in \mathbb{R}^m : \varphi \in C^*_{++} \text{ and } \varphi|M = \phi\right\},$$

where $\varphi|M$ *is the restriction of* φ *to* M,

b) P *is a convex set in* $\mathbb{R}^m$,

c) defining $N \equiv \operatorname{span}(M, y_1, y_2, \dots, y_m)$ *the dimension of the price region* P *satisfies*

$$\dim P = \operatorname{codim}_N M \tag{1}$$

which is the codimension of the marketed subspace in the enlarged marketed subspace N.

d) the no-good-deal price of y_i *is unique if and only if* y_i *is redundant, that is* $y_i \in M$

e) let K_1 *and* K_2 *be two boundedly generated sets of desirable claims,* $K_1 + \varepsilon B_1 \subset K_2$ *for some* $\varepsilon > 0$*, where* B_1 *is a unit ball in* X *in strong topology. Let* P_1 *and* P_2 *be the corresponding no-good-deal price regions. Then*

$$\mathrm{cl} P_2 \subset \mathrm{rel} - \mathrm{int} P_1 .$$

Proof. See Appendix A

[7] In practical applications it is unusual to work with abstract linear functionals. For different representations of complete market pricing rules see e.g. [9], page 104.

[8] For discussion of these results see Theorem 4.1.

3 Desirable Claims and Agent Preferences

The results derived so far were concerned with an abstract set of good deals. In this section we will show how desirable claims can be determined by agent preferences, in particular by expected utility, and examine when good deals defined in this way include all arbitrage opportunities. This approach allows to formulate a whole *range of equilibrium restrictions* as we choose K smaller or larger. We discuss two limiting cases of no-good-deal pricing – the no-arbitrage pricing and representative agent equilibrium. We only have a complete answer for X finite dimensional, so we stick to this case from the beginning, leaving the technical issues related to infinite dimension to section 4.2.

Consider a preference relation $\succ^*$ which is a) convex, in the sense that the level set $\{x \in X : x \succeq^* y\}$ is convex for all $y \in X$; b) X_{++} strictly increasing, i.e. $x - y \in X_{++}$ implies $x \succ^* y$; c) continuous, i.e. both sets $\{x \in X : x \prec^* y\}, \{x \in X : x \succ^* y\}$ are open.

For any strictly increasing utility function U the preference relation

$$x \succ^* y \Leftrightarrow \mathrm{E}[U(x + w_r)] > \mathrm{E}[U(y + w_r)] \tag{2}$$

satisfies the conditions a), b) and c). The reference point w_r is very often taken as wealth resulting from the risk-free investment, with x and y being excess returns. The analysis remains valid, however, even when reference wealth level w_r is stochastic. We may want to think of w_r, for example, as the representative agent's optimal wealth derived from investing into basis assets only.

Let $\mathbf{1}$ be a claim that pays 1 unit of the numeraire in each state of the world. Let us take a non-negative number a and define $K(a)$ as the upper level set

$$K(a) = \{x \in X : x \succeq^* a\mathbf{1}\}.$$

Thus we obtain a family of sets of desirable claims indexed by the desirability level a[9]. The quantity a is interpreted as the certainty equivalent gain over and above the reference wealth level w_r. Monotonic transformations of a define various, but in essence equivalent, reward for risk measures[10].

[9] This works well for smooth utility functions. Bernardo and Ledoit use Domar-Musgrave (piecewise linear) utility function which gives $K(a) = K(b)$ for all $a > 0, b > 0$. The widening of the set of desirable claims is not achieved by changing the parameter a but rather by changing the shape of the utility function, that is by varying the gain-loss ratio – the ratio of slopes of the two linear parts of the function.

[10] Not to be confused with 'coherent risk measures', of [1]. As noted in [12], the lower good-deal bound is a coherent risk measure in the sense of [1], whereby the set of 'acceptable risks' is identified with the set of desirable claims. See also [14] and [6].

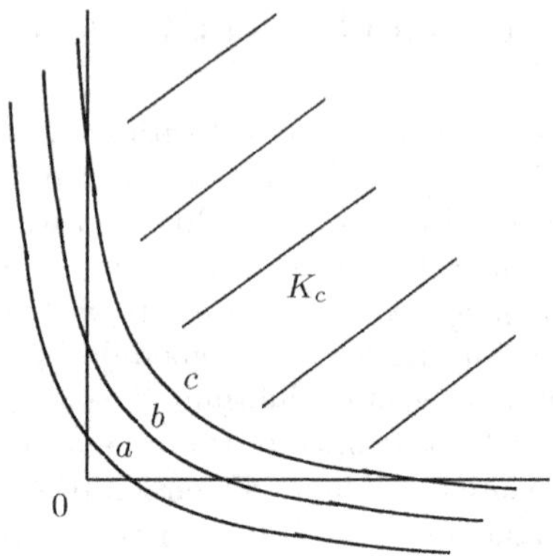

Fig. 3. Sets of desirable claims $K_a \supset K_b \supset K_c$ indexed by desirability levels $a < b < c$

Note that if a claim x is desirable then all claims $x + X_+$ are desirable too, which is a natural property that all 'good' sets of desirable claims should satisfy.

The key question is, whether, or under what assumptions, the set of desirable claims is boundedly generated. First, let us discuss situations when it is not.

Definition 3.1. The set $K(a)$ has an *asymptote* $x \in X$ if $\{\lambda x | \lambda \in \mathbb{R}\} \cap K(a) = \emptyset$ and for any $\varepsilon > 0$ $\{\lambda x | \lambda \in \mathbb{R}\} \cap K(a - \varepsilon) \neq \emptyset$.

Clearly, unless $K(a)$ is asymptote-free one cannot hope, in general, that it will be boundedly generated. With this observation in mind we proceed to examine sets of desirable claims generated by Von Neumann-Morgenstern preferences.

3.1 Arbitrage Subsumed by Good Deals

In order for no-good-deal pricing to be economically meaningful the absence of good deals must imply the absence of arbitrage. For this to be true each strictly positive claim must be virtually desirable, mathematically $C_{++}(a) \supseteq X_{++}$. In general not all arbitrage opportunities will be covered by virtually good deals. This leads us to the following definition:

Definition 3.2. We say that the preference relation $\succeq^*$is *arbitrage-sensitive* if and only if for any desirability level a and any strictly positive[11] claim x a sufficiently large scalar multiple of x is preferred to the claim $a\mathbf{1}$.

[11] Note again that strictly positive does not mean strictly positive with probability one, but rather non-negative and different from zero. See also the footnote in section 1.

In other words arbitrage sensitivity requires that a sufficiently high position in any arbitrage opportunity gives an (arbitrarily) good deal. A simple example of strictly increasing preferences that do not satisfy this requirement is given below.

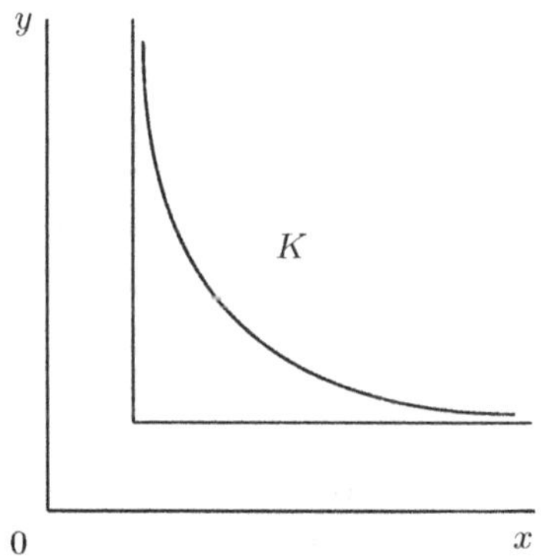

Fig. 4. Set of desirable claims K such axes x and y are not virtually desirable

The indifference curve has two asymptotes, one vertical and one horizontal. In such a case the set of virtually good deals will contain the interior of the positive quadrant but not the axes x and y.

In the case of von Neumann-Morgenstern preferences the arbitrage sensitivity condition is met by unbounded utility functions (Lemma B.3), but it is *violated by all bounded utility functions*, because strictly positive claims which pay nothing with sufficiently high probability do not constitute virtually desirable claims, see Lemma B.2. At the same time the 'inside' of positive orthant (that is all claims which are strictly positive with probability 1) is virtually desirable. Thus to prevent arbitrage opportunities one must take

$$C_+ = \mathrm{cl} \bigcup_{\lambda>0} \lambda K$$

instead of

$$C_+ = \bigcup_{\lambda>0} \lambda K.$$

Although this strengthening of no-good-deal equilibrium is purely cosmetic from the practical point of view, it highlights a different problem. Since one is not guaranteed that $C_+ \supsetneqq X_+$, the no-good-deal price bounds generated by bounded utility functions are not necessarily tighter than the no-arbitrage bounds. This problem is pointed out in [4] and it is present equally in Sharpe Ratio restrictions of [8] and generalized Sharpe Ratio analysis of [12].

3.2 Arbitrage as a Limiting Case of Good Deals

Suppose now that the preferences are arbitrage sensitive, i.e. $C_{++}(a) \supseteq X_{++}$ for all $a \in \mathbb{R}$. At the same time the sets $C_{++}(a)$ become progressively smaller as the desirability level a increases. It is interesting to see under what conditions good deals reduce to arbitrage in the limit, that is under what circumstances do we have

$$\bigcap_{a \geq 0} C_{++}(a) = X_{++}.$$

Definition 3.3. We say that the preference relation $\succeq^*$ is *downside-sensitive* if each ray generated by a non-positive claim is dominated by a claim $a\mathbf{1}$ where a is a sufficiently large positive number.

As an immediate consequence we have

Proposition 3.4. *For preferences which are arbitrage-sensitive no arbitrage is a limiting case of no-good-deal equilibria as $a \to \infty$ if and only if the preference relation is downside-sensitive.*

For von Neumann-Morgenstern preferences to be downside-sensitive the generating utility function must discount negative outcomes sufficiently heavily,

$$\lim_{x \to -\infty} \frac{x}{U(x)} = 0 \tag{3}$$

as demonstrated in Lemma B.4. This condition is satisfied by all frequently used utility functions, except for the Domar-Musgrave utility, see footnote 9.

Crucially, by virtue of Lemma B.1 the downside sensitivity property (3) guarantees that the sets of desirable claims are boundedly generated.

Theorem 3.5. *For any unbounded utility function satisfying*

$$\lim_{x \to -\infty} \frac{x}{U(x)} = 0$$

the set of desirable claims $K(a)$ is boundedly generated for any $a \in \mathbb{R}$.

Proof. See Appendix B.

3.3 Representative Agent Equilibrium as a Limiting Case

Suppose for simplicity that the reference wealth level is $w_r = 0$. As $a \to 0$ the cone of virtually good deals is getting wider and eventually becomes a hyperplane provided that the indifference surface is sufficiently smooth. At the same time the cone of complete market state prices becomes narrower until it finally collapses into the gradient of indifference surfaces.

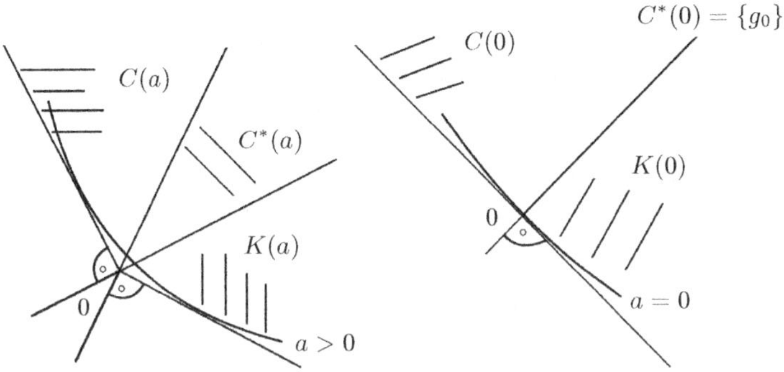

Fig. 5. Set of desirable claims $K(a)$ for $a \to 0$

In the presence of basis assets we first find the market portfolio w_M that achieves the maximum certainty equivalent gain a_M. It is clear that if we add more assets then the attainable certainty equivalent gain will be at least a_M. The condition $a \leq a_M$ is now equivalent to requiring that the new asset does not shift the efficient frontier

$$\mathrm{E}[U(Z + w_M)] - \mathrm{E}[U(w_M)] \leq 0, \tag{4}$$

where Z is the excess return of the new asset with respect to the market portfolio. For Z sufficiently small and U sufficiently smooth we have $\mathrm{E}[U(Z + w_M)] - \mathrm{E}[U(w_M)] \approx \mathrm{E}[U'(w_M)Z] = 0$, the last equality being the consequence of the no-good-deal condition (4). This implies that the new claim must be priced with the change of measure proportional to the marginal utility of the representative agent. When there are no basis assets this amounts to risk-neutral pricing[12] because $U'(R^f w_0)$ is constant. We formalize this intuition in the following section.

4 No-Good-Deal Pricing Theorem for Utility-Based Equilibrium Restrictions

4.1 Finite State Space

In a finite state model with equilibrium restrictions generated by an unbounded utility function one obtains a clear-cut characterization of no-good-deal price bounds.

Theorem 4.1. *Suppose* $\dim X < +\infty$. *Let us have a downnside-sensitive unbounded (arbitrage-sensitive) utility function* U, *which is once differentiable*

[12] Not to be confused with the pricing under risk-neutral probabilities. Here we mean the risk-neutral valuation under objective probabilities, one which is frequently used in macroeconomics.

and strictly concave. Denote the set of all claims with desirability level a or higher as $K(a)$. *Let there be a marketed subspace* M *in which there is no arbitrage and denote* M_0 *the set of marketed claims with zero price. Assume that there is a risk-free security with return* R^f. *Let there be further* m *focus assets with payoffs* $y_1, y_2, \dots, y_m$, *and let us denote* $P(a) \subset \mathbb{R}^m$ *the region of prices of the focus assets such that no claim in the extended market has desirability level exceeding* a. *Then*

a)

$$\sup_{w \in M_0} U(w) \equiv a_M < +\infty$$

is achieved in M_0. *Let us denote the unique argmax* w_M *– the market portfolio.*

b) $P(a)$ *is empty for* $a < a_M$ *and it is a non-empty convex set for* $a \geq a_M$,

c) defining $N \equiv \text{span}(M, y_1, y_2, \dots, y_m)$ *the dimension of the price region* $P(a)$ *satisfies*

$$\dim P(a) = \text{codim}_N M \tag{5}$$

for $a > a_M$, *whereas* $P(a_M)$ *is a singleton*

$$P(a_M) = \{\left(\frac{\mathrm{E}[m_M y_1]}{R^f}, \frac{\mathrm{E}[m_M y_2]}{R^f}, \dots, \frac{\mathrm{E}[m_M y_m]}{R^f}\right)\} \tag{6}$$

with $m_M = \frac{U'(w_M)}{\mathrm{E}[U'(w_M)]}$

d) if $\text{codim}_N M > 0$, *that is if at least one focus asset is non-redundant, then for all* a *and* b *such that* $a_M \leq a < b$

$$\text{cl}P(a) \subset \text{rel} - \text{int}P(b),$$

that is for $a < b$ *the no-good-deal price region* $P(a)$ *is strictly smaller than* $P(b)$

e) denoting P_{NA} *the no-arbitrage price region for the focus assets we have*

$$\lim_{a \to +\infty} P(a) = P_{NA}$$

that is $\cup_{a \in \mathbb{R}} P(a) = P_{NA}$.

f) if $\text{codim}_N M > 0$ *then for any desirability level* a *the no-good-deal price region is strictly smaller than the no-arbitrage price region.*

Remark 4.2. 1. One needs an unbounded utility function to make sure that good deals include all arbitrage opportunities. A bounded utility function leaves out strictly positive claims which are equal to zero with sufficiently high probability. For the same reason the set of good deals defined by a bounded utility function need not be asymptote-free, in which case one cannot expect property d) to hold.

2. The condition

$$\lim_{x \to -\infty} \frac{x}{U(x)} = 0$$

is necessary (and sufficient) for no-good-deal restrictions to reduce to no-arbitrage restrictions in the limit. For unbounded utility functions this condition implies that the sets of good deals are asymptote-free. In finite dimension asymptote-free set is always boundedly generated (see the proof of Theorem 4.1).
3. Existence of a risk-free security simplifies the pricing formula (6) but this assumption is not necessary. It suffices to have a marketed claim x_0 which is strictly positive, in fact desirable would suffice, see the proof of Theorem 2. The pricing formula has to be adjusted accordingly, replacing R^f with $\frac{\mathrm{E}[U'(w_M)\frac{x_0}{p_0}]}{\mathrm{E}[U'(w_M)]}$, where $p_0 > 0$ is the price of x_0.
4. Smoothness of U is necessary (and sufficient) to obtain the singleton property of $P(a_M)$. Strict concavity is sufficient but not necessary, in addition it implies uniqueness of the market portfolio w_M. The smoothness assumption is relaxed in [3], whose results imply that in general $P(a_M)$ is non-empty but not necessarily a singleton when $X = L^\infty$.

4.2 Infinite State Space

We do not know how to rephrase Theorem 5 in an infinitely dimensional state space. Let us at least summarise some of the important differences that make the problem in infinite dimension harder and more interesting.

1. **Continuity:** Expected utility in finite dimension is automatically continuous. In infinite dimension continuity is determined by the left tail of the utility function and the topology. For example, for a utility function with

$$\lim_{x \to -\infty} \frac{U'(x)}{|x|^{\delta-1}} = const$$

the expected utility is continuous in $L^p, p \geq \delta \geq 1$. With continuous expected utility Extension Theorem is available via Hahn-Banach theorem. Similarly, with U defined on the whole real line the expected utility is continuous in L^∞, this fact is used in [3].
2. **No arbitrage vs. bounded attractivness:** In finite dimension no arbitrage implies that maximum certainty equivalent gain attainable in the marketed subspace is always finite. In infinite dimension this is no longer true, one can have no arbitrage but yet the attractiveness of self-financed investment opportunities may be unbounded. Consider, for example, a complete market where the state of the world is determined by the random variable X with $\chi^2(1)$ distribution. Assume that the risk-free

rate is 0 and suppose the state prices are given by the following state price functional (change of measure)

$$m(X) = const\frac{e^{\frac{X}{2}}}{1+X}.$$

Since $\mathrm{E}\frac{e^{\frac{X}{2}}}{1+X}$ is finite the constant above can be set to satisfy

$$\mathrm{E}[m] = 1.$$

It is known, see for example [6], that the certainty equivalent gain for the negative exponential utility in a complete market is

$$a_M = \mathrm{E}[m \ln m].$$

However in this case $a_M = \infty$ as the integral

$$\mathrm{E}[m \ln m] = \int_0^\infty \frac{e^{\frac{x}{2}}}{1+x}\left[x - \ln(1+x)\right]\frac{e^{-\frac{x}{2}}}{\sqrt{x}}dx$$

diverges at the upper bound. As we pointed out above, expected utility is continuous in this case.

3. **Asymptotes:** As in finite dimension, asymptotes can only be strictly positive, and such asymptotes can exist only when utility is bounded. For unbounded utility satisfying $\lim_{x\to-\infty}\frac{x}{U(x)} = 0$ any set of desirable claims is asymptote-free. However, unlike in finite dimension, this does not imply that the set of desirable claims is boundedly generated. One can easily see this by examining a sequence of strictly positive rays which have non-zero payoff with increasingly smaller probabilities. In other words, the positive cone in infinite dimension is not boundedly generated by von Neumann-Morgenstern preferences. This is really caused by the upper tail of the utility function, thus it has nothing to do with continuity of preferences.

5 Geometric Illustration – Sharpe Ratio Restrictions

The simplest illustration of the duality between the set of desirable claims and the set of no-good-deal complete market pricing functionals comes from the mean-variance framework. The term 'good deal' was introduced by [8] in a specific situation where desirable claims are those with high Sharpe ratio of the excess return. This particular application of no-good-deal equilibrium provides a very nice geometric illustration of the theory developed in sections 1 and 2.

It is convenient to have $X \equiv L^2$. Denoting h the bound on Sharpe ratios which are acceptable in equilibrium the set of desirable claims is given as

$$K(h) = \{x \in X : \frac{\mathrm{E}[x]}{\sqrt{\mathrm{E}[x^2] - (\mathrm{E}[x])^2}} \geq h\}.$$

We note that the cone of virtually desirable claims $C_{++}(h)$ is identical to $K(h)$ and it can be rewritten more conveniently as

$$C(h) = K(h) = \{x \in X : \frac{\mathrm{E}[x]}{||x||} \geq \frac{h}{\sqrt{1+h^2}}\},$$

where $||\cdot||$ is the L^2 norm. The geometry of the cone of desirable claims is simple – it is a circular cone with the axis formed by vector $\mathbf{1} \in L^2$ and the angle at the vertex is α, such that $\cos\alpha = \frac{h}{\sqrt{1+h^2}}$, and consequently $\cot\alpha = h$, see Figure 2.

Recall that no-good-deal price functionals φ must satisfy $\varphi(K) > 0$ and that each continuous linear functional φ on L^2 is uniquely represented by a random variable $m \in L^2$ as follows

$$\varphi(x) = \mathrm{E}[mx].$$

Thus the cone of no-good-deal pricing rules can be identified with the cone of discount factors

$$\tilde{C}^*_{++}(h) = \{m \in L^2 : \mathrm{E}[mx] > 0 \text{ for all } x \in C_{++}(h)\}.$$

Note that every no-good-deal discount factor must be *at sharp angle* with every desirable claim. But since the shape of $C_{++}(h)$ is very simple we can characterize $\tilde{C}^*_{++}(h)$ explicitly.

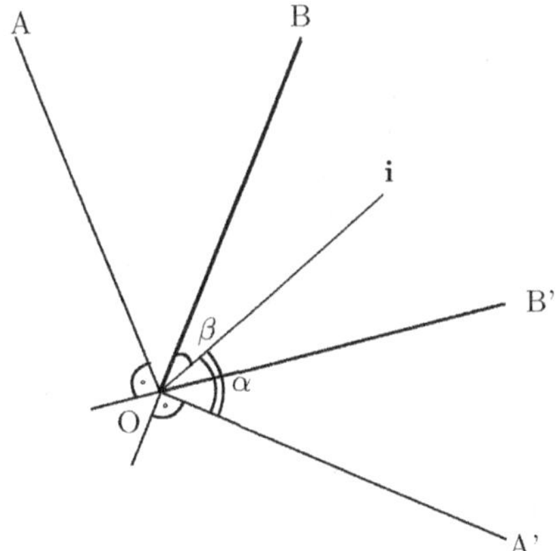

Fig. 6. Cone of good deals K (AOA') and the cone of discount factors $\tilde{C}^*$ (BOB') determined by maximum attainable Sharpe ratio $h = \cot\alpha$

As the picture shows the cone of no-good-deal discount factors is again a circular cone with the axis $\mathbf{1} \in L^2$ and with the angle at the vertex $\beta = \frac{\pi}{2} - \alpha$, that is

$$\tilde{C}^*(h) = \{m \in L^2 : \frac{\mathrm{E}[m]}{\sqrt{\mathrm{E}[m^2] - (\mathrm{E}[m])^2}} > \cot(\frac{\pi}{2} - \alpha) = \tan\alpha = \frac{1}{\cot\alpha} = \frac{1}{h}\}.$$

In other words any discount factor $m \in L^2$ that prevents Sharpe Ratios higher than h must satisfy $\frac{\mathrm{E}[m]}{\mathrm{V}(m)} \geq \frac{1}{h}$ which is the condition obtained by [11].

5.1 Preventing Arbitrage

The above relationship describes the duality between $C_{++}(h)$ and $C^*_{++}(h)$ but it does not guarantee that the functionals in $C^*(h)$ are strictly positive. To fix this problem one has to rule out both high Sharpe ratios and all arbitrage opportunities. However, one cannot take $C_{++}(h) \cup X_{++}$ as the set of desirable claims because this set is not convex and the extension property would be immediately lost. [8] therefore take convex hull of $C_{++}(h) \cup X_{++}$ as the set of desirable claims, which means they are ruling out not only high Sharpe ratios and arbitrage opportunities but also all convex combinations of the two, that are generally neither arbitrage opportunities nor high Sharpe ratios. It can be shown, however, that this set of desirable claims is generated by a truncated quadratic utility function, and that it can be associated to a level of a *generalized* Sharpe ratio, see [6].

6 Conclusions

The theory presented here shows that pricing techniques which impose equilibrium restrictions stronger than no arbitrage can be seen as a generalization of no-arbitrage pricing. We derived the Extension and Pricing Theorem in no-good-deal framework and showed that the Extension Theorem captures the trade-off between equilibrium outcomes and discount factor restrictions. We have shown that equilibrium restrictions implied by von Neumann-Morgenstern preferences contain no-arbitrage and representative agent equilibrium as the two opposite ends of a spectrum of possible restrictions. In finite state models we have settled the question of how tight are the no-good-deal price bounds generated by a utility function. It is somewhat surprising that price bounds implied by strictly increasing utility functions are not always tighter than the no-arbitrage bounds. At the same time our results moderate the Bernardo-Ledoit critique of CRRA bounds – in finite state models these are always tighter than the no-arbitrage bounds.

Acknowledgements. This is a revised version of the fourth chapter of Aleš Černý's doctoral dissertation, which was supported by the European Commission's PHARE ACE Programme 1995. Both authors wish to thank Antonio Bernardo, John Cochrane, Martin Cripps, Marcus Miller, David Oakes and the participants at the 1st World Congress of Bachelier Finance Society for helpful comments.

Appendix A

Proof of Theorem 2.3

Since K is boundedly generated, there is a closed bounded set $B \subset K$ such that $K \subset \bigcup_{\lambda>0} \lambda B$. Therefore, it is enough to strictly separate B and $M_0(p)$.

With $1 < p < \infty$ L^p is a reflexive space. By Theorem 19 C in Holmes[13] $M_0(p)$ and B can be strictly separated by a continuous linear functional, i.e. there is $\varphi \in X^*$ such that $\varphi(M_0(p)) = 0$ and $\varphi(B) > 0$. However, this implies $\varphi(C_{++}) > 0$.

By standing assumption 1 there is a marketed strictly positive claim x_0 . Define $H \triangleq M_0(\varphi)$. Because H does not intersect C_{++} , and therefore X_{++}, we have $x_0 \notin H$. Finally, because H is a hyperplane we have the spanning property $X = H \oplus Span[x_0]$, so that each claim $y \in X$ has a unique decomposition $y = y_H + \lambda_y x_0$, where $y_H \in H$. By construction $p(y) \triangleq \lambda_y p(x_0)$ is a no-good-deal price of claim y. It is easily seen now that $\tilde{p} = \frac{p(x_0)}{\varphi(x_0)}\varphi$ is an extension of the original pricing rule p. Since φ is C-strictly positive and continuous $\tilde{p}$ must be C-strictly positive and continuous which completes the proof.

Proof of Theorem 2.5

a) By Theorem 2 there is no good deal in N if and only if there is a $C \cap N$-strictly positive continuous pricing functional φ in N. It is the continuity that we are worried about. We will show that no good deal in N implies that $N_0(\varphi)$ is closed and disjoint from K. Then the assertion follows from the Extension Theorem.

Functional φ has to price correctly all claims in M, $\varphi|M = \phi$. This implies $(N \supset) N_0(\varphi) \supset M_0(\phi)$. Note, however, that $\text{codim}_N M_0(\phi) \leq m + 1$ and hence $\text{codim}_{N_0(\varphi)} M_0(\phi) \leq m + 1$. In other words $N_0(\varphi) = M_0(\phi) \oplus L$ where $\dim L$ is finite. Since ϕ is continuous and M is closed, the zero investment marketed subspace $M_0(\phi)$ is closed in X. Then also $N_0(\varphi) = M_0(\phi) \oplus L$ is closed because L is finite dimensional. Now φ is $K \cap N$-strictly positive and therefore $N_0(\varphi) \cap K = \emptyset$.

b) The convexity of the no-good-deal price region follows from the convexity of complete market no-good-deal state prices and the part a). Namely, if $p_1, p_2 \in P$ then by assertion a) there exist functionals $\varphi_1, \varphi_2 \in C^*_{++}, \varphi_i(y) =$

[13] For completeness we provide the proof of that part of the theorem which is relevant to us and which Holmes leaves as an exercise: It is known that the unit ball $U(X)$ in a normed reflexive space is weakly compact (Theorem 16 F). Furthermore for convex sets 'closed' is equivalent to 'weakly closed' (Corollary 12 A). J is closed, convex and bounded, therefore weakly compact. N is convex, closed and therefore weakly closed. The separation theorem for one closed and one compact convex set (Corollary 11 F) asserts that N and J can be strictly separated by a weakly continuous functional ψ, however such functional is continuous in the original topology on X as well (Theorem 12 A).

p_i, that price correctly all claims in M. Of course, the functional $\lambda\varphi_1 + (1-\lambda)\varphi_2 \in C^*_{++}$ prices claims in M correctly, too, and therefore by assertion a)

$$(\lambda\varphi_1 + (1-\lambda)\varphi_2)(y) = \lambda p_1 + (1-\lambda)p_2 \in P.$$

c) To prove the last statement we will first demonstrate that the cone of no-good-deal pricing functionals C^*_{++} is open. Let us take B as in the proof of Theorem 2 and denote $\kappa = \sup_{x \in B} ||x||$.

i) Take an arbitrary $\varphi \in C^*_{++}$ and denote $\varepsilon(\varphi) = \inf_{x\in B} \varphi(x)$. We claim that $\varepsilon > 0$. For the purpose of contradiction suppose that $\varepsilon = 0$. Then there is a sequence $x_n \in B$ such that $\lim \varphi(x_n) = 0$. Since B is closed, convex and bounded, from the reflexivity of X follows that B is weakly sequentially compact (Holmes, Theorem 16F and Corollary 18 A). Hence there is a subsequence x_k converging weakly to $x \in B$ implying $\varphi(x) = 0$. However, $x \in B \subset C_{++}$ contradicts $\varphi \in C^*_{++}$.

Thus $\varphi(K) \geq \varepsilon > 0$. Taking an arbitrary functional $\psi \in X^*$ such that $||\psi|| < \frac{\varepsilon}{2\kappa}$ and $x \in X_{++}$ we have

$$(\varphi + \psi)(x) > \varphi(x) - ||\psi||\,||x|| > \varepsilon - \frac{\varepsilon}{2\kappa}\kappa > 0$$

which means that $\varphi + \psi \in C^*_{++}$ whenever $||\psi|| < \frac{\varepsilon}{2\kappa}$. Since φ is arbitrary this means that C^*_{++} is open in the norm topology on X^*.

ii) Let us first assume that $\text{codim}_N M = m$. Applying Hahn-Banach theorem to the subspace $\text{span}(M \cup \{y_1, \ldots, y_{j-1}, y_{j+1}, \ldots, y_m\})$ and the point y_j one can find linear functionals $\psi_j, j = 1, \ldots, m$ such that $\psi_j(M) = 0$ for all j and $\psi_j(y_i) = \delta_{ij}$ (Kronecker's delta). By the Extension Theorem the pricing rule on M can be extended to a C-strictly positive functional φ_0 that correctly prices securities in M. Note that functionals $\varphi_0 + \lambda\psi_j$ too price these securities correctly and moreover for $|\lambda|$ sufficiently small $\varphi + \lambda\psi_j$ will be a strictly positive functional by the result in i). Thus the price vectors $\varphi_0(y) + \lambda\psi_j(y)$ give no-good-deal prices for securities $y = (y_1, \ldots, y_m)$ consistent with the predetermined prices of securities in M. By construction $\psi_j(y)$ are linearly independent vectors in $\mathbb{R}^m$ and recall that $\dim P$ is defined as the dimension of the affine hull of $P - \varphi_0(y)$ (which is a linear subspace) thus the dimension of the price region P is at least $\text{rank}(\psi_1(y), \ldots, \psi_m(y)) = m$, and of course it cannot be more than m.

iii) In a general case a certain number of vectors y_i, say $m - l$, will lie in the marketed subspace M. However, for any $\tilde{\varphi}$ that prices correctly claims in M the difference $\tilde{\varphi}(y_i) - \varphi_0(y_i)$ will be zero, hence if $y_i \in M$ it will not contribute to the dimensionality of the price region.

That leaves l claims that do not belong to the marketed subspace and these we partition into two groups; the first $\tilde{m}$ claims $c_1' = (y_1, \ldots, y_{\tilde{m}})$ that are linearly independent and the remaining $l - \tilde{m}$ claims $c_2' = (y_{\tilde{m}+1}, \ldots, y_l)$ that can be expressed as a linear combination of the first $\tilde{m}$ claims, $c_2 = Dc_1$ with $D \in \mathbb{R}^{(l-\tilde{m})\times\tilde{m}}$.

First of all it is clear that there cannot be more than $\tilde{m}$ linearly independent vectors of the type $\psi(y_1), \ldots, \psi(y_l)$. If there were more, one could find a non-trivial linear combination of these vectors that annuls the first $\tilde{m}$ coordinates, $\sum_i \lambda_i \psi_i(c_1) = 0$. However, such a linear combination annuls the remaining $l - \tilde{m}$ coordinates as well since

$$\sum_i \lambda_i \psi_i(c_2) = \sum_i \lambda_i \psi_i(Dc_1) = \sum_i \lambda_i D\psi_i(c_1) = D\left(\sum_i \lambda_i \psi_i(c_1)\right) = 0.$$

On the other hand one can find $\tilde{m}$ linearly independent prices of the desired form by the procedure described in ii). Thus $\dim P = \tilde{m}$ and by construction $\tilde{m} = \operatorname{codim}_N M$.

d) Suppose there is just one security to be priced, say security y_i and denote its no-good-deal price region $P \subset \mathbb{R}^1$. We set $N = \operatorname{span}(M \cup \{y_i\})$ and have $\operatorname{codim}_N M = 0$ if and only if $y_i \in M$. By definition the no-good-deal price of y_i is unique if and only if $\dim P = 0$. Then the assertion c) implies that y is uniquely priced if and only if y is redundant.

e) i) Define B_1, B_2 and κ_1, κ_2 in analogy to B and κ in c). We want to show that for all $\varphi_2 \in C^*_{2++}$ and for all $\psi \in X^*$ such that $||\psi|| < \frac{\varepsilon\nu}{2\kappa_1}$ we have

$$\varphi_2 + \psi \in C^*_{1++}.$$

Let us take an arbitrary $\varphi_2 \in C^*_{2++}$ which implies $\varphi_2(J_1 + \varepsilon Ball(0,1)) > 0$ and consequently

$$\varphi_2(J_1) - \varepsilon||\varphi_2|| > 0. \tag{A.1}$$

Furthermore, if φ_2 prices correctly all the basis assets then $||\varphi_2|| \geq ||\phi|| = \nu$. The standing assumption 1 implies $\nu > 0$. Taking an arbitrary $\psi \in X^*$ such that $||\psi|| < \frac{\varepsilon\nu}{2\kappa}$ and making use of (A.1) we have

$$(\varphi_2 + \psi)(J_1) \geq \varepsilon||\varphi_2|| - ||\psi||\kappa \geq \varepsilon\nu - \frac{\varepsilon\nu}{2\kappa}\kappa > 0,$$

QED.

ii) We can now construct linear functionals ψ_i as in c) ii) and denote $c \equiv \max_{i=1,\ldots,m} ||\psi_i||$. Let $\tilde{B}_\delta$ be a δ-ball in $\mathbb{R}^m$ with L^2 norm. By construction of functionals ψ_i we have

$$\{\sum_{i=1}^m \theta_i \psi_i(y) | (\theta_1, \ldots, \theta_m) \in \tilde{B}_\delta\} = \tilde{B}_\delta \tag{A.2}$$

On the other hand

$$\left\|\sum_{i=1}^m \theta_i \psi_i\right\| \leq \sum_{i=1}^m |\theta_i|\,||\psi_i|| = c\sum_{i=1}^m |\theta_i| = c||\theta||_1 \leq c\sqrt{m}||\theta||_2.$$

With $\delta = \frac{\varepsilon\nu}{2\kappa_1 c\sqrt{m}}$ therefore $\|\sum_{i=1}^m \theta_i\psi_i\| \le \frac{\varepsilon\nu}{2\kappa_1}$ for $\theta \in \tilde{B}_\delta$. Combining this result with e) i) we have for an arbitrary $\varphi_2 \in C^*_{2++}$

$$\left(\varphi_2 + \sum_{i=1}^m \theta_i\psi_i\right)(y) \in P_1 \text{ for } \theta \in \tilde{B}_\delta. \tag{A.3}$$

Let $p_2 \in \mathrm{cl}P_2$ then there is $\varphi_2 \in C^*_{2++}$ such that $\|\varphi_2(y) - p_2\|_2 < \frac{\delta}{2}$. By virtue of (A.2) and (A.3) we then have

$$p_2 + \tilde{B}_{\frac{\delta}{2}} \subset P_1.$$

QED.

Lemma A.1. *Let*

$$\begin{aligned} A &= \{(x,y,z)|z \ge \frac{1}{x+\sqrt{y}}, x+\sqrt{y} > 0, y \ge 0, x \le 1\} \\ B &= \{(0,0,z)|z \in \mathbb{R}\} \end{aligned}$$

Then A and B are two disjoint closed convex sets and they cannot be semistrictly separated.

Proof. It is easily verified that $y = 0$ is a unique separating hyperplane. Clearly this hyperplane is not disjoint from either A or B.

Appendix B

Proof of Theorem 3.5 Suppose to the contrary that the set of desirable claims $K(a)$ is not boundedly generated. Then there must be a sequence $\{x_n\}$, $\|x_n\| \to +\infty$, $x_n \in K(a)$ with the property that $\lambda x_n \notin K(a)$ for $\lambda < 1$. Nonetheless, X being finite dimensional the sequence $\frac{x_n}{\|x_n\|}$ must converge (if necessary by passing to a subsequence). Let us denote the limit z. From Lemma B.1 we know that $P(z < 0) = 0$. By virtue of Lemma B.3 for sufficiently large constant κ the claim $\kappa\alpha \in K(2a)$ and therefore, $\kappa\alpha + \delta B_1 \in K(a)$ for δ sufficiently small, because the expected utility is a continuous function on X. Furthermore, $\frac{x_n}{\|x_n\|} \to \alpha$ implies that there is n_0 such that for all $n > n_0$ $\frac{x_n}{\|x_n\|} \in \alpha + \frac{\delta}{\kappa}B_1$ and therefore

$$\frac{\kappa x_n}{\|x_n\|} \in \kappa\alpha + \delta B_1 \in K(a).$$

Since $\|x_n\| \to +\infty$ there is n_1 such that for all $n > n_1$ $\|x_n\| > \kappa$. Thus for $n > n_1$

$$\frac{\kappa}{\|x_n\|}x_n \in K(a) \text{ and } \frac{\kappa}{\|x_n\|} < 1$$

which contradicts our assumption that $\lambda x_n \notin K(a)$ for $\lambda < 1$. QED.

Proof of Theorem 4.1:

a) i) Suppose, to the contrary, that $\sup_{w\in M_0} \mathrm{E}U(w) = +\infty$. Then there is a sequence $\{w_n\} \in M_0$ such that $\{\mathrm{E}U(w_n)\}$ is increasing and unbounded from above. If $\{||w_n||\}$ were bounded then there would be a convergent subsequence $\{w_{n_k}\} \to w \in X$ and EU would not be continuous at w. Thus it must be that $\{||w_n||\}$ is unbounded. In that case, however, $\{w_n\}$ is an unbounded sequence of desirable claims and because X is finite-dimensional we can find a common direction (if necessary passing to a subsequence) $\left\{\frac{w_n}{||w_n||}\right\} \to z \in X$, in fact $z \in M_0$ because M_0 is closed. By Lemma B.1 z is strictly positive and $z \in M_0$ contradicts the assumption of no arbitrage. Q.E.D.

ii) We have shown $\sup_{w\in M_0} \mathrm{E}U(w_r + w) \triangleq \mathrm{E}U(w_r + a_M) < +\infty$. By definition of supremum there is a sequence $\{w_n\} \in M_0$ such that $\{\mathrm{E}U(w_r + w_n)\} \to \mathrm{E}U(w_r + a_M)$. Now because the sets of desirable claims are boundedly generated and M_0 is a linear subspace, we can always choose $\{w_n\}$ bounded. This implies $\{w_n\} \to w_M \in M_0$ (again using a subsequence if necessary) and by continuity of expected utility $\mathrm{E}U(w_r + w_M) = \mathrm{E}U(w_r + a_M)$. The uniqueness of w_M follows from the strict concavity of the (expected) utility function. Q.E.D.

b) The first part follows from a). For the second part of the statement it is enough to consider $a = a_M$. By Hahn-Banach Theorem one can strongly separate M_0 and the interior of $K(a_M)$ by a hyperplane N_0, and because $K(a_M)$ is closed the same hyperplane separates $K(a_M)$ and M_0 weakly. By standing assumption 1 there is a marketed strictly positive claim x with positive price $p(x) > 0$. Because N_0 does not contain internal points of $K(a_M)$ we must have $\sup_{w\in N_0} \mathrm{E}U(w_r + w) = \mathrm{E}U(w_r + a_M)$. For the same reason $N_0 \cap X_{++} = \emptyset$ and therefore $x \notin N_0$. Finally, because N_0 is a hyperplane we have the spanning property $X = N_0 \oplus Span[x]$, so that each claim $y \in X$ has a unique decomposition $y = y_0 + \lambda_y x$, where $y_0 \in N_0$. By construction $p(y) \triangleq \lambda_y p(x)$ is a no-good-deal price of claim y. Convexity of the price region follows from the argument presented in the proof of Theorem 2.5, part b).

c) The first part follows from Theorem 2.5. For the second part, it is clear from a) that the separating hyperplane has to cross $K(a_M)$ at w_M. It follows from Theorem 1.29 in [2] that EU has a unique supergradient at w_M, by direct calculation this supergradient is

$$\zeta = (\Pr(\omega_1)U'\left[w_r(\omega_1) + w_M(\omega_1)\right], \ldots, \Pr(\omega_k)U'\left[w_r(\omega_k) + w_M(\omega_k)\right]),$$

where $k = \dim X$. By definition the supergradient has the property

$$\begin{aligned}\mathrm{E}U(w_r + w_M + \triangle w) &\le \mathrm{E}U(w_r + w_M) + \zeta\triangle w \\ &= \mathrm{E}U(w_r + w_M) + \mathrm{E}U'(w_r + w_M)\triangle w.\end{aligned}$$

As long as $\mathrm{E}U'(w_r + w_M)\triangle w = 0$ for all $\triangle w \in N_0$ we have

$$\mathrm{E}U(w_r + w_M + N_0) \le \mathrm{E}U(w_r + w_M)$$

and there is no good deal in the completed market. The normalisation discussed in the proof of Theorem 2 shows that the pricing functional is

$$p(y) = p(x_0)\frac{\mathrm{E}U'(w_r + w_M)y}{\mathrm{E}U'(w_r + w_M)x_0}.$$

When x_0 is a risk-free asset this formula simplifies to

$$p(y) = \frac{\mathrm{E}\frac{U'(w_r+w_M)}{\mathrm{E}U'(w_r+w_M)}y}{R}.$$

To show uniqueness realize that by Theorem 1.30 in Beavis and Dobbs EU is continuously differentiable at $w_r + w_M$. The Taylor expansion of the form $f(x) = f(x_0)+f_x(\lambda x_0+(1-\lambda)x)(x-x_0)$ for some $0 < \lambda < 1$ with $f = \mathrm{E}U$ and $x_0 = w_r+w_M$ implies that the hyperplane defined by ζ is the only hyperplane passing through w_M that does not intersect the interior of $K(a_M)$.

d), e), f) $K(a)$ is boundedly generated by bounded closed set $B_a \subset K(a)$, likewise $K(b)$ is generated by $B_b \subset K(b)$. Because B_b is compact, EU is uniformly continuous on B_b. Therefore there is $\varepsilon > 0$ such that $B_b + Ball(0,\varepsilon) \subset K(a)$. If $B_b + Ball(0,\varepsilon) \not\subset B_a$ then we can always redefine B_a as the closed convex hull of $B_a \cup [B_b + Ball(0,\varepsilon)]$. Then the assumption of Theorem 2.5 e) is satisfied and the rest follows.

Lemma B.1. *Suppose X is a probability space,* $\dim X < \infty$, *and U is a downside-sensitive utility. If an unbounded sequence of desirable claims has a common direction, then this direction is strictly positive. Mathematically, if* $||x_n|| \to \infty$, $\frac{||x_n||}{x_n} \to z$ *and* $\mathrm{E}U(w_r + x_n) \geq \mathrm{E}U(w_r + a)$ *for a fixed* $a \in \mathbb{R}$, *then* $z \geq 0$, $\Pr(z > 0) > 0$.

Proof

Let us define $w_{\min} = \min w_r$, $w_{\max} = \max w_r$ and analogously $x_{\min}$, $x_{\max}$. Since we have finitely many states there is a state with the smallest probability $p_{\min}$.

i) For x to be a desirable claim we must have

$$\mathrm{E}U(w_r + x) - \mathrm{E}U(w_r + a) \geq 0.$$

We can rewrite this statement using conditional distribution

$$\begin{aligned}\Pr(x < 0)\mathrm{E}\left[U(w_r + x)|x < 0\right] + \Pr(x \geq 0)\mathrm{E}\left[U(w_r + x)|x \geq 0\right]& \\ \geq \mathrm{E}U(w_r + a).& \qquad \text{(B.1)}\end{aligned}$$

Let us appraise the left hand side from above. Denoting ξ the supergradient of U in $w_{\min}$ we can write

$$\begin{aligned}\mathrm{E}[U(w_r + x)|x \geq 0] &\leq U(w_{\max}) + \xi\mathrm{E}[x|x \geq 0] = \\ &= U(w_{\max}) + \xi\frac{\mathrm{E}[x^+]}{\Pr(x \geq 0)}\end{aligned}$$

Assuming that $x_{\min} < 0$ we obtain

$$\begin{aligned}\Pr(x<0)\mathrm{E}\left[U(w_r+x)|x<0\right] \leq\, & p_{\min}U(w_{\max}+x_{\min}) \\ & + \left[\Pr(x<0)-p_{\min}\right]U(w_{\max}).\end{aligned}$$

Plugging the last two expressions into the equation (B.1) we obtain

$$\begin{gathered} p_{\min}U(w_{\max}+x_{\min})+\xi\mathrm{E}[x^+]\geq\gamma \\ \gamma=\mathrm{E}[U(w_r+a)]-\Pr(x\geq 0)U(w_{\max})-\left[\Pr(x<0)-p_{\min}\right]U(w_{\max}) \\ \gamma\geq\mathrm{E}[U(w_r+a)]-U(w_{\max})\equiv c(a)\end{gathered}$$

Thus for x to be desirable we must have

$$p_{\min}U(w_{\max}+x_{\min})+\xi\mathrm{E}[x^+]\geq c(a) \tag{B.2}$$

where $p_{\min}$, $w_{\max}$, ξ, and $c(a)$ do not depend on x.

ii) Let us now take a sequence of desirable claims $||x_n||\to\infty$, $\frac{x_n}{||x_n||}\to\alpha$. If $\alpha_{\min}$ were negative then by continuity $\frac{(x_n)_{\min}}{||x_n||_1}\to\alpha_{\min}$ and hence $(x_n)_{\min}\to-\infty$. At the same time $x_n\in K(a)$ and (B.2) implies that

$$p_{\min}U(w_{\max}+(x_n)_{\min})+\xi\mathrm{E}[x_n^+]\geq c(a).$$

where $c(a)<0$ without loss of generality. After rearranging the terms we arrive at

$$\frac{w_{\max}+(x_n)_{\min}}{U(w_{\max}+(x_n)_{\min})}\frac{(x_n)_{\min}}{(w_{\max}+(x_n)_{\min})p_{\min}}\geq\frac{-(x_n)_{\min}}{\xi\mathrm{E}[x_n^+]-c(a)}.$$

The right hand side can be appraised from below

$$\frac{-(x_n)_{\min}}{\xi\mathrm{E}[x_n^+]-c(a)}\geq\frac{\mathrm{E}[x_n^-]}{\xi\mathrm{E}[x_n^+]-c(a)}.$$

The limit of the left hand side is, by the dominance condition (3), equal to zero. Because the numerator on the right hand side goes to $+\infty$, it must be that $\mathrm{E}[x_n^+]\to+\infty$. This however implies that

$$0=\lim_{n\to+\infty}\frac{\mathrm{E}[x_n^-]}{\xi\mathrm{E}[x_n^+]-c(a)}=\lim_{n\to+\infty}\frac{\mathrm{E}[x_n^-]}{\mathrm{E}[x_n^+]}\frac{1}{\xi-\frac{c(a)}{\mathrm{E}[x_n^+]}}=\frac{1}{\xi}\lim_{n\to+\infty}\frac{\mathrm{E}[x_n^-]}{\mathrm{E}[x_n^+]}$$

$$0=\lim_{n\to+\infty}\frac{\mathrm{E}[x_n^-]}{\mathrm{E}[x_n^+]}=\frac{\mathrm{E}[\alpha^-]}{\mathrm{E}[\alpha^+]}$$

which contradicts $P(\alpha<0)>0$.

Lemma B.2. *Suppose that $U:\mathbb{R}\to\mathbb{R}$ is a strictly increasing concave function such that $u=\sup_{x\in\mathbb{R}}U(x)<\infty$. Suppose further that $x\in L^p$ is strictly positive. Then*

$$\lim_{\lambda\to\infty}\mathrm{E}[U(x+\lambda y)]=u$$

if and only if $y\in L^p$ is strictly positive with probability 1.

Proof. i) Firstly, let us take $y \in L^p$ such that $P(y \leq 0) = \pi > 0$. Then we have

$$\mathrm{E}[U(x + \lambda y)] \leq (1 - \pi)u + \pi \mathrm{E}[U(x)|y \leq 0].$$

Since U is strictly increasing, we must have $U(t) < u$ for all $t \in \mathbb{R}$ and hence also $\mathrm{E}\left[U(x)|y \leq 0\right] = \tilde{u} < u$. Consequently

$$\lim_{\lambda \to \infty} \mathrm{E}[U(x + \lambda y)] \leq (1 - \pi)u + \pi \tilde{u} < u$$

ii) Now consider $y \in L^p$ such that $P(y \leq 0) = 0$. Define

$$\begin{aligned}
\pi_0 &= P(y \geq 1) \\
\pi_n &= P(\frac{1}{n} > y \geq \frac{1}{n+1}) \text{ for } n = 1, 2, \ldots \\
p_n &= \sum_0^n \pi_k \, .
\end{aligned}$$

By assumption $\lim_{n \to \infty} p_n = 1$. Now we have

$$\mathrm{E}[U(x + \lambda y)] \geq p_n U(\frac{\lambda}{n+1})$$

and therefore

$$\lim_{\lambda \to \infty} \mathrm{E}[U(x + \lambda y)] \geq p_n u \text{ for all } n$$

Since $\lim_{n \to \infty} p_n = 1$ it must be true that $\lim_{\lambda \to \infty} \mathrm{E}[U(x + \lambda y)] \geq u$.

Lemma B.3. *Suppose that $U : \mathbb{R} \to \mathbb{R}$ is strictly increasing, concave and unbounded from above. Suppose further that $x \in L^p$ is bounded below. Then*

$$\lim_{\lambda \to \infty} \mathrm{E}[U(x + \lambda y)] = \infty$$

for all strictly positive $y \in L^p$.

Proof. Define π_k as above and set $x_{\min} = \operatorname{ess\,inf} x$. Since by assumption $P(y > 0) > 0$ there must be $k \in \mathbb{N}$ such that $\pi_k > 0$. Then

$$\mathrm{E}[U(x + \lambda y)] \geq p_k U\big(x_{\min} + \frac{\lambda}{k+1}\big)$$

and letting $\lambda \to \infty$ we have

$$\lim_{\lambda \to \infty} \mathrm{E}[U(x + \lambda y)] = \infty.$$

Lemma B.4. *Von Neumann-Morgenstern preferences are downside-sensitive if and only if the generating utility function satisfies*

$$\lim_{x\to-\infty}\frac{x}{U(x)}=0.$$

Proof. First we show the 'if' part. Let us take $y\in L^p$ such that $y\notin L^p_+$, i.e. $P(y<0)=\varepsilon>0$. Then one of the numbers

$$\pi_0 = P(y<-1)$$
$$\pi_n = P(-\frac{1}{n}\le y<-\frac{1}{n+1}) \text{ for } n=1,2,\dots$$

has to be positive, otherwise $P(y<0)=0$.

Without loss of generality we can assume that $\pi_k>0$. Denoting ξ the left hand side derivative of U in zero and taking $\lambda>0$ we obtain

$$\mathrm{E}[U(\lambda y)] = \mathrm{E}\left[U(\lambda y)|y<0\right]P(y<0)+\mathrm{E}\left[U(\lambda y)|y\ge 0\right]P(y\ge 0) =$$
$$\le \pi_k U(-\frac{\lambda}{k+1})+(1-\pi_k)U(0)+\xi\lambda\mathrm{E}[y^+]\equiv v(\lambda).$$

Note that $v(\lambda)$ is a continuous function on $\mathbb{R}_+$ and thus $\sup_{\lambda\ge0}v(\lambda)=\infty$ if and only if $\lim_{\lambda\to\infty}v(\lambda)=\infty$. However, instead we have

$$\lim_{\lambda\to\infty}\frac{v(\lambda)}{\lambda}=\xi\mathrm{E}[y^+]-\lim_{x\to-\infty}\frac{\pi_k}{k+1}\frac{U(x)}{x}=-\infty$$

and hence $\lim_{\lambda\to\infty}v(\lambda)<\infty$, $\sup_{\lambda\ge0}v(\lambda)<\infty$, and consequently $\sup_{\lambda\ge0}\mathrm{E}[U(\lambda y)]<\infty$ for all $y\notin L^p_+$ which completes the proof.

The 'only if' part is shown easily once we realize that $\frac{U(x)}{x}$ is a decreasing function of x. We can take a random variable with two atoms $P(y=-1)=\pi$ and $P(y=1)=1-\pi$. Then

$$\mathrm{E}[U(\lambda y)]=\pi U(-\lambda)+(1-\pi)U(\lambda)$$

and

$$\lim_{\lambda\to\infty}\frac{\mathrm{E}[U(\lambda y)]}{\lambda}=-\pi\lim_{x\to-\infty}\frac{U(x)}{x}+(1-\pi)\lim_{x\to\infty}\frac{U(x)}{x}. \tag{B.3}$$

Now $\lim_{x\to-\infty}\frac{U(x)}{x}$ is finite and if $\lim_{x\to\infty}\frac{U(x)}{x}$ is positive one can always take π small enough so that the limit (B.3) is positive. But then $\lim_{\lambda\to\infty}\mathrm{E}[U(\lambda y)]=+\infty$.

References

1. Philippe Artzner, Freddy Delbaen, Jean-Marc Eber, and David Heath. Coherent measures of risk. *Mathematical Finance*, 9(3):203–228, 1999.

2. Brian Beavis and Ian Dobbs. *Optimization and Stability Theory for Economic Analysis.* Cambridge University Press, 1990.
3. Fabio Bellini and Marco Fritelli. On the existence of minimax martingale measures. Rapporto di Ricerca 14/2000, Universitá degli Studi Milano - Bicocca, May 2000.
4. Antonio Bernardo and Olivier Ledoit. Gain, loss and asset pricing. *Journal of Political Economy*, 108(1):144–172, 2000.
5. Antonio E. Bernardo and Olivier Ledoit. Approximate arbitrage. Working paper 18-99, Anderson School, UCLA, November 1999.
6. Aleš Černý. Generalized Sharpe Ratio and consistent good-deal restrictions in a model of continuous trading. Discussion Paper SWP9902/F, Imperial College Management School, April 1999.
7. Stephen A. Clark. The valuation problem in arbitrage price theory. *Journal of Mathematical Economics*, 22:463–478, 1993.
8. John H. Cochrane and Jesus Saá-Requejo. Beyond arbitrage: Good-deal asset price bounds in incomplete markets. *Journal of Political Economy*, 108(1):79–119, 2000.
9. Philip H. Dybvig and Stephen A. Ross. Arbitrage. In J. Eatwell, M. Milgate, and P. Newman, editors, *The New Palgrave: A Dictionary of Economics*, volume 1, pages 100–106. Macmillan, London, 1987.
10. Nicole El Karoui and Marie-Claire Quenez. Dynamic programming and pricing of contingent claims in an incomplete market. *SIAM Journal of Control and Optimization*, 33(1):29–66, 1995.
11. Lars Peter Hansen and Ravi Jagannathan. Implications of security market data for models of dynamic economies. *Journal of Political Economy*, 99(2):225–262, 1991.
12. Stewart Hodges. A generalization of the Sharpe Ratio and its application to valuation bounds and risk measures. FORC Preprint 98/88, University of Warwick, April 1998.
13. Jonathan E. Ingersoll. *Theory of Financial Decision Making.* Rowman & Littlefield Studies in Financial Economics. Rowman & Littlefield, 1987.
14. Stefan Jaschke and Uwe Kuechler. Coherent risk measures and good-deal bounds. *Finance and Stochastics*, 5(2), 2001.
15. D. Kreps. Arbitrage and equilibrium in economies with infinitely many commodities. *Journal of Mathematical Economics*, 8:15–35, 1981.
16. Robert Merton. Theory of rational option pricing. *Bell Journal of Economics and Management Science*, 4:141–183, 1973.
17. Peter H. Ritchken. On option pricing bounds. *The Journal of Finance*, 40(4):1219–1233, 1985.
18. Stephen A. Ross. The arbitrage theory of capital asset pricing. *Journal of Economic Theory*, 13:341–360, 1976.
19. Stephen A. Ross. A simple approach to the valuation of risky streams. *Journal of Business*, 51:453–475, 1978.
20. Mark Rubinstein. The valuation of uncertain income streams and the pricing of options. *The Bell Journal of Economics*, 7:407–425, 1976.
21. Walter Schachermayer. Martingale measures for discrete time processes with infinite horizon. *Mathematical Finance*, 4(1):25–55, 1994.

Spread Option Valuation and the Fast Fourier Transform

M.A.H. Dempster and S.S.G. Hong

Centre for Financial Research, Judge Institute of Management
University of Cambridge, England CB2 1AG
Email: mahd2@cam.ac.uk & gh10006@hermes.cam.ac.uk

Abstract. We investigate a method for pricing the generic spread option beyond the classical two-factor Black-Scholes framework by extending the fast Fourier Transform technique introduced by Carr & Madan (1999) to a multi-factor setting. The method is applicable to models in which the joint characteristic function of the prices of the underlying assets forming the spread is known analytically. This enables us to incorporate stochasticity in the volatility and correlation structure – a focus of concern for energy option traders – by introducing additional factors within an affine jump-diffusion framework. Furthermore, computational time does not increase significantly as additional random factors are introduced, since the fast Fourier Transform remains two dimensional in terms of the two prices defining the spread. This yields considerable advantage over Monte Carlo and PDE methods and numerical results are presented to this effect.

1 Introduction

Spread options are European derivatives with terminal payoffs of the form: $[(\boldsymbol{S}_2(T) - \boldsymbol{S}_1(T)) - K]_+$, where the two underlying processes $\boldsymbol{S}_1, \boldsymbol{S}_2$ forming the spread could refer to asset or futures prices, equity indices or (defaultable) bond yields. There is a wide variety of such options traded across different sectors of the financial markets; for example, the crack spread and crush spread options in the commodity markets [16,22], credit spread options in the fixed income markets, index spread options in the equity markets [10] and the spark (electricity/fuel) spread options in the energy markets [9,18]. They are also applied extensively in the area of real options [23] for both asset valuation and hedging a firm's production exposures. Despite their wide applicability and crucial role in managing the so-called *basis risk*, pricing and hedging of this class of options remain difficult and no consensus on a theoretical framework has emerged.

The main obstacle to a "clean" pricing methodology lies in the lack of knowledge about the distribution of the difference between two non-trivially correlated stochastic processes: the more variety we inject into the correlation structure, the less we know about the stochastic dynamics of the spread.

At one extreme, we have the *arithmetic Brownian motion* model in which $\boldsymbol{S}_1, \boldsymbol{S}_2$ are simply two Brownian motions with constant correlation [19]. The spread in this case is also a Brownian motion and an analytic solution for the spread option is thus available. This, however, is clearly an unrealistic model as it, among other things, permits negative values in the two underlying prices/rates. An alternative approach to modelling the spread directly as a *geometric Brownian motion* has also proven inadequate as it ignores the intrinsic multi-factor structure in the correlation between the spread and the underlying prices and can lead to severe misspecification of the option value when markets are volatile [13].

Going one step further we can model the individual prices as geometric Brownian motions in the spirit of Black and Scholes and assume that the two driving Brownian motions have a constant correlation [17,20,22]. The resulting spread, distributed as the difference of two lognormal random variables, does not possess an analytical expression for its density, preventing us from deriving a closed form solution to the pricing problem. We can however invoke a conditioning technique which reduces the two dimensional integral for computing the expectation under the martingale measure to a one dimensional integral, thanks to a special property of the multivariante normal distribution: conditional on one component, the remaining components are also normally distributed.

As we develop a stochastic term structure for volatilities and correlations of the underlying processes, we move out of the Gaussian world and the conditioning technique no longer applies. Furthermore, a realistic model for asset prices often requires more than two factors; for example, in the energy market, random jumps are essential in capturing the true dynamics of electricity or oil prices, and in the equity markets, stochastic volatilities are needed. Interest rate models such as the CIR or affine jump-diffusion models [11] frequently assume more than two factors and non-Gaussian dynamics for the underlying yields. However, the computational times using existing numerical techniques such as Monte Carlo or PDE methods increase dramatically as extended models take these issues into account.

In this paper we propose a new method for pricing spread options valid for the class of models which have analytic characteristic functions for the underlying asset prices or market rates. This includes the *variance gamma* (VG) model [15], the *inverse Gaussian* model [3] and numerous stochastic volatility and stochastic interest rates models in the general *affine jump-diffusion* family [1,4,6,14,21]. The method extends the *fast Fourier transform* (FFT) approach of Carr & Madan [5] to a multi-factor setting, and is applicable to options with a payoff more complex than a piecewise-linear structure. The main idea is to integrate the option payoff over approximate regions bounding the non-trivial exercise region, analogous to the method of integrating a real function by Riemann sums. As for the Riemann integral, this gives close upper and lower bounds for the spread option price which tend to the true value as we refine the discretisation.

The FFT approach is superior to existing techniques in the sense that changing the underlying diffusion models only amounts to changing the characteristic function and therefore does not alter the computational time significantly. In particular, one can introduce factors such as stochastic volatilities, stochastic interest rates and random jumps, provided the corresponding characteristic function is known, to result in a more realistic description of the market dynamics and a more sophisticated framework for managing the volatility and correlation risks involved.

We give a brief review of the FFT pricing method applied to the valuation of a simple European option on two assets in Section 2. In Section 3 our pricing scheme for a generic spread option is set out in detail. Section 4 describes the underlying models implemented for this paper and presents computational results to illustrate the advantage of the approach and the need for a non-trivial volatility and correlation structure. Section 5 concludes and describes current research directions.

2 Review of the FFT Method

To illustrate the application of the fast Fourier transform technique to the pricing of simple European style options in a multi-factor setting, in this section we derive the value of a correlation option as defined in [2] following the method and notation of [5] in the derivation of a European call on a single asset.

A *correlation option* is a two-factor analog of an European call option, with a payoff of $[\boldsymbol{S}_1(T)-K_1]_+ \cdot [\boldsymbol{S}_2(T)-K_2]_+$ at maturity T, where $\boldsymbol{S}_1, \boldsymbol{S}_2$ are the underlying asset prices. Denoting strikes and asset prices by K_1, K_2, S_1, S_2 and their logarithms by k_1, k_2, s_1, s_2, our aim is to evaluate the following integral for the option price:

$$\begin{aligned} C_T(k_1,k_2) &:= \mathbb{E}_{\mathbb{Q}}\Big[e^{-rT}\big[\boldsymbol{S}_1(T)-K_1\big]_+ \cdot \big[\boldsymbol{S}_2(T)-K_2\big]_+\Big] \\ &\equiv \int_{k_1}^{\infty}\int_{k_2}^{\infty} e^{-rT}\big(e^{s_1}-e^{k_1}\big)\big(e^{s_2}-e^{k_2}\big)q_T(s_1,s_2)ds_2 ds_1 \ , \quad (1) \end{aligned}$$

where $\mathbb{Q}$ is the risk-neutral measure and $q_T(\cdot,\cdot)$ the corresponding joint density of $\boldsymbol{s}_1(T), \boldsymbol{s}_2(T)$. The *characteristic function* of this density is defined by

$$\begin{aligned} \phi_{\mathbb{T}}(u_1,u_2) &:= \mathbb{E}_{\mathbb{Q}}\big[\exp(iu_1\boldsymbol{s}_1(T)+iu_2\boldsymbol{s}_2(T))\big] \\ &= \int_{-\infty}^{\infty}\int_{-\infty}^{\infty} e^{i(u_1 s_1+u_2 s_2)} q_T(s_1,s_2)ds_2 ds_1. \end{aligned}$$

As in [5,8], we multiply the option price (1) by an exponentially decaying term so that it is square-integrable in k_1, k_2 over the negative axes:

$$c_T(k_1,k_2) := e^{\alpha_1 k_1+\alpha_2 k_2} C_T(k_1,k_2) \quad \alpha_1, \alpha_2 > 0.$$

We now apply a Fourier transform to this modified option price:

$$\begin{aligned}
\psi_T(v_1, v_2) &:= \int_{-\infty}^{\infty}\int_{-\infty}^{\infty} e^{i(v_1 k_1 + v_2 k_2)} c_T(k_1, k_2) dk_2 dk_1 \\
&= \iint_{\mathbb{R}^2} e^{(\alpha_1 + iv_1)k_1 + (\alpha_2 + iv_2)k_2} \\
&\quad \left\{ \int_{k_2}^{\infty}\int_{k_1}^{\infty} e^{-rT}\left(e^{s_1} - e^{k_1}\right)\left(e^{s_2} - e^{k_2}\right) q_T(s_1, s_2) ds_2 ds_1 \right\} dk_2 dk_1 \\
&= \iint_{\mathbb{R}^2} e^{-rT} q_T(s_1, s_2) \\
&\quad \left\{ \int_{-\infty}^{s_2}\int_{-\infty}^{s_1} e^{(\alpha_1 + iv_1)k_1 + (\alpha_2 + iv_2)k_2} \left(e^{s_1} - e^{k_1}\right)\left(e^{s_2} - e^{k_2}\right) dk_2 dk_1 \right\} ds_2 ds_1 \\
&= \iint_{\mathbb{R}^2} \frac{e^{-rT} q_T(s_1, s_2)\, e^{(\alpha_1 + 1 + iv_1)s_1 + (\alpha_2 + 1 + iv_2)s_2}}{(\alpha_1 + iv_1)(\alpha_1 + 1 + iv_1)(\alpha_2 + iv_2)(\alpha_2 + 1 + iv_2)} ds_2 ds_1 \\
&= \frac{e^{-rT} \phi_T\left(v_1 - (\alpha_1 + 1)i, v_2 - (\alpha_2 + 1)i\right)}{(\alpha_1 + iv_1)(\alpha_1 + 1 + iv_1)(\alpha_2 + iv_2)(\alpha_2 + 1 + iv_2)}. \qquad (2)
\end{aligned}$$

Thus if the characteristic function ϕ_T is known in closed form, the Fourier transform ψ_T of the option price will also be available analytically, yielding the option price itself via an inverse transform:

$$C_T(k_1, k_2) = \frac{e^{-\alpha_1 k_1 - \alpha_2 k_2}}{(2\pi)^2} \int_{-\infty}^{\infty}\int_{-\infty}^{\infty} e^{-i(v_1 k_1 + v_2 k_2)} \psi_T(v_1, v_2) dv_2 dv_1.$$

Invoking the trapezoid rule we can approximate this Fourier integral by the following sum:

$$C_T(k_1, k_2) \approx \frac{e^{-\alpha_1 k_1 - \alpha_2 k_2}}{(2\pi)^2} \sum_{m=0}^{N-1}\sum_{n=0}^{N-1} e^{-i(v_{1,m} k_1 + v_{2,n} k_2)} \psi_T(v_{1,m}, v_{2,n}) \Delta_2 \Delta_1, \qquad (3)$$

where Δ_1, Δ_2 denote the discretization steps and

$$v_{1,m} := (m - \tfrac{N}{2})\Delta_1 \qquad v_{2,n} := (n - \tfrac{N}{2})\Delta_2 \qquad m, n = 0, \ldots, N-1. \qquad (4)$$

Recall that a two-dimensional *fast Fourier transform* (FFT) computes, for any complex (input) array, $\{X[j_1, j_2] \in \mathbb{C} \mid j_1 = 0, \ldots, N_1 - 1, j_2 = 0, \ldots, N_2 - 1\}$, the following (output) array of identical structure:

$$Y[l_1, l_2] := \sum_{j_1=0}^{N_1-1}\sum_{j_2=0}^{N_2-1} e^{-\frac{2\pi i}{N_1} j_1 l_1 - \frac{2\pi i}{N_2} j_2 l_2} X[j_1, j_2], \qquad (5)$$

for all $l_1 = 0, \ldots, N_1 - 1, l_2 = 0, \ldots, N_2 - 1$. In order to apply this algorithm to evaluate the sum in (3) above, we define a grid of size $N \times N$, set $\Lambda :=$

$\{(k_{1,p}, k_{2,q}) : 0 \le p, q \le N-1\}$, where

$$k_{1,p} := (p - \tfrac{N}{2})\lambda_1, k_{2,q} := (q - \tfrac{N}{2})\lambda_2 \ ,$$

and evaluate on it the sum

$$\Gamma(k_1, k_2) := \sum_{m=0}^{N-1} \sum_{n=0}^{N-1} e^{-i(v_{1,m}k_1 + v_{2,n}k_2)} \psi_T(v_{1,m}, v_{2,n}).$$

Choosing $\lambda_1 \Delta_1 = \lambda_2 \Delta_2 = \frac{2\pi}{N}$ gives the following values of $\Gamma(\cdot,\cdot)$ on Λ:

$$\begin{aligned}
\Gamma(k_{1,p}, k_{2,q}) &= \sum_{m=0}^{N-1} \sum_{n=0}^{N-1} e^{-i(v_{1,m}k_{1,p} + v_{2,n}k_{2,q})} \psi_T(v_{1,m}, v_{2,n}) \\
&= \sum_{m=0}^{N-1} \sum_{n=0}^{N-1} e^{-\frac{2\pi i}{N}\left[(m-N/2)(p-N/2)+(n-N/2)(q-N/2)\right]} \psi_T(v_{1,m}, v_{2,n}) \\
&= (-1)^{p+q} \sum_{m=0}^{N-1} \sum_{n=0}^{N-1} e^{-\frac{2\pi i}{N}(mp+nq)} \Big[(-1)^{m+n} \psi_T(v_{1,m}, v_{2,n})\Big].
\end{aligned}$$

This is computed by the fast Fourier transform of (5) by taking the input array as

$$X[m,n] = (-1)^{m+n} \psi_T(v_{1,m}, v_{2,n}) \quad m, n = 0, \ldots, N-1.$$

The result is an approximation for the option price at $N \times N$ different (log) strikes given by

$$C_T(k_{1,p}, k_{2,q}) \approx \frac{e^{-\alpha_1 k_{1,p} - \alpha_2 k_{2,q}}}{(2\pi)^2} \Gamma(k_{1,p}, k_{2,q}) \Delta_2 \Delta_1 \quad 0 \le p, q \le N.$$

3 FFT Pricing of the Spread Option

3.1 Pricing a Spread Option with Riemann Sums

Let us now consider the *price* of a European *spread option*, given by

$$\begin{aligned}
V(K) &:= \mathbb{E}_{\mathbb{Q}}\Big[e^{-rT}\big[\boldsymbol{S}_2(T) - \boldsymbol{S}_1(T) - K\big]_+\Big] \\
&= \int\!\!\int_{\Omega} e^{-rT}(e^{s_2} - e^{s_1} - K) q_T(s_1, s_2) ds_2 ds_1 \\
&= \int_{-\infty}^{\infty} \int_{\log(e^{s_1}+K)}^{\infty} e^{-rT}(e^{s_2} - e^{s_1} - K) q_T(s_1, s_2) ds_2 ds_1,
\end{aligned}$$

where the *exercise region* is defined as

$$\Omega := \Big\{(s_1, s_2) \in \mathbb{R}^2 \Big| e^{s_2} - e^{s_1} - K \ge 0\Big\}.$$

Transforming the option price with respect to the log of the strike K no longer gives the same kind of simple relationship with the characteristic function as in (2) of the previous section as a consequence of the simple shape of the exercise region Ω of the correlation option. If the boundaries of Ω are made up of straight edges, an appropriate affine change of variables can be introduced to make the method in the previous section applicable. This will not work for the pricing of spread options for which the exercise region is by nature non-linear (see Figure 1).

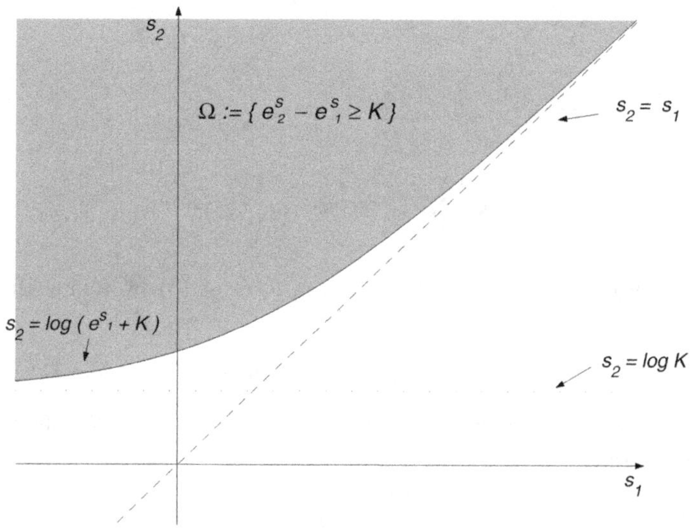

Fig. 1. Exercise region of a spread option in logarithmic variables

Notice however from above that the FFT option pricing method gives $N \times N$ prices simultaneously in one transform, that is, integrals of the payoff over $N \times N$ different regions. By subtracting and collecting the correct pieces, we can form tight upper and lower bounds for an integral over a non-polygonal region analogous to integrating by Riemann sums. More specifically, we consider the following modified exercise region:

$$\Omega_\lambda := \left\{ (s_1, s_2) \in \left[-\tfrac{1}{2}N\lambda, \tfrac{1}{2}N\lambda \right) \times \mathbb{R} \;\middle|\; e^{s_2} - e^{s_1} - K \geq 0 \right\}$$

and construct two "sandwiching" regions $\underline{\Omega} \subset \Omega_\lambda \subset \overline{\Omega}$ out of rectangular strips with vertices on the grid of the inverse transform (see Figures 2 and 3).

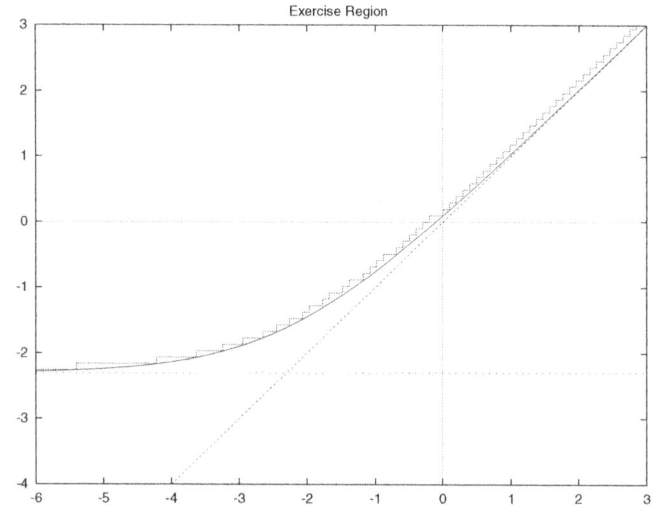

Fig. 2. Construction of the boundary of the approximate region $\underline{\Omega}$

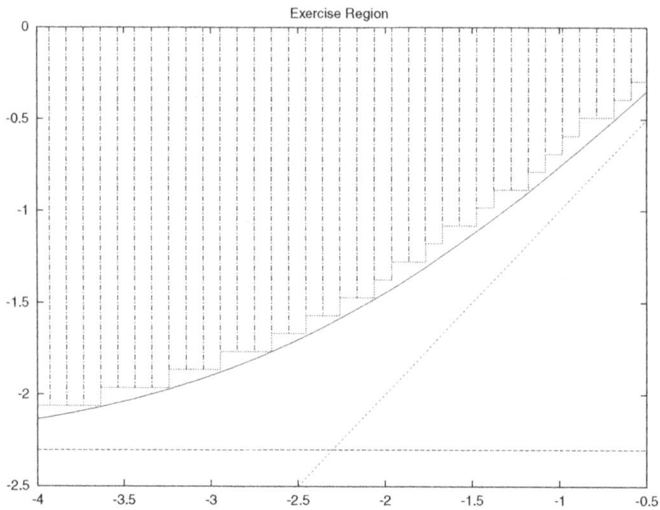

Fig. 3. Approximation of the exercise region with rectangular strips

Take as before an $N \times N$ equally spaced grid $\Lambda_1 \times \Lambda_2$, where

$$\Lambda_1 := \left\{k_{1,p} \in \mathbb{R} \mid k_{1,p} := (p - \tfrac{1}{2}N)\lambda,\ \ p = 0, \cdots, N-1\right\}$$
$$\Lambda_2 := \left\{k_{2,q} \in \mathbb{R} \mid k_{2,q} := (q - \tfrac{1}{2}N)\lambda,\ \ q = 0, \cdots, N-1\right\}.$$

For each $p = 0, \ldots, N-1$, define

$$\underline{k}_2(p) := \min_{0 \le q \le N-1} \left\{k_{2,q} \in \Lambda_2 \middle| e^{k_{2,q}} - e^{k_{1,p+1}} \ge K\right\}$$
$$\overline{k}_2(p) := \max_{0 \le q \le N-1} \left\{k_{2,q} \in \Lambda_2 \middle| e^{k_{2,q}} - e^{k_{1,p}} < K\right\}$$

to be the s_2-coordinates of the lower edges of the following rectangular strips,

$$\underline{\Omega}_p := [k_{1,p}, k_{1,p+1}) \times [\underline{k}_2(p), \infty)$$
$$\overline{\Omega}_p := [k_{1,p}, k_{1,p+1}) \times [\overline{k}_2(p), \infty).$$

Putting these together we obtain two regions bounding Ω_λ:

$$\underline{\Omega} := \bigcup_{p=0}^{N-1} \underline{\Omega}_p\,, \quad \overline{\Omega} := \bigcup_{p=0}^{N-1} \overline{\Omega}_p.$$

Since $\underline{\Omega} \subset \Omega_\lambda$ and the spread option payoff is positive over Ω_λ, we have a lower bound for its integral with the pricing kernel over this region:

$$\begin{aligned} V(K) &:= \int\!\!\int_{\Omega_\lambda} e^{-rT}(e^{s_2} - e^{s_1} - K) q_T(s_1, s_2) ds_2 ds_1 \\ &\gtrsim \int\!\!\int_{\underline{\Omega}} e^{-rT}(e^{s_2} - e^{s_1} - K) q_T(s_1, s_2) ds_2 ds_1. \end{aligned} \tag{6}$$

Establishing the upper bound is a trickier issue since the integrand is not positive over the entire region $\overline{\Omega}$. In fact, the payoff is strictly negative over $\overline{\Omega} \setminus \Omega_\lambda$ by the definition of Ω_λ. To overcome this, we shall pick some $\epsilon > 0$ such that

$$\overline{\Omega} \subset \left\{(s_1, s_2) \in \mathbb{R}^2 \middle| e^{s_2} - e^{s_1} - K \ge -\epsilon\right\}.$$

We then have

$$\begin{aligned} V(K) &= e^{-rT}\Big[\int\!\!\int_{\Omega_\lambda} \big(e^{s_2} - e^{s_1} - (K - \epsilon)\big) q_T(s_1, s_2) ds_2 ds_1 \\ &\qquad - \int\!\!\int_{\Omega_\lambda} \epsilon \cdot q_T(s_1, s_2) ds_2 ds_1\Big] \\ &\lesssim e^{-rT}\Big[\int\!\!\int_{\overline{\Omega}} \big(e^{s_2} - e^{s_1} - (K - \epsilon)\big) q_T(s_1, s_2) ds_2 ds_1 \\ &\qquad - \int\!\!\int_{\underline{\Omega}} \epsilon \cdot q_T(s_1, s_2) ds_2 ds_1\Big] \\ &= e^{-rT}\Big[\int\!\!\int_{\overline{\Omega}} (e^{s_2} - e^{s_1}) q_T(s_1, s_2) ds_2 ds_1 - (K - \epsilon)\int\!\!\int_{\overline{\Omega}} q_T(s_1, s_2) ds_2 ds_1 \\ &\qquad - \epsilon \int\!\!\int_{\underline{\Omega}} q_T(s_1, s_2) ds_2 ds_1\Big]. \end{aligned} \tag{7}$$

By breaking (6) and (7) into two components we can obtain these bounds by integrating $(e^{s_2} - e^{s_1}) \cdot q_T(s_1, s_2)$ and the density $q_T(s_1, s_2)$ over $\underline{\Omega}$ and $\overline{\Omega}$, using the fast Fourier transform method described in the previous section. Set

$$\underline{\Pi}_1 := \int\int_{\underline{\Omega}} (e^{s_2} - e^{s_1}) q_T(s_1, s_2) ds_2 ds_1 \quad \underline{\Pi}_2 := \int\int_{\underline{\Omega}} q_T(s_1, s_2) ds_2 ds_1$$
$$\overline{\Pi}_1 := \int\int_{\overline{\Omega}} (e^{s_2} - e^{s_1}) q_T(s_1, s_2) ds_2 ds_1 \quad \overline{\Pi}_2 := \int\int_{\overline{\Omega}} q_T(s_1, s_2) ds_2 ds_1.$$

Equations (6) and (7) can now be written as

$$e^{-rT} [\, \underline{\Pi}_1 - K \underline{\Pi}_2 \,] \lesssim V(K) \lesssim e^{-rT} [\, \overline{\Pi}_1 - (K - \epsilon) \overline{\Pi}_2 - \epsilon \underline{\Pi}_2 \,] \tag{8}$$

for suitable $\epsilon > 0$.

3.2 Computing the Sums by FFT

We now demonstrate in detail how to compute, by performing two fast Fourier transforms, the four components $\underline{\Pi}_1, \underline{\Pi}_2, \overline{\Pi}_1, \overline{\Pi}_2$ in the approximate pricing equations (8) and hence the spread option prices across different strikes. (In fact, if one only wishes to approximate the option price from below, a single transform is sufficient.) The argument is set out explicitly for $\underline{\Pi}_1$ below and the other three cases follow similarly.

$$\begin{aligned}
\underline{\Pi}_1 &:= \int\int_{\underline{\Omega}} (e^{s_2} - e^{s_1}) q_T(s_1, s_2) ds_2 ds_1 = \sum_{p=0}^{N-1} \int\int_{\underline{\Omega}_p} (e^{s_2} - e^{s_1}) q_T(s_1, s_2) ds_2 ds_1 \\
&= \sum_{p=0}^{N-1} \Bigg[\int_{k_{1,p}}^{\infty} \int_{\underline{k}_2(p)}^{\infty} (e^{s_2} - e^{s_1}) q_T(s_1, s_2) ds_2 ds_1 \\
&\qquad - \int_{k_{1,p+1}}^{\infty} \int_{\underline{k}_2(p)}^{\infty} (e^{s_2} - e^{s_1}) q_T(s_1, s_2) ds_2 ds_1 \Bigg] \\
&= \sum_{p=0}^{N-1} \big(\Pi_1(k_{1,p}, \underline{k}_2(p)) - \Pi_1(k_{1,p+1}, \underline{k}_2(p)) \big),
\end{aligned} \tag{9}$$

where for all $(k_1, k_2) \in \mathbb{R}^2$

$$\Pi_1(k_1, k_2) := \int_{k_1}^{\infty} \int_{k_2}^{\infty} (e^{s_2} - e^{s_1}) q_T(s_1, s_2) ds_2 ds_1.$$

As before we apply the Fourier transform to the following modified integrand:

$$\pi_1(k_1, k_2) := e^{\alpha_1 k_1 + \alpha_2 k_2} \Pi_1(k_1, k_2), \quad \alpha_1, \alpha_2 > 0$$

whose transform χ_1 bears a simple relationship with the characteristic function ϕ_T:

$$\begin{aligned}\chi_1(v_1, v_2) &:= \int_{-\infty}^{\infty}\int_{-\infty}^{\infty} e^{i(v_1k_1+v_2k_2)}\pi_1(k_1,k_2)dk_2dk_1 \\ &= \int_{-\infty}^{\infty}\int_{-\infty}^{\infty} e^{(\alpha_1+iv_1)k_1+(\alpha_2+iv_2)k_2} \\ &\quad \cdot \int_{k_2}^{\infty}\int_{k_1}^{\infty} (e^{s_2}-e^{s_1})q_T(s_1,s_2)ds_2ds_1\ dk_2dk_1 \\ &= \int_{-\infty}^{\infty}\int_{-\infty}^{\infty} (e^{s_2}-e^{s_1})q_T(s_1,s_2) \\ &\quad \cdot \int_{-\infty}^{s_2}\int_{-\infty}^{s_1} e^{(\alpha_1+iv_1)k_1+(\alpha_2+iv_2)k_2}dk_2dk_1\ ds_2ds_1 \\ &= \int_{-\infty}^{\infty}\int_{-\infty}^{\infty} (e^{s_2}-e^{s_1})q_T(s_1,s_2)\frac{e^{(\alpha_1+iv_1)s_1+(\alpha_2+iv_2)s_2}}{(\alpha_1+iv_1)(\alpha_2+iv_2)}ds_2ds_1 \\ &= \frac{\phi_T(v_1-\alpha_1 i, v_2-(\alpha_2+1)i)-\phi_T(v_1-(\alpha_1+1)i, v_2-\alpha_2 i)}{(\alpha_1+iv_1)(\alpha_2+iv_2)}. \quad (10)\end{aligned}$$

Discretising as in the previous section with

$$\begin{aligned}\lambda_1\cdot\Delta_1 = \lambda_2\cdot\Delta_2 = \tfrac{2\pi}{N} \\ v_{1,m} := (m-\tfrac{N}{2})\Delta_1 \qquad v_{2,n} := (n-\tfrac{N}{2})\Delta_2,\end{aligned}$$

we now have via an (inverse) fast Forier transform values of $\Pi_1(\cdot,\cdot)$ on all $N\times N$ vertices of the grid $\Lambda_1\times\Lambda_2$ given by

$$\begin{aligned}\Pi_1(k_{1,p},k_{2,q}) &= \frac{e^{-\alpha_1k_{1,p}-\alpha_2k_{2,q}}}{(2\pi)^2}\int_{-\infty}^{\infty}\int_{-\infty}^{\infty} e^{-i(v_1k_{1,p}+v_2k_{2,q})}\chi_1(v_1,v_2)dv_2dv_1 \\ &\approx \frac{e^{-\alpha_1k_{1,p}-\alpha_2k_{2,q}}}{(2\pi)^2}\sum_{m=0}^{N-1}\sum_{n=0}^{N-1} e^{-i(v_{1,m}k_{1,p}+v_{2,n}k_{2,q})}\chi_1(v_{1,m},v_{2,n})\Delta_2\Delta_1 \\ &= \frac{(-1)^{p+q}\cdot e^{-\alpha_1k_{1,p}-\alpha_2k_{2,q}}}{(2\pi)^2}\Delta_2\Delta_1 \\ &\quad \cdot\sum_{m=0}^{N-1}\sum_{n=0}^{N-1} e^{-\frac{2\pi i}{N}(mp+nq)}\left[(-1)^{m+n}\chi_1(v_{1,m},v_{2,n})\right] \quad (11)\end{aligned}$$

and hence the values of the $2\cdot N$ required components in (9). Repeating the same procedure for the other components in (8) gives the bounds for the spread option value $V(K)$.

4 Numerical Performance

4.1 Underlying Models

Previous works on spread options have concentrated on the two-factor *geometric Brownian motion* (GBM) model in which the risk-neutral dynamics of the underlying assets are given by

$$\begin{aligned} d\boldsymbol{S}_1 &= \boldsymbol{S}_1((r-\delta_1)dt + \sigma_1 d\boldsymbol{W}_1) \\ d\boldsymbol{S}_2 &= \boldsymbol{S}_2((r-\delta_2)dt + \sigma_2 d\boldsymbol{W}_2), \end{aligned}$$

where $\mathbb{E}_{\mathbb{Q}}[d\boldsymbol{W}_1 d\boldsymbol{W}_2] = \rho dt$ and r, δ_i, σ_i denote the risk-free rate, dividend yields and volatilities respectively. Working with the log prices, $\boldsymbol{s}_i := \log \boldsymbol{S}_i$, one has the following pair of SDEs:

$$\begin{aligned} d\boldsymbol{s}_1 &= (r-\delta_1-\tfrac{1}{2}\sigma_1^2)dt + \sigma_1 d\boldsymbol{W}_1 \\ d\boldsymbol{s}_2 &= (r-\delta_2-\tfrac{1}{2}\sigma_2^2)dt + \sigma_2 d\boldsymbol{W}_2 \ . \end{aligned}$$

We shall now extend this model to include a third factor, a mean-revert stochastic volatility for the two underlying processes:

$$\begin{aligned} d\boldsymbol{s}_1 &= (r-\delta_1-\tfrac{1}{2}\sigma_1^2\boldsymbol{\nu})dt + \sigma_1\sqrt{\boldsymbol{\nu}}\, d\boldsymbol{W}_1 \\ d\boldsymbol{s}_2 &= (r-\delta_2-\tfrac{1}{2}\sigma_2^2\boldsymbol{\nu})dt + \sigma_2\sqrt{\boldsymbol{\nu}}\, d\boldsymbol{W}_2 \\ d\boldsymbol{\nu} &= \kappa(\mu-\boldsymbol{\nu})dt + \sigma_\nu\sqrt{\boldsymbol{\nu}}\, d\boldsymbol{W}_\nu \ , \end{aligned} \tag{12}$$

where

$$\begin{aligned} \mathbb{E}_{\mathbb{Q}}[d\boldsymbol{W}_1 d\boldsymbol{W}_2] &= \rho\, dt \\ \mathbb{E}_{\mathbb{Q}}[d\boldsymbol{W}_1 d\boldsymbol{W}_\nu] &= \rho_1\, dt \\ \mathbb{E}_{\mathbb{Q}}[d\boldsymbol{W}_2 d\boldsymbol{W}_\nu] &= \rho_2\, dt. \end{aligned}$$

This is a direct generalisation of the single-asset stochastic volatility model [14,21] and is considered for the case of correlation options in [2]. Applying Ito's lemma and solving the resulting PDE, one obtains an analytical expression for its characteristic function:

$$\begin{aligned} \phi_{\mathrm{SV}}(u_1,u_2) &:= \mathbb{E}_{\mathbb{Q}}\Big[\exp\big(iu_1\boldsymbol{s}_1(T) + iu_2\boldsymbol{s}_2(T)\big)\Big] \\ &= \exp\Bigg[iu_1\cdot s_1(0) + iu_2\cdot s_2(0) + \left(\frac{2\zeta(1-e^{-\theta T})}{2\theta-(\theta-\gamma)(1-e^{-\theta T})}\right)\cdot\nu(0) \\ &\quad + \sum_{j=1,2} u_j(r-\delta_j)T - \frac{\kappa\mu}{\sigma_\nu^2}\left[2\cdot\log\left(\frac{2\theta-(\theta-\gamma)(1-e^{-\theta T})}{2\theta}\right) + (\theta-\gamma)T\right]\Bigg] , \end{aligned} \tag{13}$$

where

$$\zeta := -\frac{1}{2}\Big[\big(\sigma_1^2 u_1^2 + \sigma_2^2 u_2^2 + 2\rho\sigma_1\sigma_2 u_1 u_2\big) + i\big(\sigma_1^2 u_1 + \sigma_2^2 u_2\big)\Big]$$
$$\gamma := \kappa - i\big(\rho_1\sigma_1 u_1 + \rho_2\sigma_2 u_2\big)\sigma_\nu$$
$$\theta := \sqrt{\gamma^2 - 2\sigma_\nu^2\zeta}.$$

Notice that as we let the parameters of the *stochastic volatility* process approach the limits

$$\kappa, \mu, \sigma_\nu \to 0,\ \nu(0) \to 1,$$

the three-factor stochastic volatility (SV) model degenerates into the two-factor GBM model and the characteristic function simplifies to that of a bivariate normal distribution:

$$\phi_{\mathrm{GBM}}(u_1, u_2) = \exp\Big[\ iu_1 \cdot s_1(0) + iu_2 \cdot s_2(0) + \zeta\, T + \sum_{j=1,2} u_j(r - \delta_j)T\Big].$$

We shall use these two characteristic functions to compute the spread option prices under the GBM and SV model. In the former case the prices computed by the FFT method are compared to the analytic option value obtained by a one dimensional integration based on the conditioning technique. This fails when we introduce a stochastic volatility factor and thus a Monte Carlo pricing method is used as a benchmark for the SV model.

Prices are also compared for the two diffusion models. Given a set of parameter values for the SV model, one can compute from the characteristic function the mean vector and covariance matrix of $s_1(T), s_2(T)$ under the stochastic volatility assumption. We can infer for these the parameter values of the two-factor GBM model needed to produce the same moments. Option values may then be computed and compared to the three factor SV prices.

The code is written in C++ and includes the fast Fourier transform routine FFTW (the *Fastest Fourier Transform in the West*), written by M. Frigo and S.G. Johnson [12]. The experiments were conducted on an Athlon 650 MHz machine running under Linux with 512 MB RAM.

4.2 Computational Results

Table 1 documents the spread option prices across a range of strikes under the two factor geometric Brownian motion model [22], computed by three different techniques: one-dimensional integration (analytic), the fast Fourier transform and the Monte Carlo method. The values for the FFT methods shown are the "lower" prices, computed over $\underline{\Omega}$, regions that approach the true exercise region from inside and are therefore all less than the analytic price in the first column. 80000 simulation paths were used to produce the Monte Carlo prices and the average standard errors are recorded in brackets at the bottom. Note that if one is only interested in computing prices in the two factor world, it

Table 1. Prices computed by alternative methods under the 2-factor GBM model

	Analytic	Fast Fourier Transform				Monte Carlo	
		No. Discretisation N				Time Steps	
Strikes K		512	1024	2048	4096	1000	2000
0.0	8.513201	8.509989	8.511891	8.512981	8.513079	8.500949	8.516613
0.4	8.312435	8.311424	8.311995	8.312370	8.312385	8.300180	8.315818
0.8	8.114964	8.113877	8.114304	8.114901	8.114916	8.102730	8.118328
1.2	7.920790	7.919520	7.920173	7.920712	7.920741	7.908614	7.924135
1.6	7.729903	7.728471	7.729268	7.729810	7.729852	7.717831	7.733193
2.0	7.542296	7.540686	7.541637	7.542185	7.542242	7.530322	7.545496
2.4	7.357966	7.356278	7.357288	7.357830	7.357901	7.346038	7.361136
2.8	7.176888	7.175080	7.176185	7.176734	7.176818	7.164956	7.180054
3.2	6.999052	6.997200	6.998345	6.998881	6.998979	6.987070	7.002243
3.6	6.824451	6.822477	6.823721	6.824259	6.824371	6.812353	6.827700
4.0	6.653060	6.651047	6.652306	6.652852	6.652976	6.640874	6.656364
						(0.018076)	(0.018184)

$S_1(0) = 96$ $\delta_1 = 0.05$ $\sigma_1 = 0.1$ $S_2(0) = 100$ $\delta_2 = 0.05$ $\sigma_2 = 0.2$
$r = 0.1$ $T = 1.0$ $K = 4.0$ $\rho = 0.5$
Note: 80000 simulations have been used in the Monte Carlo method

is not actually necessary to discretise the time horizon $[0, T]$ and simulate price paths as was done here. Since we know the terminal joint distribution of the two asset prices are bivariate normal, they can be simulated directly and one single time step is sufficient. However, the point of this exercise is to acquire an intuition into how the computational time and accuracy varies as one changes the underlying assumptions, since the introduction of extra factors into a model inevitably involves generating the whole time paths of these factors.

The average errors of the two methods are computed and recorded in Table 2. First we note that integrating over $\underline{\Omega}$ from inside Ω is more accurate than integrating over $\overline{\Omega}$, as one can expect from the less straightforward pro-

Table 2. Accuracy of alternative methods for the 2-factor GBM model (b.p.)

Fast Fourier Transform			Monte Carlo		
Discretisation			Number of	Time Steps	
Number	Lower	Upper	Simulations	1000	2000
512	4.44	25.60	10000	129.15 (0.051839)	70.81 (0.050949)
1024	1.13	13.90	20000	22.34 (0.036225)	40.67 (0.035899)
2048	0.32	7.20	40000	7.44 (0.025737)	7.63 (0.025733)
4096	0.10	3.65	80000	18.34 (0.018076)	4.94 (0.018184)

$S_1(0) = 96$ $\delta_1 = 0.05$ $\sigma_1 = 0.1$ $S_2(0) = 100$ $\delta_2 = 0.05$ $\sigma_2 = 0.2$
$r = 0.1$ $T = 1.0$ $K = 4.0$ $\rho = 0.5$

cedure for constructing the upper bound. For $N = 1024$ the lower bound has an error of roughly one basis point, whereas $N = 2048$ takes us well below this error level. From Table 3 the inner approximation take 4.28 and 18.46 seconds respectively, clearly outperforming the Monte-Carlo method. For the same level of accuracy, one would require far more than 80000 simulations, which already take 304.95 seconds (606.40 seconds for the case of 2000 time steps) to generate.

Although the Monte Carlo code employed uses no variance reduction technique other than antithetic variates and its speed could be significantly improved, the method is still unlikely to beat the FFT method in performance.

A close examination of Table 3 reveals the real strength of the FFT method. As we introduce a stochastic volatility factor, the Monte Carlo technique needs to generate this value at each time step, which is then multiplied with the increments $d\boldsymbol{W}_1, d\boldsymbol{W}_2$ of the Brownian motions to give the asset prices in the next period. As indicated across the columns this increases the computational time by almost a factor of 4. Recalling the FFT method described in the previous section, we notice that only a different characteristic function is substituted when more factors are included, and the transform

Table 3. Computing times of alternative methods

Fast Fourier Transform				
Discretisation	10 Strikes		100 Strikes	
Number	GBM	SV	GBM	SV
512	1.04	1.11	1.10	1.20
1024	4.28	4.64	4.48	4.83
2048	18.46	19.54	18.42	19.74
4096	74.45	81.82	76.47	81.27

Monte Carlo: 1000 Time Steps				
Number of	10 Strikes		100 Strikes	
Simulation	GBM	SV	GBM	SV
10000	38.2	144.87	41.95	151.75
20000	76.22	288.09	83.81	303.31
40000	152.5	576.25	168.48	606.53
80000	304.95	1152.9	335.20	1212.76

Monte Carlo: 2000 Time Steps				
Number of	10 Strikes		100 Strikes	
Simulation	GBM	SV	GBM	SV
10000	75.57	287.41	79.83	295.21
20000	157.28	574.18	159.08	590.23
40000	303.37	1149.25	317.49	1184.32
80000	606.40	2298.37	636.33	2359.05

remains two dimensional. Comparing the times for the GBM and SV models, we observe only a 5 to 9 percent increase which is dropping as we increase the discretisation number. The extra computing time is due to the more complex expression of the characteristic function with a larger set of parameters.

For both methods however, increasing the number of strikes does not result in dramatic increases in the computational times.

Table 4 shows the spread option prices for different strikes under the three factor SV model. The Monte Carlo prices with a discretisation of 2000 time steps oscillate around those computed by the FFT method. Since we observe that in the two factor case the errors of the Monte Carlo method remain high even for 80000 simulations, more experiments need to be conducted for a conclusive judgment on this point.

Finally, Figure 4 plots the difference in the spread option values under the 3-factor stochastic volatility model and the corresponding 2-factor geometric Brownian motion model with the same means and covariances at exercise. Under the SV model, knowing the characteristic function of $\boldsymbol{s}_1, \boldsymbol{s}_2$, we can calculate their means and covariance matrix, which can then be used as the implied parameters $r - \delta_i$ and σ_i, $i = 1, 2$, and ρ for the GBM model. We repeat the procedure for different values of ρ_1, ρ_2, the correlation parameters between the Brownian motions $\boldsymbol{W}_i, i = 1, 2$, driving the asset prices and $\boldsymbol{W}_\nu$ driving the stochastic volatility factor $\boldsymbol{\nu}$. When ρ_1, ρ_2 are high, a large

Table 4. Prices computed by alternative methods under the 3-factor SV model

	Fast Fourier Transform		Monte Carlo			
	No. of Discretisations		No. of Simulations			
Strikes K	512	2048	10000	20000	40000	80000
2.0	7.546895	7.543618	7.514375	7.567536	7.572211	7.523968
2.2	7.451878	7.452998	7.421861	7.475093	7.479742	7.431489
2.4	7.357703	7.363377	7.330142	7.383470	7.388080	7.339813
2.6	7.264298	7.274876	7.239209	7.292616	7.297218	7.248961
2.8	7.171701	7.186990	7.149234	7.202571	7.207191	7.158919
3.0	7.079987	7.099819	7.060043	7.113303	7.117954	7.069687
3.2	6.989008	7.013731	6.971625	7.024808	7.029515	6.981272
3.4	6.898826	6.928373	6.884026	6.937119	6.941875	6.893664
3.6	6.809471	6.843671	6.797246	6.850283	6.854984	6.806859
3.8	6.720957	6.759903	6.711328	6.764275	6.768886	6.720859
4.0	6.633232	6.676768	6.626221	6.679076	6.683587	6.635661
			(0.052702)	(0.036984)	(0.025739)	(0.018206)

$r = 0.1$ $T = 1.0$ $\rho = 0.5$
$S_1(0) = 96$ $\delta_1 = 0.05$ $\sigma_1 = 0.5$ $\rho_1 = 0.25$
$S_2(0) = 100$ $\delta_2 = 0.05$ $\sigma_2 = 1.0$ $\rho_1 = -0.5$
$\nu(0) = 0.04$ $\kappa = 1.0$ $\mu = 0.04$ $\sigma_\nu = 0.05$
Note: 2000 time steps have been used for the Monte Carlo simulation.

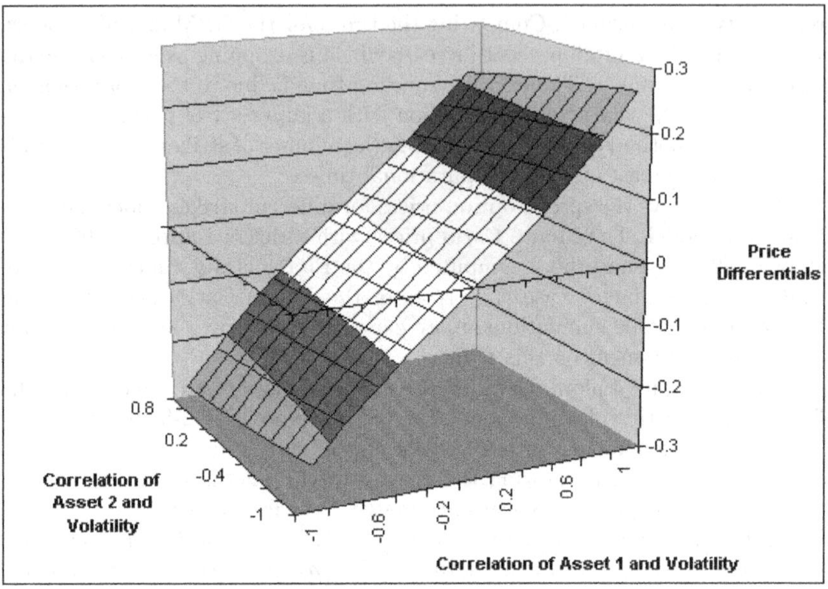

Fig. 4. Price difference between SV model and the GBM model with implied parameters

increment $\boldsymbol{W}_\nu$ in (12) is more likely to induce simultaneously large values of $\boldsymbol{W}_i, i = 1, 2$, and $d\boldsymbol{\nu}$. This increases the volatilities of both $\boldsymbol{s}_1$ and $\boldsymbol{s}_2$ and hence the spread and the spread option value. Compared with the two factor GBM model, the SV model of (12) obviously exhibits a richer structure for the spread option value which can be used by traders with forward views on the term structures of volatilities and correlations of the components of the spread [16].

5 Conclusions and Future Directions

We have described and implemented an efficient method of computing, via the construction of suitable approximate exercise regions, the value of a generic spread option under models for which the characteristic function of the two underlying asset prices is known in closed form. This takes us well beyond the two factor constant correlation Gaussian framework found in the existing literature, which is commonly assumed only for its tractability. In particular, one can now price spread options under many multi-factor models in the affine jump-diffusion family. For example, an index spread option in the equity markets can be priced under stochastic volatility models. Spark and crack spread options in the energy markets can now be valued with asset price spikes and random volatility jumps, with major implications for trading, as well as for asset and real option valuation.

Furthermore, switching between alternative diffusion models only amounts to substituting a different characteristic function for the underlying prices/rates, leaving the dimension of the transform and the summation procedure unchanged. As more factors are introduced more time is devoted to the inexpensive evaluation of the more complex characteristic function, but *not* to the fast Fourier Transform algorithm. This significantly cuts down the increase of computational times expected when one applies the generic PDE or Monte Carlo approaches to such a high dimensional option pricing problem.

The computational advantage of the approach is demonstrated with numerical experiments for both the two factor geometric Brownian motion and the three factor stochastic volatility models. Price differentials between the models as one varies the parameters of the volatility process confirm the significance of a non-trivial correlation structure in the model dynamics.

One possible direction to enrich the volatility and correlation structure further is to assume a four factor model with two correlated stochastic volatility processes [7]. The calibration issue also remains to be resolved in detail, where the focus of concern will be on an efficient procedure for backing out a implied correlation surface from observed option prices.[1]

References

1. Bakshi, G. and Z. Chen (1997). An alternative valuation model for contingent claims. *Journal of Financial Economics* **44** (1) 123–165.
2. Bakshi, G. and D. Madan (2000). Spanning and derivative-security valuation. *Journal of Financial Economics* **55** 205–238.
3. Barndorff-Nielsen, O. (1997). Processes of normal inverse Gaussian type. *Finance and Stochastics* **2** 41–68.
4. Bates, D. (1996). Jumps and stochastic volatility: Exchange rate process implicit in Deutschmark options. *Review of Financial Studies* **9** 69–108.
5. Carr, P. and D. B. Madan (1999). Option valuation using the fast Fourier transform. *The Journal of Computational Finance* **2** (4) 61–73.
6. Chen, R. and L. Scott (1992). Pricing interest rate options in a two-factor Cox-Ingersoll-Ross model of the term structure. *Review of Financial Studies* **5** 613–636.
7. Clewlow, L. and C. Strickland (1998). *Implementing Derivatives Models.* John Wiley & Sons Ltd.
8. Dempster, M. A. H. and J. P. Hutton (1999). Pricing American stock options by linear programming. *Mathematical Finance* **9** (3) 229–254.
9. Deng, S. (1999). Stochastic models of energy commodity prices and their applications: mean-reversion with jumps and spikes. Working paper, Georgia Institute of Technology, October.

[1] More ambitions still would be to attempt to combine the FFT techniques of this paper with the linear programming methods of [8] in order to produce fast valuation of *American* spread options.

10. Duan, J.-C. and S. R. Pliska (1999). Option valuation with co-integrated asset prices. Working paper, Department of Finance, Hong Kong University of Science and Technology, January.
11. Duffie, D., J. Pan and K. Singleton (1999). Transform analysis and asset pricing for affine jump-diffusions. Working paper, Graduate School of Business, Stanford University, August.
12. Frigo, M. and S. G. Johnson (1999). *FFTW user's manual.* MIT, May.
13. Garman, M. (1992). Spread the load. *RISK* **5** (11) 68–84.
14. Heston, S. (1993). A closed-form solution for options with stochastic volatility, with applications to bond and currency options. *Review of Financial Studies* **6** 327–343.
15. Madan, D., P. Carr and E. Chang (1998). The variance gamma process and option pricing. *European Finance Review* **2** 79–105.
16. Mbanefo, A. (1997). Co-movement term structure and the valuation of crack energy spread options. In *Mathematics of Derivatives Securities.* M. A. H. Dempster and S. R. Pliska, eds. Cambridge University Press, 89-102.
17. Pearson, N. D. (1995). An efficient approach for pricing spread options. *Journal of Derivatives* **3**, Fall, 76–91.
18. Pilipovic, D. and J. Wengler (1998). Basis for boptions. *Energy and Power Risk Management,* December, 28–29.
19. Poitras, G. (1998). Spread options, exchange options, and arithmetic Brownian motion. *Journal of Futures Markets* **18** (5) 487–517.
20. Ravindran, K. (1993). Low-fat spreads. *RISK* **6** (10) 56–57.
21. Scott, L. O. (1997). Pricing stock options in a jump-diffusion model with stochastic volatility and interest rates: Applications of Fourier inversion methods. *Mathematical Finance* **7** (4) 413–426.
22. Shimko, D. C. (1994). Options on futures spreads: hedging, speculation, and valuation. *The Journal of Futures Markets* **14** (2) 183–213.
23. Trigeorgis, L. (1996). *Real Options - Managerial Flexibility and Strategy in Resource Allocation.* MIT Press, Cambridge, Mass.

The Law of Geometric Brownian Motion and its Integral, Revisited; Application to Conditional Moments

Catherine Donati-Martin[1], Hiroyuki Matsumoto[2] and Marc Yor[3]

[1]Laboratoire de Statistique et Probabilités, Université Paul Sabatier, 118, route de Narbonne, 31062 Toulouse Cedex 04, France,
[2]School of Informatics and Sciences, Nagoya University, Chikusa-ku, Nagoya 464-8601, Japan,
[3]Laboratoire de Probabilités, Université de Paris IV, 175, rue du Chevaleret, 75013 Paris, France

Introduction

I.a. Part A of this paper is a transcript of the third author's lecture at the Bachelier Conference, July 1st, 2000; it is a summary of our joint works on this topic.

The most encountered stochastic process in Mathematical Finance is $e^{(\mu)} = \{e_t^{(\mu)} = \exp(B_t+\mu t), t \geqq 0\}$, the so-called geometric Brownian motion, with parameter $\mu \in \mathbf{R}$. The exponential function transforms the independence of increments into independence of ratios, but more importantly, a main difference with the Brownian motion $B_t^{(\mu)} = B_t + \mu t$ itself is that the quadratic variation

$$\langle e^{(\mu)} \rangle_t = \int_0^t \exp(2B_s^{(\mu)})\, ds \equiv A_t^{(\mu)}$$

is not deterministic and, in fact, provides as much information as $e^{(\mu)}$ itself. A large part of the results presented below consists in constructing one-dimensional diffusion processes from the two-dimensional one $\{(e_t^{(\mu)}, \langle e^{(\mu)} \rangle_t), t \geqq 0\}$.

I.b. In Part B, we apply the results of Part A to obtain some closed formulae for the conditional moments

$$E[(A_t)^r | B_t = x]$$

for $r \in \mathbf{R}$.

Part A: Important features of the law of $(e_t^{(\mu)}, A_t^{(\mu)})$

1 Lamperti's Representation of Geometric Brownian Motions

Lamperti [13] showed that the exponential of every Lévy process $\{\xi_t, t \geqq 0\}$ may be represented implicitly as follows:

$$\exp(\xi_t) = X_{\int_0^t \exp(\xi_s)ds}, \qquad t \geqq 0,$$

where $\{X_u, u \geqq 0\}$ is a semi-stable Markov process, i.e., a Markov process which satisfies the scaling property

$$(\{X_{cu}, u \geqq 0\}, X_0 = x) \overset{\text{(law)}}{=} (\{cX_u, u \geqq 0\}, X_0 = x/c).$$

More generally, for any $\alpha \in \mathbf{R}$, $\alpha \neq 0$, it holds that

$$\exp(\xi_t) = X^{(1/\alpha)}_{\int_0^t \exp(\alpha\xi_s)ds},$$

where $\{X_u^{(1/\alpha)}, u \geqq 0\}$ has the scaling property:

$$(\{X^{(1/\alpha)}_{c^\alpha u}, u \geqq 0\}, X_0^{(1/\alpha)} = x) \overset{\text{(law)}}{=} (\{cX_u^{(1/\alpha)}, u \geqq 0\}, X_0^{(1/\alpha)} = x/c).$$

In the case of $\xi_t = B_t^{(\nu)} \equiv B_t + \nu t$, we choose systematically $\alpha = 2$, and consequently, $X^{(1/2)}$ is a Bessel process of index ν or of dimension $2(\nu+1)$ with infinitesimal generator

$$\frac{1}{2}\frac{d^2}{dx^2} + \frac{2\nu+1}{2x}\frac{d}{dx}, \qquad \text{on } C_c^\infty(0,\infty).$$

We write

$$\exp(B_t^{(\nu)}) = R^{(\nu)}_{A_t^{(\nu)}}, \qquad \text{where} \quad A_t^{(\nu)} = \int_0^t \exp(2B_s^{(\nu)})\, ds.$$

We also note that

$$(A_t^{(\nu)} < u) = (t < C_u^{(\nu)}),$$

where $C_u^{(\nu)} = \int_0^u (R_s^{(\nu)})^{-2} ds$ is the Bessel clock featured, when ν is a half integer, in the skew product decomposition of a $d = 2(\nu+1)$ dimensional Brownian motion $B^{(d)} = \{B_t^{(d)}\}$: there exists a Brownian motion $\{\Sigma_u, u \geqq 0\}$ on the $(d-1)$-sphere S^{d-1}, independent of $\{R_t^{(\nu)}, t \geqq 0\}$, such that

$$B_t^{(d)} = R_t^{(\nu)} \Sigma_{C_t^{(\nu)}}.$$

For this see Itô–McKean [11], Section 7.15; see also Pauwels–Rogers [19] for a more complete panorama of skew product decompositions.

2 Some Consequences

2.a. For $\nu < 0$, we have $A_\infty^{(\nu)} = \inf\{u; R_u^{(\nu)} = 0\}$, which yields

$$A_\infty^{(\nu)} \stackrel{\text{(law)}}{=} \frac{1}{2\gamma_{(-\nu)}}, \tag{2.1}$$

where $\gamma_{(\mu)}$ denotes a Gamma(μ) random variable. See [6] and [24] for details; see also [3] for a more general set of results concerning perpetuities, that is, the random variables of the form $A_\infty(\xi) \equiv \int_0^\infty \exp(\xi_s)ds$ for a Lévy process $\xi = \{\xi_s\}$.

2.b. With the help of Lamperti's representation, one can show the following (cf. [9], [25]):

$$A_{T_\lambda}^{(\nu)} \stackrel{\text{(law)}}{=} \frac{1 - U^{1/a}}{2\gamma_b}, \tag{2.2}$$

where T_λ is an exponentially distributed random variable with parameter $\lambda > 0$, which is independent of B; on the other hand, U and γ_b are respectively a standard uniform and a Gamma(b) independent random variables, and

$$a = \frac{\mu + \nu}{2}, \qquad b = \frac{\mu - \nu}{2}, \qquad \mu = \sqrt{2\lambda + \nu^2}.$$

2.c. Comparing the laws found in (2.2) for $\nu > 0$ and $\nu < 0$, Dufresne [7] remarked that, for fixed t and $\mu > 0$,

$$\frac{1}{A_t^{(-\mu)}} \stackrel{\text{(law)}}{=} \frac{1}{A_t^{(\mu)}} + \frac{1}{\tilde{A}_\infty^{(-\mu)}}. \tag{2.3}$$

holds, where $\tilde{A}_\infty^{(-\mu)}$ is a copy of $A_\infty^{(-\mu)} \equiv \int_0^\infty \exp(2B_s^{(-\mu)})ds$ independent of $\{A_t^{(\mu)}\}$. In Section 6 we shall show that this relationship holds at the process level, that is, for all t's simultaneously. For more details, see Matsumoto–Yor [18].

2.d. This is quite remarkable, and should be contrasted with Bougerol's identity:

$$\sinh(B_t) \stackrel{\text{(law)}}{=} \int_0^t \exp(B_s)d\beta_s \stackrel{\text{(law)}}{=} \beta_{A_t} \stackrel{\text{(law)}}{=} \sqrt{A_t}\beta_1 \qquad \text{for fixed } \ t > 0, \tag{2.4}$$

where $\{\beta_s, s \geqq 0\}$ is a Brownian motion independent of B. This time, the identity in law does not hold at the process level.

We note here that (2.4) implies

$$N^2 A_t \stackrel{\text{(law)}}{=} (\sinh B_t)^2 \tag{2.5}$$

for a standard Gaussian random variable N, independent of B. Identity (2.5) characterizes the law of A_t for fixed t.

2.e. Identity (2.2) may also be compared with explicit results at fixed time [25]:

$$\frac{d}{du}\frac{d}{dx}P(A_t \leqq u, B_t \leqq x) = \frac{1}{u}\exp(-\frac{1}{2u}(1+e^{2x}))\theta_{e^x/u}(t), \tag{2.6}$$

where $\theta_r(t), t \geqq 0$, is Hartman's density given by the Laplace transform

$$I_{|\nu|}(r) = \int_0^\infty \exp(-\frac{\nu^2 t}{2})\theta_r(t)\,dt,$$

where I_μ denotes the usual modified Bessel function of index μ. In fact, (2.6) follows from

$$E[\exp(-\frac{\nu^2}{2}C_t^{(0)})|R_t^{(0)} = \rho] = (\frac{I_{|\nu|}}{I_0})(\frac{\rho}{t}).$$

More generally, for $\mu \geqq 0$,

$$E[\exp(-\frac{\nu^2}{2}C_t^{(\mu)})|R_t^{(\mu)} = \rho] = (\frac{I_\lambda}{I_\mu})(\frac{\rho}{t}),$$

where $\lambda = \sqrt{\mu^2+\nu^2}$. For a general discussion of the interplay between the inverse processes $A^{(\mu)}$ and $C^{(\mu)}$, see, e.g., Yor [26].

3 A Parallel Between the Laws of $(B_t^{(\mu)}, \sup_{s\leqq t} B_s^{(\mu)})$ and $(e_t^{(\mu)}, A_t^{(\mu)})$

3.a. Let $B = \{B_s, s \geqq 0\}$ be a standard Brownian motion starting from 0 and consider also a Brownian motion with drift $B^{(\mu)}$ given by $B_t^{(\mu)} = B_t + \mu t$. The conjunction of the well known approximation result(the Laplace method),

$$\frac{1}{c}\log\Big(\int_0^t \exp(cB_s^{(\mu)})ds\Big) \to M_t^{(\mu)} \equiv \sup_{s\leqq t} B_s^{(\mu)}, \qquad \text{as } c\to\infty,$$

Lévy's theorem $\{(M_t - B_t, M_t), t \geqq 0\} \overset{(\text{law})}{=} \{(|B_t|, L_t), t \geqq 0\}$, L_t being the local time at 0 of B, and its extensions to Brownian motion with drift $B^{(\mu)}$ make it natural to consider the Markov processes

$$\{\frac{1}{c}\log\Big(\int_0^t \exp(cB_s^{(\mu)})\,ds\Big) - B_t^{(\mu)}, t \geqq 0\}$$

or equivalently

$$\{\exp(-cB_t^{(\mu)})\int_0^t \exp(cB_s^{(\mu)})ds, t \geqq 0\}.$$

3.b. We may now describe both joint laws of (M_t, B_t) and $(A_t, \exp(B_t))$.

Theorem 3.1 (Bachelier, Lévy). *For any $t > 0$, it holds that*

$$P(M_t \in dm, B_t \in db) = \sqrt{\frac{2}{\pi t^3}}(2m - b)\exp(-\frac{(2m-b)^2}{2t})\, db\, dm, \tag{3.1}$$

for $m \geqq b^+ \equiv \max(b, 0)$. In other terms, one has

$$(2M_t - B_t, M_t) \overset{(\text{law})}{=} (|B_t| + L_t, L_t) \overset{(\text{law})}{=} (R_t^{(1/2)}, U R_t^{(1/2)}),$$

where U denotes a uniform random variable on $[0,1]$, independent of $R_t^{(1/2)}$, the value at time t of a three-dimensional Bessel process, and

$$P(L_t + |x| \in dy | B_t = x) = \frac{y}{t}\exp(-\frac{y^2 - x^2}{2t})\, dy. \tag{3.2}$$

For a detailed proof of (3.1) and a motivation from Mathematical Finance, see Lamberton–Lapeyre [12], Exercise 18, Chapter 3. See also Revuz–Yor [20], Chapter VI, and Williams [23], Chapter II.

The rest of Part A shall be devoted to proving the following result in a "natural" way.

Theorem 3.2. *Let $\mathbf{e}$ denote an exponential random variable with mean 1, independent of B. Then, for any fixed t, it holds that*

$$(\mathbf{e}A_t, B_t) \overset{(\text{law})}{=} (e^{B_t}(\cosh(|B_t| + L_t) - \cosh(B_t)), B_t). \tag{3.3}$$

As interesting applications, following Dufresne, one may obtain closed form formulae for moments of $A_t^{(\mu)}$ or even of $A_t^{(\mu)}$ conditioned on $B_t^{(\mu)} = x$ (see, e.g., [4], [5]), the most striking being

$$E[(\int_0^1 \exp(\alpha b(u))\, du)^{-1}] = 1, \qquad \text{for all} \quad \alpha, \tag{3.4}$$

where $\{b(u), 0 \leqq u \leqq 1\}$ denotes the standard Brownian bridge. D.Hobson showed how this identity follows from Vervaat's relationship between the standard Brownian bridge and excursion. See Chaumont–Hobson–Yor [2] for an extended discussion, based on cyclic exchangeability properties. A more computational approach to these conditional moments will be provided in Part B below.

4 Lévy's and Pitman's Theorems

4.a. Lévy's theorem may be presented as

$$\{(L_t - |B_t|, L_t), t \geqq 0\} \overset{(\text{law})}{=} \{(B_t, M_t), t \geqq 0\},$$

whereas Pitman's theorem may be presented as

$$\{(L_t+|B_t|,L_t),t\geqq 0\}\overset{(\text{law})}{=}\{(R_t^{(1/2)},\inf_{s\geqq t}R_s^{(1/2)}),t\geqq 0\},$$

or, in terms of $\{M_t,t\geqq 0\}$ instead of $\{L_t,t\geqq 0\}$,

$$\{(M_t-B_t,M_t),t\geqq 0\}\overset{(\text{law})}{=}\{(|B_t|,L_t),t\geqq 0\}$$

and

$$\{(2M_t-B_t,M_t),t\geqq 0\}\overset{(\text{law})}{=}\{(R_t^{(1/2)},\inf_{s\geqq t}R_s^{(1/2)}),t\geqq 0\}.$$

Note that this second identity gives an explanation at the process level of the form of the joint law for fixed t of $(2M_t-B_t,M_t)$, as presented in Theorem 3.1.

4.b. Both Lévy's and Pitman's theorems extend to Brownian motions with drift. See [20], Exercise (3.19), Chapter VIII, pp.361–362. We only present the extension of Pitman's theorem, which is due to Rogers–Pitman: setting $M_t^{(\mu)}=\max_{0\leqq s\leqq t}B_s^{(\mu)}$ and letting $L_t^{(\mu)}$ be the local time of $B^{(\mu)}$ at 0,

$$\{2M_t^{(\mu)}-B_t^{(\mu)},t\geqq 0\}\overset{(\text{law})}{=}\{(|B_t^{(\mu)}|+L_t^{(\mu)},t\geqq 0\}\overset{(\text{law})}{=}\{Y_t^{(\mu)},t\geqq 0\},$$

where $\{Y_t^{(\mu)},t\geqq 0\}$ is a diffusion process with infinitesimal generator

$$\frac{1}{2}\frac{d^2}{dx^2}+\mu\coth(\mu x)\frac{d}{dx}.$$

5 Exponential Counterparts of Lévy's and Pitman's Theorems

5.a Going back to the parallel between the laws of $\{(M_t,B_t)\}$ and $\{(A_t,B_t)\}$, let us consider the following two-parameter family of exponential type processes (μ being fixed):

$$X_t^{c,b}=\exp(-cB_t^{(\mu)})\{x+\int_0^t\exp(bB_s^{(\mu)})ds\}.$$

Itô's formula yields

$$X_t^{c,b}=x-c\int_0^t X_s^{c,b}\,dB_s^{(\mu)}+\frac{c^2}{2}\int_0^t X_s^{c,b}\,ds+\int_0^t\exp((b-c)B_s^{(\mu)})\,ds. \quad (5.1)$$

Thus, when $b=c$, $\{X_t^{c,c},t\geqq 0\}$ is a diffusion process on $[0,\infty)$ generated by the second order differential operator

$$\frac{c^2}{2}x^2\frac{d^2}{dx^2}+\left((\frac{c^2}{2}-c\mu)x+1\right)\frac{d}{dx}.$$

Lévy's theorem is recovered by letting $c \to \infty$.

More generally, we can show that, if ξ and η are two independent Lévy processes, then

$$X_t = \exp(-\xi_t)(x + \int_0^t \exp(\xi_s)\, d\eta_s)$$

is a Markov process.

5.b. Here is now an extension of the Rogers–Pitman theorem, which corresponds to $b = 2c$.

Theorem 5.1. *The stochastic processes* $\{Z_t^{(-\mu)} = \exp(-B_t^{(-\mu)})A_t^{(-\mu)}, t \geqq 0\}$ *and* $\{Z_t^{(\mu)} = \exp(-B_t^{(\mu)})A_t^{(\mu)}, t \geqq 0\}$ *have the same distribution, that of a diffusion process on* $(0,\infty)$ *with generator (on* C_c^2 *functions)*

$$\frac{1}{2}z^2\frac{d^2}{dz^2} + \left((\frac{1}{2} - \mu)z + (\frac{K_{1+\mu}}{K_\mu})(\frac{1}{z})\right)\frac{d}{dz}. \tag{5.2}$$

The modified Bessel function $K_\mu(1/z)$ found in (5.2) is best understood as the normalizing factor of the GiG distribution $\mathrm{GiG}(\mu, 1/z)(dx)$ whose density is given by

$$\frac{1}{2K_\mu(1/z)}x^{\mu-1}\exp\left(-\frac{1}{2z}(x + \frac{1}{x})\right). \tag{5.3}$$

The heart of the proof of Theorem 5.1 is the following.

Theorem 5.2. *For any* $\mu \in \mathbf{R}$, *the conditional law of* $e_t^{(\mu)} \equiv \exp(B_t^{(\mu)})$ *given* $\mathcal{Z}_t^{(\mu)} \equiv \sigma\{Z_s^{(\mu)}, s \leqq t\}$ *and* $Z_t^{(\mu)} = z$, *is* $\mathrm{GiG}(\mu; 1/z)(dx)$, *whose density is given by* (5.3).

Theorem 5.1 is easily proved by using Theorem 5.2. In fact, from (5.1), taken for $c = 1$ and $b = 2$, we know that

$$Z_t^{(\mu)} - (\frac{1}{2} - \mu)\int_0^t Z_s^{(\mu)}\, ds - \int_0^t \exp(B_s^{(\mu)})\, ds$$

is a $(\mathcal{B}_t)$-martingale, where $\mathcal{B}_t = \sigma\{B_s, s \leqq t\}$. Hence, taking the conditional expectation of $\exp(B_s^{(\mu)})$ given $\mathcal{Z}_s^{(\mu)}$, we obtain from Theorem 5.2 that

$$Z_t^{(\mu)} - (\frac{1}{2} - \mu)\int_0^t Z_s^{(\mu)}\, ds - \int_0^t (\frac{K_{\mu+1}}{K_\mu})(\frac{1}{Z_s^{(\mu)}})\, ds$$

is a $(\mathcal{Z}_t^{(\mu)})$-martingale, which implies Theorem 5.1.

6 Extension of Dufresne's Affine Identity and Proof of Theorem 5.2

6.a. From the result (2.1) and enlarging the natural filtration $(\mathcal{B}_t, t \geqq 0)$ of $\{B_t, t \geqq 0\}$ with the variable $A_\infty^{(-\mu)}$ (we set $\hat{\mathcal{B}}_t^{(-\mu)} = \mathcal{B}_t \vee \sigma\{A_\infty^{(-\mu)}\}$), one obtains the following decomposition:

$$B_t^{(-\mu)} = \hat{B}_t^{(\mu)} - \int_0^t \frac{\exp(2B_s^{(-\mu)})}{A_\infty^{(-\mu)} - A_s^{(-\mu)}} ds, \tag{6.1}$$

where $\{\hat{B}_t^{(\mu)}, t \geqq 0\}$ is a Brownian motion with drift μ with respect to $(\hat{\mathcal{B}}_t^{(-\mu)})$, hence is independent of $A_\infty^{(-\mu)}$.

Identity (6.1) may be considered as an equation for $\{B_t^{(-\mu)}\}$ and solved explicitly

$$B_t^{(-\mu)} = \hat{B}_t^{(\mu)} - \log\left(1 + \frac{\hat{A}_t^{(\mu)}}{A_\infty^{(-\mu)}}\right), \qquad t \geqq 0, \tag{6.2}$$

or equivalently

$$\left\{\frac{1}{A_t^{(-\mu)}}, t \geqq 0\right\} \stackrel{(\text{law})}{=} \left\{\frac{1}{A_t^{(\mu)}} + \frac{1}{\tilde{A}_\infty^{(-\mu)}}, t \geqq 0\right\}, \tag{6.3}$$

where $\hat{A}_t^{(\mu)} = \int_0^t \exp(2\hat{B}_s^{(\mu)}) ds$ and $\tilde{A}_\infty^{(-\mu)}$ is a copy of $A_\infty^{(-\mu)}$, independent of $\{A_t^{(\mu)}, t \geqq 0\}$. This is precisely (2.3) extended at the process level.

6.b. In this subsection we give a proof of Theorem 5.2. Differentiating both hand sides of (6.3), we obtain

$$\left(\left\{\frac{1}{(Z_t^{(-\mu)})^2}, t \geqq 0\right\}, \frac{1}{A_\infty^{(-\mu)}}\right) \stackrel{(\text{law})}{=} \left(\left\{\frac{1}{(Z_t^{(\mu)})^2}, t \geqq 0\right\}, \frac{1}{\tilde{A}_\infty^{(-\mu)}},\right),$$

which proves at the same time that

$$\{Z_t^{(-\mu)}, t \geqq 0\} \stackrel{(\text{law})}{=} \{Z_t^{(\mu)}, t \geqq 0\},$$
$$\{Z_t^{(-\mu)}, t \geqq 0\} \text{ is independent of } A_\infty^{(-\mu)}.$$

To prove Theorem 5.2, we condition $B^{(-\mu)}$ with respect to $\mathcal{Z}_t^{(-\mu)}$, $Z_t^{(-\mu)} = z$; let $Q_{\omega,t}^z$ denote the regular conditional laws of $B_t^{(-\mu)}$. Under $Q_{\omega,t}^z$, one has

$$A_\infty^{(-\mu)} = A_t^{(-\mu)} + (e_t^{(-\mu)})^2 \tilde{A}_\infty^{(-\mu)} = e_t^{(-\mu)} z + (e_t^{(-\mu)})^2 \tilde{A}_\infty^{(-\mu)},$$

where $A_\infty^{(-\mu)}$ is distributed as $1/2\gamma_\mu$ under $Q_{\omega,t}^z$. Using (2.1), we need to solve the stochastic equation

$$AX^2 + zX \stackrel{(\text{law})}{=} A \tag{6.4}$$

with a random variable A which is distributed as $1/2\gamma_\mu$ and is independent of X. We make the following conjecture.

Conjecture *Equation* (6.4) *admits only one solution given by*

$$X \overset{\text{(law)}}{=} \text{GiG}(-\mu, 1/z).$$

In the Appendix, we present our attempts to solve this conjecture. However, in the Brownian context, we need only prove the formula for $\mu = 1/2$, which we shall give below, because we obtain the result for general μ by applying the Cameron–Martin theorem. Note that the result is also meaningful for $\mu = 0$.

Proof for the case $\mu = 1/2$. We note

$$\frac{1}{2\gamma_{1/2}} \overset{\text{(law)}}{=} \frac{1}{N^2} \overset{\text{(law)}}{=} T,$$

where N is a standard normal random variable and T is a stable(1/2) random variable. Hence the equation

$$\exp(-\lambda) = E[\exp(-\frac{\lambda^2}{2}T)] = E[\exp(-\frac{\lambda^2}{2}(X^2T + zX))], \qquad \lambda > 0,$$

becomes

$$\exp(-\lambda) = E[\exp(-(\lambda + \frac{\lambda^2 z}{2})X)],$$

which yields

$$E[\exp(-\theta X)] = \exp(-\frac{1}{z}(\sqrt{2\theta z + 1} - 1)).$$

Hence, the distribution of X is determined uniquely and is GiG$(-1/2, 1/z)$ by virtue of the uniqueness of Laplace transform. We note that, in this case, this is a "standard" inverse Gaussian distribution with parameter $1/z$.

7 Proof of Theorem 3.2

In this section we give a proof of Theorem 3.2 and its immediate consequence.

From Theorem 5.2 or (2.6), we obtain

$$P(B_t \in dx | Z_t = z) = \frac{1}{2K_0(1/z)} \exp(-\frac{\cosh(x)}{z})\, dx,$$

hence, from the Bayes formula,

$$P(Z_t \in dz | B_t = x) = \sqrt{2\pi t} \exp(\frac{x^2}{2t} - \frac{\cosh(x)}{z}) \frac{P(Z_t \in dz)}{2K_0(1/z)},$$

which, in particular, characterizes $P(Z_t \in dz)$. From this, we deduce

$$\begin{aligned} E[f(\mathrm{e}A_t)|B_t = x] &= E[f(\mathrm{e}e^x Z_t)|B_t = x] \\ &= \int_{|x|}^{\infty} f(e^x(\cosh(y) - \cosh(x)))\frac{y}{t}\exp(-\frac{y^2 - x^2}{2t})\,dy, \end{aligned} \tag{7.1}$$

which, with the help of (3.2), proves the identity (3.3). □

Setting $f(u) = u^r$ in (7.1), we obtain the following semi-closed expression for the conditional moments of A_t given $B_t = x$.

Proposition 7.1. *For any $r > -1$, one has*

$$E[(A_t)^r|B_t = x] = \frac{\exp(rx)}{\Gamma(r+1)}\int_{|x|}^{\infty} \frac{y}{t}\exp(-\frac{y^2 - x^2}{2t})(\cosh(y) - \cosh(x))^r\,dy.$$

Again, we postpone to Part B a more detailed discussion of these quantities.

8 Some Symmetric Diffusions

In this section we show that the key property in our study, which is expressed by Theorem 5.2, is closely related to the following symmetry property of the Markovian semigroup $\{\Pi_t^{(\mu)}\}$, $t \geqq 0$ for any $\mu \in \mathbf{R}$, of the exponential type process:

$$X_t^{(\mu),x} = \exp(-2B_t^{(\mu)})(x + \int_0^t \exp(2B_s^{(\mu)})\,ds)$$

already considered, in fact in greater generality, in Subsection 5.a.

Proposition 8.1. *For every $\mu \in \mathbf{R}$ and Borel functions $f, g : \mathbf{R}_+ \to \mathbf{R}_+$, there is the following identity:*

$$\int_0^{\infty} (\Pi_t^{(\mu)} f)(x)g(x)\frac{1}{x^{\mu+1}}e^{-1/2x}\,dx = \int_0^{\infty} f(x)(\Pi_t^{(\mu)} g)(x)\frac{1}{x^{\mu+1}}e^{-1/2x}\,dx. \tag{8.1}$$

Proof. We may assume that f and g are smooth functions with compact supports. Then some straightforward computation shows

$$\int_0^{\infty} (L^{(\mu)} f)(x)g(x)\frac{1}{x^{\mu+1}}e^{-1/2x}\,dx = \int_0^{\infty} f(x)(L^{(\mu)} g)(x)\frac{1}{x^{\mu+1}}e^{-1/2x}\,dx, \tag{8.2}$$

where $L^{(\mu)}$ is the generator (on smooth functions) of the diffusion process $\{X_t^{(\mu),x}\}$ which from Section 5 has been shown to be

$$L^{(\mu)} = 2x^2 \frac{d^2}{dx^2} + (2(1-\mu)x + 1)\frac{d}{dx}. \qquad \square$$

With the help of the Cameron–Martin relationship, it is not difficult to deduce from (8.1) the following identity:

$$\int_0^\infty E[H(e^{-2B_t}(x+A_t), xe^{-B_t}, x)] \frac{e^{-1/2x}}{x}\, dx = \int_0^\infty E[H(x, xe^{-B_t}, e^{-2B_t}(x+A_t))] \frac{e^{-1/2x}}{x}\, dx$$

for any non-negative Borel function H on $\mathbf{R}_+^3$. Using Fubini's theorem and making the change of variable $y = x\exp(-B_t)$ on both hand sides yield

$$E[\int_0^\infty H(ye^{-B_t} + e^{-2B_t}A_t, y, ye^{B_t}) \exp(-\frac{e^{-B_t}}{2y}) \frac{dy}{y}] = E[\int_0^\infty H(ye^{B_t}, y, ye^{-B_t} + e^{-2B_t}A_t) \exp(-\frac{e^{-B_t}}{2y}) \frac{dy}{y}].$$

Thus we obtain, for a generic Borel function $G : \mathbf{R}_+^2 \to \mathbf{R}_+$,

$$E[G(ye^{-B_t} + e^{-2B_t}A_t, ye^{B_t}) \exp(-\frac{e^{-B_t}}{2y})] = E[G(ye^{B_t}, ye^{-B_t} + e^{-2B_t}A_t) \exp(-\frac{e^{-B_t}}{2y})]$$

for every $y > 0$. Another elementary transformation shows that, writing $e_t = \exp(B_t)$ for clarity, we have

$$E[\exp(-\frac{1}{2ye_t})\tilde{G}(Z_t, ye_t)] = E[\exp(-\frac{1}{2ye_t})\tilde{G}(Z_t, \frac{y+Z_t}{e_t})],$$

where $Z_t = (e_t)^{-1}A_t$ and $\tilde{G}$ is another generic function, so that, for every Borel function $h : \mathbf{R}_+ \to \mathbf{R}_+$,

$$E[h(ye_t)\exp(-\frac{1}{2ye_t})|Z_t = z] = E[h(\frac{y+z}{e_t})\exp(-\frac{1}{2ye_t})|Z_t = z]. \tag{8.3}$$

Now assume for simplicity that the conditional distribution of e_t given $Z_t = z$ has a continuous density $q(u)$ with respect to the Lebesgue measure. Then, by (8.3), it is easy to show that q satisfies the identity

$$\frac{1}{y}\exp(-\frac{1}{2v})q(\frac{v}{y}) = \frac{y+z}{v^2}\exp(-\frac{v}{2y(y+z)})q(\frac{y+z}{v}), \qquad y, z, v > 0.$$

Setting $x = v^{-1}(y+z)$ and $\eta = v/y$, we obtain

$$q(x) = \eta \exp(\frac{\eta^2+1}{2z\eta}) q(\eta) \cdot \frac{1}{x} \exp(-\frac{1}{2z}(x+\frac{1}{x}))$$

for every $x, \eta, z > 0$. Since q is a probability density, integration with respect to x yields

$$\eta \exp(\frac{\eta^2+1}{2z\eta}) q(\eta) = \frac{1}{2K_0(1/z)}$$

and we obtain an important part of Theorem 5.2:

$$P(e_t \in dx | Z_t = z) = \frac{1}{2K_0(1/z)} x^{-1} \exp(-\frac{1}{2z}(x+\frac{1}{x}))\, dx, \qquad x > 0,$$

from which, with the help of the Cameron–Martin relationship, the full statement of Theorem 5.2 follows.

Part B. (Conditional) Moments of A_t

In the second part of this paper, we show that the identity (2.6) yields a number of interesting results. In particular, we obtain closed expressions of the conditional moments $E[(A_t)^r | B_t = x]$ for every negative integer r and the striking identity (3.4).

9 Negative Moments of A_t

At first we give a closed form formula for the Laplace transform (with respect to the parameter r) of the function $\theta_r(t)$ featured in (2.6).

Proposition 9.1. *Let $t > 0$ be fixed. Then one has*

$$\int_0^\infty e^{-xr} \theta_r(t)\, dr = \frac{1}{\sqrt{2\pi t^3}} a_c(x) a_c'(x) \exp(-\frac{(a_c(x))^2}{2t}) \tag{9.1}$$

or, equivalently,

$$\int_0^\infty e^{-xr} \theta_r(t)\, \frac{dr}{r} = \frac{1}{\sqrt{2\pi t}} \exp(-\frac{(a_c(x))^2}{2t}),$$

where $a_c(x) = \mathrm{Argcosh}(x) = \log(x + \sqrt{x^2-1}), x \geqq 1$.

We can prove (9.1) from the uniqueness of Laplace transform once we have recalled the well known identities

$$\int_0^\infty \exp(-u\cosh(r)) I_\nu(u)\, \frac{du}{u} = \frac{1}{\nu} e^{-\nu r}, \qquad r \geqq 0,$$

(see Watson [22], p.388, for a more general formula) and

$$\int_0^\infty \frac{1}{\sqrt{2\pi t^3}} \exp\left(-\left(\frac{a^2}{2t} + \frac{\lambda^2 t}{2}\right)\right) dt = \frac{1}{a} e^{-\lambda a}, \qquad a > 0, \lambda \geqq 0.$$

The latter common expression is equal to $a^{-1}E[\exp(-\lambda^2 T_a/2)]$, where $T_a = \inf\{t; B_t = a\}$.

From Proposition 9.1, we easily deduce the following. We set $\varphi_x(\lambda) = a_c(\lambda \exp(-x) + \cosh(x))$ and

$$\phi_x(\lambda) = \frac{(\varphi_x(\lambda))^2 - x^2}{2t}.$$

Proposition 9.2. *The joint law of (A_t, B_t) may be characterized by the following formula: for $t > 0$ and $\lambda \geqq 0$,*

$$E[\exp(-\frac{\lambda}{A_t})|B_t = x] = \exp(-\phi_x(\lambda)). \tag{9.2}$$

By differentiating both hand sides of (9.2) in λ and setting $\lambda = 0$, we obtain the expressions for the conditional moments of negative orders of A_t given B_t.

Proposition 9.3. *For any $n \in \mathbf{N}$, one has*

$$(-1)^n E[(A_t)^{-n}|B_t = x] = \psi_n(0),$$

where $\{\psi_n(\lambda)\}_{n=0}^\infty$ is the sequence of functions defined by the recurrence formula

$$\begin{cases} \psi_{n+1}(\lambda) = \psi_n'(\lambda) - \phi_x'(\lambda)\psi_n(\lambda), & n \geqq 0, \\ \psi_0(\lambda) = 1. \end{cases} \tag{9.3}$$

In particular, for $n = 1$, one has

$$E[\frac{1}{A_t}|B_t = x] = \frac{2x}{t(\exp(2x) - 1)}. \tag{9.4}$$

Now, recalling that $b_t(s) = B_s - sB_t/t$, $s \leqq t$, is a standard Brownian bridge with length t, we obtain the following, which shows (3.4) as a particular case.

Proposition 9.4. *Let $\{b(s), 0 \leqq s \leqq 1\}$ be a standard Brownian bridge. Then the identity*

$$E[\left(\int_0^1 \exp(\alpha b(s) + sy)ds\right)^{-1}] = \frac{y}{\exp(y) - 1} \tag{9.5}$$

holds for all $\alpha \in \mathbf{R}$ and $y \in \mathbf{R}$.

10 Expressions in Terms of the Jacobi Theta Function

(10.a) We now recall the famous series development:

$$\frac{y}{\exp(y)-1} = 1 - \frac{y}{2} + \sum_{k=1}^{\infty}(-1)^{k+1}B_k\frac{y^{2k}}{(2k)!}, \tag{10.1}$$

where, on the right hand side, the B_k's are the Bernoulli numbers, i.e.,

$$B_1 = 1/6,\ B_2 = 1/30,\ B_3 = 1/42,\ B_4 = 1/30,\ B_5 = 5/66,\ \ldots$$

(see further classical developments in Serre [21], p.147). Setting $\delta_n = d_n/d_0$, we then write

$$\int_0^1 \exp(\alpha b(s)du + sy)\,du = \sum_{n=0}^{\infty}\frac{y^n}{n!}\int_0^1 \exp(\alpha b(s))\ s^n\ ds$$
$$= d_0 + \sum_{n=1}^{\infty} d_n y^n = d_0(1+\sum_{n=1}^{\infty}\delta_n y^n).$$

We also write

$$\frac{1}{1+\sum_{n=1}^{\infty}\delta_n y^n} = 1 + \sum_{n=1}^{\infty}\eta_n y^n \tag{10.2}$$

Then, from (9.5), (10.1), (10.2), we can derive (at least formally)

$$E[\,\frac{1}{d_0}\,] = 1,\quad E[\,\frac{\eta_1}{d_0}\,] = -1/2,\quad E[\,\frac{\eta_{2k}}{d_0}\,] = (-1)^{k+1}\frac{B_k}{(2k)!},\quad E[\,\frac{\eta_{2k+1}}{d_0}\,] = 0.$$

Obviously, we need to relate the variables $\{\eta_n\}$ and $\{\delta_n\}$, which is easy since by definition we have

$$(1+\sum_{n=1}^{\infty}\delta_n y^n)(1+\sum_{n=1}^{\infty}\eta_n y^n) = 1.$$

Thus, we obtain

$$\eta_1 = -\delta_1,\quad \eta_2 = -\delta_2 + \delta_1^2,\quad \eta_3 = -\delta_1^3 + 2\delta_1\delta_2 - \delta_3,$$

from which we deduce

$$E[\,\frac{d_1}{d_0^2}\,] = \frac{1}{2},\quad E[\,-\frac{d_2}{d_0^2} + \frac{d_1^2}{d_0^3}\,] = \frac{B_1}{2},\quad E[\,\frac{d_3}{d_0^2} - 2\frac{d_1 d_2}{d_0^3} + \frac{d_1^3}{d_0^4}\,] = 0.$$

(10.b) We now exploit (9.4) to obtain closed expressions for $E[(A_t)^{-1}\varphi(B_t)]$ for certain functions φ. We first present some well known relations between the coth function and the Jacobi theta function Θ:

$$x\coth(x) = \frac{x^2}{2}\int_0^{\infty}\Theta(\frac{\pi t}{2})\exp(-\frac{x^2 t}{2})\,dt, \tag{10.3}$$

where

$$\Theta(t) = \sum_{n=-\infty}^{\infty} \exp(-n^2 \pi t). \tag{10.4}$$

Below, we shall exploit the Jacobi theta function property of symmetry:

$$\sqrt{t}\Theta(t) = \Theta(\frac{1}{t}). \tag{10.5}$$

It will be helpful to use the following auxiliary functions:

$$\Theta_{\alpha,\beta}(u) = \int_0^\infty \frac{\Theta(v)}{v^\beta (1+uv)^\alpha}\, dv,$$

which may be expressed in terms of the confluent hypergeometric functions of second type (see [14, p. 268]) given by

$$\Psi(\alpha, \gamma; z) = \frac{1}{\Gamma(\alpha)} \int_0^\infty \exp(-zv) v^{\alpha-1} (1+v)^{\gamma-\alpha-1}\, dv.$$

Indeed, it is immediate to show

$$\begin{aligned}\Theta_{\alpha,\beta}(u) &= \int_0^\infty \frac{dt}{t^\beta (1+ut)^\alpha} + 2u^{\beta-1} \sum_{n=1}^{\infty} \Psi(1-\beta, 2-\beta-\alpha; \frac{n^2\pi}{u}) \\ &= u^{\beta-1}\{B(\alpha+\beta-1, 1-\beta) + 2\sum_{n=1}^{\infty} \Psi(1-\beta, 2+\beta-\alpha; \frac{n^2\pi}{u})\}\end{aligned}$$

Thanks to (10.5), $\Theta_{\alpha,\beta}$ satisfies

$$\Theta_{\alpha,\beta}(u) = \frac{1}{u^\alpha} \Theta_{\alpha,3/2-\alpha-\beta}(\frac{1}{u}). \tag{10.6}$$

It is not difficult to prove, using (10.5) in particular, that $\Theta_{\alpha,\beta}(u) < \infty$ for $u > 0$, if and only if $\beta < 1/2$ and $\alpha + \beta > 1$. Needless to say, these constraints are also satisfied by $(\alpha, 3/2 - \alpha - \beta)$.

Now we set

$$u_k(t) = E[\frac{1}{A_t} B_t^k],$$

for which we obtain some explicit expression. By (9.4), we have

$$u_k(t) = \frac{1}{t}\, E[B_t^{k+1}(\coth(B_t) - 1)].$$

Proposition 10.1. *For any $t > 0$, one has*

$$u_{2k+1}(t) = -\frac{1}{t} E[B_t^{2k+2}] = -t^k \frac{(2k+1)!}{2^k k!} \tag{10.7}$$

and

$$u_{2k}(t) = \frac{t^k}{2}\frac{(2k+1)!}{2^k k!}\int_0^\infty \Theta(\frac{\pi u}{2})\frac{du}{(1+ut)^{k+3/2}} \tag{10.8}$$
$$= \frac{t^k}{\pi}\frac{(2k+1)!}{2^k k!}\Theta_{k+3/2,0}(2t/\pi)$$

In particular, $u_0(t) = u(t)$ *satisfies*

$$u(t) = \left(\frac{\pi}{2t}\right)^{3/2} u\left(\frac{\pi^2}{4t}\right). \tag{10.9}$$

Proof. (10.7) follows from the symmetry of the law of B_t and the scaling property. From the representation (10.3) of $x \coth x$, we obtain

$$u_{2k}(t) = \frac{1}{t}E[B_t^{2k+1}\coth B_t]$$
$$= \frac{1}{t}E\left[\frac{t^{k+1}N^{2k+2}}{2}\int_0^\infty \Theta(\frac{\pi u}{2})\exp(-\frac{utN^2}{2})\right] du,$$

where N denotes a standard Gaussian variable. Therefore (10.8) easily follows. (10.9) is a consequence of (10.6). □

Remark 10.2. (i) The identity (10.9) was already obtained in [5], where it is related to some computations involving the 3-dimensional Bessel bridge as in [27], Part B. More generally, in [5], we also compare (3.3) with some formulae obtained by physicists.
(ii) As we are discussing some applications of the identity (9.4), it may be of some interest to mention that it yields some (very partial!) verification of Dufresne's identity

$$\frac{1}{A_t^{(-\mu)}} \overset{\text{(law)}}{=} \frac{1}{A_t^{(\mu)}} + \frac{1}{\tilde{A}_\infty^{(-\mu)}},$$

for fixed $t > 0$ and $\mu > 0$, mentioned in Subsection 2.c. Indeed, we may verify from (9.4) that the expectations of both hand sides are equal. Using the Cameron-Martin formula and the identity (9.4), it is sufficient for this purpose to check

$$E[B_t(\coth(B_t)-1)\exp(-\mu B_t)]$$
$$= E[B_t(\coth(B_t)-1)\exp(\mu B_t)] + 2\mu t\exp(\frac{\mu^2 t}{2}).$$

By symmetry of the law of B, we obtain

$$E[B_t\coth(B_t)\exp(-\mu B_t)] = E[B_t\coth(B_t)\exp(\mu B_t)]$$

and the "remaining" equality $E[B_t\sinh(\mu B_t)] = \mu t\exp(\mu^2 t/2)$ follows from the differentiation of both hand sides of $E[\cosh(\mu B_t)] = \exp(\mu^2 t/2)$ with respect to μ.

11 A General "Recurrence" Formula

In this section, we consider a regular (C^∞, say) function $h : \mathbf{R} \to \mathbf{R}$ and set

$$A_t^h = \int_0^t h(B_s)\, ds.$$

The function h will be fixed throughout the following discussion, whereas $f : \mathbf{R}_+ \to \mathbf{R}$ is a generic "test" function. The aim of this section is to derive some relationship between the space-time functions defined by

$$\hat{f}(t,x) = E[f(A_t^h)|B_t = x].$$

We also consider, for the derivative f' of f,

$$\widehat{f'}(t,x) = E[f'(A_t^h)|B_t = x]$$

and we are most interested in the relationship between $\hat{f}$ and $\widehat{f'}$. The recurrence formula invoked in the title of this section is the following expression for $\widehat{f'}$ in terms of $\hat{f}$.

Theorem 11.1. *The following formula holds:*

$$h(x)\, \widehat{f'}(t,x) = (\hat{f})'_t(t,x) + \frac{x}{t}\, (\hat{f})'_x(t,x) - \frac{1}{2}(\hat{f})''_{xx}(t,x), \tag{11.1}$$

where $(\hat{f})'_t, (\hat{f})'_x, (\hat{f})''_{xx}$ *are the derivatives of the function* $\hat{f}$ *in the obvious sense.*

Proof. We shall obtain this formula by writing Itô's formula for each side of the following equality:

$$E[f(A_t^h)\, \varphi(B_t)] = E[\hat{f}(t,B_t)\, \varphi(B_t)], \tag{11.2}$$

where φ is a regular (C_c^∞) function. On the right hand side, we obtain

$$\frac{d}{dt}\Big(E[\hat{f}(t,B_t)\, \varphi(B_t)]\Big) = E[(\frac{1}{2}\{(\hat{f})''_{xx}\, \varphi + 2(\hat{f})'_x\, \varphi' + \hat{f}\, \varphi''\} + (\hat{f})'_t\, \varphi)(t,B_t)]. \tag{11.3}$$

On the left hand side, we obtain

$$\begin{aligned} \frac{d}{dt}\left(E[f(A_t^h)\, \varphi(B_t)]\right) &= E[h(B_t)\, f'(A_t^h)\, \varphi(B_t)] + \frac{1}{2}E[f(A_t^h)\, \varphi''(B_t)] \\ &= E[h(B_t)\, \widehat{f'}(t,B_t)\, \varphi(B_t)] + \frac{1}{2}E[\hat{f}(t,B_t)\, \varphi''(B_t)]. \end{aligned} \tag{11.4}$$

Thus, comparing (11.3) and (11.4), we obtain

$$E[h(B_t)\ \widehat{f}'(t,B_t)\ \varphi(B_t)] = E[(\frac{1}{2}(\hat{f})''_{xx}\ \varphi + (\hat{f})'_x\ \varphi' + (\hat{f})'_t\ \varphi)(t,B_t)]. \quad (11.5)$$

In order to derive formula (11.1) from (11.5), we still need to use an integration by parts formula to transform the term

$$E[((\hat{f})'_x\ \varphi')(t,B_t)]$$

This is done in the next well known lemma, which is really one of the very first steps of Malliavin's stochastic calculus of variations. □

Lemma 11.2. *Let $q : \mathbf{R} \to \mathbf{R}$ be a regular function. Then, for any $\varphi : \mathbf{R} \to \mathbf{R}$ in C_c^∞, one has*

$$E[(q\ \varphi')(B_t)] = E[(\frac{B_t}{t}q(B_t) - q'(B_t))\varphi(B_t)] \quad (11.6)$$

Proof of Lemma 11.2. We set

$$p_t(x) = \frac{1}{\sqrt{2\pi t}}\exp\bigl(-\frac{x^2}{2t}\bigr).$$

The left hand side of (11.6) equals

$$\int_{-\infty}^{\infty} p_t(x)q(x)\varphi'(x)\,dx = -\int_{-\infty}^{\infty} (p_t(x)q(x))'\varphi(x)\,dx,$$

which easily yields the right hand side of (11.6). □

End of the proof of Theorem 11.1. Using Lemma 11.2, we obtain

$$E[((\hat{f})'_x\ \varphi')(t,B_t)] = E[(\frac{x}{t}\ (\hat{f})'_x - (\hat{f})''_{xx})(t,B_t)\ \varphi(B_t)].$$

Hence, we may now write the right hand side of (11.5) as

$$E[((\hat{f})'_t + \frac{x}{t}(\hat{f})'_x - \frac{1}{2}(\hat{f})''_{xx})(t,B_t)\ \varphi(B_t)]. \quad (11.7)$$

Comparing formula (11.7) with the left hand side of (11.5) yields formula (11.1). □

Corollary 11.3. *Let $n \in \mathbf{N}$, and set*

$$k_n(t,x) = E[\frac{1}{A_t^n}|B_t = x].$$

Then the functions k_n's satisfy the recurrence formula :

$$-ne^{2x}k_{n+1}(t,x) = \frac{\partial}{\partial t}k_n(t,x) + \frac{x}{t}\frac{\partial}{\partial x}k_n(t,x) - \frac{1}{2}\frac{\partial^2}{\partial x^2}k_n(t,x). \quad (11.8)$$

Proof. Immediate from formula (11.1) in which we take $f(x) = x^{-n}$. □

We now apply formula (11.8) for $n = 1$ which yields, with the help of formula (9.4), the following expression:

$$e^{2x} E[\frac{1}{A_t^2} | B_t = x]$$
$$= \frac{1}{t^2} (\frac{2x}{\exp(2x) - 1}) - \frac{x}{t^2} \frac{d}{dx} (\frac{2x}{\exp(2x) - 1}) + \frac{1}{2t} \frac{d^2}{dx^2} (\frac{2x}{\exp(2x) - 1}).$$

Straightforward computations lead to

$$E[\frac{1}{A_t^2} | B_t = x] = \frac{4}{(\exp(2x) - 1)^2} (\frac{x^2}{t^2} - \frac{1}{t}(1 - x \coth x)). \tag{11.9}$$

This expression coincides with the formula given in Proposition 9.3, where we computed $\psi_2(0)$ by using (9.3).

We now transform this formula in a similar one concerning the variance of

$$X_y^{(\alpha)} \equiv (a_y^{(\alpha)})^{-1} = \left(\int_0^1 \exp(\alpha b(u) + uy) \, du \right)^{-1}.$$

We set

$$\varphi(y) = \frac{1}{(\exp(y) - 1)^2} (\frac{y}{2} \coth(\frac{y}{2}) - 1) \qquad \text{and} \qquad A(y) = \frac{\exp(y) - 1}{y}.$$

Proposition 11.4. (i) *The following formula holds:*

$$E[(X_y^{(\alpha)})^2] = A(y)^{-2} + \alpha^2 \varphi(y). \tag{11.10}$$

(ii) *Setting*

$$X(y) = \int_0^1 e^{yu} b(u) \, du \qquad \textit{and} \qquad Y(y) = \int_0^1 e^{yu} b(u)^2 \, du,$$

we have

$$\varphi(y) = \frac{1}{A(y)^4} E[X(y)^2] = \frac{1}{2A(y)^3} E[Y(y)]. \tag{11.11}$$

Proof. (i) From (11.9), we proceed as in the proof of Proposition 9.4.
(ii) Since $E[b(u)b(s)] = u \wedge s - us$, an easy computation leads to (11.11). □

Remark 11.5. A Taylor expansion in α of $E[(X_y^{(\alpha)})^2]$ gives a priori the following:

$$E[(X_y^{(\alpha)})^2] = \frac{1}{A(y)^2} + (\frac{3}{A(y)^4} E[X(y)^2] - \frac{1}{A(y)^3} E[Y(y)]) \alpha^2 + o(\alpha^2) \tag{11.12}$$

as $\alpha \to 0$. In fact, we know from (11.10) that the error term $o(\alpha^2)$ on the right hand side of (11.12) is identically 0 and this fact is a convenient check for (11.11).

We now extend the discussion made in Proposition 11.4 to higher moments:

$$M_n(\alpha, y) = E[(X_y^{(\alpha)})^n], \quad n \geqq 1.$$

Proposition 11.6. *The sequence of moments $\{M_n(\alpha, y), n \geqq 1\}$ satisfies the following recurrence formula:*

$$M_1(\alpha, y) = \frac{y}{\exp(y) - 1}$$

$$M_{n+1}(\alpha, y) = e^{-y} M_n(\alpha, y) + \frac{e^{-y}}{n} \left\{ \frac{\alpha^2}{2} \frac{\partial^2}{\partial y^2} M_n(\alpha, y) \right. \tag{11.13}$$
$$\left. -y \frac{\partial}{\partial y} M_n(\alpha, y) - \frac{\alpha}{2} \frac{\partial}{\partial \alpha} M_n(\alpha, y) \right\}.$$

Proof. Again, the argument used in the proof of Proposition 9.4 yields:

$$E[\frac{1}{A_t^n} | B_t = x] = \frac{1}{t^n} M_n(2\sqrt{t}, 2x).$$

As a consequence, the recurrence formula (11.13) follows from the recurrence formula (11.8). □

Corollary 11.7. *For each $n \geqq 1$, the function $M_n(\alpha, y)$ is of the form:*

$$M_n(\alpha, y) = \mu_{n,0}(y) + \alpha^2 \mu_{n,1}(y) + \cdots + \alpha^{2(n-1)} \mu_{n,n-1}(y)$$

and the functions $\{\mu_{n,k}; 0 \leqq k \leqq n-1\}$ satisfy the following recurrence relations:

$$\mu_{n+1,0}(y) = e^{-y} \mu_{n,0}(y) - \frac{e^{-y}}{n} y \, \mu'_{n,0}(y)$$

$$\mu_{n+1,k}(y) = e^{-y}(1 - \frac{k}{n}) \mu_{n,k}(y) - \frac{e^{-y}}{n} (y \mu'_{n,k}(y) - \frac{1}{2} \mu''_{n,k-1}(y))$$

$$\mu_{n+1,n}(y) = \frac{e^{-y}}{2n} \mu''_{n,n-1}(y)$$

with

$$\mu_{1,0}(y) = \frac{y}{\exp(y) - 1} \quad \textit{and} \quad \mu_{n,0}(y) = \left(\frac{y}{\exp(y) - 1}\right)^n.$$

Acknowledgements. We are very grateful to D. Dufresne, whose preprint [8] has been an important stimulation for the present paper. We also thank N.O'Connell who, motivated by studies of Brownian tandem queues [10], strongly suggested Proposition 8.1.

Appendix: On the Quadratic Equation (6.4)

In this appendix we present two approaches to the stochastic equation (6.4), which we first recall:

$$AX^2 + zX \overset{\text{(law)}}{=} A, \tag{6.4}$$

where z is a fixed positive number and A is a random variable which is distributed as $1/2\gamma_\mu$ for a $Gamma(\mu)$ random variable γ_μ and is independent of X.

a) As a first approach, we show the following.

Proposition A.1. *Assume that the probability law of X admits a continuous density $g(u) = \varphi(u)f(u)$ with respect to the Lebesgue measure du, where f is the density of* GiG$(-\mu, 1/z)$ *distribution given by*

$$f(u) = \frac{1}{2K_\mu(1/z)} u^{-\mu-1} \exp\big(-\frac{1}{2z}(u + \frac{1}{u})\big).$$

Then, X solves (6.4) *if and only if the function $\phi(v) = \varphi(1/v)$ satisfies*

$$\int_0^\infty f(u)\phi(u+x)\, du = 1 \tag{A.1}$$

for every $x > 0$.

Proof. We rewrite the equation (6.4) as

$$\frac{1}{z}\big(\frac{1}{X} - \frac{1}{X + 2z\gamma_\mu}\big) \overset{\text{(law)}}{=} 2\gamma_\mu .$$

Then, noting that, for $x > 0$, the left hand side is less than x if and only if $X > h_{z,x}(\gamma_\mu)$, where

$$h_{z,x}(c) = -zc + \sqrt{\frac{2c}{x} + z^2c^2},$$

we obtain

$$\int_0^\infty c^{\mu-1}e^{-c}\, dc \int_{h_{z,x}(c)}^\infty g(u)\, du = \int_0^{x/2} c^{\mu-1}e^{-c}\, dc.$$

Differentiating both hand sides, we get

$$\int_0^\infty c^\mu e^{-c} g(h_{z,x}(c))\big(\frac{2c}{x} + z^2c^2\big)^{-1/2} dc = 2^{-\mu}x^{\mu+1}e^{-x/2}.$$

Moreover, changing the variable in the integration by $h_{z,x}(c) = (zx+v)^{-1}$, we obtain

$$\int_0^\infty \big(\frac{1}{v(v+zx)}\big)^{\mu+1} g\big(\frac{1}{v+zx}\big) \exp\big(-\frac{x}{2v(v+zx)}\big)\, dv = e^{-x/2}.$$

Now, setting $g(u) = \varphi(u)f(u)$, we easily arrive at (A.1). □

If we can deduce $\phi \equiv 1$ from $(A.1)$, we will obtain a proof of the conjecture.

In the theory of statistics (see, e.g., [15]), a family $\mathcal{P}$ of probability distributions P is called complete if $E^P[\phi(X)] = 1$ for all $P \in \mathcal{P}$ implies $\phi = 1$, *a.e.* $\mathcal{P}$. Since

$$\int_0^\infty f(u)\phi(u+x)\, du = \int_0^\infty I_{(u>x)} f(u-x)\phi(u)\, du,$$

the conjecture is equivalent to the completeness of the probability distributions on $\mathbf{R}_+$ given by $I_{(u>x)}f(u-x)du, x>0$, for the density f of $\mathrm{GiG}(-\mu, 1/z)$.

After showing Proposition A.1, we arrive at a more general problem: for which probability measure μ on $[0,\infty)$ with density f, if a bounded continuous function k satisfies

$$\int_0^\infty f(u)k(u+x)\,du = 1$$

for all $x>0$, then it implies that $k\equiv 1$? For example, when $f(u)=e^{-u}$, it is easy to see this is the case.

b) Next we consider the Laplace transform approach. For any $\lambda>0$, we set $\mathcal{L}_A(\lambda)=E[\exp(-\lambda A)]$. Then, letting $\mu(dx)$ be the probability law of X, we have from (6.4)

$$\int_0^\infty \mathcal{L}_A(\lambda x^2)e^{-\lambda zx}\mu(dx)=\mathcal{L}_A(\lambda). \tag{A.2}$$

Therefore, if we can show that the functions $u_{A,\lambda}: x\mapsto \mathcal{L}_A(\lambda x^2)\exp(-\lambda zx), \lambda>0$, form a total set in $C_0(\mathbf{R}_+)$, the space of continuous functions on $\mathbf{R}_+$ which vanish at infinity, we will obtain a proof of the conjecture. When $\mu=1/2$, we gave a proof by using the fact $\mathcal{L}_A(\lambda)=\exp(-\sqrt{2\lambda})$ and the uniqueness of Laplace transform. This argument is very close to the above totality criterion: indeed, in the case $\mu=1/2$, we have $u_{A,\lambda}(x)=\exp(-(\sqrt{2\lambda}+\lambda z)x), \lambda>0$, which is clearly a total family in $C_0(\mathbf{R}_+)$.

References

1. L. Alili, D. Dufresne et M. Yor, Sur l'identité de Bougerol pour les fonctionnelles exponentielles du mouvement brownien avec drift, in Exponential Functionals and Principal Values related to Brownian Motion, A collection of research papers, Ed. by M. Yor, Biblioteca de la Revista Matemática Iberoamericana, 1997.
2. L. Chaumont, D.G. Hobson and M. Yor, Some consequences of the cyclic exchangeability property for exponential functionals of Lévy processes, Sém. Prob. XXXV, Lec. Notes Math. 1755, 334–347, Springer-Verlag, 2001.
3. P. Carmona, F. Petit and M. Yor, Exponential functionals of Lévy processes, in Lévy Processes: Theory and Applications, ed. by O.E. Barndorff-Nielsen, T. Mikosch and S.I. Resnick, 41–56, Birkhäuser, 2001.
4. C. Donati-Martin, H. Matsumoto and M. Yor, On striking identities about the exponential functionals of the Brownian bridge and Brownian motion, Periodica Math. Hung., **41** (2000), 103–119.
5. C. Donati-Martin, H. Matsumoto and M. Yor, On positive and negative moments of the integral of geometric Brownian motions, Stat. Prob. Lett., **49** (2000), 45–52.
6. D. Dufresne, The distribution of perpetuity, with applications to risk theory and pension funding, Scand. Act. J.,1990, 39–79.
7. D. Dufresne, An affine property of the reciprocal Asian option process, Osaka J. Math. **38** (2001), 379–381.
8. D. Dufresne, Laguerre series for Asian and other options, Math. Finance, **10** (2000), 407–428.

9. H. Geman and M. Yor, Bessel processes, Asian options, and perpetuities, Math. Finance, **3** (1993), 349–375.
10. N. O'Connell and M. Yor, Brownian analogues of Burke's theorem, to appear in Stoch. Proc. Appl., 2001.
11. K. Itô and H.P. McKean, Jr., Diffusion Processes and Their Sample Paths, Springer-Verlag, Berlin, 1965.
12. D. Lamberton and B. Lapeyre, Introduction to Stochastic Calculus Applied to Finance, Chapman & Hall, London, 1996.
13. J. Lamperti, Semi-stable Markov processes I, Z.W., **22** (1972), 205–255.
14. N.N. Lebedev, Special Functions and their Applications, Dover, New York, 1972.
15. E.H. Lehmann, Testing Statistical Hypotheses, 2nd Ed., Wiley, New York, 1986.
16. H. Matsumoto and M. Yor, On Bougerol and Dufresne's identities for exponential Brownian functionals, Proc. Japan Acad., **74** Ser.A (1998), 152–155.
17. H. Matsumoto and M. Yor, A version of Pitman's $2M-X$ theorem for geometric Brownian motions, C. R. Acad. Sc. Paris, **328** (1999), 1067–1074.
18. H. Matsumoto and M. Yor, A relationship between Brownian motions with opposite drifts via certain enlargements of the Brownian filtration, Osaka J. Math. **38** (2001), 383–398.
19. E.J. Pauwels and L.C.G. Rogers, Skew-product decompositions of Brownian motions, Geometry of random motion, 237–262, Contemp. Math., **73**, AMS, Providence, 1988.
20. D. Revuz and M. Yor, Continuous Martingales and Brownian Motion, 3rd. Ed., Springer-Verlag, Berlin, 1999.
21. J.P. Serre, Cours d'arithmétique, PUF, 1970.
22. G.N. Watson, A Treatise on the Theory of Bessel Functions, 2nd ed., Cambridge Univ. Press, Cambridge, 1944.
23. D. Williams, Diffusions, Markov Processes and Martingales, vol. 1: Foundations, Wiley and Sons, New York, 1979.
24. M. Yor, Sur certaines fonctionnelles exponentielles du mouvement brownien réel, J. Appl. Prob., **29** (1992), 202–208.
25. M. Yor, On some exponential functionals of Brownian motion, Adv. Appl. Prob., **24** (1992), 509–531.
26. M. Yor, From planar Brownian windings to Asian options, Insurance Math. Econom., **13** (1993), 23–34.
27. M. Yor (ed.), Exponential Functionals and Principal Values related to Brownian Motion, A collection of research papers, Biblioteca de la Revista Matemática Iberoamericana, 1997.

The Generalized Hyperbolic Model: Financial Derivatives and Risk Measures

Ernst Eberlein[1] and Karsten Prause[2]

[1] Institute for Mathematical Stochastics and Freiburg Center for Data Analysis and Modelling (FDM), University of Freiburg, Eckerstr. 1, D-79104 Freiburg, Germany

[2] HypoVereinsbank, Credit Risk Controlling, Arabellastr. 12, D-81925 München, Germany

Abstract. Statistical analysis of data from the financial markets shows that generalized hyperbolic (GH) distributions allow a more realistic description of asset returns than the classical normal distribution. GH distributions contain as subclasses hyperbolic as well as normal inverse Gaussian (NIG) distributions which have recently been proposed as basic ingredients to model price processes. GH distributions generate in a canonical way Lévy processes, i.e. processes with stationary and independent increments. We introduce a model for price processes which is driven by generalized hyperbolic Lévy motions. This GH model is a generalization of the hyperbolic model developed by Eberlein and Keller (1995). It is incomplete. We derive an option pricing formula for GH driven models using the Esscher transform as martingale measure and compare the prices with classical Black-Scholes prices. The objective of this study is to examine the consistency of our model assumptions with the empirically observed price processes for underlyings and derivatives. Finally we present a simplified approach to the estimation of high-dimensional GH distributions and their application to measure risk in financial markets.

1 Introduction

Generalized hyperbolic (GH) distributions were introduced by Ole E. Barndorf-Nielsen (1977) in the context of the sand project as a variance-mean mixture of normal and generalized inverse Gaussian (GIG) distributions.

These distributions seem to be tailor-made to describe the statistical behaviour of asset returns. Analyzing financial time series such as stock prices, indices, FX-rates or interest rates, one gets empirical distributions with a rather typical shape. They place substantial probability mass near the origin, have slim flanks and a number of observations far out in the tails. The normal distribution on which the classical models in finance are based, fails in all three aspects. How far this deviation from normality goes, depends on the time scale of the underlying data sets.

For long term studies based on weekly or even monthly data points the empirical distributions are close to the normal. But using scarce data sets effectively ignores a lot of information. Daily data is the minimum one has to consider for most purposes. Analyzing intraday data, i.e. looking at price movements on a microscopic scale leads to a deeper understanding of the relevant processes.

Definition 1.1. For $x \in \mathbb{R}$ the density of the *generalized hyperbolic distribution* is defined as

$$\begin{aligned} \mathrm{gh}(x;\lambda,\alpha,\beta,\delta,\mu) &= a(\lambda,\alpha,\beta,\delta)\,\big(\delta^2+(x-\mu)^2\big)^{(\lambda-1/2)/2} \\ &\quad\times K_{\lambda-1/2}\big(\alpha\sqrt{\delta^2+(x-\mu)^2}\,\big)\exp\big(\beta(x-\mu)\big) \\ a(\lambda,\alpha,\beta,\delta) &= \frac{(\alpha^2-\beta^2)^{\lambda/2}}{\sqrt{2\pi}\,\alpha^{\lambda-1/2}\,\delta^{\lambda}K_{\lambda}\big(\delta\sqrt{\alpha^2-\beta^2}\,\big)}, \end{aligned}$$

where K_ν denotes the modified Bessel function of the third kind with index ν. The domain of variation of the parameters is $0 \le |\beta| < \alpha$, $\mu, \lambda \in \mathbb{R}$ and $\delta > 0$.

Thus GH distributions are characterized by the five parameters $(\lambda, \alpha, \beta, \delta, \mu)$. Alternative parameters used in the literature are

$$\begin{aligned} \zeta &= \delta\sqrt{\alpha^2-\beta^2}, \qquad \rho = \beta/\alpha, \\ \xi &= (1+\zeta)^{-1/2}, \qquad \chi = \xi\rho, \\ \bar{\alpha} &= \alpha\delta, \qquad \bar{\beta} = \beta\delta. \end{aligned}$$

These alternative parameters are scale- and location-invariant, i.e. they do not change under affine transformations $Y = aX + b$ with $a \neq 0$ of a given variable X. Let η (resp. ν) denote the expectation (resp. the variance) of the distribution given by the density above. It can be shown that the mapping $(\lambda, \alpha, \beta, \delta, \mu) \to (\lambda, \xi, \chi, \nu, \eta)$ is bijective. Therefore $(\lambda, \xi, \chi, \nu, \eta)$ where $0 \le |\chi| < \xi < 1$ represents a parametrization with a rather intuitive interpretation. λ is a class parameter, ξ and χ are invariant shape parameters whereas ν (resp. η) are the variance (resp. the expectation), i.e. they are the scale and the location parameter.

The properties of the Bessel function K_λ (Abramowitz and Stegun (1968)) allow one to find simpler expressions for the Lebesgue density if $\lambda \in {}^1\!/_2\,\mathbb{Z}$. For $\lambda = 1$ we get the hyperbolic distribution which is characterized by the fact that the log-density is a hyperbola. This subclass has the simplest representation of all GH laws, which is favourable from a numerical point of view. For $\lambda = -1/2$ we get the normal inverse Gaussian (NIG) distribution. This subclass is closed under convolution for fixed parameters α and β. See Eberlein and Keller (1995), Eberlein, Keller, and Prause (1998), Barndorff-Nielsen (1998), Barndorff-Nielsen and Prause (1999) for statistical results concerning the subclasses of hyperbolic (resp. NIG) distributions.

2 Estimation of Densities

We estimate generalized hyperbolic, hyperbolic and normal inverse Gaussian distributions from daily as well as from high-frequency data. The algorithm and the results concerning German stock prices and NYSE indices are described in detail in Prause (1997, 1999). Analogous results are obtained for the DAX, the German stock index (see Figure 1). Let $(S_t)_{t\geq 0}$ be the price process for a given financial instrument. We define the *return* of this instrument for a given time interval Δt, e.g. one trading day, as

$$X_t = \log S_t - \log S_{t-\Delta t}. \tag{1}$$

Thus the return during n periods is the sum of the one period returns. The numerical estimates for the GH distribution and the subclasses are given in Table 1.

Table 1. Generalized hyperbolic parameter estimates for the daily returns of the DAX from December 15, 1993 to November 26, 1997. The parameter λ is fixed for the estimation of the hyperbolic and the NIG distribution

	λ	α	β	δ	μ	Log-Likelihood
GH	−2.018	46.82	−24.91	0.0163	0.00336	3138.28
Hyperbolic	1	158.87	−29.02	0.0059	0.00374	3135.15
NIG	−0.5	105.96	−26.15	0.0112	0.00348	3137.33

Figure 1 (top) provides a typical plot of empirical and estimated GH densities. The plot of the densities shows that the GH, hyperbolic and NIG distributions are more peaked and have more mass in the tails than the normal distribution. Consequently they are much closer to the empirical distribution of asset returns. Although the difference between GH, hyperbolic and the NIG distribution is small, it is clear that the generalized hyperbolic distributions are superior to those of the subclasses.

Value-at-Risk (VaR) has become a major tool in the modelling of risk inherent in financial markets. Essentially VaR is defined as the potential loss given a level of probability $\alpha \in (0,1)$

$$P[X_t < -\mathrm{VaR}_\alpha] = \alpha. \tag{2}$$

The quantity defined here has to be transformed in the proper way if one wants to express VaR in currency units. The plot of VaR as a function of α could also be used to visualize the tail behaviour of distributions. Note, that the concept of VaR applied only for a single α is not satisfactory: VaR does not identify extreme risks appearing with a probability smaller than α. Figure 1 (bottom) shows that the tails of the generalized hyperbolic distributions are heavier than the tails of the normal distribution and therefore VaR

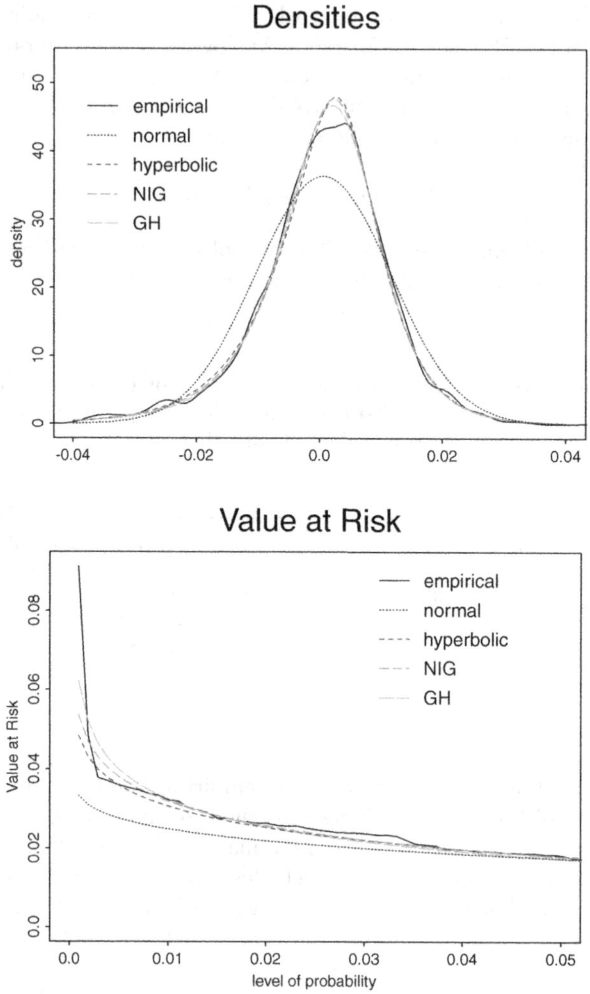

Fig. 1. DAX from December 15, 1993 to November 26, 1997, daily prices at 12:00h, IBIS data (Karlsruher Kapitalmarktdatenbank)

computed parametrically for the GH distribution and its subclasses is closer to the empirically observed Value-at-Risk.

In the global foreign exchange (FX) market it is particularly important to look at price movements on an intraday basis. Many traders close their positions over night and try to make a profit from intraday trading only. Therefore we examine 6 hours returns of USD/DEM exchange rates from the

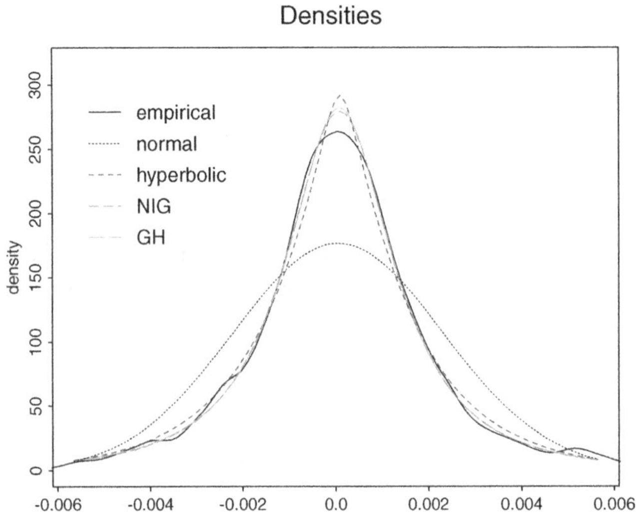

Fig. 2. USD/DEM exchange rate from January 1 to December 31, 1996, 6 hours returns, HFDF96 data set (Olsen & Associates, Zürich)

HFDF96 data set provided by Olsen & Associates. See also J.P. Morgan and Reuters (1996, p. 65) for some remarks concerning the leptokurtosis of daily USD/DEM returns. For high-frequency data we follow Guillaume, Dacorogna, Davé, Müller, Olsen, and Pictet (1997) in the definition of the log-price

$$p(t_i) = [\log p_{ask}(t_i) + \log p_{bid}(t_i)]/2 \tag{3}$$

and the corresponding return

$$r(t_i) = p(t_i) - p(t_i - \Delta t). \tag{4}$$

We estimate the GH parameters for the increments $r(t_i)$ after removing all zero-returns. Although this is only a provisional approach to focus on time periods where trading takes place, the results as plotted in Figure 2 provide a clear picture: The excellent fit of generalized hyperbolic distributions and the typical difference to the normal distribution observed for daily returns is repeated for high-frequency data (see also Barndorff-Nielsen and Prause (1999)).

3 The Generalized Hyperbolic Model

We follow Eberlein and Keller (1995) in the design of the price process $(S_t)_{t\geq 0}$ and the derivation of an option pricing formula. First we construct the driving process. Generalized hyperbolic distributions are infinitely divisible (Barndorff-Nielsen and Halgreen (1977)). Therefore they generate a Lévy process $(X_t)_{t\geq 0}$, i.e. a process with stationary and independent increments, such that the distribution of X_1 and thus of $X_t - X_{t-1}$ is generalized hyperbolic. We call this process $(X_t)_{t\geq 0}$ the *generalized hyperbolic Lévy motion*. It depends on the five parameters $(\lambda, \alpha, \beta, \delta, \mu)$ and is purely discontinuous. This property follows from the explicit form of the Lévy-Khintchine representation of the characteristic function of generalized hyperbolic distributions which is given in the appendix. The exponent consists only of a drift term and the integral representing the jumps, but has no Gaussian term $-{}^{c}\!/_{2}\, u^2$. The new model for the price process itself is defined by

$$S_t = S_0 \exp(X_t). \tag{5}$$

Let us emphasize that (5) is only the basic model which replaces the classical geometric Brownian motion introduced by Osborne and Samuelson. During the last 40 years this classical Gaussian model, which can also be defined via the diffusion equation

$$dS_t = S_t(\mu dt + \sigma dB_t), \tag{6}$$

has been generalized and refined in many directions. In its most sophisticated generalization (see e.g. Bakshi, Cao, and Chen (1997)) jumps are added through a Poisson process, the constant volatility σ is replaced by a diffusion process driven by a different Brownian motion and a stochastic interest rate is considered, which is typically given in the form of a Cox-Ingersoll-Ross model. Taking correlations between the various driving processes into acount one has to consider more than ten parameters. Calibration of such a model is not an easy task.

Essentially every extension which has been considered for the geometric Brownian motion can be applied to the exponential Lévy model (5) as well. The extension we consider to be crucial and which improves the model considerably is stochastic volatility. In (5) this can be done by writing X_t in the form $\mu t + \sigma L_t$ where $(L_t)_{t\geq 0}$ is a standardized Lévy process, that is one with mean zero and variance one. In this form σ can be replaced by any of the standard models for stochastic volatility such as diffusion models or the Ornstein-Uhlenbeck-based models considered by Barndorff-Nielsen and Shephard (2001) and Nicolato and Prause (1999) or any member of the ARCH and GARCH-family. A detailed discussion of this issue supported by a number of empirical results will be given in Eberlein, Kallsen, and Kristen (2001).

The key property of our model–besides its simplicity–is that taking log-returns in (5) one obtains the corresponding increment of the driving Lévy process $(X_t)_{t\geq 0}$. For time intervals of length 1 its distribution is by construction the generating generalized hyperbolic distribution. Thus the model produces for time intervals of length 1 exactly that distribution which one gets from fitting data. If, for example, the underlying data set consists of daily

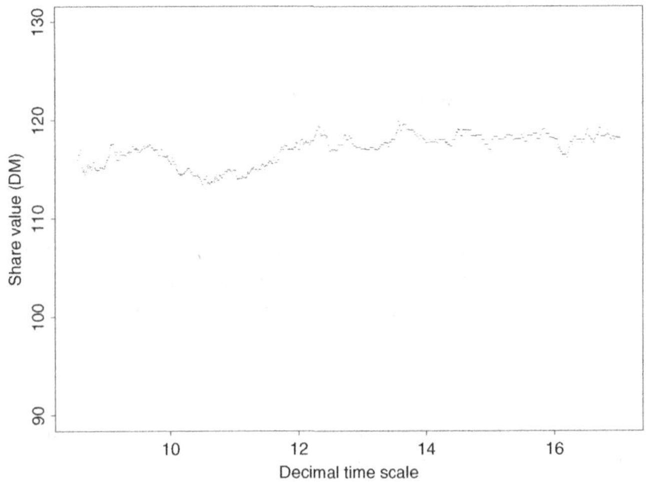

Fig. 3. Siemens, Xetra data from August 28, 1998

prices, one trading day in real time corresponds to a time interval of length 1 in the model. It is not only this reproduction of observed distributions which makes (5) attractive, the model is also consistent in a much deeper sense. If one calibrates a model using daily data a natural question is whether the distribution produced by the model for a weekly horizon is close to the distribution one obtains by fitting the corresponding weekly data. This turns out to be the case to a certain degree of accuracy. Of course the same should hold if one goes in the other direction, namely from daily to intraday hourly data. Recall that the classical Gaussian model produces normal log-returns along any time interval Δt. Detailed results on this consistency property in both directions will appear in a forthcoming joint paper with Fehmi Özkan.

In this context let us clarify that the generalized hyperbolic model (5) does not have anything in common with the hyperbolic diffusion model introduced by Bibby and Sørensen (1997) and discussed further in Rydberg (1999). The latter is a classical diffusion model with completely different statistical as well as path properties.

The price process (5) has purely discontinuous paths as has the driving Lévy process. In order to give the reader an idea of what the paths of such a process look like, we show in Figure 3 a sample of the intraday price behaviour of stocks. To model the microstructure of asset prices, purely discontinuous processes are more appropriate than the classical or the hyperbolic diffusion processes with continuous paths.

Since we are in an incomplete setting, we have to select a specific equivalent martingale measure. Arbitrage free prices are obtained as expectations under these measures (Delbaen and Schachermayer (1994)). Note, that it is possible to obtain every price in the full no-arbitrage interval by chosing the proper equivalent martingale measure (Eberlein and Jacod (1997)). We choose the Esscher equivalent martingale measure P^θ given by

$$\mathrm{d}P^\theta = \exp\big(\theta X_t - t \log M(\theta)\big)\mathrm{d}P. \tag{7}$$

The parameter θ is the solution of $r = \log M(\theta+1) - \log M(\theta)$ where M is the moment generating function given in the Appendix and r is the constant interest rate. The equation for θ ensures that the discounted price process is in fact a P^θ-martingale. Chan (1999) remarked that in a model very similar to the exponential Lévy model (5), the Esscher transform is the minimal martingale measure in the sense of Föllmer and Schweizer (1991). A much deeper motivation for the choice of this particular martingale measure came out of several recent papers, where via duality theory it was shown that the choice of a minimal martingale measure corresponds to maximizing expected utility. More precisely, taking the Esscher transform corresponds to maximizing utility with respect to the power utility function $u(x) = x^p/p$. One among several good references for this application of duality theory to finance is Goll and Rüschendorf (2000).

Following the arbitrage pricing theory, the price of an option with time to expiration T and payoff function $H(S_T)$ is given by $e^{-rT}\mathrm{E}^\theta[H(S_T)]$. In particular, for a call option with strike K whose payoff is $H(S_T) = (S_T - K)^+$ we obtain the price formula

$$S_0 \int_\gamma^\infty \mathrm{gh}^{*T}(x;\theta+1)\,\mathrm{d}x - e^{-rT}K \int_\gamma^\infty \mathrm{gh}^{*T}(x;\theta)\,\mathrm{d}x, \tag{8}$$

where $\gamma = \ln(K/S_0)$ and $\mathrm{gh}^{*t}(\,\cdot\,;\theta)$ is the density of the distribution of X_t under the risk-neutral measure. The density $\mathrm{gh}^{*t}(\,\cdot\,)$ of the t-fold convolution of the generalized hyperbolic distribution can be computed by applying the Fourier inversion formula to the characteristic function. In the case of NIG distributions one should of course use the property that this subclass is closed under convolution.

Figure 4 shows that the difference of the generalized hyperbolic option prices to those from the Black-Scholes model resembles the W-shape which was observed for hyperbolic option prices. Note that for options with short maturities the W-shape is more pronounced in the case of the NIG and the GH model.

4 Rescaling of Generalized Hyperbolic Distributions

For the computation of implicit volatilities in the GH model we need to rescale the generalized hyperbolic distribution while keeping the shape fixed. An analogous problem occurs when computing GH option prices for a given volatility,

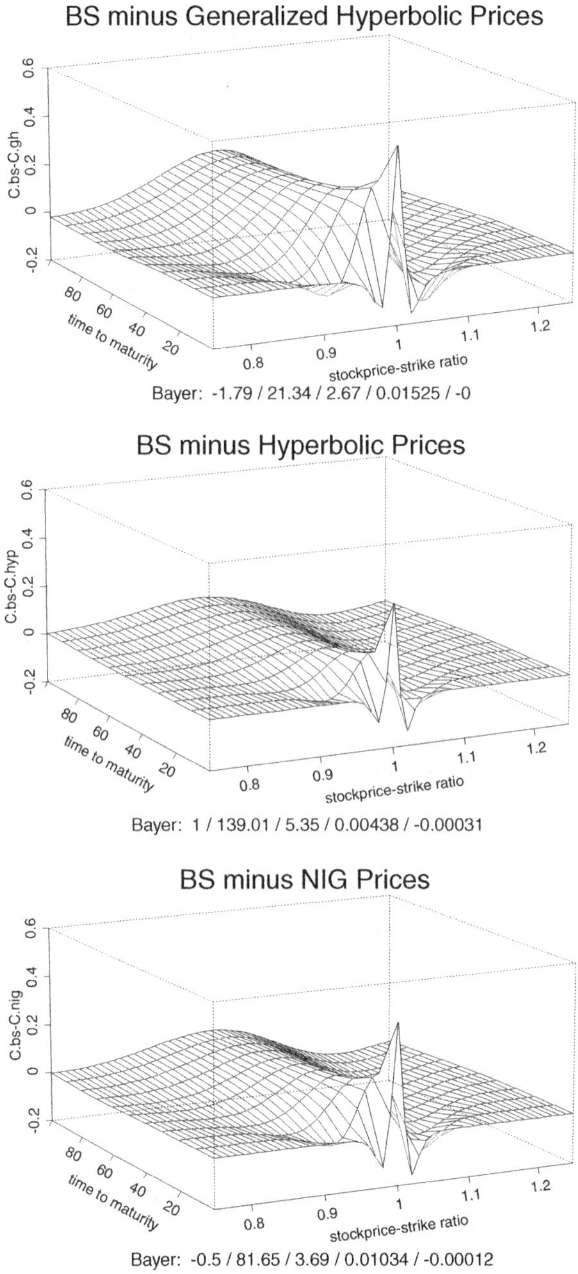

Fig. 4. Black-Scholes minus GH prices (Bayer parameters, strike K=1000)

e.g. a volatility estimated from historical stock returns. In this section we also give some insights into the structure of GH distributions. The rescaling of generalized hyperbolic distributions is based on the following property concerning scale- and location-invariance.

Lemma 4.1. *The terms λ, $\alpha\delta$ and $\beta\delta$ are scale- and location-invariant parameters of the univariate generalized hyperbolic distribution. The very same holds for the alternative parametrizations $(\zeta,\ \rho)$ and $(\xi,\ \chi)$.*

Proof. According to Blæsild (1981) a linear transformation $Y = aX + b$ of a GH distributed variable X is again GH-distributed with parameters $\lambda^+ = \lambda$, $\alpha^+ = \alpha/|a|$, $\beta^+ = \beta/|a|$, $\delta^+ = \delta|a|$ and $\mu^+ = a\mu + b$. Obviously $\alpha^+\delta^+ = \alpha\delta$ and $\beta^+\delta^+ = \beta\delta$.

A consequence of Lemma 4.1 is that the variance of the generalized hyperbolic distribution has the linear structure $\mathrm{Var}[X_1] = \delta^2 C_\zeta$ in δ^2 where C_ζ depends only on the shape, i.e. the scale- and location-invariant parameters (Barndorff-Nielsen and Blæsild (1981)). Therefore one can also use δ as a scaling parameter. To rescale the distribution for a given variance $\widehat{\sigma}^2$ one obtains the new $\widetilde{\delta}$ as

$$\widetilde{\delta} = \widehat{\sigma}\left[\frac{K_{\lambda+1}(\widehat{\zeta})}{\widehat{\zeta}K_\lambda(\widehat{\zeta})} + \frac{\widehat{\beta}^2}{\widehat{\alpha}^2 - \widehat{\beta}^2}\left(\frac{K_{\lambda+2}(\widehat{\zeta})}{K_\lambda(\widehat{\zeta})} - \left(\frac{K_{\lambda+1}(\widehat{\zeta})}{K_\lambda(\widehat{\zeta})}\right)^2\right)\right]^{-1/2} \tag{9}$$

where $(\widehat{\alpha}, \widehat{\beta}, \widehat{\delta})$ and consequently $\widehat{\zeta}$ are estimated from a longer time series. To fix the shape of the distribution while rescaling with a new $\widetilde{\delta}$, one has to change the other parameters in the following way

$$\widetilde{\lambda} = \widehat{\lambda}, \quad \widetilde{\alpha} = \frac{\widehat{\alpha}\,\widehat{\delta}}{\widetilde{\delta}}, \quad \widetilde{\beta} = \frac{\widehat{\beta}\,\widehat{\delta}}{\widetilde{\delta}} \text{ and } \widetilde{\mu} = \widehat{\mu}. \tag{10}$$

Note, that the term in the square brackets is scale- and location-invariant. In order to value German stock options we use shape parameters estimated from stock prices from January 1, 1988 to May 24, 1994 and we rescale the estimated generalized hyperbolic distributions while feeding in volatility estimates from shorter time periods.

Figure 5 shows the densities and the corresponding log-densities of hyperbolic distributions. In the first row we fix the shape estimated from Bayer stock prices and rescale the distribution as described in (10). The second row of Figure 5 reveals that ζ describes the kurtosis of the distribution. For increasing ζ the density becomes less peaked and converges to the Gaussian distribution. Log-densities give some insight into the tail behaviour of the density. The log-density of the hyperbolic distribution is a hyperbola whereas the normal log-density is a parabola. Therefore hyperbolic distributions possess substantially heavier tails than the normal distribution. Nevertheless, in contrast to those of stable distributions, excluding the normal distribution, all moments of GH distributions do exist.

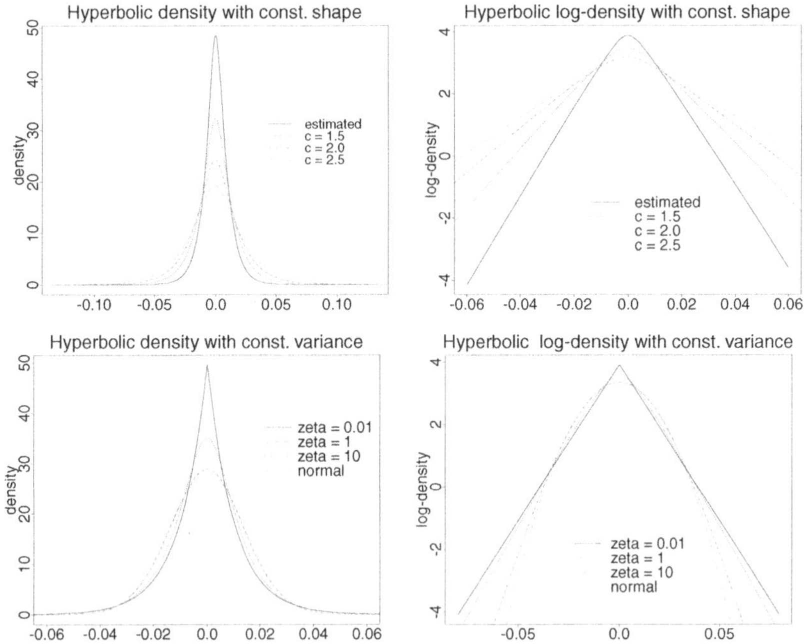

Fig. 5. Rescaled hyperbolic densities ($\delta' = c\,\widehat{\delta}$ with constant shape parameters $\widehat{\zeta} = 0.608$, $\widehat{\rho} = 0.0385$ estimated from Bayer stock prices) and hyperbolic densities with constant variance and different shapes

5 Smile Reduction

The comparison of generalized hyperbolic prices with Black-Scholes prices in Section 3 hints at the possibility to correct the well-known smiles which appear in Black-Scholes implicit volatilities. Implicit volatilities are computed from observed option prices by inverting the corresponding pricing formula with respect to the volatility parameter. Usually all parameters necessary for option pricing are known to traders except the volatility. In the GH model we rely on the rescaling mechanism described in Section 4 to obtain the volatility parameter. In this section we compute the implicit volatilities. The study is based on intraday option and stock market data of Bayer, Daimler Benz, Deutsche Bank, Siemens and Thyssen from July 1992 to August 1994. The option data set contains all trades reported by the Deutsche Terminbörse (since 1998 Eurex Germany) during the period above. The preparation of the data sets is described in detail in Eberlein, Keller, and Prause (1998, Chapter IV). The latter article includes also a discussion of implicit volatilities in the hyperbolic model and of the different approaches to reduce the smile.

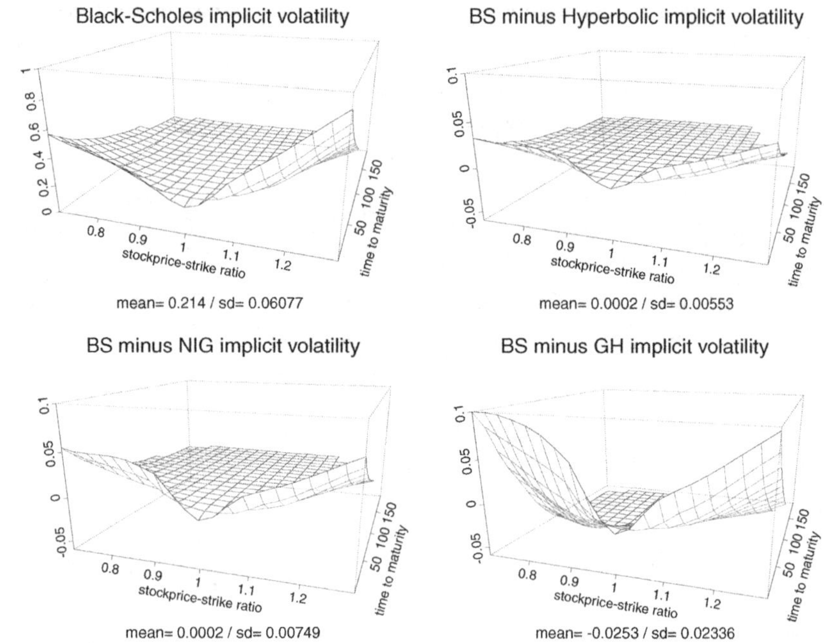

Fig. 6. Black-Scholes implicit volatilities and comparison of the implicit volatilities of Black-Scholes, hyperbolic, NIG, and GH prices (Daimler Benz calls from July 1992 to August 1994, 62504 observations)

Implicit volatilities in the Black-Scholes model typically follow a pattern denoted as *smile*, i.e. they are low for options at the money and the highest implicit volatilities are observed for options with short maturities in and out of the money. Figure 6 (top left) shows the implicit volatilities of Daimler Benz calls in the Black-Scholes model. To compare these with implicit volatilities in the GH model, we computed the differences and plotted them in Figure 6. The pattern reflects the W-shapes from Figure 4. Obviously we observe a more pronounced correction of the smile effect in the GH model–due to the heavier tails of the distribution.

A different approach to analyse the smile behaviour of a particular option pricing model is to fit a linear model for the implicit volatilities of the form

$$\sigma_{\mathrm{Imp},i} = b_0 + b_1 T_i + b_2(\rho_i - 1)^2 / T_i + e_i, \tag{11}$$

where e_i is the random error term, ρ_i the stockprice-strike ratio S/K and i the number of the trade in the option data set. The cross-term $(\rho - 1)^2/T$ reflects the degression of the smile effect with increasing time to maturity T. Table 2 shows the regression coefficients for the Black-Scholes, the GH models and the respective symmetric centered versions of each. The values of

Table 2. Fitted coefficients for call options from July 1992 to August 1994. SC marks the results for the symmetric centered versions of the models

		b_0	b_1	b_2	R^2
Daimler Benz	Black-Scholes	0.2177	−0.00029	40.53	0.5416
	Hyperbolic	0.2186	−0.0003	36.89	0.4972
	Hyperbolic SC	0.2184	−0.000293	36.33	0.4951
	NIG	0.2191	−0.000305	35.11	0.4746
	NIG SC	0.2189	−0.000296	34.48	0.4716
	GH	0.2207	−0.000321	32.81	0.4378
	GH SC	0.2201	−0.000306	31.98	0.4343

the coefficient for $(\rho-1)^2/T$ are smaller for hyperbolic, NIG and GH prices compared to those from Black-Scholes prices. Hence these new models reduce the smile effect. The largest correction is observed for the symmetric centered GH model.

6 Multivariate Generalized Hyperbolic Distributions

In the previous sections we have discussed univariate generalized hyperbolic distributions as the basic ingredient for a stock price model focussing on option pricing. We shall now look into the estimation of multivariate GH distributions and its application to risk measurement.

Definition 6.1. For $x \in \mathbb{R}^d$, the d-dimensional *generalized hyperbolic distribution* (GH_d) is defined by its Lebesgue density, which is given by

$$\mathrm{gh}_d(x) = a_d \frac{K_{\lambda-d/2}\big(\alpha\sqrt{\delta^2+(x-\mu)'\Delta^{-1}(x-\mu)}\big)}{\big(\alpha^{-1}\sqrt{\delta^2+(x-\mu)'\Delta^{-1}(x-\mu)}\big)^{d/2-\lambda}} \exp(\beta'(x-\mu)),$$

$$a_d = a_d(\lambda,\alpha,\beta,\delta,\Delta) = \frac{\big(\sqrt{\alpha^2-\beta'\Delta\beta}/\delta\big)^{\lambda}}{(2\pi)^{d/2}K_\lambda\big(\delta\sqrt{\alpha^2-\beta'\Delta\beta}\big)}$$

The parameters have the following domain of variation[1]: $\lambda \in \mathbb{R}$, $\beta,\mu \in \mathbb{R}^d$, $\delta > 0$, $\beta'\Delta\beta < \alpha^2$. The positive definite matrix $\Delta \in \mathbb{R}^{d\times d}$ has a determinant $|\Delta| = 1$.

For $\lambda = (d+1)/2$ we obtain the *multivariate hyperbolic* and for $\lambda = -1/2$ the *multivariate normal inverse Gaussian* distribution. Generalized hyperbolic distributions are symmetric iff $\beta = (0,\ldots,0)'$.

[1] We omitted the limiting distributions obtained at the boundary of the parameter space; see e.g. Blæsild and Jensen (1981).

Blæsild and Jensen (1981) introduced alternative parameters ζ, π, S where $\zeta = \delta\sqrt{\alpha^2 - \beta'\Delta\beta}$, $\pi = \beta\Delta^{1/2}(\alpha^2 - \beta'\Delta\beta)^{-1/2}$ and $S = \delta^2\Delta$. Generalized hyperbolic distributions are closed under forming marginals, conditioning and affine transformations (Blæsild (1981)). For the mean and the variance of $X \sim \mathrm{GH}_d$ one obtains

$$\mathrm{E}X = \mu + \delta R_\lambda(\zeta)\pi\Delta^{1/2}, \tag{12}$$

$$\mathrm{Var}\, X = \delta^2 \left(\zeta^{-1} R_\lambda(\zeta)\Delta + S_\lambda(\zeta)(\pi\Delta^{1/2})'(\pi\Delta^{1/2})\right), \tag{13}$$

where in order to simplify notation we introduced $R_\lambda(x) = K_{\lambda+1}(x)/K_\lambda(x)$ and $S_\lambda(x) = \left[K_{\lambda+2}(x)K_\lambda(x) - K^2_{\lambda+1}(x)\right]/K^2_\lambda(x)$.

A maximum likelihood estimation of all parameters in higher dimensions is computationally too demanding since the number of parameters $3 + d(d+5)/2$ increases rapidly with the number of dimensions. Therefore we propose a simplified algorithm for symmetric GH distributions which allows for an efficient estimation also in higher dimensions. The first step of the estimation follows a method of moments approach: we estimate the sample mean $\widehat{\mu} \in \mathbb{R}^d$ and the sample dispersion matrix Σ using canonical estimators. Since $\pi = 0$ in the symmetric case, $\mathrm{E}X = \mu$, and from (13) we get the following estimate for Δ

$$\widehat{\Delta} = \frac{\zeta}{\delta^2 R_\lambda(\zeta)}\,\Sigma. \tag{14}$$

Consequently we may compute $\widehat{\Delta}$ by norming the sample dispersion matrix such that $|\Delta| = 1$. The second step is to compute

$$y_i = (x_i - \widehat{\mu})'\widehat{\Delta}^{-1}(x_i - \widehat{\mu}) \tag{15}$$

from observations $x_i \in \mathbb{R}^d$, $1 \le i \le n$. Then the log-likelihood function is given as

$$\begin{aligned} L(x;\lambda,\alpha,\delta) &= n\Big(\lambda\log(\alpha/\delta) - \frac{d}{2}\log(2\pi) - \log K_\lambda(\delta\alpha)\Big) \\ &+ \sum_{i=1}^n \log K_{\lambda-(d/2)}\big(\alpha\sqrt{\delta^2 + y_i}\,\big) + \Big(\lambda - \frac{d}{2}\Big)\sum_{i=1}^n \log\big(\sqrt{\delta^2 + y_i}/\alpha\big). \end{aligned} \tag{16}$$

The last step is to maximize this log-likelihood function with respect to $(\lambda, \alpha, \delta)$. We have developed efficient estimation algorithms for hyperbolic and NIG distributions, i.e. for fixed $\lambda = (d+1)/2$ and $\lambda = -1/2$. In the case of arbitrary λ one may encounter numerical problems due to extremely small values of the Bessel functions K_λ. As in the univariate case the log-likelihood function simplifies for $\lambda \in \frac{1}{2}\mathbb{Z}$. For NIG distributions, i.e. $\lambda = -1/2$, the number of Bessel functions K_λ which have to be computed for the log-likelihood function is reduced by one. In the case of hyperbolic and hyperboloid distributions we have to compute only one Bessel function instead of $n+1$. Since the evaluation of Bessel functions is the time-consuming part of the third step, computation is much simpler for hyperbolic distributions. For fixed λ it is also possible to estimate only ζ in the second step. Nevertheless, we have chosen to estimate the covariance structure in the first "method of moments" step and the parameters (α, δ) characterizing the kurtosis and the scale in the likelihood step.

For a price process $S_t \in \mathbb{R}^d$ we define *relative returns* $x_t \in \mathbb{R}^d$ by

$$x_t^{(i)} = \left[S_t^{(i)} - S_{t-\Delta t}^{(i)}\right] / S_{t-\Delta t}^{(i)} \approx \log S_t^{(i)} - \log S_{t-\Delta t}^{(i)}, \quad 1 \le i \le d, \tag{17}$$

which are approximated by the log-returns defined in (1). The motivation to choose this definition is that the return of a portfolio described by a vector $h \in \mathbb{R}^d$ is then simply given by $h'x_t$. See J.P. Morgan and Reuters (1996, Section 4.1) for a discussion of temporal and cross-section aggregation of asset returns.

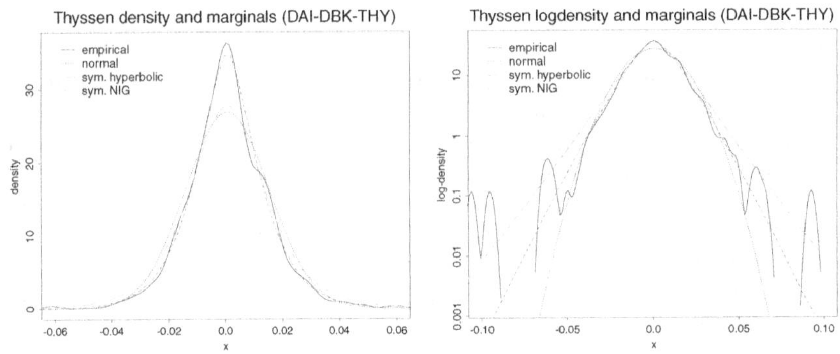

Fig. 7. Marginal density for Thyssen obtained from the 3-dimensional estimate for Daimler Benz–Deutsche Bank–Thyssen

The marginal densities of the GH distributions can be derived using a theorem of Blæsild (1981). Typically we obtain the pattern shown in Figure 7 for the densities and log-densities: The marginal distributions of hyperbolic and NIG distributions are closer to the empirical distribution than the normal distribution. In the center, marginals of hyperbolic distributions are closer to the empirical distribution but in the tails, marginals of NIG distributions provide a better fit.

7 Market Risk Measurement

Let us start with a general result on densities.

Theorem 7.1. *Let X be a d-dimensional random variable with symmetric generalized hyperbolic distribution, i.e. with $\beta = (0,\dots,0)'$, and let $h \in \mathbb{R}^d$ where $h \neq (0,\dots,0)'$. The distribution of $h'X$ is univariate generalized hyperbolic $\mathrm{GH}_d(\lambda^\times, \alpha^\times, \beta^\times, \delta^\times, \mu^\times)$, where $\lambda^\times = \lambda$, $\alpha^\times = \alpha|h'\Delta h|^{-1/2}$, $\beta^\times = 0$, $\delta^\times = \delta|h'\Delta h|^{1/2}$ and $\mu^\times = h'\mu$.*

Proof. Let $h_1 \neq 0$ without loss of generality. Apply Theorem Ic) of Blæsild (1981) with

$$A = \begin{pmatrix} h_1 & h_2 & \cdots & h_d \\ 0 & 1 & & 0 \\ \vdots & & \ddots & \\ 0 & 0 & & 1 \end{pmatrix} \quad \text{and } B = \begin{pmatrix} 0 \\ \vdots \\ 0 \end{pmatrix}. \tag{18}$$

Then project the d-dimensional GH distribution onto the first coordinate using Theorem Ia).

The latter theorem may be used to calculate risk measures for a portfolio of d assets with investments given by a vector $h \in \mathbb{R}^d$. As an example we look at a portfolio consisting of three German stocks: Daimler Benz, Deutsche Bank and Thyssen from January 1, 1988 to May 24, 1994. We choose $h = (1,1,1)'$ and show the empirical density of the returns $h'x_t$ of the portfolio in Figure 8. The previous theorem gives the corresponding densities obtained from the d-dimensional estimates of symmetric hyperbolic and symmetric NIG distributions. Figure 8 shows also the direct estimate of the univariate GH distribution from $h'x_t$.

The densities and log-densities in Figure 8 indicate that symmetric GH distributions enable one to perform more precise modelling of the return distribution of the portfolio. As a consequence one can get more realistic risk measures than the traditional ones based on the normal distribution. Figure 9 shows a risk measure over a 1-day horizon with respect to a level of probability $\alpha \in (0,1)$, namely the shortfall which we define as

$$\mathrm{Shortfall}_{\alpha,t} = -\mathrm{E}\big[h'x_t \big| h'x_t < q(\alpha)\big], \tag{19}$$

where $q : [0,1] \to \mathbb{R}$ is the corresponding quantile function.

Note that the shortfall goes clearly beyond the concept of VaR because it takes into account the extreme negative returns. The log-density of the empirical distribution in Figure 8 shows the magnitude of the negative returns of multi-asset portfolios in relation to the more frequent small returns.

The Basle Committee on Banking Supervision (1995, IV.23) has proposed a *backtesting* procedure to test the quality of Value-at-Risk estimators. We

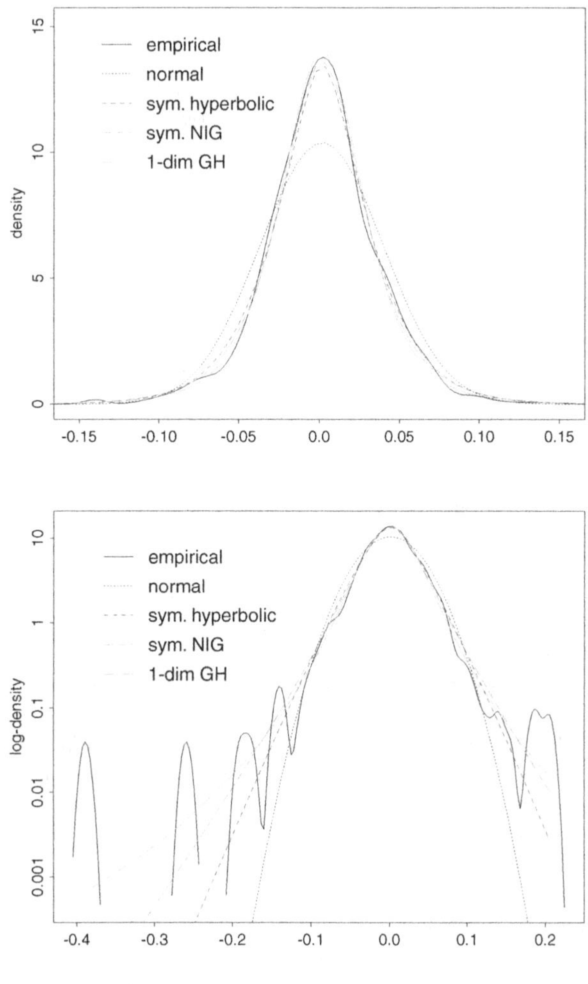

Fig. 8. Distribution of the returns of a portfolio consisting of Daimler Benz, Deutsche Bank, and Thyssen (equal weights)

follow this procedure to compare standard VaR estimation approaches with VaR estimators based on GH distributions.[2] After computing the VaR for

[2] The Basle Committee on Banking Supervision (1995, IV.3) recommends a holding period of 10 days. Nevertheless, we consider a 1-day horizon only because the increased number of returns in the observation period allows more accurate statistical results. Note, that we would not upscale a 1-day VaR by multiplying it

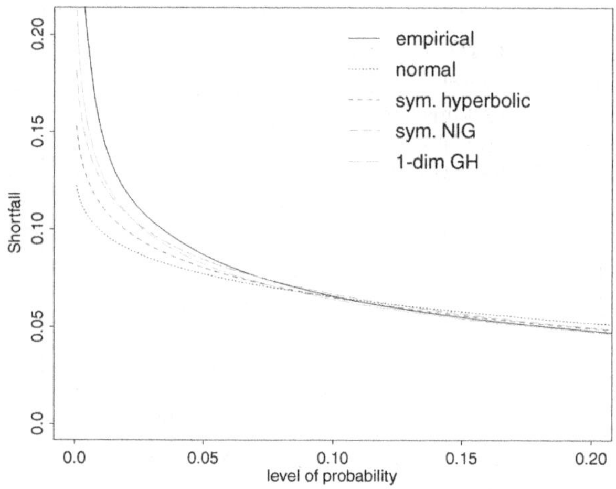

Fig. 9. Shortfall for Daimler Benz–Deutsche Bank–Thyssen (portfolio with equal weights)

each day in the time period from January 1, 1989 to May 24, 1994 we count the observed losses greater than the Value-at-Risk. Since Value-at-Risk is essentially a quantile, the percentage of excess losses should correspond to the level of probability α. One standard method to compute VaR is to simulate the return of the portfolio by the preceding 250 observed returns and to take the quantile of this empirical distribution. This is historical simulation. A second simulation technique proposed to forecast VaR is Monte Carlo simulation. It is computationally intensive for large portfolios. We have not applied this method here because distinct differences to the Variance-Covariance approach are only obtained for nonlinear portfolios (Bühler, Korn, and Schmidt (1998)). However, a full valuation approach based on GH distributions, for instance for portfolios with derivative contracts, is easily implemented (Prause (1999)). The mixture representation of GH distributions allows to generate random numbers efficiently. We also apply the Variance-Covariance approach which is based on the multivariate normal distribution.

The results given in Table 3 show that the standard estimators for Value-at-Risk underestimate the risk of extreme losses on the relevant level of 1%. This effect is visible in the percentage of excess losses in the historical simulation. In the Variance-Covariance approach we observe too high values for the level of probability $\alpha = 1\%$ and too small values for $\alpha = 5\%$. The percentages

with $\sqrt{10}$. Instead we would use the distribution corresponding to 10 days in our model.

of realized losses greater than VaR are closer to the level of probability in both cases $\alpha = 1\%$ and $\alpha = 5\%$ for the symmetric hyperbolic and the symmetric NIG distribution.

Table 3. Ex post evaluation of risk measures: percentage of losses greater than VaR. Each trading day the Value-at-Risk for a holding period of one day is estimated from the preceding 250 trading days (Daimler Benz, Deutsche Bank, and Thyssen from January 1, 1989 to May 24, 1994, Investment of 1DM in each asset)

VaR Estimation Method	$\alpha = 1\%$	$\alpha = 5\%$
Historical Simulation	2.08	5.79
Variance-Covariance	1.63	4.45
RiskMetrics / IGARCH	1.34	4.75
Symmetric hyperbolic	1.48	4.9
Symmetric NIG	1.26	4.75
Symmetric hyperbolic, long-term ζ	1.26	4.45
Symmetric NIG, long-term ζ	1.04	4.9
Hyperbolic IGARCH, long-term ζ	1.11	4.82
NIG IGARCH, long-term ζ	1.11	5.34
1-dimensional Hyperbolic	1.41	4.9
1-dimensional NIG	1.41	4.97

An approach similar to the rescaling mechanism proposed above for the univariate case is to estimate the shape from a longer time period and to use an up-to-date covariance matrix Σ. This allows one to incorporate the risk of extreme events, even if they do not occur in the preceding 250 trading days, which is the minimum time period proposed by the Basle Committee (1995). Therefore we have to choose a subclass, i.e. a parameter $\lambda \in \mathbb{R}$, and to fix a long-term estimate for ζ. We compute the matrix S in the alternative parametrization by

$$S = \delta^2 \Delta = \frac{\zeta}{R_\lambda(\zeta)} \Sigma. \tag{20}$$

A further refinement is possible by choosing an appropriate estimate for the covariance matrix. We select the multivariate IGARCH model of Nelson (1990) in which variance $\sigma^2_{1,t}$ and covariance $\sigma^2_{12,t}$ are given by

$$\sigma^2_{1,t} = (1-\lambda) \sum_{t \geq 1} \lambda^{t-1} (r_t - \bar{r}), \tag{21}$$

$$\sigma^2_{12,t} = (1-\lambda) \sum_{t \geq 1} \lambda^{t-1} (r_{1,t} - \bar{r}_1)(r_{2,t} - \bar{r}_2), \tag{22}$$

where $0 < \lambda < 1$ is a decay factor, $r_t, r_{1,t}, r_{2,t}$ returns of financial assets and $\bar{r}, \bar{r}_1, \bar{r}_2$ the corresponding mean values. To allow for a comparison, we have used the decay factor $\lambda = 0.94$ applied in J.P. Morgan and Reuters (1996) for daily returns.

Finally we propose to reduce the risk measurement problem to one dimension by computing quantiles for the return $h'x_t$ of the whole portfolio. We estimate hyperbolic and NIG distributions and derive the corresponding quantiles.

The results of the study for a linear portfolio are shown in Table 3. Taken together, the use of a long-term shape parameter incorporates the possibility of extreme events, even if there was no crash in the preceding 250 trading days, whereas the GH-IGARCH approach describes the volatility clustering observed in financial markets. This yields more accurate results for GH-based models in the ex-post evaluation of the risk measures.

8 Conclusion

In the first part of this paper we presented generalized hyperbolic distributions resp. their subclasses and estimation results concerning daily as well as high-frequency returns. The greater flexibility of this class of distributions allows an almost perfect fit to empirical asset return distributions. Based on the Lévy processes generated by these infinitely divisible distributions we introduced in section 3 the generalized hyperbolic model as a new way to describe asset prices. It is a rather natural model, since it reproduces exactly those distributions which one observes in the data. An option pricing formula can be derived using the Esscher transform as in Eberlein and Keller (1995). Using the rescaling mechanism of generalized hyperbolic distributions we analyzed implicit volatilities and prices obtained in the GH model. We observed a correction of the smile effect in the GH model.

Risk measures are used in financial institutions with two objectives. Internally they give the management a possibility to allocate risk capital.

> Setting limits in terms of risk helps business managers to allocate risk to those areas which they feel offer the most potential, or in which their firms' expertise is greatest. This motivates managers of multiple risk activities to favor risk reducing diversification strategies.[3]

On the other hand regulators as well as the management want to reduce the probability of default. Therefore they set limits to the exposure to market risk relative to the capital of the firm.

Is Value-at-Risk the adequate measure for this purpose? Quantile-based methods like VaR have the disadvantage that they do not consider losses occuring with a probability below a given level of probability. Stress testing offers a partial solution to this problem focussing on extreme scenarios. To

[3] J.P. Morgan and Reuters (1996, p. 33), see also Chart 3.1.

quantify risk properly one has to forecast the whole profit and loss distribution. Regulators should use other risk measures than VaR as well. In this context we also would like to mention the axiomatic concept of coherent risk measures developed by Artzner, Delbaen, Eber, and Heath (1999).

In the last two sections we have shown that it is possible to estimate generalized hyperbolic distributions in an efficient way and to construct more accurate risk measures for multivariate price processes. Symmetric hyperbolic and symmetric NIG distributions are characterized by the covariance matrix and a shape parameter. This simple structure allows a further sophistication of GH risk measures by fixing a long-term shape parameter, which describes the probability of rare events, and choosing a short-term estimate for the covariance matrix. A study in accordance with the backtesting concept required by the Basle Committee on Banking Supervision reconfirms the excellent results concerning VaR estimation for multivariate price processes. Moreover, we have shown that generalized hyperbolic distributions are also the proper building block for risk measures beyond VaR.

Acknowledgements. We thank Deutsche Börse AG, Frankfurt for a number of data sets concerning stock and option prices. We also used IBIS data from the Karlsruher Kapitalmarktdatenbank and the high-frequency data set HFDF96 provided by Olsen & Associates, Zürich.

Appendix

Moment Generating and Characteristic Function

Lemma A.1. *The moment generating function of the generalized hyperbolic distribution is*

$$M(u) = e^{u\mu}\left(\frac{\alpha^2-\beta^2}{\alpha^2-(\beta+u)^2}\right)^{\lambda/2}\frac{K_\lambda(\delta\sqrt{\alpha^2-(\beta+u)^2}\,)}{K_\lambda(\delta\sqrt{\alpha^2-\beta^2}\,)}, \quad |\beta+u|<\alpha.$$

Lemma A.2. *The characteristic function of the generalized hyperbolic distribution is*

$$\phi(u) = e^{i\mu u}\left(\frac{\alpha^2-\beta^2}{\alpha^2-(\beta+iu)^2}\right)^{\lambda/2}\frac{K_\lambda(\delta\sqrt{\alpha^2-(\beta+iu)^2}\,)}{K_\lambda(\delta\sqrt{\alpha^2-\beta^2}\,)}.$$

Theorem A.3. *The Lévy-Khintchine representation of $\phi(u)$ is*

$$\ln\phi(u) = iu\mu + \int\left(e^{iux}-1-iux\right)g(x)\mathrm{d}x, \tag{23}$$

with density

$$g(x) = \frac{e^{\beta x}}{|x|}\left(\int_0^\infty \frac{\exp(-\sqrt{2y+\alpha^2}\,|x|)}{\pi^2 y\big(J_\lambda^2(\delta\sqrt{2y})+Y_\lambda^2(\delta\sqrt{2y})\big)}\,\mathrm{d}y + \mathbf{1}_{\{\lambda\geq 0\}}\,\lambda e^{-\alpha|x|}\right). \tag{24}$$

Here J_λ and Y_λ denote Bessel functions of the first and second kind.

References

1. Abramowitz, M. and Stegun, I. A. (1968) *Handbook of Mathematical Functions*. Dover Publ., New York
2. Artzner, P., Delbaen, F., Eber, J.-M., and Heath, D. (1999) Coherent measures of risk. *Mathematical Finance*, **9**, 203–228
3. Bakshi, G., Cao, C., and Chen, Z. (1997) Empirical performance of alternative option pricing models. *Journal of Finance*, **70**, 2003–2049
4. Barndorff-Nielsen, O. E. (1977) Exponentially decreasing distributions for the logarithm of particle size. *Proceedings of the Royal Society London A*, **353**, 401–419
5. Barndorff-Nielsen, O. E. (1998) Processes of normal inverse Gaussian type. *Finance & Stochastics*, **2**, 41–68
6. Barndorff-Nielsen, O. E. and Blæsild, P. (1981) Hyperbolic distributions and ramifications: Contributions to theory and application. In Taillie, C., Patil, G., and Baldessari, B., editors, *Statistical Distributions in Scientific Work*, volume 4, pages 19–44. Reidel, Dordrecht
7. Barndorff-Nielsen, O. E., Blæsild, P., Jensen, J. L., and Sørensen, M. (1985) The fascination of sand. In Atkinson, A. C. and Fienberg, S. E., editors, *A Celebration of Statistics*. Springer, New York
8. Barndorff-Nielsen, O. E. and Halgreen, O. (1977) Infinite divisibility of the hyperbolic and generalized inverse Gaussian distributions. *Zeitschrift für Wahrscheinlichkeitstheorie und verwandte Gebiete*, **38**, 309–312
9. Barndorff-Nielsen, O. E. and Prause, K. (2001) Apparent scaling. *Finance & Stochastics*, **5**, 103–113
10. Barndorff-Nielsen, O. E. and Shephard, N. (2001) Non-Gaussian Ornstein-Uhlenbeck-based models and some of their uses in financial economics. *J.R. Statist. Soc. B*, **63**, 1–42
11. Basle Committee on Banking Supervision (1995) *An internal model-based approach to market risk capital requirements*. Bank for International Settlements, Basle
12. Bates, D. S. (1996) Testing option pricing models. In Maddala, G. S. and Rao, C. R., editors, *Handbook of Statistics*, volume 14, pages 567–611. Elsevier Science, Amsterdam
13. Bibby, B. M. and Sørensen, M. (1997) A hyperbolic diffusion model for stock prices. *Finance & Stochastics*, **1**, 25–41
14. Black, F. and Scholes, M. (1973) The pricing of options and corporate liabilities. *Journal of Political Economy*, **81**, 637–654
15. Blæsild, P. (1981) The two-dimensional hyperbolic distribution and related distributions, with an application to Johannsen's bean data. *Biometrika*, **68**, 251–263
16. Blæsild, P. and Jensen, J. L. (1981) Multivariate distributions of hyperbolic type. In Taillie, C., Patil, G., and Baldessari, B., editors, *Statistical distributions in scientific work*, volume 4, pages 45–66. Reidel, Dordrecht
17. Bühler, W., Korn, O., and Schmidt, A. (1998) Ermittlung von Eigenkapitalanforderungen mit internen Modellen. Eine empirische Studie zur Messung von Zins-, Währungs- und Optionsrisiken mit Value-at-Risk Ansätzen. *Die Betriebswirtschaft*, **58**(1), 64–85
18. Chan, T. (1999) Pricing contingent claims on stocks driven by Lévy processes. *Annals of Applied Probability*, **9**, 504–528

19. Delbaen, F. and Schachermayer, W. (1994) A general version of the fundamental theorem of asset pricing. *Mathematische Annalen*, **300**, 463–520
20. Eberlein, E. and Jacod, J. (1997) On the range of options prices. *Finance & Stochastics*, **1**, 131–140
21. Eberlein, E. and Keller, U. (1995) Hyperbolic distributions in finance. *Bernoulli*, **1**, 281–299
22. Eberlein, E., Kallsen, J., and Kristen, J. (2001) Risk management based on stochastic volatility. FDM-Preprint, **72**, University of Freiburg
23. Eberlein, E., Keller, U., and Prause, K. (1998) New insights into smile, mispricing and value at risk: the hyperbolic model. *Journal of Business*, **71**, 371–405
24. Eberlein, E. and Prause, K. (1998) The generalized hyperbolic model: financial derivatives and risk measures. FDM Preprint 56, University of Freiburg
25. Föllmer, H. and Schweizer, M. (1991) Hedging of contingent claims under incomplete information. In Davis, M. H. A. and Elliott, R. J., editors, *Applied Stochastic Analysis*, volume 5 of *Stochastics Monographs*, pages 389–414. Gordon and Breach, New York
26. Goll, Th. and Rüschendorf, L. (2001) Minimax and minimal distance martingale measures and their relationship to portfolio optimization. to appear in *Finance & Stochastics*
27. Guillaume, D. M., Dacorogna, M. M., Davé, R. D., Müller, U. A., Olsen, R. B., and Pictet, O. V. (1997) From the bird's eye to the microscope: a survey of new stylized facts of the intra-daily foreign exchange markets. *Finance & Stochastics*, **1**, 95–129
28. J. P. Morgan and Reuters. (1996) RiskMetrics – Technical document. New York
29. Keller, U. (1997) *Realistic modelling of financial derivatives*. Dissertation, University of Freiburg
30. Lo, A. W. (1986) Statistical tests of contingent-claims asset-pricing models. A new methodology. *Journal of Financial Economics*, **17**, 143–173
31. Nelson, D. B. (1990) Stationarity and persistence in the GARCH(1,1) model. *Econometric Theory*, **6**, 318–334
32. Nicolato, E. and Prause, K. (1999) Derivative pricing in stochastic volatility models of the Ornstein-Uhlenbeck type. *Working Paper, University of Århus*
33. Olsen & Associates. (1998) HFDF96. High frequency data in preparation for the Second International Conference on High Frequency Data in Finance (HFDF II), Zürich, April 1–3, 1998.
34. Prause, K. (1997) Modelling financial data using generalized hyperbolic distributions. FDM Preprint 48, University of Freiburg
35. Prause, K. (1999) *The generalized hyperbolic model: Estimation, financial derivatives, and risk measures*. Dissertation, University of Freiburg
36. Rydberg, T. H. (1999) Generalized hyperbolic diffusion processes with applications towards finance. *Mathematical Finance*, **9**, 183–201

Using the Hull and White Two Factor Model in Bank Treasury Risk Management

Robert J. Elliott[1,2] and John van der Hoek[2]

[1] Department of Mathematical Sciences, University of Alberta, Edmonton, Alberta, Canada T6G 2G1
[2] Department of Applied Mathematics, University of Adelaide, Adelaide, South Australia 5005

1 Introduction

In order to manage interest rank risk exposure, bank treasurers allocate their portfolio of assets and liabilities to what they call standard "buckets". In order words the portfolio of cash flows from assets and liabilities at various times are re-expressed as a portfolio of cash flows at certain specified standard times. These specified times usually correspond to important dates like expiry dates of hedging instruments (bank bills, bank bill futures, interest rate swaps, and so on). The exact choice of these dates will not concern us in this paper, but will be specified by the users of these results. Of course, this procedure can only be done in an approximate way. Traditionally, the real cash flows are assigned so that the portfolio of assets and liabilities of the bank portfolio match the portfolio of bucketed assets with respect to present value and (Macaulay) duration (or other risk measures like Reddington convexity), which could hold for a short time interval. We will not review the various methods used by practitioners, but present a method based on the use of Wiener Chaos expansions, (with respect to suitable forward probability measures), of assets and liabilities and to select bucketing in order to match various orders of the chaos expansion in the assets and liabilities portfolio and bucketed portfolio. In fact matching the first three orders of the chaos expansion is closely related to present value, duration, and convexity matching, (see Brace and Musiela [1]). In this framework it is possible to study and give results about optimal assignments, so that matching may hold over a working time horizon (a day, a week, and so on). Once assets and liabilities have been bucketed, the interest rate or exposure can be managed by the use of standard treasury products. This can involve a similar procedure to the one we have described already, by which we represent the bucketed cash flows in terms of the cash flows of a portfolio of hedging instruments.

We shall present the results in the framework of the Hull and White two factor interest rate model [6], in which it is possible to make very explicit calculations. The Hull and White two factor model generalizes the Vasicek and the continuous Ho and Lee one factor models. For these, even more explicit results are available. We derive and present various explicit formulae for bucketing calculations. Further results on the bucketing of the cash flows from options, the use of hedging instruments, and calibration issues of the Hull and White model will be presented in a sequel to this paper.

This work complements the studies of Jarrow and Turnbull [8], Musiella, Turnbull and Wakeman [11], on risk management, and Brace and Musiela [1], LaCoste [10] on the use of Wiener Chaos expansions. In our presentation, the computations of chaos expansions are achieved by the theory of stochastic flows (see Elliott [2], Kunita [9], Elliott and Kohlmann [3].

2 The Hull and White Two-Factor Model

In this model we assume that the short interest rate r has dynamics given (for $t \geq 0$) by:

$$dr(t) = [\theta(t) + u(t) - ar(t)]dt + \sigma_1 dW_1(t) \tag{2.1}$$

$$du(t) = -bu(t)dt + \sigma_2[\rho dW_1(t) + \sqrt{1-\rho^2}\, dW_2(t)] \tag{2.2}$$

where W_1 and W_2 are independent (one-dimensional) Brownian motions on some probability space $(\Omega, \mathcal{F}, \boldsymbol{P})$ and $(\mathcal{F}_t)$ is the usual filtration generated by $W = (W_1, W_2)$. We assume that $r(0)$, $u(0)$ are given. Also a, b, σ_1, σ_2 and $|\rho| \leq 1$ are all constants (which need to be estimated) and $\theta(t)$ can be selected so that the model is consistent with an initial term structure. The model expressed in (2.1) and (2.2) is assumed to be the model in the risk-neutral setting. For further details of this model see Hull and White [6].

The price at time $t \leq T$ of a contingent claim $H \in \mathcal{L}^2(\Omega, \mathcal{F}_T, \boldsymbol{P})$ is given by

$$\boldsymbol{E}\Big[\exp\Big(-\int_t^T r(s)ds\Big)H \,\Big|\, \mathcal{F}_t\Big]. \tag{2.3}$$

Here $\boldsymbol{E}$ denotes expectation with respect to probability $\boldsymbol{P}$. In particular, the price of a zero coupon bond at time t, with maturity T is given by

$$P(t,T) = \boldsymbol{E}\Big[\exp\Big(-\int_t^T r(s)ds\Big) \,\Big|\, \mathcal{F}_t\Big]. \tag{2.4}$$

It is sometimes useful to express the explicit dependence of $P(t,T)$ on the spot values of r and u as follows

$$P\big(t,T,(r,u)\big) = \boldsymbol{E}\Big[\exp\Big(-\int_t^T r(s)ds\Big) \,\Big|\, r(t) = r,\ u(t) = u\Big]. \tag{2.5}$$

It is shown in Hull and White [6], and by a different argument in Elliott and van der Hoek [4] that

$$P\big(t,T,(r,u)\big) = \exp\big(A(t,T) - B(t,T)r - C(t,T)u\big) \tag{2.6}$$

where

$$B(t,T) = \frac{1}{a}\left(1 - e^{-a(T-t)}\right) \tag{2.7}$$

$$C(t,T) = \frac{1}{a(a-b)}e^{-a(T-t)} - \frac{1}{b(a-b)}e^{-b(T-t)} - \frac{1}{ab} \tag{2.8}$$

and $A(t,T)$ is the solution of

$$\frac{\partial A}{\partial t} - B\theta + \frac{1}{2}\left(\sigma_1^2 B^2 + \sigma_2^2 C^2 + 2\rho\sigma_1\sigma_2 BC\right) = 0 \tag{2.9}$$

subject to $A(T,T) = 0$, and for which an explicit expression can be written down (see Hull and White [6]).

3 Forward Measures

For any $T > 0$ we can introduce a forward measure $\boldsymbol{P}^T$ on $\mathcal{F}_T$ by setting

$$\left.\frac{d\boldsymbol{P}^T}{d\boldsymbol{P}}\right|_{\mathcal{F}_T} = \Lambda_T = \frac{\exp\left(-\int_0^T r(s)ds\right)}{P(0,T)} \tag{3.1}$$

and write

$$\Lambda_t = \boldsymbol{E}[\Lambda_T \mid \mathcal{F}_t] = \frac{\exp\left(-\int_0^t r(s)ds\right)P(t,T)}{P(0,T)}. \tag{3.2}$$

Let $\boldsymbol{E}^T$ denote expectation with respect to $\boldsymbol{P}^T$. Under this forward probability measure, the process

$$\left\{\frac{P(t,U)}{P(t,T)} \,\middle|\, t \geq 0\right\}$$

is a martingale for any $U > 0$. This can be shown directly using Bayes' rule

$$\boldsymbol{E}^T[X \mid \mathcal{F}_t] = \frac{\boldsymbol{E}[\Lambda_T X \mid \mathcal{F}_t]}{\Lambda_t} \tag{3.3}$$

for any $\mathcal{F}_T$ measurable X. Under P^T we can find independent Brownian motions W_1^T, W_2^T on $\left(\Omega, \mathcal{F}, \boldsymbol{P}^T, (\mathcal{F}_t)\right)$. In fact they satisfy

$$dW_1^T(t) = dW_1(t) + \phi_1(t)dt \tag{3.4}$$

$$dW_2^T(t) = dW_2(t) + \phi_2(t)dt \tag{3.5}$$

with

$$\phi_1(t) = \sigma_1 B(t,T) + \rho\sigma_2 C(t,T) \tag{3.6}$$

$$\phi_2(t) = \sigma_2\sqrt{1-\rho^2}\,C(t,T). \tag{3.7}$$

Now if we introduce the notation

$$B(s,T,U) \equiv B(s,T) - B(s,U) \tag{3.8}$$

$$C(s,T,U) \equiv C(s,T) - C(s,U) \tag{3.9}$$

then a direct calculation shows that

$$d\left[\frac{P(t,U)}{P(t,T)}\right] = \left[\frac{P(t,U)}{P(t,T)}\right] dZ^U(t) \tag{3.10}$$

where

$$Z^U(t) \equiv \int_0^t \Big\{ \big(\sigma_1 B(s,T,U) + \rho\sigma_2 C(s,T,U)\big) dW_1^T(s) + \sigma_2 \sqrt{1-\rho^2}\, C(s,T,U) dW_2^T(s) \Big\}. \tag{3.11}$$

4 Wiener Chaos with Respect to Forward Measures

For a fixed T, we are interested in the Wiener Chaos expansion of the zero coupon prices $P(T,U)$ for various choices of $U > 0$. We do not require that $U \geq T$. That is, we wish to express

$$\begin{aligned} P(T,U) &= \boldsymbol{E}^T[P(T,U)] + \int_0^T \sum_{i=1}^2 \gamma^i(s_1) dW_i^T(s_1) \\ &+ \int_0^T \int_0^{s_1} \sum_{i,j=1}^2 \gamma^{ij}(s_1,s_2) dW_i^T(s_2) dW_j^T(s_1) \\ &+ \dots \end{aligned} \tag{4.1}$$

where the successive multiple integrals have deterministic integrands. The terms in (4.1) are called the $0^{\text{th}}, 1^{\text{st}}, 2^{\text{nd}}, \dots$ order chaos terms of the expansion. These expansions originate with Wiener [12], but see also Itô [7]. For the Hull and White two factor model the terms in (4.1) can be calculated explicitly.

In fact define

$$V\big(s, r(s), u(s)\big) \equiv \boldsymbol{E}^T[P(T,U) \,|\, \mathcal{F}_s] \tag{4.2}$$

for $0 \leq s \leq T$. Then

$$\begin{aligned} V\big(s, r(s), u(s)\big) &= \boldsymbol{E}^T\left[\frac{P(T,U)}{P(T,T)} \,\Big|\, \mathcal{F}_s\right] \\ &= \frac{P(s,U)}{P(s,T)} \end{aligned} \tag{4.3}$$

for $0 \le s \le T$. In particular

$$\boldsymbol{E}^T[P(T,U)] = \frac{P(0,U)}{P(0,T)} \,. \tag{4.4}$$

By (3.10),

$$V\big(s,r(s),u(s)\big) = V\big(0,r(0),u(0)\big) + \int_0^s V\big(\tau,t,(\tau),u(\tau)\big)dZ^U(\tau). \tag{4.5}$$

In particular

$$\begin{aligned}
P(T,U) &= \boldsymbol{E}^T[P(T,U)] + \int_0^T V\big(\tau,r(\tau),u(\tau)\big)dZ^U(\tau) \\
&= \boldsymbol{E}^T[P(T,U)] \\
&\quad + \int_0^T \Big[V\big(0,r(0),u(0)\big) + \int_0^\tau V\big(s,r(s),u(s)\big)dZ^U(s)\Big]dZ^U(\tau) \\
&= \boldsymbol{E}^T[P(T,U)] + \boldsymbol{E}^T[P(T,U)]\int_0^T dZ^U(\tau) \\
&\quad + \int_0^T \Big[\int_0^\tau V\big(s,r(s),u(s)\big)dZ^U(s)\Big]dZ^U(\tau)
\end{aligned}$$

which leads to the basic identity

$$P(T,U) = \frac{P(0,U)}{P(0,T)}[1 + Z^U(T)] + \int_0^T \Big[\int_0^\tau \frac{P(s,U)}{P(s,T)}\, dZ^U(s)\Big]dZ^U(\tau). \tag{4.6}$$

We can now iterate on (4.6) or (4.5) to obtain chaos expansions of any order. In fact

$$\begin{aligned}
P(T,U) &= \frac{P(0,U)}{P(0,T)}\Big[1 + \int_0^T dZ^U(s_1) + \int_0^T \int_0^{s_1} dZ^U(s_2)dZ^U(s_1) + \dots\Big] \\
&= \frac{P(0,U)}{P(0,T)}\,\Theta^U(T)
\end{aligned} \tag{4.7}$$

where

$$\Theta^U(t) = \exp\Big[Z^U(t) - \frac{1}{2}[Z^U,Z^U](t)\Big] \tag{4.8}$$

and

$$\begin{aligned}
[Z^U,Z^U](t) = \int_0^t \{&\sigma_1^2 B(s,T,U)^2 + 2\rho\sigma_1\sigma_2 B(s,T,U)C(s,T,U) \\
&+\sigma_2^2 C(s,T,U)^2\}ds
\end{aligned} \tag{4.9}$$

or more generally,

$$[Z^U, Z^V](t) = \int_0^t \{[\sigma_1 B(s,T,U) + \rho\sigma_2 C(s,T,U)] \\ \times[\sigma_1 B(s,T,V) + \rho\sigma_2 C(s,T,V)] \\ +\sigma_2^2(1-\rho^2)C(s,T,U)C(s,T,V)\}ds. \quad (4.10)$$

Whereas the chaos expansion can be written down explicitly for the zero coupon bond, only the first few terms can be written down for the chaos expansions for options on bonds. Calculations for these are provided in the second part of this paper.

5 Bucketting of Cash Flows

We now think of $T > 0$ as expressing the horizon over which we wish to work. It could express the time between the rehedging of a bank's book. This will be chosen by the practitioner. We shall consider without loss of generality a cash flow of 1 at time U, whose value at time T will be $P(T,U)$. We will let the standard times be

$$0 < T_1 < T_2 < T_3 \cdots < T_N$$

and we shall select cash flows c_i for $i = 1, 2, 3, \ldots, N$, so that

$$P(T,U) \approx \sum_{i=1}^{N} c_i P(T,T_i). \quad (5.1)$$

We will select c_i for $i = 1, 2, 3, \ldots, N$ so that both sides of equation (5.1) have equal 0th and 1st order chaos terms (if possible) and the residual

$$R(T) = P(T,U) - \sum_{i=1}^{N} c_i P(T,T_i)$$

has minimum value for

$$\boldsymbol{E}^T[R(T)^2].$$

As pointed out by Brace and Musiela [1], this choice generalizes the idea of equating the present values and duration of both sides of (4.6) and minimizing the convexity of the residual.

Because of the special structure of the Hull and White two factor model, this task can be carried out explicitly. In fact by (4.6)

$$\begin{aligned} R(T) &= \frac{1}{P(0,T)}\left[P(0,U) - \sum_{i=1}^{N} c_i P(0,T_i)\right] \\ &+ \frac{1}{P(0,T)}\left[P(0,U)Z^U(T) - \sum_{i=1}^{N} c_i P(0,T_i) Z^{T_i}(T)\right] \\ &+ \int_0^T \int_0^\tau \left[\frac{P(s,U)}{P(s,T)} dZ^U(s) dZ^U(\tau) - \sum_{i=1}^{N} c_i \frac{P(s,T_i)}{P(s,T)} dZ^{T_i}(s) dZ^{T_i}(\tau)\right] \end{aligned} \tag{5.2}$$

and the first two terms on the right hand side of (5.2) can be made zero by requiring

$$P(0,U)B(s,T,U) = \sum_{i=1}^{N} c_i B(s,T,T_i) P(0,T_i) \tag{5.3}$$

$$P(0,U)C(s,T,U) = \sum_{i=1}^{N} c_i C(s,T,T_i) P(0,T_i) \tag{5.4}$$

$$P(0,U) = \sum_{i=1}^{N} c_i P(0,T_i) \tag{5.5}$$

for all $0 \leq s \leq T$. Because of the special structure of $B(s,T,U)$ and $C(s,T,U)$, namely

$$\begin{aligned} B(s,T,U) &= \left[-\frac{1}{a} e^{-a(T-s)}\right] + \left[\frac{1}{a} e^{as}\right] e^{-aU} \\ C(s,T,U) &= \left[\frac{e^{-a(T-s)}}{a(a-b)} - \frac{e^{-b(T-s)}}{b(a-b)}\right] - \left[\frac{e^{as}}{a(a-b)}\right] e^{-aU} + \left[\frac{e^{bs}}{b(a-b)}\right] e^{-bU}, \end{aligned}$$

equations (5.3), (5.4) and (5.5) are equivalent to (5.5) plus the following two identities

$$P(0,U)e^{-aU} = \sum_{i=1}^{N} c_i P(0,T_i) e^{-aT_i} \tag{5.6}$$

$$P(0,U)e^{-bU} = \sum_{i=1}^{N} c_i P(0,T_i) e^{-bT_i}. \tag{5.7}$$

We will therefore assume that (5.5), (5.6) and (5.7) are constraints that hold while we minimize the expression $\boldsymbol{E}^T[R(T)^2]$.

Remark: It can be shown that equations (5.5), (5.6) and (5.7) are equivalent to (5.5) and the requirements:

$$\frac{d}{dr_0} P(0,U) = \frac{d}{dr_0} \sum_{i=1}^{N} c_i P(0,T_i) \tag{5.8}$$

$$\frac{d}{du_0} P(0,U) = \frac{d}{du_0} \sum_{i=1}^{N} c_i P(0,T_i). \tag{5.9}$$

so the equating of 0th and 1st order chaos expansions generalizes the equating of present values and Macaulay durations. This observation is also true in other interest rate models where the spot rate or the forward rates (as in the HJM models) have deterministic volatility structure. This will be made more explicit in a sequel to this paper where we derive chaos expansions (for options) using the theory of flows (as in Elliott and Kohlmann [3]). With other volatility structures (as for the Cox-Ingersoll-Ross model) this equivalence no longer holds, and we have a true extension of Macaulay duration. In this more general context there are now a number of competing definitions for duration measures for such stochastic interest rate models. With all this said, an equivalent statement for our program is to minimize $\boldsymbol{E}^T[R(T)^2]$ subject to (5.5), (5.8) and (5.9).

In the present context $\boldsymbol{E}^T[R(T)^2]$ can be calculated explicitly as follows. By (4.7)

$$R(T) = \frac{P(0,U)}{P(0,T)} \Theta^U(T) - \sum_{i=1}^{N} c_i \frac{P(0,T_i)}{P(0,T)} \Theta^{T_i}(T)$$

and hence

$$\begin{aligned} \boldsymbol{E}^T[R(T)^2] = \left[\frac{P(0,U)}{P(0,T)}\right]^2 & \boldsymbol{E}^T[\Theta^U(T)^2] \\ & -2\sum_{i=1}^{N} c_i \frac{P(0,T_i)P(0,U)}{P(0,T)^2} \boldsymbol{E}^T[\Theta^U(T)\Theta^{T_i}(T)] \\ & + \sum_{i,j=1}^{N} c_i c_j \frac{P(0,T_i)P(0,T_j)}{P(0,T)^2} \boldsymbol{E}^T[\Theta^{T_i}(T)\Theta^{T_j}(T)] \end{aligned} \tag{5.10}$$

and a direct calculation shows that for any positive U and V,

$$\boldsymbol{E}^T[\Theta^U(t)\Theta^V(t)] = \exp\big[[Z^U, Z^V](t)\big]. \tag{5.11}$$

Now setting variables

$$x_i \equiv \frac{P(0,T_i)}{P(0,T)} c_i$$

we can write

$$\boldsymbol{E}^T[R(T)^2] = \left[\frac{P(0,U)}{P(0,T)}\right]^2 \Big[\sum_{i,j=1}^{N} A_{ij} x_i x_j - 2\sum_{i=1}^{N} B_i x_i + \Gamma\Big] \tag{5.12}$$

where

$$\begin{aligned} A_{ij} &= \exp\left[[Z^{T_i}, Z^{T_j}](T) \right] \\ B_i &= \exp\left[[Z^{T_i}, Z^{U}](T) \right] \\ \Gamma &= \exp\left[[Z^{U}, Z^{U}](T) \right]. \end{aligned}$$

It is possible to show that the matrix (A_{ij}) is symmetric and positive definite. In fact a direct calculation shows that if $(\xi_1, \xi_2, \ldots, \xi_N) \in \mathcal{R}^N$, then

$$\sum_{i,j=1}^{N} A_{ij}\xi_i\xi_j = \boldsymbol{E}^T\left[\left(\sum_{i=1}^{N} \xi_i \Theta^{T_i}(T) \right)^2 \right] \geq 0$$

and if

$$\sum_{i,j=1}^{N} A_{ij}\xi_i\xi_j = 0$$

then almost surely

$$\sum_{i=1}^{N} \xi_i \Theta^{T_i}(T) = 0$$

and hence (by conditioning)

$$\sum_{i=1}^{N} \xi_i \Theta^{T_i}(t) = 0$$

for all $0 \leq t \leq T$, whence $\xi_1 = \xi_2 = \cdots = \xi_N = 0$. We now solve the following quadratic linear programming problem:

$$\frac{1}{2}\sum_{i,j=1}^{N} A_{ij}x_i x_j - \sum_{i=1}^{N} B_i x_i + \frac{1}{2}\Gamma = \text{ minimum} \tag{5.13}$$

subject to

$$1 = \sum_{i=1}^{N} x_i \tag{5.14}$$

$$e^{-aU} = \sum_{i=1}^{N} x_i e^{-aT_i} \tag{5.15}$$

$$e^{-bU} = \sum_{i=1}^{N} x_i e^{-bT_i}. \tag{5.16}$$

The solution (using Lagrange multipliers) is

$$\boldsymbol{x} = A^{-1}B - \lambda_0 A^{-1}\mathbf{1} - \lambda_a A^{-1}e^{-a\boldsymbol{T}} - \lambda_b A^{-1}e^{-b\boldsymbol{T}} \tag{5.17}$$

where

$$\begin{aligned}
\boldsymbol{x} &= (x_1, x_2, \ldots, x_N)' \\
A &= (A_{ij}) \\
B &= (B_1, B_2, \ldots, B_N)' \\
\mathbf{1} &= (1, 1, \ldots, 1)' \\
e^{-a\boldsymbol{T}} &= (e^{-aT_1}, e^{-aT_2}, \ldots, e^{-aT_N})' \\
e^{-b\boldsymbol{T}} &= (e^{-bT_1}, e^{-bT_2}, \ldots, e^{-bT_N})'.
\end{aligned}$$

The Lagrange multipliers $(\lambda_0, \lambda_a, \lambda_b)$ satisfy the equation

$$\begin{bmatrix} \mathbf{1}'A^{-1}\mathbf{1} & \mathbf{1}'A^{-1}e^{-a\boldsymbol{T}} & \mathbf{1}'A^{-1}e^{-b\boldsymbol{T}} \\ e^{a\boldsymbol{T}\prime}A^{-1}\mathbf{1} & \mathbf{1}'A^{-1}e^{-a\boldsymbol{T}} & e^{-a\boldsymbol{T}\prime}A^{-1}e^{-b\boldsymbol{T}} \\ e^{-b\boldsymbol{T}\prime}A^{-1}\mathbf{1} & e^{-b\boldsymbol{T}\prime}A^{-1}e^{-a\boldsymbol{T}} & e^{-b\boldsymbol{T}\prime}A^{-1}e^{-b\boldsymbol{T}} \end{bmatrix} \begin{bmatrix} \lambda_0 \\ \lambda_a \\ \lambda_b \end{bmatrix}$$

$$= \begin{bmatrix} \mathbf{1}'A^{-1}B - \mathbf{1} \\ e^{-a\boldsymbol{T}\prime}A^{-1}B - e^{-aU} \\ e^{-b\boldsymbol{T}\prime}A^{-1}B - e^{-bU} \end{bmatrix}. \tag{5.18}$$

The matrix on the left hand side of (5.18) is symmetric and positive definite provided $N \geq 3$. The solution to the bucketing problem is then

$$c_i = \frac{P(0,U)}{P(0,T_i)}\, x_i \tag{5.19}$$

for $i = 1, 2, \ldots, N$. Each cash flow can be handled in the same way. As the matrix A and the matrix on the left hand side of (5.18) do not involve U, the inversion of the respective matrices need be performed only once in each application. We note that U enters the solution through B in (5.17) and the right hand side of (5.18) and in the rescaling in (5.19).

6 Special Examples

6.1 The Continuous Ho and Lee Model

This makes the choices:

$$a = b = 0, \quad \sigma_2 = 0, \quad \sigma_1 = \sigma, \quad u \equiv 0, \quad W = W_1$$

in which case

$$\begin{aligned} B(t,T) &= T - t \\ B(s,T,U) &= T - U \\ dW^T(t) &= dW(t) + \sigma(T-t)dt \\ Z^U(t) &= \sigma(T-U)W^T(t) \\ [Z^U, Z^U] &= \sigma^2(T-U)(T-V)t \\ A_{ij} &= \exp\left[\sigma^2(T-T_i)(T-T_j)T\right] \\ B_i &= \exp\left[\sigma^2(T-T_i)(T-U)T\right] \\ \Gamma &= \exp\left[\sigma^2(T-U)(T-U)T\right]. \end{aligned}$$

Equation (5.6) becomes:

$$UP(0,U) = \sum_{i=1}^{N} c_i T_i P(0,T_i).$$

6.2 Extended Vasicek Model

This is also a Hull and White one factor model [5]. This makes the choices

$$b = 0, \quad \sigma_2 = 0, \quad \sigma_1 = \sigma, \quad u \equiv 0 \; W = W_1$$

in which case we get the same analysis as the two factor model except that all the C and W_2 terms are missing.

Acknowledgements. This work was carried out during the visit of RJE to Adelaide. He wishes to thank the Department of Applied Mathematics for its hospitality. RJE also wishes to thank SSHRCC for its support. Both authors acknowledge the useful discussions with two practitioners, Dr. Bob Arnold (senior quantitative analyst, treasury systems and reporting, SA Government Financing Authority) and Guy Pitman (actuary and software developer for Global Information Solutions).

References

1. A. Brace and M. Musiela: Duration, convexity and Wiener chaos. Working paper (1995) University of NSW.
2. R.J. Elliott: *Stochastic Calculus and Applications.* Springer Verlag 1982.
3. R.J. Elliott and M. Kohlmann: Integration by parts, homogeneous chaos expansions and smooth densities. Ann. Probab. **17** (1989) 194-207.
4. R.J. Elliott and J. van der Hoek: Stochastic flows and the forward measure. Working paper (1998).
5. J. Hull and A. White: Numerical procedures for implementing term structure models I. Single factor models. Journal of Derivatives **2** (Fall 1994) 7-16.
6. J. Hull and A. White: Numerical procedures for implementing term structure models II. Two factor models. Journal of Derivatives **2** (Winter 1994) 37-48.

7. K. Itô: Multiple Wiener integrals. J. Math. Soc. Japan **3** (1951) 157-169.
8. R. Jarrow and S.M. Turnbull: Delta, gamma and bucket hedging of interest rate derivatives. Applied Mathematical Finance **1** (1994) 21-48.
9. H. Kunita: Stochastic partial differential equations connected with non-linear filtering, in *Lecture Notes in Mathematics*, vol. 972. Springer, Berlin 1982.
10. V. Lacoste: Wiener chaos: A new approach to option hedging. Mathematical Finance **6** (April 1996) 197-213.
11. M. Musiela, S.M. Turnbull and L.M. Wakeman: Interest rate risk management. Review of Futures Markets **12** (1992) 221-261.
12. N. Wiener: The homogeneous chaos. Am. J. Math. **55** (1938) 897-936.

Default Risk and Hazard Process

Monique Jeanblanc[1] and Marek Rutkowski[2]

[1] M.J. Equipe d'Analyse et Probabilités, Université d'Evry Val d'Essonne
Boulevard François Mitterrand, 91025 Evry Cedex, France
jeanbl@maths.univ-evry.fr

[2] M.R. Faculty of Mathematics and Information Science
Warsaw University of Technology
00-661 Warszawa, Poland
markrut@mini.pw.edu.pl

1 Introduction

The so-called intensity-based approach to the modelling and valuation of defaultable securities has attracted a considerable attention of both practitioners and academics in recent years; to mention a few papers in this vein: Duffie [8], Duffie and Lando [9], Duffie et al. [10], Jarrow and Turnbull [13], Jarrow et al. [14], Jarrow and Yu [15], Lando [21], Madan and Unal [23]. In the context of financial modelling, there was also a renewed interest in the detailed analysis of the properties of random times; we refer to the recent papers by Elliott et al. [12] and Kusuoka [20] in this regard. In fact, the systematic study of stopping times and the associated enlargements of filtrations, motivated by a purely mathematical interest, was initiated in the 1970s by the French school, including: Brémaud and Yor [4], Dellacherie [5], Dellacherie and Meyer [7], Jeulin [16], and Jeulin and Yor [17]. On the other hand, the classic concept of the intensity (or the hazard rate) of a random time was also studied in some detail in the context of the theory of Cox processes, as well as in relation to the theory of martingales. The interested reader may consult, in particular, the monograph by Last and Brandt [22] for the former approach, and by Brémaud [3] for the latter. It seems to us that no single comprehensive source focused on the issues related to default risk modelling is available to financial researchers, though. Furthermore, it is worth noting that some challenging mathematical problems associated with the modelling of default risk remain still open. The aim of this text is thus to fill the gap by furnishing a relatively concise and self-contained exposition of the most relevant – from the viewpoint of financial modelling – results related to the analysis of random times and their filtrations. We also present some recent developments and we indicate the directions for a further research. Due to the limited space, the proofs of some results were omitted; a full version of the working paper [19] is available from the authors upon request.

2 Hazard Process Γ of a Random Time

Let τ be a non-negative random variable on a probability space $(\Omega, \mathcal{G}, \mathbf{P})$, such that $\mathbf{P}(\tau = 0) = 0$ and $\mathbf{P}(\tau > t) > 0$ for any $t \geq 0$. We introduce a right-continuous process D by setting $D_t = 1\!\!1_{\{\tau \leq t\}}$, and we denote by $\mathbf{D}$ the filtration generated by D; that is, $\mathcal{D}_t = \sigma(D_u : u \leq t)$.
Setup 1. Suppose that $\mathbf{F} = (\mathcal{F}_t)_{t \geq 0}$ is a given (but arbitrary) filtration[1] on $(\Omega, \mathcal{G}, \mathbf{P})$. Let us consider the joint filtration $\mathbf{G} := \mathbf{D} \vee \mathbf{F}$; that is, we set $\mathcal{G}_t = \mathcal{D}_t \vee \mathcal{F}_t$ for every $t \in I\!\!R_+$. Our first goal is to find a representation of the conditional expectation $\mathbf{E}(Y \mid \mathcal{G}_t)$ in terms of the random time τ and the 'hazard process' of τ with respect to $\mathbf{F}$. Subsequently, we analyse the martingale representation of the process $\mathbf{E}(Y \mid \mathcal{G}_t)$, $t \in I\!\!R_+$, and we examine the behaviour of the hazard process under an equivalent change of probability measure. The properties of the reference filtration $\mathbf{F}$ appear to play a crucial role in this study.
Setup 2. Suppose that we are given an arbitrary filtration $\mathbf{G}$ such that $\mathbf{D} \subset \mathbf{G}$. It is important to observe that the equality $\mathbf{G} = \mathbf{D} \vee \mathbf{F}$ does not specify uniquely the 'complementary' filtration $\mathbf{F}$. For instance, when $\mathcal{G}_t = \mathcal{D}_t$, we may take the trivial filtration (as, e.g., in [9,12,18]), but also $\mathbf{F} = \mathbf{D}$ (or indeed any other sub-filtration of $\mathbf{D}$). At the intuitive level, given the random time τ and the 'enlarged' filtration $\mathbf{G}$, one might be interested in searching for a filtration which represents an additional flow of informations. Formally, one looks for a 'minimal' filtration $\widetilde{\mathbf{F}}$ such that the equality $\mathbf{G} = \mathbf{D} \vee \widetilde{\mathbf{F}}$ is valid. As soon as the minimal complementary filtration $\widetilde{\mathbf{F}}$ is determined, we are back in Setup 1.

Remark 2.1. In both cases, the σ-field $\mathcal{G}_t$ is assumed to represent all observations available at time t. Since obviously $\mathcal{D}_t \subset \mathcal{G}_t$ for any t, τ is a stopping time with respect to $\mathbf{G}$; τ is not necessarily a stopping time with respect to $\mathbf{F}$, however. In financial interpretation, the filtration $\mathbf{F}$ is assumed to model the flow of observations available to investors prior to the *default time* τ. In case of several default times (see Section 5), the filtration $\mathbf{F}$ may also include some events related to some of them. We do not pretend that the default time τ is not observed, but we are interested in the valuation of (defaultable) contingent claims only strictly prior to the default time.

For any $t \in I\!\!R_+$, we write $F_t = \mathbf{P}(\tau \leq t \mid \mathcal{F}_t)$, so that $1 - F_t = \mathbf{P}(\tau > t \mid \mathcal{F}_t)$. It is easily seen that F is a bounded, non-negative, $\mathbf{F}$-submartingale. We may thus deal with the right-continuous modification of F. The following definition is standard.

Definition 2.2. The $\mathbf{F}$-*hazard process* of τ, denoted by Γ, is defined through the formula $1 - F_t = e^{-\Gamma_t}$. Equivalently, $\Gamma_t = -\ln(1 - F_t)$ for every $t \in I\!\!R_+$.

[1] In most applications, $\mathbf{F}$ is the natural filtration of a certain stochastic process. Filtrations $\mathbf{D}$ and $\mathbf{F}$ are not independent, in general.

Unless otherwise explicitly stated, it is assumed throughout that the inequality $F_t < 1$ holds for every t, and thus the $\mathbf{F}$-hazard process of τ given by Definition 2.2 exists. The special case when τ is an $\mathbf{F}$-stopping time (in other words, the case when $\mathbf{F} = \mathbf{G}$) is not analysed in detail in the present work.

2.1 Conditional Expectation With Respect to G

We make throughout the technical assumption that all filtrations are $(\mathbf{P}, \mathcal{G})$-completed. In addition, we assume also that the enlarged filtration $\mathbf{G} = \mathbf{D} \vee \mathbf{F}$ satisfies the 'usual conditions.' The first result examines the structure of the enlarged filtration $\mathbf{G}$.

Lemma 2.3. *We have $\mathcal{G}_t \subset \mathcal{G}_t^*$, where*

$$\mathcal{G}_t^* := \big\{ A \ \in \mathcal{G} \,|\, \exists B \in \mathcal{F}_t \ \ A \ \cap \{\tau > t\} = B \cap \{\tau > t\} \big\}.$$

PROOF: Observe that $\mathcal{G}_t = \mathcal{D}_t \vee \mathcal{F}_t = \sigma(\mathcal{D}_t, \mathcal{F}_t) = \sigma(\{\tau \le u\}, u \le t, \mathcal{F}_t)$. Also, it is easily seen that the class $\mathcal{G}_t^*$ is a sub-σ-field of $\mathcal{G}$. Therefore, it is enough to check that if either $A = \{\tau \le u\}$ for some $u \le t$ or $A \in \mathcal{F}_t$, then there exists an event $B \in \mathcal{F}_t$ such that $A \cap \{\tau > t\} = B \cap \{\tau > t\}$. Indeed, in the former case we may take $B = \emptyset$, in the latter $B = A$. △

The next result provides the key formula (see, e.g., Dellacherie [5]) which relates the conditional expectation with respect to $\mathcal{G}_t$ to the conditional expectation with respect to $\mathcal{F}_t$.

Lemma 2.4. *For any $\mathcal{G}$-measurable[2] random variable Y we have, for any $t \in I\!\!R_+$,*

$$\mathbf{E}\,(1\!\!1_{\{\tau>t\}} Y | \mathcal{G}_t) = 1\!\!1_{\{\tau>t\}} \frac{\mathbf{E}\,(1\!\!1_{\{\tau>t\}} Y | \mathcal{F}_t)}{\mathbf{P}(\tau > t | \mathcal{F}_t)} = 1\!\!1_{\{\tau>t\}} e^{\Gamma_t} \mathbf{E}\,(1\!\!1_{\{\tau>t\}} Y | \mathcal{F}_t). \quad (2.1)$$

PROOF: Let us fix $t \in I\!\!R_+$. In view of Lemma 2.3, any $\mathcal{G}_t$-measurable random variable coincides on the set $\{\tau > t\}$ with some $\mathcal{F}_t$-measurable random variable. Therefore,

$$\mathbf{E}\,(1\!\!1_{\{\tau>t\}} Y \,|\, \mathcal{G}_t) = 1\!\!1_{\{\tau>t\}} \mathbf{E}\,(Y \,|\, \mathcal{G}_t) = 1\!\!1_{\{\tau>t\}} X,$$

where X is an $\mathcal{F}_t$-measurable random variable. Taking conditional expectation with respect to $\mathcal{F}_t$, we obtain

$$\mathbf{E}\,(1\!\!1_{\{\tau>t\}} Y \,|\, \mathcal{F}_t) = \mathbf{P}(\tau > t \,|\, \mathcal{F}_t)\, X. \qquad \triangle$$

Proposition 2.5. *Let Z be a (bounded) $\mathbf{F}$-predictable process. Then for any $t < s \le \infty$*

$$\mathbf{E}\,(1\!\!1_{\{t<\tau\le s\}} Z_\tau | \mathcal{G}_t) = 1\!\!1_{\{\tau>t\}} e^{\Gamma_t}\, \mathbf{E}\left(\int_{]t,s]} Z_u \, dF_u \,\Big|\, \mathcal{F}_t \right). \quad (2.2)$$

[2] Of course, it is also assumed throughout that Y is $\mathbf{P}$-integrable.

PROOF: We start by assuming that Z is a stepwise **F**-predictable process, so that (we are interested only in values of Z for $u \in]t,s]$)

$$Z_u = \sum_{i=0}^{n} Z_{t_i} \mathbb{1}_{]t_i,t_{i+1}]}(u),$$

where $t_0 = t < \cdots < t_{n+1} = s$ and the random variable Z_{t_i} is $\mathcal{F}_{t_i}$-measurable. In view of (2.1), for any i we have

$$\begin{aligned}\mathbf{E}\left(\mathbb{1}_{\{t_i<\tau\le t_{i+1}\}} Z_\tau \,|\, \mathcal{G}_t\right) &= \mathbb{1}_{\{\tau>t\}} e^{\Gamma_t}\, \mathbf{E}\left(\mathbb{1}_{\{t_i<\tau\le t_{i+1}\}} Z_{t_i} \,|\, \mathcal{F}_t\right) \\ &= \mathbb{1}_{\{\tau>t\}} e^{\Gamma_t}\, \mathbf{E}\left(Z_{t_i}(F_{t_{i+1}} - F_{t_i}) \,|\, \mathcal{F}_t\right).\end{aligned}$$

In the second step we approximate an arbitrary (bounded) **F**-predictable process by a sequence of stepwise **F**-predictable processes. △

Let us remark that Proposition 2.5 remains valid if $\mathbf{F} = \mathbf{G}$; that is, when τ is an **F**-stopping time. However, in this case, it does not provide a non-trivial formula. Indeed, the left-hand member of (2.2) is then $\mathbf{E}\,(\mathbb{1}_{\{t<\tau\le s\}} Z_\tau |\, \mathcal{F}_t)$. On the other hand, since $F_t = \mathbb{1}_{\{\tau\le t\}}$, the random variable e^{Γ_t} is equal to 1 on the set $\{\tau > t\}$, and thus the right-hand side of (2.2) is also equal to $\mathbf{E}\,(\mathbb{1}_{\{t<\tau\le s\}} Z_\tau |\, \mathcal{F}_t)$. Notice also that any **G**-predictable process coincides with a (unique) **F**-predictable process up to time τ; this shows that there is no point in dealing in Proposition 2.5 with **G**-predictable processes.

Corollary 2.6. *Let Y be a $\mathcal{G}$-measurable random variable. Then, for $t \le s$,*

$$\mathbf{E}\,(\mathbb{1}_{\{\tau>s\}} Y \,|\, \mathcal{G}_t) = \mathbb{1}_{\{\tau>t\}} \mathbf{E}\,(\mathbb{1}_{\{\tau>s\}} e^{\Gamma_t} Y \,|\, \mathcal{F}_t). \tag{2.3}$$

Furthermore, for any $\mathcal{F}_s$-measurable random variable Y we have

$$\mathbf{E}\,(\mathbb{1}_{\{\tau>s\}} Y \,|\, \mathcal{G}_t) = \mathbb{1}_{\{\tau>t\}} \mathbf{E}\,\big(Y e^{\Gamma_t - \Gamma_s} \,|\, \mathcal{F}_t\big). \tag{2.4}$$

*If F (and thus also Γ) is a continuous increasing process then for any **F**-predictable (bounded) process Z we have*

$$\mathbf{E}\,(\mathbb{1}_{\{t<\tau\le s\}} Z_\tau |\, \mathcal{G}_t) = \mathbb{1}_{\{\tau>t\}} \mathbf{E}\left(\int_t^s Z_u e^{\Gamma_t - \Gamma_u}\, d\Gamma_u \,\Big|\, \mathcal{F}_t\right). \tag{2.5}$$

PROOF: In view of (2.1), to show that (2.3) holds it is enough to observe that $\mathbb{1}_{\{\tau>s\}} = \mathbb{1}_{\{\tau>t\}} \mathbb{1}_{\{\tau>s\}}$. Equality (2.4) is a straightforward consequence of (2.3). Formula (2.5) follows immediately from (2.2), since, when F is increasing, $dF_u = e^{-\Gamma_u}\, d\Gamma_u$. △

2.2 Valuation of Defaultable Claims

Let us fix $t \le T$, and let δ be a constant. In what follows, we implicitly assume that **P** is the martingale measure for a financial market model. For any $\mathcal{G}$-measurable random variable Y we have

$$\mathbf{E}\,(\mathbb{1}_{\{t<\tau\le T\}} \delta + \mathbb{1}_{\{\tau>T\}} Y \,|\, \mathcal{G}_t) = \delta\, \mathbf{P}(t<\tau\le T \,|\, \mathcal{G}_t) + \mathbf{E}\,(\mathbb{1}_{\{\tau>T\}} Y \,|\, \mathcal{G}_t).$$

In particular if Y is an $\mathcal{F}_T$-measurable random variable, using (2.4) we may rewrite the formula as follows, on the set $\{\tau > t\}$

$$\mathbf{E}\,(\mathbb{1}_{\{t<\tau\le T\}}\delta + \mathbb{1}_{\{\tau>T\}}Y \mid \mathcal{G}_t) = \delta\,\mathbf{E}\,(1 - e^{\Gamma_t-\Gamma_T} \mid \mathcal{F}_t) + \mathbf{E}\,(e^{\Gamma_t-\Gamma_T}\,Y \mid \mathcal{F}_t).$$

In financial interpretation, the random variable Y represents the *promised payoff* of a defaultable claim, and δ is the *recovery payoff* – that is, the amount received if default occurs prior to the claim's maturity date T. If the constant δ is replaced by a random payoff Z_τ, where Z is an $\mathbf{F}$-predictable process, referred to as the *recovery process*, and F is an increasing continuous process, then (2.4)-(2.5) yield

$$\mathbf{E}\,(\mathbb{1}_{\{t<\tau\le T\}}Z_\tau + \mathbb{1}_{\{\tau>T\}}Y|\mathcal{G}_t) = \mathbb{1}_{\{\tau>t\}}e^{\Gamma_t}\mathbf{E}\left(\int_t^T Z_u e^{-\Gamma_u}\,d\Gamma_u + e^{-\Gamma_T}Y\,\Big|\,\mathcal{F}_t\right).$$

In the formulae above, we have made an implicit assumption that the interest rate equals zero. In general, we introduce the savings account process B; that is, a strictly positive, $\mathbf{F}$-adapted process of finite variation. For any $t \le T$, we set $B(t,T) := B_t\,\mathbf{E}\,(B_T^{-1} \mid \mathcal{F}_t)$, and we refer to $B(t,T)$ as the price of a unit default-free zero-coupon bond of maturity T. First,

$$B_t\,\mathbf{E}\,(\mathbb{1}_{\{\tau>T\}}B_T^{-1}Y \mid \mathcal{G}_t) = \mathbb{1}_{\{\tau>t\}}\hat{B}_t\,\mathbf{E}\,(\hat{B}_T^{-1}Y \mid \mathcal{F}_t),$$

where we write $\hat{B}_t := B_t e^{\Gamma_t}$. Next, if the process F is increasing and continuous, then

$$B_t\,\mathbf{E}\,(\mathbb{1}_{\{t<\tau\le T\}}B_\tau^{-1}Z_\tau| \mathcal{G}_t) = \mathbb{1}_{\{\tau>t\}}\,\hat{B}_t\,\mathbf{E}\left(\int_t^T \hat{B}_u^{-1}Z_u\,d\Gamma_u\,\Big|\,\mathcal{F}_t\right).$$

The last two formulae lead to the following result.

Corollary 2.7. *Assume that F is an increasing continuous process. Consider a defaultable contingent claim with default time τ, the promised payoff Y and the recovery process Z. The pre-default value V of this claim, defined as*

$$V_t := B_t\,\mathbf{E}\,(\mathbb{1}_{\{t<\tau\le T\}}B_\tau^{-1}Z_\tau + \mathbb{1}_{\{\tau>T\}}B_T^{-1}Y \mid \mathcal{G}_t),$$

satisfies

$$V_t = \mathbb{1}_{\{\tau>t\}}\,\hat{B}_t\,\mathbf{E}\left(\int_t^T \hat{B}_u^{-1}Z_u\,d\Gamma_u + \hat{B}_T^{-1}Y\,\Big|\,\mathcal{F}_t\right). \tag{2.6}$$

Example 2.8. Assume that F is an increasing continuous process. Consider a *defaultable zero-coupon bond* with the nominal value N, maturity T, and the recovery process $Z_t = h(t)$ for some function $h : [0,T] \to I\!R$. The *pre-default bond price* $D(t,T)$ is defined through the formula

$$D(t,T) := B_t\,\mathbf{E}\,(\mathbb{1}_{\{t<\tau\le T\}}B_\tau^{-1}h(\tau) + \mathbb{1}_{\{\tau>T\}}B_T^{-1}N \mid \mathcal{G}_t).$$

Setting $Z_t = h(t)$ and $Y = N$ in (2.6), we obtain the following representation for $D(t,T)$

$$D(t,T) = \mathbb{1}_{\{\tau>t\}} \hat{B}_t \left(\int_t^T \hat{B}_u^{-1} h(u)\, d\Gamma_u + \hat{B}_T^{-1} N \,\Big|\, \mathcal{F}_t \right).$$

Notice that manifestly $D(t,T) = 0$ after default time τ, so that $D(t,T)$ does not give the right value of a defaultable bond after default in case of non-zero recovery (i.e., when h is nonvanishing). The correct expression for the price of a defaultable bond reads

$$\mathbb{1}_{\{\tau>t\}} D(t,T) + \mathbb{1}_{\{\tau\le t\}} h(\tau) B^{-1}(\tau,T) B(t,T), \quad \forall\, t \in [0,T],$$

if the recovery payoff $h(\tau)$ received at default time τ is invested in default-free bonds of maturity T. Let us introduce the 'credit-risk-adjusted' bond price $\hat{B}(t,T)$ by setting $\hat{B}(t,T) := \hat{B}_t\, \mathbf{E}\,(\hat{B}_T^{-1} \mid \mathcal{F}_t)$. Then

$$D(t,T) = \mathbb{1}_{\{\tau>t\}} \hat{B}_t\, \mathbf{E} \left(\int_t^T \hat{B}_u^{-1} h(u)\, d\Gamma_u \,\Big|\, \mathcal{F}_t \right) + \mathbb{1}_{\{\tau>t\}} N \hat{B}(t,T). \tag{2.7}$$

Assume, in addition, that $dB_t = r_t B_t\, dt$ for some process r representing the short-term interest rate, and that the process F is absolutely continuous, so that

$$1 - F_t = 1 - \int_0^t f_u\, du = e^{-\Gamma_t} = \exp\left(- \int_0^t \gamma_u\, du \right),$$

where the *intensity of default* γ satisfies $\gamma_u = f_u (1 - F_u)^{-1}$. Then $\hat{B}_t = \exp\left(\int_0^t \hat{r}_u\, du \right)$, where $\hat{r}_u := r_u + \gamma_u$ is the 'credit-risk-adjusted' short-term interest rate. Using (2.7), we conclude that in the case of $h = 0$ (i.e., under zero recovery) to value a defaultable zero-coupon bond prior to default, it is enough to discount its nominal value N using the 'credit-risk-adjusted' rate $\hat{r}$. In the special case when the filtration $\mathbf{F}$ is trivial, and the probability law of τ under $\mathbf{P}$ admits the density function f, formula (2.7) becomes (we write $R_t = 1/B_t$ and $G(t) = 1 - F(t)$)

$$R_t G(t) D(t,T) = \mathbb{1}_{\{\tau>t\}} \left(\int_t^T R_u h(u) f(u)\, du + R_T G(T) N \right).$$

Remark 2.9. Using the technique described above, it is possible to solve the problem studied extensively in Duffie and Lando [9] (see Elliott et al. [12] for further comments).

2.3 Semimartingale Representation of the Stopped Process

In the next auxiliary result we assume that m is an arbitrary $\mathbf{F}$-martingale, and we examine the semimartingale decomposition with respect to the enlarged filtration $\mathbf{G} = \mathbf{D} \vee \mathbf{F}$ of the stopped process $\widetilde{m}_t = m_{t\wedge\tau}$ (we sometimes use the standard notation m^τ for the process m stopped at τ).

Lemma 2.10. *Assume that the process m is a continuous $\mathbf{F}$-martingale.*
(i) *If F is a continuous increasing process then the stopped process $\widetilde{m}_t = m_{t\wedge\tau}$ is a $\mathbf{G}$-martingale.*
(ii) *If F is a continuous submartingale then the process*

$$\hat{m}_t = \widetilde{m}_t - \int_0^{t\wedge\tau} e^{\Gamma_u}\, d\langle m, F\rangle_u = m_{t\wedge\tau} - \int_0^{t\wedge\tau} (1-F_u)^{-1}\, d\langle m, F\rangle_u \quad (2.8)$$

is a $\mathbf{G}$-martingale.

PROOF: We establish only the second statement, which implies the first one. For $s \geq t$,

$$\mathbf{E}\left(m_{s\wedge\tau} - m_{t\wedge\tau} \,|\, \mathcal{G}_t\right) = \mathbb{1}_{\{\tau>t\}}\left(\mathbf{E}\left(m_{s\wedge\tau} \,|\, \mathcal{G}_t\right) - m_t\right).$$

The process m is $\mathbf{F}$-predictable, hence, on the set $\{\tau > t\}$, from (2.2)

$$\begin{aligned}\mathbf{E}\left(m_{s\wedge\tau} \,|\, \mathcal{G}_t\right) &= e^{\Gamma_t}\mathbf{E}\left(\int_t^\infty m_{s\wedge u}\, dF_u \,\Big|\, \mathcal{F}_t\right)\\ &= e^{\Gamma_t}\mathbf{E}\left(\int_t^s m_u\, dF_u + m_s(1-F_s) \,\Big|\, \mathcal{F}_t\right).\end{aligned}$$

Now, the integration by parts formula leads to

$$\int_t^s m_u\, dF_u = F_s m_s - F_t m_t - \int_t^s F_u\, dm_u - \langle m, F\rangle_s + \langle m, F\rangle_t.$$

From the $\mathbf{F}$-martingale property of m, it follows that

$$\mathbf{E}\left(m_{s\wedge\tau} - m_{t\wedge\tau} \,|\, \mathcal{G}_t\right) = \mathbb{1}_{\{\tau>t\}} e^{\Gamma_t}\left(\langle m, F\rangle_t - \mathbf{E}\left(\langle m, F\rangle_s \,|\, \mathcal{F}_t\right)\right).$$

Using (2.2) again, and introducing

$$\tilde{M}_t = \int_0^{t\wedge\tau} (1-F_u)^{-1}\, d\langle m, F\rangle_u = \int_0^{t\wedge\tau} e^{\Gamma_u} d\langle m, F\rangle_u,$$

we obtain $\mathbf{E}(\tilde{M}_s - \tilde{M}_t \,|\, \mathcal{F}_t) = \mathbb{1}_{\{\tau>t\}} e^{\Gamma_t} J_t^s$, where

$$J_t^s = \mathbf{E}\left(\int_t^s dF_u \int_t^u e^{\Gamma_v} d\langle m, F\rangle_v + (1-F_s)\int_t^s e^{\Gamma_v} d\langle m, F\rangle_v \,\Big|\, \mathcal{F}_t\right).$$

Using Fubini's theorem,

$$J_t^s = \mathbf{E}\left(\int_t^s e^{\Gamma_v}(F_s - F_v)\, d\langle m, F\rangle_v + (1-F_s)\int_t^s e^{\Gamma_v}\, d\langle m, F\rangle_v \,\Big|\, \mathcal{F}_t\right).$$

The result follows from $J_t^s = \mathbf{E}\left(\int_t^s d\langle m, F\rangle_v \,\Big|\, \mathcal{F}_t\right)$. △

Remark 2.11. All these results have been established in a general setting. Under some additional assumptions imposed on τ (namely, when τ is an *honest time* – i.e. the end of a predictable set). the decomposition of the semimartingale m in the filtration $\mathbf{G}$ can be given (for details, see [17,16,26]). We can not avoid the pleasure to quote from Dellacherie et Meyer (see Page 137 in [6]) "If X is an adapted continuous process, the random variable $\tau = \inf\{s \in [0,T] : X_s = \sup_{0 \le u \le T} X_u\}$ is honest. For example, if X represents the dynamics of an asset price, τ would be the best time to sell. All speculators try to obtain some knowledge on τ, but they cannot succeed; that's why this variable is named honest." Notice that formula (2.8) is similar to Girsanov's transformation; things are not so easy, though. In a typical case, when F has a nonzero martingale part, this process is equal to 1 at time $\tau-$, and the "drift" term goes to infinity.

2.4 Martingales Associated With the Hazard Process Γ

Lemma 2.12. *The process*

$$L_t := \mathbb{1}_{\{\tau > t\}} e^{\Gamma_t} = (1 - D_t) e^{\Gamma_t} = \frac{1 - D_t}{1 - F_t}$$

is a $\mathbf{G}$-martingale. Moreover, for any $\mathbf{F}$-martingale m, the product Lm is a $\mathbf{G}$-martingale.

PROOF: We establish the second statement which implies the first one. In view of (2.4), for $t \le s$

$$\mathbf{E}\,(\mathbb{1}_{\{\tau > s\}} e^{\Gamma_s} m_s \,|\, \mathcal{G}_t) = \mathbb{1}_{\{\tau > t\}} e^{\Gamma_t} \mathbf{E}\,(e^{-\Gamma_s} e^{\Gamma_s} m_s \,|\, \mathcal{F}_t) = L_t m_t.$$

△

In the next result we assume in addition that Γ is an increasing continuous process.

Proposition 2.13. *Assume that the $\mathbf{F}$-hazard process Γ of τ is an increasing continuous process. Then:*
(i) *the process $M_t = D_t - \Gamma_{t \wedge \tau}$ is a $\mathbf{G}$-martingale, more specifically,*

$$M_t = -\int_{]0,t]} e^{-\Gamma_u}\, dL_u. \tag{2.9}$$

Furthermore, L satisfies

$$L_t = 1 - \int_{]0,t]} L_{u-}\, dM_u. \tag{2.10}$$

(ii) *If an $\mathbf{F}$-martingale m is also a $\mathbf{G}$-martingale then the product Mm is a $\mathbf{G}$-martingale.*

PROOF: The martingale property of M and equalities (2.9)-(2.10) can be shown using Itô's lemma combined with Lemma 2.12. To establish (ii), notice that from Lemma 2.12 we know that Lm is a **G**-martingale, so that $[L, m]$ is a **G**-martingale. Therefore, using the equality $dM_t = -e^{-\Gamma_t} dL_t$, we get that

$$d(mM)_t = m_{t-}dM_t + M_{t-}dm_t + d[M, m]_t = m_{t-}dM_t + M_{t-}dm_t - e^{-\Gamma_t} d[L, m]_t$$

and (ii) follows. △

Remark 2.14. In general, Γ has a nonzero martingale part and we have

$$L_t = (1 - D_t)e^{\Gamma_t} = 1 + \int_0^t e^{\Gamma_u} \left[(1 - D_u) \left(d\Gamma_u + (1/2)\langle \Gamma \rangle_u \right) - dD_u \right]$$

so that M is no longer a **G**-martingale (but the process $D_t - \Gamma_{t\wedge\tau} - (1/2)\langle \Gamma \rangle_{t\wedge\tau}$ is clearly a **G**-martingale).

2.5 Intensity of a Random Time

Let us consider the classic case of an absolutely continuous, increasing, **F**-hazard process Γ. We assume that $\Gamma_t = \int_0^t \gamma_u \, du$ for some **F**-progressively measurable process γ, referred to as the **F**-*intensity* of a random time τ. By virtue of Proposition 2.13, the process M, which is given by the formula

$$M_t = D_t - \int_0^{t\wedge\tau} \gamma_u \, du = D_t - \int_0^t 1\!\!1_{\{\tau > u\}} \gamma_u \, du,$$

follows a **G**-martingale. The property above is frequently used in the financial literature as the definition of the **F**-intensity of a random time. The intuitive meaning of the **F**-intensity γ as the "intensity of survival given the information flow **F**" becomes apparent from the following corollary.

Corollary 2.15. *If the* **F**-*hazard process* Γ *of* τ *is absolutely continuous then for any* $t \le s$

$$\mathbf{P}(t < \tau \le s \,|\, \mathcal{G}_t) = 1\!\!1_{\{\tau > t\}} \mathbf{E}\left(1 - e^{-\int_t^s \gamma_u du} \,\middle|\, \mathcal{F}_t\right).$$

Remark 2.16. The **F**-hazard function Γ is not well defined when τ is a **F**-stopping time (that is, when $\mathbf{D} \subset \mathbf{F}$ so that $\mathbf{G} = \mathbf{F}$), and thus Corollary 2.15 cannot be directly applied in this case. It appears, however, that for a certain class of a **G**-stopping times we can find an increasing **G**-predictable process Λ such that for any $t \le s$

$$\mathbf{P}(\tau > s \,|\, \mathcal{G}_t) = 1\!\!1_{\{\tau > t\}} \mathbf{E}\left(e^{\Lambda_t - \Lambda_s} \,\middle|\, \mathcal{G}_t\right) = 1\!\!1_{\{\tau > t\}} \mathbf{E}\left(e^{-\int_t^s \lambda_u du} \,\middle|\, \mathcal{G}_t\right),$$

where the second equality holds provided that the process Λ is absolutely continuous. It seems natural to conjecture that the *martingale hazard* process Λ, which is formally defined in Section 4 below, coincides with the $\hat{\mathbf{F}}$-hazard process of τ for some filtration $\hat{\mathbf{F}}$, such that τ is not an $\hat{\mathbf{F}}$-stopping time. Let us again emphasise that the existence and the value of the intensity depend strongly on the choice of $\mathbf{F}$. If τ is an $\mathbf{F}$-predictable stopping time (so that, in particular, $\mathbf{G} = \mathbf{F}$), it does not admit an intensity with respect to $\mathbf{F}$. The intensity of τ with respect to the trivial filtration exists, however, provided that $F(t) = \mathbf{P}(\tau \leq t) < 1$ for every t and F admits a density function.

2.6 Tradable Contingent Claims in a Defaultable Market

Blanchet-Scalliet and Jeanblanc [2] give the following suggestion to study a defaultable market. Assume that every $\mathcal{F}_T$-square-integrable contingent claim X is hedgeable, i.e., there exists a real number x and a square-integrable $\mathbf{F}$-predictable process θ such that[3]

$$R_T X = x + \int_0^T \theta_u \, d\widetilde{S}_u,$$

where $\widetilde{S} = RS$; we shall refer to this property as the completeness of the $\mathcal{F}_T$-market. The t-time price of X is equal to V_t, where

$$R_t V_t = \mathbf{E_Q}(R_T X \,|\, \mathcal{F}_t) = x + \int_0^t \theta_u \, d\widetilde{S}_u$$

and $\mathbf{Q}$ the unique equivalent martingale measure (e.m.m.) for $\widetilde{S}$ with respect to the filtration $\mathbf{F}$.

Assume that the $\mathcal{F}_T$-measurable claims X are available in the defaultable market – that is, it is possible to get a payoff equal to X, no matter whether the default has occurred prior to T or not. Then these claims are obviously also hedgeable in the $\mathcal{G}_T$-market (with the same hedging portfolio θ), and thus the discounted price of X must be equal to $\mathbf{E_R}(XR_T \,|\, \mathcal{G}_t)$, where $\mathbf{R}$ is any e.m.m. with respect to $\mathbf{G}$ (we assume that the $\mathbf{G}$-market is arbitrage-free; more precisely, that there exists at least one e.m.m.) The uniqueness of the price of a hedgeable claim yields

$$\mathbf{E_Q}(XR_T \,|\, \mathcal{F}_t) = \mathbf{E_R}(XR_T \,|\, \mathcal{G}_t) = x + \int_0^t \theta_u \, d\widetilde{S}_u.$$

This means, in particular, that $\mathbf{E_Q}(Z) = \mathbf{E_R}(Z)$ for any $Z \in \mathcal{F}_T$ (take $t = 0$ and $X = ZR_T^{-1}$), and thus the restriction of any e.m.m. $\mathbf{R}$ to the σ-field $\mathcal{F}_T$ coincides with $\mathbf{Q}$. Moreover, since any square-integrable $\mathbf{F}$-martingale can be written as $\mathbf{E_R}(X \,|\, \mathcal{F}_t)$ under any e.m.m. $\mathbf{R}$, we conclude that any square-integrable $\mathbf{F}$-martingale is also a $\mathbf{G}$-martingale.

[3] Recall that the process R equals $1/B$, where B is assumed to represent the savings account.

2.7 Hypothesis (H)

Let us introduce the following hypothesis.

(H) Every **F** square-integrable martingale is a **G** square-integrable martingale.

Hypothesis (H) (which implies, in particular, that any **F**-Brownian motion remains a Brownian motion in the enlarged filtration) was studied by Brémaud and Yor [4], Mazziotto and Szpirglas [24], and in the financial context by Kusuoka [20]. Hypothesis (H) can also be expressed directly in terms of filtrations **F** and **G**; namely, the following condition (H*) is equivalent to (H):

(H*) For any t, the σ-fields $\mathcal{F}_\infty$ and $\mathcal{G}_t$ are conditionally independent given $\mathcal{F}_t$.

Lemma 2.17. *If* $\mathbf{G} = \mathbf{F} \vee \mathbf{D}$ *then* (H) *is equivalent to any of the two equivalent conditions* (H′)

$$\forall\, s \le t, \quad \mathbf{P}(\tau \le s \,|\, \mathcal{F}_\infty) = \mathbf{P}(\tau \le s \,|\, \mathcal{F}_t),$$

$$\forall\, t \in I\!R_+, \quad \mathbf{P}(\tau \le t \,|\, \mathcal{F}_\infty) = \mathbf{P}(\tau \le t \,|\, \mathcal{F}_t). \tag{2.11}$$

PROOF: The proof of this result can be found in [7] (see also [19]). △

Remark 2.18. (i) Equality (2.11) appears in several papers devoted to default risk, usually with no explicit reference to condition (H). For instance, the main theorem in Madan and Unal [23] follows from the fact that (2.11) holds (see the proof of B9 in the appendix of this paper). This is also the case for Wong's [25] setup.
(ii) If τ is $\mathcal{F}_\infty$-measurable and (2.11) holds, then τ is an **F**-stopping time. If τ is an **F**-stopping time, equality (2.11) holds.
(iii) Though condition (H) is not always satisfied, it holds when τ is constructed through a standard approach (see Section 4.4 below). This hypothesis is quite natural under the historical probability, and it is stable under some changes of the underlying probability measure. However, Kusuoka [20] provides a simple example in which (H) is satisfied under the historical probability, but fails to hold after an equivalent change of a probability measure. This counter-example is linked to the *default correlations* across various firms (see Section 5.4 below).

3 Martingale Representation Theorems

It is well known that the concept of replication of contingent claims is closely related to martingale representation theorems. In the case of defaultable markets this problem of hedging of contingent claims becomes more delicate than in the classic case of default-free markets.

3.1 Martingale Representation Theorem: General Case

We shall first consider a general setup; that is, we do not assume that the filtration $\mathbf{F}$ supports only continuous martingales. We shall postulate that the process F is increasing and continuous, however. We reproduce here without proof the result due to Blanchet-Scalliet and Jeanblanc [2], a useful tool for hedging purposes.

Proposition 3.1. *Assume that the* $\mathbf{F}$*-hazard process* Γ *of* τ *is an increasing continuous process. Let* Z *be an* $\mathbf{F}$*-predictable process such that the random variable* Z_τ *is integrable. Then the* $\mathbf{G}$*-martingale* $M_t^Z := \mathbf{E}(Z_\tau \,|\, \mathcal{G}_t)$ *admits the following decomposition*

$$M_t^Z = m_0 + \int_{]0,t\wedge\tau]} L_{u-}\,dm_u + \int_{]0,t\wedge\tau]} (Z_u - M_{u-}^Z)\,dM_u,$$

where m *is an* $\mathbf{F}$*-martingale*

$$m_t := \mathbf{E}\left(\int_0^\infty Z_u e^{-\Gamma_u} d\Gamma_u \,\Big|\, \mathcal{F}_t\right) = \mathbf{E}\left(\int_0^\infty Z_u\,dF_u \,\Big|\, \mathcal{F}_t\right).$$

As usual $M_t = D_t - \Gamma_{t\wedge\tau}$. Note also that $m_0 = M_0^Z$.

3.2 Martingale Representation Theorem: Case of a Brownian Filtration

We shall now consider the case of the Brownian filtration; that is, we assume that $\mathbf{F} = \mathbf{F}^W$ for some Brownian motion W. We postulate that W remains a martingale (and thus a Brownian motion) with respect to the enlarged filtration $\mathbf{G}$ (see Section 2.7). The next result is borrowed from Kusuoka [20].

Proposition 3.2. *Assume that the hazard process* Γ *is an increasing continuous process. Then for any* $\mathbf{G}$*-martingale* N *we have*

$$N_t = N_0 + \int_0^t \xi_u dW_u + \int_{]0,t]} \zeta_u dM_u = N_0 + \widetilde{N}_t + \hat{N}_t,$$

where ξ *and* ζ *are* $\mathbf{G}$*-predictable stochastic processes, and* M *is given by (2.9). Moreover, the continuous* $\mathbf{G}$*-martingale* $\widetilde{N}$ *and the purely discontinuous* $\mathbf{G}$*-martingale* $\hat{N}$ *are mutually orthogonal.*

The proposition above shows that in order to get a complete market with defaultable securities, a default-free asset and a defaultable asset (for instance, a defaultable zero-coupon bond) should be taken as hedging instruments to mimic the price process of a defaultable claim (i.e., to generate an arbitrary discontinuous $\mathbf{G}$-martingale).

4 Martingale Hazard Process Λ of a Random Time

In this section, the case of $\mathbf{D} \subset \mathbf{F}$ (i.e., the case when $\mathbf{F} = \mathbf{G}$) is not excluded. Put another way, the case when τ is an $\mathbf{F}$-stopping time is also covered by the foregoing results.

Definition 4.1. An $\mathbf{F}$-predictable right-continuous increasing process Λ is called an $(\mathbf{F}, \mathbf{G})$-*martingale hazard process* of a random time τ if and only if the process $\widetilde{M}_t := D_t - \Lambda_{t\wedge\tau}$ is a $\mathbf{G}$-martingale. In addition, $\Lambda_0 = 0$.

For the sake of brevity, we shall frequently refer to the $(\mathbf{F}, \mathbf{G})$-martingale hazard process as the $\mathbf{F}$-*martingale hazard process,* if the meaning of $\mathbf{G}$ is clear from the context.

Remark 4.2. We assume here that the reference filtration $\mathbf{F}$ is given a priori. It seems natural, however, to search for the 'minimal' filtration such that $\hat{\mathbf{F}} \subset \mathbf{F}$ and the $(\hat{\mathbf{F}}, \mathbf{G})$-martingale hazard process is actually $\hat{\mathbf{F}}$-adapted (we do not insist that the equality $\mathbf{G} = \mathbf{D} \vee \hat{\mathbf{F}}$ needs to hold). Though this problem is rather difficult to solve in general, in particular cases the filtration $\hat{\mathbf{F}}$ emerges in the calculation of the $\mathbf{F}$-martingale hazard process (see, e.g., Example 4.1 in [19]).

4.1 Evaluation of Λ: Special Case

Our first goal will be to examine a special case when the $\mathbf{F}$-martingale hazard process Λ can be expressed by means of Γ. We find it convenient to introduce the following condition (notice that (H) implies (G)):
(G) F is an increasing process.

Proposition 4.3. *Assume that* (G) *holds. If the process* Λ *given by the formula*

$$\Lambda_t = \int_{]0,t]} \frac{dF_u}{1 - F_{u-}} = \int_{]0,t]} \frac{d\mathbf{P}(\tau \le u \,|\, \mathcal{F}_u)}{1 - \mathbf{P}(\tau < u \,|\, \mathcal{F}_u)} \tag{4.12}$$

is $\mathbf{F}$*-predictable, then* Λ *is the* $\mathbf{F}$*-martingale hazard process of a random time* τ.

PROOF: It suffices to check that $D_t - \Lambda_{t\wedge\tau}$ is a $\mathbf{G}$-martingale, where $\mathbf{G} = \mathbf{D} \vee \mathbf{F}$. Using (2.1), we obtain for $t < s$

$$\mathbf{E}\,(D_s - D_t \,|\, \mathcal{G}_t) = 1\!\!1_{\{\tau > t\}} \frac{\mathbf{P}(t < \tau \le s \,|\, \mathcal{F}_t)}{\mathbf{P}(\tau > t \,|\, \mathcal{F}_t)} = 1\!\!1_{\{\tau > t\}} \frac{\mathbf{E}\,(F_s \,|\, \mathcal{F}_t) - F_t}{1 - F_t}.$$

On the other hand,

$$J_t^s := \mathbf{E}\left(\Lambda_{s\wedge\tau} - \Lambda_{t\wedge\tau} \,|\, \mathcal{G}_t\right) = \mathbf{E}\left(1\!\!1_{\{\tau > s\}}(\Lambda_s - \Lambda_t) \,|\, \mathcal{G}_t\right) + \mathbf{E}\left(1\!\!1_{\{t < \tau \le s\}}\, \widetilde{\Lambda}_\tau \,|\, \mathcal{G}_t\right),$$

where, for a fixed t, we write $\widetilde{\Lambda}_u = (\Lambda_u - \Lambda_t)1\!\!1_{]t,\infty[}(u)$ (so that $\widetilde{\Lambda}$ is an **F**-predictable process). Therefore, an application of formula (2.2) gives

$$\mathbf{E}\left(1\!\!1_{\{t<\tau\le s\}}\,\widetilde{\Lambda}_\tau \,|\, \mathcal{G}_t\right) = 1\!\!1_{\{\tau>t\}}e^{\Gamma_t}\,\mathbf{E}\left(\int_{]t,s]}(\Lambda_u-\Lambda_t)\,dF_u\,\Big|\,\mathcal{F}_t\right).$$

Furthermore, (2.3) yields

$$\mathbf{E}\left(1\!\!1_{\{\tau>s\}}(\Lambda_s-\Lambda_t)\,|\,\mathcal{G}_t\right) = 1\!\!1_{\{\tau>t\}}e^{\Gamma_t}\,\mathbf{E}\left(1\!\!1_{\{\tau>s\}}\,(\Lambda_s-\Lambda_t)\,|\,\mathcal{F}_t\right).$$

Combining the formulae above, we get, on the set $\{\tau > t\}$

$$\begin{aligned} J_t^s &= e^{\Gamma_t}\,\mathbf{E}\left(1\!\!1_{\{\tau>s\}}\,(\Lambda_s-\Lambda_t)+\int_{]t,s]}(\Lambda_u-\Lambda_t)\,dF_u\,\Big|\,\mathcal{F}_t\right)\\ &= e^{\Gamma_t}\,\mathbf{E}\left((1-F_s)(\Lambda_s-\Lambda_t)+\int_{]t,s]}(\Lambda_u-\Lambda_t)\,dF_u\,\Big|\,\mathcal{F}_t\right), \end{aligned}$$

where the last equality follows from the conditioning with respect to the σ-field $\mathcal{F}_s$. To conclude, it is enough to apply Itô's integration by parts formula (see [19] for details). △

4.2 Evaluation of Λ: General Case

Assume now that either (G) is not satisfied (so that F is not an increasing process), or (G) holds, but the increasing process F is not **F**-predictable.[4] The **F**-martingale hazard process Λ can nevertheless be found through a suitable modification of formula (4.12).

We write $\widetilde{F}$ to denote the **F**-compensator of the bounded **F**-submartingale F. This means that $\widetilde{F}$ is the unique **F**-predictable, increasing process, with $\widetilde{F}_0 = 0$, and such that the process $U = F - \widetilde{F}$ is an **F**-martingale (the existence and uniqueness of $\widetilde{F}$ is a consequence of the Doob-Meyer decomposition theorem). In the next result it is not assumed that (G) holds.

Proposition 4.4. (i) *The* **F**-*martingale hazard process of a random time* τ *is given by the formula*

$$\Lambda_t = \int_{]0,t]}\frac{d\widetilde{F}_u}{1-F_{u-}}. \tag{4.13}$$

(ii) *If* $\widetilde{F}_t = \widetilde{F}_{t\wedge\tau}$ *for every* $t \in I\!\!R_+$ *(that is, the process* $\widetilde{F}$ *is stopped at* τ*) then* $\Lambda = \widetilde{F}$.

[4] For instance, τ can be an **F**-stopping time, which is not **F**-predictable. If τ is an **F**-stopping time, we have simply $F = D$, and the process D is not **F**-predictable, unless the stopping time τ is **F**-predictable.

PROOF: It is clear that the process Λ given by (4.13) is predictable. Therefore, one needs only to verify that the process $\widetilde{M}_t = D_t - \Lambda_{t\wedge\tau}$ is a $\mathbf{G}$-martingale. To this end, it is enough to proceed along the same lines as in the proof of Proposition 4.3 (see, e.g., [19]).

We shall now prove part (ii). We assume that $\widetilde{F}_{t\wedge\tau} = \widetilde{F}_t$ for every $t \in \mathbb{R}_+$. This means, in particular, that the process $F_t - \widetilde{F}_{t\wedge\tau}$ is an $\mathbf{F}$-martingale. We wish to show that the process $D_t - \widetilde{F}_{t\wedge\tau}$ is a $\mathbf{G}$-martingale – that is, for any $t \leq s$

$$\mathbf{E}\,(D_s - \widetilde{F}_{s\wedge\tau}\,|\,\mathcal{G}_t) = D_t - \widetilde{F}_{t\wedge\tau},$$

or equivalently,

$$\mathbf{E}\,(D_s - D_t\,|\,\mathcal{G}_t) = \mathbf{E}\,(\widetilde{F}_{s\wedge\tau} - \widetilde{F}_{t\wedge\tau}\,|\,\mathcal{G}_t).$$

By virtue of (2.1), we have

$$\mathbf{E}\,(D_s - D_t\,|\,\mathcal{G}_t) = (1 - D_t)\,\frac{\mathbf{E}\,(D_s - D_t\,|\,\mathcal{F}_t)}{\mathbf{E}\,(1 - D_t\,|\,\mathcal{F}_t)}. \tag{4.14}$$

On the other hand,

$$\begin{aligned}\mathbf{E}\,(\widetilde{F}_{s\wedge\tau} - \widetilde{F}_{t\wedge\tau}\,|\,\mathcal{G}_t) &= (1 - D_t)\,\frac{\mathbf{E}\,(\widetilde{F}_{s\wedge\tau} - \widetilde{F}_{t\wedge\tau}\,|\,\mathcal{F}_t)}{\mathbf{E}\,(1 - D_t\,|\,\mathcal{F}_t)}\\ &= (1 - D_t)\,\frac{\mathbf{E}\,(F_s - F_t\,|\,\mathcal{F}_t)}{\mathbf{E}\,(1 - D_t\,|\,\mathcal{F}_t)} = (1 - D_t)\,\frac{\mathbf{E}\,(D_s - D_t\,|\,\mathcal{F}_t)}{\mathbf{E}\,(1 - D_t\,|\,\mathcal{F}_t)},\end{aligned}$$

where the second equality follows from (2.1), and the third is a consequence of our assumption that the process $F_t - \widetilde{F}_{t\wedge\tau}$ is an $\mathbf{F}$-martingale. △

Remark 4.5. In standard examples, τ is a totally inaccessible $\mathbf{F}$-stopping time, and $\widetilde{F}$ is an $\mathbf{F}$-adapted process with continuous increasing sample paths. For example, if τ is the first jump time of a Poisson process N, and $\mathbf{F} = \mathbf{F}^N$ is the natural filtration of this process, then clearly $F_t = D_t$ and $\widetilde{F}_t = \lambda t$, where λ is the (constant) intensity of N. Let us mention that if $\mathbf{F}$ is the Brownian filtration, the process $\widetilde{F}$ is continuous if and only if for any $\mathbf{F}$-stopping time U we have: $\mathbf{P}(\tau = U) = 0$.

Remark 4.6. Under Hypothesis (H), the process $\widetilde{F}$ is never stopped at τ, unless τ is an $\mathbf{F}$-stopping time. To show this assume, on the contrary, that $\widetilde{F}_t = \widetilde{F}_{t\wedge\tau}$. Let us stress that if (H) holds, the process $F_t - \widetilde{F}_{t\wedge\tau}$ is not only an $\mathbf{F}$-martingale, but also a $\mathbf{G}$-martingale. Since by virtue of part (ii) in Proposition 4.4 the process $D_t - \widetilde{F}_{t\wedge\tau}$ is a $\mathbf{G}$-martingale, we see that $D - F$ also is a $\mathbf{G}$-martingale. In view of the definition of F, the last property reads

$$\mathbf{E}\,\big(D_s - \mathbf{E}\,(D_s\,|\,\mathcal{F}_s)\,\big|\,\mathcal{G}_t\big) = D_t - \mathbf{E}\,(D_t\,|\,\mathcal{F}_t),$$

for $t \le s$, or equivalently

$$\mathbf{E}(D_s - D_t \,|\, \mathcal{G}_t) = \mathbf{E}\big(\mathbf{E}(D_s \,|\, \mathcal{F}_s) \,|\, \mathcal{G}_t\big) - \mathbf{E}(D_t \,|\, \mathcal{F}_t) = I_1 - I_2. \tag{4.15}$$

Under (H), we have

$$I_1 = \mathbf{E}(\mathbf{P}(\tau \le s \,|\, \mathcal{F}_\infty) \,|\, \mathcal{F}_t \vee \mathcal{D}_t) = \mathbf{E}(\mathbf{P}(\tau \le s \,|\, \mathcal{F}_\infty) \,|\, \mathcal{F}_t)$$

since the random variable $\mathbf{P}(\tau \le s \,|\, \mathcal{F}_\infty)$ is obviously $\mathcal{F}_\infty$-measurable, and the σ-fields $\mathcal{F}_\infty$ and $\mathcal{D}_t$ are conditionally independent given $\mathcal{F}_t$. Consequently, $I_1 = \mathbf{E}(\mathbf{E}(D_s \,|\, \mathcal{F}_\infty) \,|\, \mathcal{F}_t) = \mathbf{E}(D_s \,|\, \mathcal{F}_t)$. Therefore, (4.15) can be rewritten as follows

$$\mathbf{E}(D_s - D_t \,|\, \mathcal{G}_t) = \mathbf{E}(D_s \,|\, \mathcal{F}_t) - \mathbf{E}(D_t \,|\, \mathcal{F}_t).$$

Furthermore, applying (4.14) to the left-hand side of the last equality, we obtain

$$(1 - D_t)\, \frac{\mathbf{E}(D_s - D_t \,|\, \mathcal{F}_t)}{\mathbf{E}(1 - D_t \,|\, \mathcal{F}_t)} = \mathbf{E}(D_s - D_t \,|\, \mathcal{F}_t).$$

By letting s tend to ∞, we obtain $D_t = \mathbf{E}(D_t \,|\, \mathcal{F}_t)$ or more explicitly, $\mathbf{P}(\tau \le t \,|\, \mathcal{F}_t) = 1\!\!1_{\{\tau \le t\}}$ for every $t \in I\!\!R_+$. We conclude that τ is a $\mathbf{F}$-stopping time.

The theory of the compensator proves that the process $\widetilde{F}$ enjoys the property that for any $\mathbf{F}$-predictable bounded process Z we have

$$\mathbf{E}(Z_\tau \,|\, \mathcal{G}_t) = Z_\tau 1\!\!1_{\{\tau \le t\}} + 1\!\!1_{\{\tau > t\}} e^{\Gamma_t} \mathbf{E}\Big(\int_t^\infty Z_u \, d\widetilde{F}_u \,\Big|\, \mathcal{F}_t\Big).$$

This property appears to be useful, for instance, in the computation of the value of a rebate (recovery payoff at default)

$$\mathbf{E}\big(1\!\!1_{\{\tau \le T\}} Z_\tau\big) = \mathbf{E}\Big(\int_0^T Z_u \, d\widetilde{F}_u\Big).$$

Let us examine the relationship between the concept of an $\mathbf{F}$-martingale hazard process Λ of τ and the classic concept of a compensator.

Definition 4.7. The *compensator* of a $\mathbf{G}$-stopping time τ is a process A which satisfies: (i) A is a $\mathbf{G}$-predictable right-continuous increasing process, with $A_0 = 0$, (ii) the process $D - A$ is a $\mathbf{G}$-martingale.

It is well known that for any random time τ and any filtration $\mathbf{G}$ such that τ is a $\mathbf{G}$-stopping time there exists a unique $\mathbf{G}$-compensator A of τ. Moreover, $A_t = A_{t \wedge \tau}$, that is, A is stopped at τ. In the next proposition, $\mathbf{F}$ is an arbitrary filtration such that $\mathbf{G} = \mathbf{D} \vee \mathbf{F}$.

Proposition 4.8. (i) *Let Λ be an $\mathbf{F}$-martingale hazard process of τ. Then the process $A_t = \Lambda_{t \wedge \tau}$ is the $\mathbf{G}$-compensator of τ.*
(ii) *Let A be the $\mathbf{G}$-compensator of τ. Then there exists an $\mathbf{F}$-martingale hazard process Λ such that $A_t = \Lambda_{t \wedge \tau}$.*

Remark 4.9. For a given filtration $\mathbf{G}$ and a given $\mathbf{G}$-stopping time τ, the condition $\mathbf{G} = \mathbf{D} \vee \mathbf{F}$ does not specify uniquely the filtration $\mathbf{F}$, in general. Assume that $\mathbf{G} = \mathbf{D} \vee \mathbf{F}^1 = \mathbf{D} \vee \mathbf{F}^2$, and denote by Λ^i the $\mathbf{F}^i$-martingale hazard process of τ. Then obviously $\Lambda^1_{t\wedge\tau} = A_{t\wedge\tau} = \Lambda^2_{t\wedge\tau}$ so that the stopped hazard processes coincide.

4.3 Relationships Between Hazard Processes Γ and Λ

Let us assume that the $\mathbf{F}$-hazard process Γ is well defined (in particular, τ is not an $\mathbf{F}$-stopping time). Recall that for any $\mathcal{F}_s$-measurable random variable Y we have (cf. (2.4))

$$\mathbf{E}\,(\mathbb{1}_{\{\tau>s\}}\,Y \mid \mathcal{G}_t) = \mathbb{1}_{\{\tau>t\}}\mathbf{E}\left(Ye^{\Gamma_t-\Gamma_s} \mid \mathcal{F}_t\right). \tag{4.16}$$

The natural question which arises in this context is: can we substitute Γ with the $\mathbf{F}$-martingale hazard function Λ in the formula above? Of course, the answer is trivial when it is known that the equality $\Lambda = \Gamma$ is valid, for instance, when condition (G) are satisfied and F is a continuous process. More precisely, we have the following result (see [19] for the proof).

Proposition 4.10. *Under assumption* (G) *the following assertions are valid.*
(i) *If the* $\mathbf{F}$*-hazard process* Γ *is continuous, then the* $\mathbf{F}$*-martingale hazard process* Λ *is also continuous, and both processes coincide, namely,*

$$\Gamma_t = \Lambda_t = -\ln(1-F_t), \quad \forall\, t \in I\!R_+.$$

(ii) *If the* $\mathbf{F}$*-hazard process* Γ *is a discontinuous process then the equality* $\Lambda = \Gamma$ *is never satisfied. More precisely, we have*

$$e^{-\Gamma_t} = e^{-\Lambda^c_t} \prod_{0<u\le t} (1-\Delta\Lambda_u),$$

where Λ^c *is the continuous component of* Λ *– that is,* $\Lambda^c_t = \Lambda_t - \sum_{0\le u\le t} \Delta\Lambda_u$.

We shall now examine the following question: does the continuity of the $\mathbf{F}$-martingale hazard process Λ imply the equality $\Lambda = \Gamma$? The next result provides only a partial solution to this problem.

Proposition 4.11. *Under* (G), *assume that any* $\mathbf{F}$*-martingale is continuous. If the* $\mathbf{F}$*-martingale hazard process* Λ *is a continuous process then for arbitrary* $t \le s$ *and any bounded* $\mathcal{F}_s$*-measurable random variable* Y *we have*

$$\mathbb{1}_{\{\tau>t\}}\mathbf{E}\,(Ye^{\Gamma_t-\Gamma_s} \mid \mathcal{F}_t) = \mathbb{1}_{\{\tau>t\}}\mathbf{E}\,(Ye^{\Lambda_t-\Lambda_s} \mid \mathcal{F}_t). \tag{4.17}$$

PROOF: We shall show that for any $t \le s$

$$\mathbf{E}\,(\mathbb{1}_{\{\tau>s\}}Y \mid \mathcal{G}_t) = \mathbb{1}_{\{\tau>t\}}\mathbf{E}\,(Ye^{\Lambda_t-\Lambda_s} \mid \mathcal{F}_t). \tag{4.18}$$

Let us introduce the **F**-martingale $m_t = \mathbf{E}\,(Ye^{-\Lambda_s}\,|\,\mathcal{F}_t)$, where Y is a bounded $\mathcal{F}_s$-measurable random variable. Also let $\widetilde{L}_t = 1\!\!1_{\{\tau>t\}}\,e^{\Lambda_t}$. The **G**-martingale property of $\widetilde{L}$ follows easily from Itô's lemma and the continuity of Λ. Indeed,

$$d\widetilde{L}_t = (1-D_{t-})e^{\Lambda_t}\,d\Lambda_t - e^{\Lambda_t}\,dD_t = -e^{\Lambda_t}\,d\widetilde{M}_t = -\widetilde{L}_{t-}\,d\widetilde{M}_t. \tag{4.19}$$

By virtue of part (i) in Lemma 2.10 the stopped process $\widetilde{m}_t = m_{t\wedge\tau}$ is a continuous **G**-martingale, so that it is orthogonal to the **G**-martingale $Z_t = \widetilde{L}_{t\wedge s}$ (which is obviously of finite variation). Therefore the product $\widetilde{m}Z$ is a **G**-martingale, and thus

$$\mathbf{E}\,(\widetilde{m}_s Z_s\,|\,\mathcal{G}_t) = \widetilde{m}_t Z_t = 1\!\!1_{\{\tau>t\}}\,m_t e^{\Lambda_t} = 1\!\!1_{\{\tau>t\}}\mathbf{E}\,(Ye^{\Lambda_t-\Lambda_s}\,|\,\mathcal{F}_t).$$

Furthermore,

$$\widetilde{m}_s Z_s = 1\!\!1_{\{\tau>s\}}\,m_{s\wedge\tau}e^{\Lambda_s} = 1\!\!1_{\{\tau>s\}}\,Ym_s e^{\Lambda_s} = Y1\!\!1_{\{\tau>s\}}.$$

This shows that (4.18) is indeed satisfied. Combining (4.18) with (4.16), we get (4.17). △

It appears that under the assumptions of Proposition 4.11 we can establish the equality $\Gamma = \Lambda$, as the following result shows.

Proposition 4.12. *Under* (G), *assume that any* **F***-martingale is continuous. Then:*
(i) *if Λ is a continuous process, then Γ is also continuous and $\Lambda = \Gamma$,*
(ii) *if Λ is a discontinuous process, then Γ is also a discontinuous process, and $\Lambda \neq \Gamma$.*

PROOF: We know that the **F**-martingale hazard process Λ is given by (4.13). Therefore, if Λ is continuous then also $\widetilde{F}$ is continuous, and thus also $F = \widetilde{M} + \widetilde{F}$ is an increasing continuous process. Consequently, Λ is given by (4.12) and thus $\Lambda_t = -\ln(1-F_t) = \Gamma_t$. The second statement follows by similar arguments. △

To the best of our knowledge, the problem whether the continuity of Λ implies the continuity of Γ remains open in general (under (G), say). The following conjecture seems to be natural.

Conjecture (A). If (G) holds and the **F**-martingale hazard process Λ is continuous, then $\Gamma = \Lambda$.

In view of Proposition 4.3, it would be enough to show that Γ is a continuous process, and the equality $\Gamma = \Lambda$ would then follow. The following example[5] shows that Conjecture (A) is false, in general, when hypothesis (G) fails to hold.

[5] More examples of this kind can be found in [18].

Example 4.13. Let $(W_t,\, t \geq 0)$ be a standard Brownian motion on $(\Omega, \mathbf{F}, \mathbf{P})$, where $\mathbf{F} = \mathbf{F}^W$ is the natural filtration of W. Let S_t denote the maximum of W on $[0,t]$; that is, $S_t = \sup\{W_s : s \leq t\}$. A random time $\widetilde{\tau}$ is defined by setting $\widetilde{\tau} = \inf\{t \leq 1 : W_t = S_1\}$ (notice that $\widetilde{\tau}$ is the honest time of Meyer mentioned in Remark 2.11). We set $\mathbf{G} = \mathbf{D} \vee \mathbf{F}$. Notice that

$$\{\widetilde{\tau} \leq t\} = \{\sup_{s \leq t} W_s \geq \sup_{t \leq u \leq 1} W_u\} = \{S_t - W_t \geq \sup_{0 \leq s \leq 1-t} \hat{W}_s\},$$

where $(\hat{W}_s := W_{s+t} - W_t\,, s \geq 0)$ is a standard Brownian motion independent of $\mathcal{F}_t$. Therefore,

$$\mathbf{P}(\widetilde{\tau} \leq t|\mathcal{F}_t) = \mathbf{P}\big(\sup_{0 \leq s \leq 1-t} \hat{W}_s \leq S_t - W_t \,\big|\, \mathcal{F}_t\big) = \mathbf{P}\big(\sup_{0 \leq s \leq 1-t} \hat{W}_s \leq x\big)_{|x = S_t - W_t}.$$

Using the standard relationships $\mathbf{P}\big(\sup_{0 \leq s \leq 1-t} \hat{W}_s \leq x\big) = 1 - 2\mathbf{P}(\hat{W}_{1-t} \geq x) = \mathbf{P}(|\hat{W}_{1-t}| \leq x)$ which are valid for any $x \geq 0$, we finally obtain, for $t \in [0,1)$,

$$F_t = \mathbf{P}(\tilde{\tau} \leq t \,|\, \mathcal{F}_t) = \widetilde{\Phi}(t, S_t - W_t) = \Phi\left(\frac{S_t - W_t}{\sqrt{1-t}}\right),$$

where (G stands here for a random variable with the standard Gaussian law under $\mathbf{P}$)

$$\widetilde{\Phi}(t,x) := \mathbf{P}(|\hat{W}_{1-t}| \leq x) = \mathbf{P}(\sqrt{1-t}\,|G| \leq x) = \sqrt{\frac{2}{\pi}} \int_0^{x/\sqrt{1-t}} e^{-\frac{y^2}{2}}\, dy$$

and

$$\Phi(x) := \sqrt{\frac{2}{\pi}} \int_0^x e^{-\frac{y^2}{2}}\, dy, \quad \forall\, x \in I\!R_+.$$

Since the $\mathbf{F}$-hazard process of $\widetilde{\tau}$ equals $\Gamma_t = -\ln(1 - F_t)$, it is apparent that it follows a process of infinite variation. The next goal is to find the canonical decomposition of the submartingale F. Let us denote by Z the non-negative continuous semimartingale, for $t \in [0,1)$,

$$Z_t = \frac{S_t - W_t}{\sqrt{1-t}}.$$

Since clearly $F_t = \Phi(Z_t)$, using Itô's formula, we obtain

$$\begin{aligned} F_t &= \int_0^t \Phi'(Z_u)\, dZ_u + \frac{1}{2}\int_0^t \Phi''(Z_u)\, \frac{du}{1-u} \\ &= -\int_0^t \Phi'(Z_u) \frac{dW_u}{\sqrt{1-u}} + \sqrt{\frac{2}{\pi}} \int_0^t \frac{dS_u}{\sqrt{1-u}} - \frac{1}{2}\sqrt{\frac{2}{\pi}} \int_0^t Z_u e^{-Z_u^2/2} \frac{du}{1-u}, \end{aligned}$$

where in the second equality we have used the fact that $\Phi'(0) = \sqrt{2/\pi}$, and that the process S_t, $t < 1$, increases only on the set $\{t \in [0,1) : S_t = W_t\} =$

$\{t \in [0,1) : Z_t = 0\}$. By virtue of Proposition 4.4, the martingale hazard process of $\widetilde{\tau}$ equals, for $t \leq 1$ (cf. (4.13))

$$\Lambda_t = \int_0^t \frac{d\widetilde{F}_u}{1 - F_u} = \int_0^t \frac{d\widetilde{F}_u}{1 - \Phi(Z_u)}.$$

Of course, processes Γ and Λ do not coincide (recall that Γ is not of finite variation).

As noted in [1], such a model would admit arbitrage opportunities if the occurrence of the random time $\widetilde{\tau}$ could be observed by an investor. From a financial point of view, they are obvious. Consider any date t. If $\widetilde{\tau}$ is smaller than t, the investor should not enter the market. If $\widetilde{\tau} > t$, the investor knows that the supremum is not yet attained. Therefore, he should buy the asset and wait until the asset price attains some higher value (this will happen with probability one). Finally, he should sell the asset at some time when his profit[6] is strictly positive (for example, at time $\widetilde{\tau}$).

Let us now take a purely mathematical point of view. Suppose that there exists a probability $\mathbf{Q}$, equivalent to the historical probability $\mathbf{P}$, and such that the stopped process $W_{t \wedge \widetilde{\tau}}$ is a $\mathbf{G}$-martingale. Let a be a constant such that the event $A = \{\widetilde{\tau} > \frac{1}{2}, W_{1/2} < a\}$ has a positive probability under $\mathbf{P}$. Since $A \in \mathcal{G}_{1/2}$, we have

$$0 = \mathbf{E_Q}\big(1\!\!1_A W_{\widetilde{\tau}}\big) - \mathbf{E_Q}\big(1\!\!1_A W_{1/2}\big) = \mathbf{E_Q}\big(1\!\!1_A (W_{\widetilde{\tau}} - W_{1/2})\big).$$

We conclude that $\mathbf{Q}(A) = 0$, and thus $\mathbf{Q}$ and $\mathbf{P}$ are not mutually equivalent.

Remark 4.14. The example above shows that the "stochastic intensity" does not characterize the default time, in general. It is well known (see the next section) that it is possible to construct a random time τ with a given hazard processes $\Gamma = \Lambda$. Unlike $\widetilde{\tau}$, the random time τ defined below satisfies (H).

4.4 Random Time with a Given Hazard Process

We shall now examine the 'standard' construction of a random time for a given 'hazard process' Ψ. In the 'standard' construction of τ, the following properties hold:
(i) Ψ coincides with the $\mathbf{F}$-hazard process Γ of τ,
(ii) Ψ is the $\mathbf{F}$-martingale hazard process of a random time τ,
(iii) Ψ is a $\mathbf{G}$-martingale hazard process of τ considered as a $\mathbf{G}$-stopping time.
Let us notice that the random time τ defined below is not a stopping time with respect to the filtration $\mathbf{F}$, but it is a totally inaccessible stopping time with respect to the enlarged filtration $\mathbf{G}$. Let Ψ be an $\mathbf{F}$-adapted, continuous,

[6] For simplicity, we assume that the interest rate is zero. Otherwise, we should take into account the cost of borrowed money.

increasing process, defined on a filtered probability space $(\widetilde{\Omega}, \mathbf{F}, \widetilde{\mathbf{P}})$ such that $\Psi_0 = 0$ and $\Psi_\infty = +\infty$. For instance, it can be given by the formula

$$\Psi_t = \int_0^t \psi_u \, du, \quad \forall \, t \in I\!R_+, \tag{4.20}$$

where ψ is a non-negative $\mathbf{F}$-progressively measurable process. Our goal is to construct a random time τ, on an enlarged probability space $(\Omega, \mathcal{G}, \mathbf{P})$, in such a way that Ψ is an $\mathbf{F}$-(martingale) hazard process of τ. To this end, we assume that ξ is a random variable on some probability space[7] $(\hat{\Omega}, \hat{\mathcal{F}}, \hat{\mathbf{P}})$, with the uniform probability law on $[0,1]$. We may take the product space $\Omega = \widetilde{\Omega} \times \hat{\Omega}$, $\mathcal{G} = \mathcal{F}_\infty \otimes \hat{\mathcal{F}}$ and $\mathbf{P} = \widetilde{\mathbf{P}} \otimes \hat{\mathbf{P}}$. We introduce the random time τ by setting

$$\tau = \inf \{ \, t \in I\!R_+ : e^{-\Psi_t} \le \xi \, \} = \inf \{ \, t \in I\!R_+ : \Psi_t \ge - \ln \xi \, \}.$$

As usual, we set $\mathcal{G}_t = \mathcal{D}_t \vee \mathcal{F}_t$ for every t. We shall now check that properties (i)-(iii) also hold.
Indeed, since clearly $\{\tau > t\} = \{e^{-\Psi_t} > \xi\}$, we get $\mathbf{P}(\tau > t \,|\, \mathcal{F}_\infty) = e^{-\Psi_t}$. Consequently,

$$1 - F_t = \mathbf{P}(\tau > t|\mathcal{F}_t) = \mathbf{E}\left(\mathbf{P}(\tau > t|\mathcal{F}_\infty)|\mathcal{F}_t\right) = e^{-\Psi_t} = \mathbf{P}(\tau > t|\mathcal{F}_\infty) \tag{4.21}$$

and thus F is an $\mathbf{F}$-adapted continuous increasing process. We conclude that Ψ coincides with the $\mathbf{F}$-hazard process Γ. Since (H) is valid (cf. (4.21) and (2.11)), using Proposition 4.3, we conclude that the $\mathbf{F}$-martingale hazard process Λ of τ coincides with Γ. To be more specific, we have $\Psi_t = \Lambda_t = \Gamma_t = -\ln(1 - F_t)$ and thus (ii) is valid. Furthermore, the process $D_t - \Psi_{t \wedge \tau}$ is indeed a $\mathbf{G}$-martingale so that (iii) holds.

Remark 4.15. If, in addition, Ψ satisfies (4.20) then

$$\mathbf{P}(t < \tau \le s \,|\, \mathcal{G}_t) = 1\!\!1_{\{\tau > t\}} \, \mathbf{E}\left(1 - e^{-\int_t^s \psi_u \, du} \,\Big|\, \mathcal{F}_t\right).$$

In particular, the cumulative distribution function of τ equals (we write γ^0 to denote the unique intensity function of τ with respect to the trivial filtration)

$$\mathbf{P}(\tau \le t) = 1 - \mathbf{E}\left(e^{-\int_0^t \psi_u \, du}\right) = 1 - e^{-\int_0^t \gamma^0(u) \, du}.$$

Remark 4.16. If Hypothesis (H) is satisfied, there exists an $\mathbf{F}$-adapted increasing process ζ such that $\mathbf{P}(\tau > t \,|\, \mathcal{F}_\infty) = e^{-\zeta_t}$. The variable $\xi := e^{-\zeta_\tau}$ is independent of $\mathcal{F}_\infty$, it is uniformly distributed on $[0,1]$ and $\tau = \inf\{t : \zeta_t \ge -\ln \xi\}$. See [11].

[7] In principle, it is enough to assume that there exists a random variable ξ on $(\Omega, \mathcal{G}, \mathbf{P})$ such that ξ is uniformly distributed on $[0,1]$, and it is independent of the process Ψ (we then set $\hat{\mathcal{F}} = \sigma(\xi)$).

5 Analysis of Several Random Times

Assume that we are given random times $\tau_1, \ldots, \tau_n$, defined on a common probability space $(\Omega, \mathcal{G}, \mathbf{P})$ endowed with a filtration $\mathbf{F}$. For $i = 1, \ldots, n$ we set $D_t^i = 1\!\!1_{\{\tau_i \le t\}}$, and we denote by $\mathbf{D}^i$ the filtration generated by the process D^i. We introduce the enlarged filtration $\mathbf{G} := \mathbf{D}^1 \vee \cdots \vee \mathbf{D}^n \vee \mathbf{F}$. It is thus evident that $\tau_1, \ldots, \tau_n$ are stopping times with respect to the filtration $\mathbf{G}$.

5.1 Ordered Random Times

Consider the two $\mathbf{F}$-adapted increasing continuous processes, Ψ^1 and Ψ^2, which satisfy $\Psi_0^2 = \Psi_0^1 = 0$ and $\Psi_t^1 > \Psi_t^2$ for every $t \in I\!R_+$. Let ξ be a random variable uniformly distributed on $[0,1]$, independent of the processes $\Psi^i, i = 1, 2$. We introduce random times satisfying $\tau_1 < \tau_2$ with probability 1 by setting

$$\tau_i = \inf\{\, t \in I\!R_+ : e^{-\Psi_t^i} \le \xi \,\} = \inf\{\, t \in I\!R_+ : \Psi_t^i \ge -\ln \xi \,\}.$$

We shall write $\mathbf{G}^i = \mathbf{D}^i \vee \mathbf{F}$, for $i = 1, 2$, and $\mathbf{G} = \mathbf{D}^1 \vee \mathbf{D}^2 \vee \mathbf{F}$. An analysis of each random time τ_i with respect to its 'natural' enlarged filtration $\mathbf{G}^i$ can be done along the same lines as in the previous section.

From results of Section 4.4, it follows that for each i the process Ψ^i represents: (i) the $(\mathbf{F}, \mathbf{G}^i)$-hazard process Γ^i of τ_i, (ii) the $(\mathbf{F}, \mathbf{G}^i)$-martingale hazard process Λ^i of τ_i, and (iii) the $\mathbf{G}^i$-martingale hazard process of τ_i when τ_i is considered as a $\mathbf{G}^i$-stopping time.

We find it convenient to introduce the following notation:[8] $\mathbf{F}^i = \mathbf{D}^i \vee \mathbf{F}$, so that $\mathbf{G} = \mathbf{D}^1 \vee \mathbf{F}^2$ and $\mathbf{G} = \mathbf{D}^2 \vee \mathbf{F}^1$. Let us start by an analysis of τ_1. We search for the $(\mathbf{F}^2, \mathbf{G})$-hazard process $\widetilde{\Gamma}^1$ of τ_1 and for the $(\mathbf{F}^2, \mathbf{G})$-martingale hazard process $\widetilde{\Lambda}^1$ of τ_1. We shall first check that $\widetilde{\Gamma}^1 \neq \Gamma^1$. Indeed, by virtue of the definition of a hazard process we have, for $t \in I\!R_+$,

$$e^{-\Gamma_t^1} = \mathbf{P}(\tau_1 > t \,|\, \mathcal{F}_t) = e^{-\Psi_t^1}.$$

and

$$e^{-\widetilde{\Gamma}_t^1} = \mathbf{P}(\tau_1 > t \,|\, \mathcal{F}_t^2) = \mathbf{P}(\tau_1 > t \,|\, \mathcal{F}_t \vee \mathcal{D}_t^2).$$

Equality $\widetilde{\Gamma}^1 = \Gamma^1$ would thus imply the following equality, for every $t \in I\!R_+$,

$$\mathbf{P}(\tau_1 > t \,|\, \mathcal{F}_t \vee \mathcal{D}_t^2) = \mathbf{P}(\tau_1 > t \,|\, \mathcal{F}_t). \tag{5.22}$$

The equality above is not valid, however. Indeed, the condition $\tau_2 \le t$ implies $\tau_1 \le t$, an thus on the set $\{\tau_2 \le t\} \in \mathcal{D}_t^2$ we obtain $\mathbf{P}(\tau_1 > t \,|\, \mathcal{F}_t \vee \mathcal{D}_t^2) = 0$. The last equality contradicts (5.22) since the right-hand side in (5.22) is non-zero.

[8] Though $\mathbf{F}^i = \mathbf{G}^i$ in the present setup, this double notation will prove useful in what follows.

Notice also that the $(\mathbf{F}^2, \mathbf{G})$-hazard process $\widetilde{\Gamma}^1$ is well defined only strictly before τ_2.

On the other hand, one can check that the process $D_t^1 - \Psi^1_{t\wedge\tau_1}$, which is obviously stopped at τ_1, is not only a $\mathbf{G}^1$-martingale, but also a $\mathbf{G}$-martingale. To check this, let us consider an arbitrary $\mathbf{G}^1$-martingale M stopped at τ_1. To show that M is a $\mathbf{G}$-martingale, it is enough to check that for any bounded $\mathbf{G}$-stopping time τ we have $\mathbf{E}\,(M_\tau) = M_0$. Since M is stopped at τ_1, it is clear that $\mathbf{E}\,(M_\tau) = \mathbf{E}\,(M_{\tau\wedge\tau_1})$. Furthermore, for any bounded $\mathbf{G}$-stopping time τ, the random time $\tilde{\tau} = \tau \wedge \tau_1$ is a bounded $\mathbf{G}^1$-stopping time. To see that, take an arbitrary t, and consider the event $\{\tilde{\tau} \le t\}$. We have

$$\{\tilde{\tau} \le t\} = \{\tau \wedge \tau_1 \le t\} = \{\tau_1 \le t\} \cup (\{\tau \le t\} \cap \{\tau_1 > t\}) = A \cup B .$$

Clearly $A = \{\tau_1 \le t\} \in \mathcal{D}_t^1 \subset \mathcal{F}_t \vee \mathcal{D}_t^1 = \mathcal{G}_t^1$. Since $\tau_1 \le \tau_2$, we have

$$B = \{\tau \le t\} \cap \{\tau_1 > t\} = \{\tau \le t\} \cap \{\tau_1 > t\} \cap \{\tau_2 > t\}.$$

Since $\mathcal{G}_t = \mathcal{G}_t^1 \vee \mathcal{D}_t^2$, there exists a set $C \in \mathcal{G}_t^1$ such that $B = C \cap \{\tau_2 > t\}$. Thus

$$B = C \cap \{\tau_2 > t\} = C \cap \{\tau_1 > t\} \cap \{\tau_2 > t\} = C \cap \{\tau_1 > t\} \in \mathcal{G}_t^1 .$$

By assumption, M is a $\mathbf{G}^1$-martingale, and thus $\mathbf{E}\,(M_{\tilde{\tau}}) = M_0$ for any bounded $\mathbf{G}^1$-stopping time $\tilde{\tau}$. Combining the properties above, we get the equality $\mathbf{E}\,(M_\tau) = M_0$ for any bounded $\mathbf{G}$-stopping time τ. Notice that since $D_t^1 - \Psi^1_{t\wedge\tau_1}$ is a $\mathbf{G}$-martingale, Ψ^1 coincides with the $(\mathbf{F}^2, \mathbf{G})$-martingale hazard process $\widetilde{\Lambda}^1$ of τ_1. Furthermore, Ψ^1 represents also the $\mathbf{G}$-martingale hazard process $\hat{\Lambda}^1$ of τ_1.

As expected, the properties of τ_2 with respect to the filtration $\mathbf{F}^1$ are different. First, by definition of the $(\mathbf{F}^1, \mathbf{G})$-martingale hazard process of τ_2 we have

$$e^{-\widetilde{\Gamma}_t^2} = \mathbf{P}(\tau_2 > t \,|\, \mathcal{F}_t^1) = \mathbf{P}(\tau_2 > t \,|\, \mathcal{F}_t \vee \mathcal{D}_t^1).$$

We claim that $\widetilde{\Gamma}^2 \neq \Gamma^2$; that is, that the equality

$$\mathbf{P}(\tau_2 > t \,|\, \mathcal{F}_t \vee \mathcal{D}_t^1) = \mathbf{P}(\tau_2 > t \,|\, \mathcal{F}_t) \tag{5.23}$$

does not hold, in general. Indeed, the inequality $\tau_1 > t$ implies $\tau_2 > t$, and thus on set $\{\tau_1 > t\} \in \mathcal{D}_t^1$ we have $\mathbf{P}(\tau_2 > t \,|\, \mathcal{F}_t \vee \mathcal{D}_t^1) = 1$, in contradiction with (5.23). Notice that the process $\widetilde{\Gamma}^2$ is not well defined after time τ_1.

Furthermore, the process $D_t^2 - \Psi^2_{t\wedge\tau_2}$ is a $\mathbf{G}^2$-martingale; it does not follow a $\mathbf{G}$-martingale, however (otherwise, the equality $\widetilde{\Gamma}_t^2 = \Gamma_t^2 = \Psi_t^2$ would be true for $t < \tau_2$, but this is not the case). The explicit formula for the $(\mathbf{F}^1, \mathbf{G})$-martingale hazard process $\widetilde{\Lambda}^2$ of τ_2 is not easily available (it seems plausible that $\widetilde{\Lambda}^2$ has discontinuity at τ_1).

Let us observe that τ_1 is a totally inaccessible stopping time not only with respect to $\mathbf{G}^1$, but also with respect to $\mathbf{G}$. On the other hand, τ_2 is a totally inaccessible stopping time with respect to $\mathbf{G}^1$, but it is a predictable stopping time with respect to $\mathbf{G}$. Indeed, we may easily find an announcing sequence τ_2^n of $\mathbf{G}$-stopping times, for instance,

$$\tau_2^n = \inf\{\, t \geq \tau_1 \, : \, \Psi_t^2 \geq -\ln \xi - \tfrac{1}{n} \,\}.$$

Therefore, the $\mathbf{G}$-martingale hazard process $\hat{\Lambda}^2$ of τ_2 coincides with the $\mathbf{G}$-predictable process D^2.

Let us set $\tau = \tau_1 \wedge \tau_2$. In the present setup, it is evident that $\tau = \tau_1$, and thus the $\mathbf{G}$-martingale hazard process $\hat{\Lambda}$ of τ is equal to Ψ^1. It is also equal to the sum of $\mathbf{G}$-martingale hazard processes $\hat{\Lambda}^i$ of τ_i, $i = 1, 2$, stopped at τ. Indeed, we have

$$\hat{\Lambda}_{t\wedge\tau} = \Psi^1_{t\wedge\tau} = \Psi^1_{t\wedge\tau} + D^2_{t\wedge\tau} = \hat{\Lambda}^1_{t\wedge\tau} + \hat{\Lambda}^2_{t\wedge\tau}.$$

We shall see in the next section that this property is universal (notice that it is of limited use though).

5.2 Properties of the Minimum of Several Random Times

The exposition here is partially based on Duffie [8] and Kusuoka [20]. We shall examine the following problem: given a finite family of random times τ_i, $i = 1, \ldots, n$, and the associated hazard processes, find the hazard process of the random time $\tau = \tau_1 \wedge \cdots \wedge \tau_n$. The problem above cannot be solved in such a generality; that is, without the knowledge of the joint law of $(\tau_1, \ldots, \tau_n)$. Indeed, the solution depends on specific assumptions on random times and the choice of filtrations.

When the reference filtration $\mathbf{F}$ is trivial, the hazard process is a deterministic function, known as the hazard function: $\Gamma(t) = -\ln \mathbf{P}(\tau > t)$. The next simple result deals with the hazard function of the minimum of mutually independent random times.

Lemma 5.1. *Let τ_i, $i = 1, \ldots, n$, be n random times defined on a common probability space $(\Omega, \mathcal{G}, \mathbf{P})$. Assume that τ_i admits the hazard function Γ^i. If τ_i, $i = 1, \ldots, n$, are mutually independent random variables, the hazard function Γ of τ is equal to the sum of hazard functions Γ^i, $i = 1, \ldots, n$.*

PROOF: For any $t \in \mathbb{R}_+$ we have

$$e^{-\Gamma(t)} = \mathbf{P}(\tau_1 \wedge \cdots \wedge \tau_n > t) = \prod_{i=1}^n (1 - F_i(t)) = e^{-\sum_{i=1}^n \Gamma^i(t)}. \qquad \triangle$$

Conversely, if the hazard function of $\tau = \tau_1 \wedge \cdots \wedge \tau_n$ satisfies $\Lambda(t) = \Gamma(t) = \sum_{i=1}^n \Gamma^i(t) = \sum_{i=1}^n \Lambda^i(t)$ for every t then we obtain

$$\mathbf{P}(\tau_1 > t, \ldots, \tau_n > t) = \prod_{i=1}^n \mathbf{P}(\tau_i > t), \quad \forall\, t \in \mathbb{R}_+.$$

Lemma 5.1 admits a rather trivial extension to the general case. Let τ_i, $i = 1, \ldots, n$, be n random times defined on a common probability space $(\Omega, \mathcal{G}, \mathbf{P})$. Assume that τ_i admits the $\mathbf{F}^i$-hazard process Γ^i. If, for every $t \in I\!\!R_+$, the events $\{\tau_i < t\}$, $i = 1, \ldots, n$, are conditionally independent with respect to $\mathcal{F}_t$, then the hazard process Γ of τ is equal to the sum of hazard processes Γ^i, $i = 1, \ldots, n$.

We borrow from Duffie [8] the following simple result (see Lemma 1 in [8]).

Lemma 5.2. *Let τ_i, $i = 1, \ldots, n$, be $\mathbf{G}$-stopping times such that $\mathbf{P}(\tau_i = \tau_j) = 0$ for $i \neq j$. Then the $(\mathbf{F}, \mathbf{G})$-martingale hazard process Λ of $\tau = \tau_1 \wedge \cdots \wedge \tau_n$ is equal to the sum of $(\mathbf{F}, \mathbf{G})$-martingale hazard processes Λ^i; that is, $\Lambda_t = \sum_{i=1}^n \Lambda_t^i$ for $t \in I\!\!R_+$. If Λ is a continuous process then the process $\widetilde{L}$ given by the formula $\widetilde{L}_t = (1 - D_t)e^{\Lambda_t}$ is a $\mathbf{G}$-martingale.*

We consider once again the case of the Brownian filtration; that is, we assume that $\mathbf{F} = \mathbf{F}^W$ for some Brownian motion W. We postulate that W remains a martingale (and thus a Brownian motion) with respect to the enlarged filtration $\mathbf{G} = \mathbf{D}^1 \vee \cdots \vee \mathbf{D}^n \vee \mathbf{F}$. In view of the martingale representation property of the Brownian filtration this means, of course, that any $\mathbf{F}$-local martingale is also a local martingale with respect to $\mathbf{G}$ (or indeed with respect to any enlargement of the filtration $\mathbf{F}$), so that (H) holds. It is worthwhile to stress that the case of a trivial filtration $\mathbf{F}$ is also covered by the results of this section.

Our next goal is to generalize the martingale representation property established in Proposition 3.2. Recall that in Proposition 3.2 we have assumed that the $\mathbf{F}$-hazard process Γ of a random time τ is an increasing continuous process. Also, by virtue of results of Section 4.3 (see Proposition 4.10) under the assumptions of Corollary 3.2 we have $\Gamma = \Lambda$; that is, the $\mathbf{F}$-hazard process Γ and the $(\mathbf{F}, \mathbf{G})$-martingale hazard process Λ coincide.

In the present setup, we prefer to make assumptions directly about the $(\mathbf{F}, \mathbf{G})$-martingale hazard processes Λ^i of random times τ_i, $i = 1, \ldots, n$. We assume throughout that the processes Λ^i, $i = 1, \ldots, n$ are continuous. As before, we assume that $\mathbf{P}(\tau_i = \tau_j) = 0$ for $i \neq j$. Recall that by virtue of the definition of the $(\mathbf{F}, \mathbf{G})$-martingale hazard process Λ^i of a random time τ_i the process $\widetilde{M}_t^i = D_t^i - \Lambda_{t \wedge \tau_i}^i$ is a $\mathbf{G}$-martingale. Notice that the process $\widetilde{L}_t^i = (1 - D_t^i)e^{\Lambda_t^i}$ is also a $\mathbf{G}$-martingale, since clearly (cf. (4.19))

$$\widetilde{L}_t^i = 1 - \int_{]0,t]} \widetilde{L}_{u-}^i \, d\widetilde{M}_u^i.$$

It is easily seen that $\widetilde{L}^i$ and $\widetilde{L}^j$ are mutually orthogonal $\mathbf{G}$-martingales for any $i \neq j$ (a similar remark applies to $\widetilde{M}^i$ and $\widetilde{M}^j$).

For a fixed k with $0 \leq k \leq n$, we introduce the filtration $\widetilde{\mathbf{G}} = \mathbf{D}^1 \vee \cdots \vee \mathbf{D}^k \vee \mathbf{F}$. Then obviously $\widetilde{\mathbf{G}} = \mathbf{G}$ if $k = n$, and by convention $\widetilde{\mathbf{G}} = \mathbf{F}$ for $k = 0$. It is clear that for any fixed k and arbitrary $i \leq k$ processes $\widetilde{L}^i$

and $\widetilde{M}^i$ are $\widetilde{\mathbf{G}}$-adapted. More specifically, $\widetilde{L}^i$ and $\widetilde{L}^j$ are mutually orthogonal $\widetilde{\mathbf{G}}$-martingales for $i,j \leq k$ provided that $i \neq j$. A trivial modification of Lemma 5.2 shows that the $(\mathbf{F},\widetilde{\mathbf{G}})$-martingale hazard process of the random time $\widetilde{\tau} := \tau_1 \wedge \cdots \wedge \tau_k$ equals $\widetilde{\Lambda} = \sum_{i=1}^k \Lambda^i$. In other words, the process $\widetilde{D}_t - \sum_{i=1}^k \Lambda^i_{t\wedge\widetilde{\tau}}$ is a $\widetilde{\mathbf{G}}$-martingale, where we set $\widetilde{D}_t = 1\!\!1_{\{\widetilde{\tau}\leq t\}}$. For a fixed k with $0 \leq k \leq n$, we set $\widetilde{\mathbf{F}} := \mathbf{D}^{k+1} \vee \cdots \vee \mathbf{D}^n \vee \mathbf{F}$. The next two results are due to Kusuoka [20].

Proposition 5.3. *Assume that the* $\mathbf{F}$*-Brownian motion* W *remains a Brownian motion with respect to the enlarged filtration* $\mathbf{G}$. *Let* Y *be a bounded* $\mathcal{F}_T$*-measurable random variable, and let* $\widetilde{\tau} = \tau_1 \wedge \cdots \wedge \tau_k$. *Then for any* $t \leq s \leq T$ *we have*

$$\mathbf{E}\,(1\!\!1_{\{\widetilde{\tau}>s\}}Y \,|\, \mathcal{G}_t) = \mathbf{E}\,(1\!\!1_{\{\widetilde{\tau}>s\}}Y \,|\, \widetilde{\mathcal{G}}_t) = 1\!\!1_{\{\widetilde{\tau}>t\}}\mathbf{E}\,\big(Ye^{\widetilde{\Lambda}_t-\widetilde{\Lambda}_s} \,|\, \mathcal{F}_t\big).$$

In particular,

$$\mathbf{P}(\widetilde{\tau} > s \,|\, \mathcal{G}_t) = \mathbf{P}(\widetilde{\tau} > s \,|\, \widetilde{\mathcal{G}}_t) = 1\!\!1_{\{\widetilde{\tau}>t\}}\mathbf{E}\,(e^{\widetilde{\Lambda}_t-\widetilde{\Lambda}_s} \,|\, \mathcal{F}_t).$$

Let us set $\Lambda = \sum_{i=1}^n \Lambda^i$. *Then for* $\tau = \tau_1 \wedge \cdots \wedge \tau_n$ *we have*

$$\mathbf{P}(\tau > s \,|\, \mathcal{G}_t) = 1\!\!1_{\{\tau>t\}}\mathbf{E}\,(e^{\Lambda_t-\Lambda_s} \,|\, \mathcal{F}_t).$$

Proposition 5.4. *Assume that the* $\mathbf{F}$*-Brownian motion* W *remains a Brownian motion with respect to* $\mathbf{G}$, *and that for each* $i = 1,\dots,n$ *the* $\mathbf{F}$*-martingale hazard process* Λ^i *is continuous. Then any* $\widetilde{\mathbf{F}}$*-martingale* N *admits the integral representation*

$$N_t = N_0 + \int_0^t \xi_u\,dW_u + \sum_{i=k+1}^n \int_{]0,t]} \zeta^i_u\,d\widetilde{M}^i_u, \tag{5.24}$$

where ξ *and* ζ^i, $i = k+1,\dots,n$ *are* $\widetilde{\mathbf{F}}$*-predictable processes.*

It is interesting to observe that the $\mathcal{F}_T$-measurability of Y can be replaced by the $\widetilde{\mathcal{F}}_T$-measurability of Y in Proposition 5.3. Indeed, it is clear that $\widetilde{\Lambda} = \sum_{i=1}^k \Lambda^i$ is also the $(\widetilde{\mathbf{F}},\mathbf{G})$-martingale hazard process of τ. Furthermore, Proposition 5.4 shows that the process $\hat{Y}$, given by the formula

$$\hat{Y}_t = \mathbf{E}\,(Ye^{-\widetilde{\Lambda}_s} \,|\, \widetilde{\mathcal{F}}_t), \quad \forall\, t \in [0,T],$$

where Y is an $\widetilde{\mathcal{F}}_T$-measurable random variable, admits the following integral representation

$$\hat{Y}_t = \hat{Y}_0 + \int_0^t \xi_u\,dW_u + \sum_{i=k+1}^n \int_{]0,t]} \zeta^i_u\,d\widetilde{M}^i_u,$$

where ξ and ζ^i, $i = k+1, \dots, n$ are $\widetilde{\mathbf{F}}$-predictable processes. We conclude that $\hat{Y}$ follows a $\mathbf{G}$-martingale orthogonal to the $\mathbf{G}$-martingale U, which is given by

$$U_t = (1 - \widetilde{D}_{t\wedge s})e^{\widetilde{\Lambda}_{t\wedge s}} = \prod_{i=1}^{k} \widetilde{L}^i_{t\wedge s}.$$

Arguing in a much the same way as in the proof of Proposition 5.3, we obtain the following result (see [19] for the details).

Corollary 5.5. *Let Y be a bounded $\widetilde{\mathcal{F}}_T$-measurable random variable. Let $\widetilde{\tau} = \tau_1 \wedge \dots \wedge \tau_k$. Then for any $t \leq s \leq T$ we have*

$$\mathbf{E}\left(\mathbb{1}_{\{\widetilde{\tau}>s\}} Y \mid \mathcal{G}_t\right) = \mathbb{1}_{\{\widetilde{\tau}>t\}} \mathbf{E}\left(Y e^{\widetilde{\Lambda}_t - \widetilde{\Lambda}_s} \mid \widetilde{\mathcal{F}}_t\right).$$

5.3 Change of a Probability Measure

In this section – in which we follow Kusuoka [20] – it is assumed throughout that the filtration $\mathbf{F}$ is generated by a Brownian motion W, which is also a $\mathbf{G}$-martingale (the case of a trivial filtration $\mathbf{F}$ is also covered by the results of this section, though). For a fixed $T > 0$, we shall examine the properties of $\widetilde{\tau}$ under a probability measure $\mathbf{P}^*$ equivalent to $\mathbf{P}$ on $(\Omega, \mathcal{G}_t)$, for every $t \in \mathbb{R}_+$. We introduce the $\mathbf{G}$-martingale η by setting

$$\eta_t := \frac{d\mathbf{P}^*}{d\mathbf{P}}\bigg|_{\mathcal{G}_t}, \quad \mathbf{P}\text{-a.s.},$$

By virtue of Proposition 5.4 (with $k = 0$), the Radon-Nikodým density process η admits the integral representation

$$\eta_t = 1 + \int_0^t \xi_u \, dW_u + \sum_{i=1}^{n} \int_{]0,t]} \zeta^i_u \, d\widetilde{M}^i_u, \tag{5.25}$$

where ξ and ζ^i, $i = 1, \dots, n$ are $\mathbf{G}$-predictable stochastic processes. It can be shown that η is a strictly positive process, so that we may rewrite (5.25) as follows

$$\eta_t = 1 + \int_{]0,t]} \eta_{u-}\big(\beta_u \, dW_u + \sum_{i=1}^{n} \kappa^i_u \, d\widetilde{M}^i_u\big), \tag{5.26}$$

where β and $\kappa^i > -1$, $i = 1, \dots, n$ are $\mathbf{G}$-predictable processes. For the proof of the next result, we refer to [20] or [19].

Proposition 5.6. *Let $\mathbf{P}^*$ be a probability measure which is equivalent to $\mathbf{P}$ on $(\Omega, \mathcal{G}_t)$, for every $t \in \mathbb{R}_+$. If the Radon-Nikodým density of $\mathbf{P}^*$ with respect to $\mathbf{P}$ on $(\Omega, \mathcal{G}_t)$ is given by (5.26), then the process*

$$W^*_t = W_t - \int_0^t \beta_u \, du,$$

is a **G***-Brownian motion under* $\mathbf{P}^*$*, and for each* $i = 1, \dots, n$ *the process*

$$M_t^{i*} := \widetilde{M}_t^i - \int_{]0,t\wedge\tau_i]} \kappa_u^i \, d\Lambda_u^i = D_t^i - \int_{]0,t\wedge\tau_i]} (1+\kappa_u^i) \, d\Lambda_u^i,$$

is a **G***-martingale orthogonal to* W^* *under* $\mathbf{P}^*$*. Moreover, processes* M^{i*} *and* M^{j*} *follow mutually orthogonal* **G***-martingales under* $\mathbf{P}^*$ *for any* $i \neq j$*.*

Though the process M^{i*} is a **G**-martingale under $\mathbf{P}^*$, it should be stressed that the process $\int_{]0,t]} (1+\kappa_u^i)\, d\Lambda_u^i$ is not necessarily the $(\mathbf{F}, \mathbf{G})$-martingale hazard process of τ_i under $\mathbf{P}^*$, since it is not **F**-adapted, but merely **G**-adapted, in general. To circumvent this, we choose, for any fixed i, a suitable version of the process κ^i. Specifically, we take a process κ^{i*}, which coincides with κ^i on a random interval $[0, \tau_i]$, and which is predictable with respect to the enlarged filtration $\mathbf{F}^{i*} := \mathbf{D}^1 \vee \dots \vee \mathbf{D}^{i-1} \vee \mathbf{D}^{i+1} \vee \dots \mathbf{D}^n \vee \mathbf{F}$. Since

$$D_t^i - \int_{]0,t\wedge\tau_i]} (1+\kappa_u^{i*}) \, d\Lambda_u^i = D_t^i - \int_{]0,t\wedge\tau_i]} (1+\kappa_u^i) \, d\Lambda_u^i,$$

we conclude that for each fixed i the process

$$\Lambda_t^{i*} = \int_{]0,t]} (1+\kappa_u^{i*}) \, d\Lambda_u^i$$

represents the $(\mathbf{F}^{i*}, \mathbf{G})$-martingale hazard process of τ_i under $\mathbf{P}^*$. This does not mean, however, that the equality

$$\mathbf{P}^*(\tau_i > s \,|\, \mathcal{G}_t) = \mathbb{1}_{\{\tau_i > t\}} \mathbf{E}_{\mathbf{P}^*}(e^{\Lambda_t^{i*} - \Lambda_s^{i*}} \,|\, \mathcal{F}_t^{i*})$$

is valid for every $s \le t$. We prefer to examine the validity of the last formula in a slightly more general setting. For a fixed $k \le n$, we set $\widetilde{\tau} = \tau_1 \wedge \dots \wedge \tau_k$, and we write $\widetilde{\mathbf{F}} = \mathbf{D}^{k+1} \vee \dots \vee \mathbf{D}^n \vee \mathbf{F}$. For any $i = 1, \dots, n$ we denote by $\widetilde{\kappa}^i$ ($\widetilde{\beta}$, resp.) the $\widetilde{\mathbf{F}}$-predictable process such that $\widetilde{\kappa}^i = \kappa^i$ ($\widetilde{\beta} = \beta$, resp.) on the random set $[0, \widetilde{\tau}]$.

Lemma 5.7. *The* $(\widetilde{\mathbf{F}}, \mathbf{G})$*-martingale hazard process of the random time* $\widetilde{\tau}$ *under* $\mathbf{P}^*$ *is given by the formula*

$$\Lambda_t^* = \sum_{i=1}^k \int_{]0,t]} (1+\widetilde{\kappa}_u^i) \, d\Lambda_u^i. \tag{5.27}$$

PROOF: Let us set

$$\widetilde{W}_t^* = W_t - \int_0^t \widetilde{\beta}_u \, du,$$

and

$$\widetilde{M}_t^{i*} = D_t^i - \int_{]0,t\wedge\tau_i]} (1+\widetilde{\kappa}_u^i) \, d\Lambda_u^i$$

for $i = 1, \ldots, n$. The processes $\widetilde{W}^*$ and $\widetilde{M}^{i*}$ follow $\mathbf{G}$-martingales under $\mathbf{P}^*$, provided that they are stopped at $\tilde{\tau}$ (since $\widetilde{W}^*_{t\wedge\tilde{\tau}} = W^*_{t\wedge\tilde{\tau}}$ and $\widetilde{M}^{i*}_{t\wedge\tilde{\tau}} = M^{i*}_{t\wedge\tilde{\tau}}$). Consequently, the process

$$\widetilde{D}_t - \sum_{i=1}^{k} \int_{]0,t\wedge\tilde{\tau}]} (1 + \tilde{\kappa}^i_u)\, d\Lambda^i_u = \sum_{i=1}^{k} (\widetilde{M}^{i*}_t)^{\tilde{\tau}}$$

is a $\mathbf{G}$-martingale. △

In view of Corollary 5.5 and Lemma 5.7, it would be natural to conjecture that, for any bounded $\widetilde{\mathcal{F}}_T$-measurable random variable Y, and every $t \le s \le T$, we have

$$\mathbf{E}_{\mathbf{P}^*}(\mathbb{1}_{\{\tilde{\tau}>s\}} Y \,|\, \mathcal{G}_t) = \mathbb{1}_{\{\tilde{\tau}>t\}}\, \mathbf{E}_{\mathbf{P}^*}\big(Y e^{\Lambda^*_t - \Lambda^*_s} \,|\, \widetilde{\mathcal{F}}_t\big). \tag{5.28}$$

It appears, however, that the last formula is not valid, in general, unless we substitute the probability measure $\mathbf{P}^*$ in the right-hand side of (5.28) with some related probability measure. To this end, we introduce the following auxiliary processes $\hat{\eta}^\ell$, for $\ell = 1, 2, 3$,

$$\hat{\eta}^1_t = 1 + \int_{]0,t]} \hat{\eta}^1_{u-} \Big(\tilde{\beta}_u\, dW_u + \sum_{i=k+1}^{n} \tilde{\kappa}^i_u\, d\widetilde{M}^i_u\Big), \tag{5.29}$$

$$\hat{\eta}^2_t = 1 + \int_{]0,t]} \hat{\eta}^2_{u-} \Big(\tilde{\beta}_u\, dW_u + \sum_{i=1}^{n} \tilde{\kappa}^i_u\, d\widetilde{M}^i_u\Big),$$

and

$$\hat{\eta}^3_t = 1 + \int_{]0,t]} \hat{\eta}^3_{u-} \Big(\tilde{\beta}_u\, dW_u + \sum_{i=1}^{k} \kappa^i_u\, d\widetilde{M}^i_u + \sum_{i=k+1}^{n} \tilde{\kappa}^i_u\, d\widetilde{M}^i_u\Big).$$

It is worth noting that the process $\hat{\eta}^1$ is $\widetilde{\mathbf{F}}$-adapted (since, in particular, each process $\widetilde{M}^i$ is adapted to the filtration $\mathbf{D}^i \vee \mathbf{F}$). On the other hand, processes $\hat{\eta}^2$ and $\hat{\eta}^3$ are merely $\mathbf{G}$-adapted, but not necessarily $\widetilde{\mathbf{F}}$-adapted, in general.

For $\ell = 1, 2, 3$, we define a probability measure $\widetilde{\mathbf{P}}_\ell$, equivalent to $\mathbf{P}$ on $(\Omega, \mathcal{G}_t)$, by setting

$$\hat{\eta}^\ell_t := \frac{d\widetilde{\mathbf{P}}_\ell}{d\mathbf{P}}\Big|_{\mathcal{G}_t}, \quad \mathbf{P}\text{-a.s.}$$

The following proposition, which generalizes a result of Kusuoka [20], is a counterpart of Corollary 5.5. We refer to [19] for the proof.

Proposition 5.8. *Let Y be a bounded $\widetilde{\mathcal{F}}_T$-measurable random variable, and let Λ^* be given by (5.27). Then for any $t \le s \le T$ and any $\ell = 1, 2, 3$, we have*

$$\mathbf{E}_{\mathbf{P}^*}(\mathbb{1}_{\{\tilde{\tau}>s\}} Y \,|\, \mathcal{G}_t) = \mathbb{1}_{\{\tilde{\tau}>t\}}\, \mathbf{E}_{\widetilde{\mathbf{P}}_\ell}\big(Y e^{\Lambda^*_t - \Lambda^*_s} \,|\, \widetilde{\mathcal{F}}_t\big).$$

5.4 Kusuoka's Example

We shall now examine a purely mathematical example, due to Kusuoka [20]; the calculations presented in [20] are not complete, though (see [19] for details). In this example, the hazard process is not increasing, but the market is still arbitrage-free. Under the real-world probability $\mathbf{P}$, the random times τ_i, $i = 1, 2$ are mutually independent random variables with the exponential law with constant parameters λ_1 and λ_2, respectively. The joint law of (τ_1, τ_2) under $\mathbf{P}$ has thus the density $f(x, y) = \lambda_1 \lambda_2 e^{-(\lambda_1 x + \lambda_2 y)}$ for $(x, y) \in \mathbb{R}^2_+$. We denote by $M^i_t = D^i_t - \lambda_i t$ the martingales associated with these random times. Let α_1 and α_2 be strictly positive real numbers, and let the probability measure $\mathbf{Q}$ on $(\Omega, \mathcal{G})$ be given on $\mathcal{G}_t$ by the formula

$$\frac{d\mathbf{Q}}{d\mathbf{P}} = \eta_t, \quad \mathbf{P}\text{-a.s.},$$

with

$$\eta_t = 1 + \sum_{i=1}^{2} \int_{]0,t]} \eta_{u-} \kappa^i_u \, dM^i_u,$$

where in turn

$$\kappa^1_t = 1\!\!1_{\{\tau_2 < t\}} \Big(\frac{\alpha_1}{\lambda_1} - 1 \Big), \quad \kappa^2_t = 1\!\!1_{\{\tau_1 < t\}} \Big(\frac{\alpha_2}{\lambda_2} - 1 \Big).$$

Assume that the filtration in the default-free market is $\mathcal{D}^2_t$, and that τ_1 represents the default time. We have

$$\mathbf{Q}(\tau_1 > t \,|\, \mathcal{D}^2_t) = (1 - D^2_t) \, \frac{\mathbf{Q}(\tau_1 > t, \tau_2 > t)}{\mathbf{Q}(\tau_2 > t)} + D^2_t \mathbf{Q}(\tau_1 > t \,|\, \tau_2).$$

Rather tedious calculations show that (see [20] or [19])

$$\mathbf{Q}(\tau_1 > t, \tau_2 > t) = e^{-(\lambda_1 + \lambda_2)t},$$

$$\mathbf{Q}(\tau_2 > t) = \frac{1}{\lambda_1 + \lambda_2 - \alpha_2} \Big(\lambda_1 e^{-\alpha_2 t} + (\lambda_2 - \alpha_2) e^{-(\lambda_1 + \lambda_2)t} \Big),$$

and

$$\mathbf{Q}(\tau_1 > t \,|\, \tau_2 = u) = \frac{(\lambda_1 + \lambda_2 - \alpha_2)\lambda_2 e^{-(\lambda_1 + \lambda_2)u t} e^{-\alpha_1 (t-u)}}{\lambda_1 \alpha_2 e^{-\alpha_2 u} + (\lambda_2 - \alpha_2)(\lambda_1 + \lambda_2) e^{-(\lambda_1 + \lambda_2)u}}.$$

Consequently, we obtain

$$1 - F_t = 1\!\!1_{\{t < \tau_2\}} \frac{c}{\lambda_1 e^{ct} + \lambda_2 - \alpha_2} + 1\!\!1_{\{\tau_2 \le t\}} \frac{c \lambda_2 e^{-\alpha_1 (t - \tau_2)}}{\lambda_1 \alpha_2 e^{c\tau_2} + (\lambda_2 - \alpha_2)(\lambda_1 + \lambda_2)},$$

where $c = \lambda_1 + \lambda_2 - \alpha_2$. The two terms in the right-hand side are decreasing functions, the jump in τ_2 equals

$$\Delta = \frac{c}{\lambda_1 e^{ct} + \lambda_2 - \alpha_2} - \frac{c \lambda_2}{\lambda_1 \alpha_2 e^{c\tau_2} + (\lambda_2 - \alpha_2)(\lambda_1 + \lambda_2)}$$

and thus it is negative if and only if $\lambda_2 \le \alpha_2$. We conclude that Hypothesis (G) is not satisfied.

References

1. Blanchet-Scalliet, C.: Doctoral dissertation. In preparation.
2. Blanchet-Scalliet, C. and Jeanblanc, M.: Hazard process and hedging defaultable contingent claims. Working paper, 2001.
3. Brémaud, P.: *Point Processes and Queues. Martingale Dynamics.* Springer-Verlag, Berlin, 1981.
4. Brémaud, P. and Yor, M.: Changes of filtrations and of probability measures. *Z. Wahrsch. Verw. Gebiete* 45, 269-295, 1978.
5. Dellacherie, C.: Un exemple de la théorie générale des processus. Séminaire de Probabilités IV, Lecture Notes in Math. 124, Springer-Verlag, Berlin, 1970, pp. 60-70.
6. Dellacherie, C., Maisonneuve, B. and Meyer, P.A.: *Probabilités et potentiel. Processus de Markov. Compléments de calcul stochastique.* Hermann, Paris, 1992.
7. Dellacherie, C. and Meyer, P.A.: A propos du travail de Yor sur les grossissements des tribus. Séminaire de Probabilités XII, Lect. Notes in Math. 649, Springer-Verlag, Berlin, 1978, pp. 69-78.
8. Duffie, D.: First-to-default valuation. Working paper, 1998.
9. Duffie, D. and Lando, D.: Term structure of credit spreads with incomplete accounting information. Forthcoming in *Econometrica,* 2001.
10. Duffie, D., Schroder, M. and Skiadas, C.: Recursive valuation of defaultable securities and the timing of resolution of uncertainty. *Annals of Applied Probability* 6, 1075-1090, 1997.
11. El Karoui, N.: Modélisation de l'information, CEA-EDF-INRIA, Ecole d'été, 1999.
12. Elliott, R.J., Jeanblanc, M. and Yor, M.: On models of default risk. *Mathematical Finance* 10, 179-195, 2000.
13. Jarrow, R.A. and Turnbull, S.M.: Pricing derivatives on financial securities subject to credit risk. *Journal of Finance* 50, 53-85, 1995.
14. Jarrow, R.A., Lando, D. and Turnbull, S.M.: A Markov model for the term structure of credit risk spreads. *Review of Financial Studies* 10, 481-523, 1997.
15. Jarrow, R.A. and Yu, F.: Counterparty risk and the pricing of defaultable securities. Working paper, 1999.
16. Jeulin, T.: *Semi-martingales et grossissement de filtration.* Lecture Notes in Math. 833, Springer-Verlag, Berlin, 1980.
17. Jeulin, T. and Yor, M.: Grossissement d'une filtration et semi-martingales: formules explicites. Séminaire de Probabilités XII, Lecture Notes in Math. 649, Springer-Verlag, Berlin, 1978, pp. 78-97.
18. Jeanblanc, M. and Rutkowski, M.: Modelling of default risk: An overview. *Mathematical Finance: Theory and Practice,* Higher Education Press, Beijing, 2000, pp. 171-269.
19. Jeanblanc, M. and Rutkowski, M.: Modelling of default risk: Mathematical tools. Workshop *Fixed Income and Credit Risk Modeling and Management,* New York University, Stern School of Business, May 5, 2000.
20. Kusuoka, S.: A remark on default risk models. *Advances in Mathematical Economics* 1, 69-82, 1999.
21. Lando, D.: On Cox processes and credit-risky securities. *Review of Derivatives Research* 2, 99-120, 1998.
22. Last, G. and Brandt, A.: *Marked Point Processes on the Real Line. The Dynamic Approach.* Springer-Verlag, Berlin, 1995.

23. Madan, D. and Unal, H.: Pricing the risk of default. *Review of Derivatives Research* 2, 121-160, 1998.
24. Mazziotto, G. and Szpirglas, J.: Modèle général de filtrage non linéaire et équations différentielles stochastiques associées. *Ann. Inst. Henri Poincaré* 15, 147-173, 1979.
25. Wong, D.: A unifying credit model. Scotia Capital Markets Group, 1998.
26. Yor, M.: *Some Aspects of Brownian Motion. Part II: Some Recent Martingale Problems.* Lectures in Mathematics, ETH Zürich. Birkhäuser, 1997.

Utility-Based Derivative Pricing in Incomplete Markets

Jan Kallsen*

Universität Freiburg i. Br., Institut für Mathematische Stochastik
Eckerstraße 1, D-79104 Freiburg i. Br., Germany
e-mail: kallsen@stochastik.uni-freiburg.de

Abstract. In recent years various suggestions concerning contingent claim valuation in incomplete markets have been made. We argue that some of them can be naturally interpreted in terms of *neutral derivative prices* which occur if derivative demand and supply are balanced. Secondly, we introduce the notion of *consistent derivative pricing* which is a way of constructing market models that are consistent with initially observed derivative quotations.

Keywords

fundamental theorem of utility maximization, local utility, neutral pricing, consistent pricing

1 Introduction

Portfolio optimization, hedging, and derivative pricing constitute three fundamental quantitative problems in mathematical finance. Since martingale methods have been introduced by [33] and [34], these questions can be addressed very elegantly in complete models. As the key tool they come up with the unique *equivalent martingale measure (EMM)*. Computing conditional expectations of some random variables under this auxiliary measure yields important parts of the solution: the unique arbitrage-free contingent claim value in the derivative pricing problem, the value of its replicating portfolio in the hedging problem, and the optimal wealth process in utility maximization problems (cf. [34], [58], [48], [7]). An important early reference in this context is [4].

In incomplete markets the situation is more involved because one is facing a mathematical and a conceptual problem: Mathematically, the use of martingale methods is complicated by the fact that there exists more than one EMM. This may explain why they have been applied only in the last decade.

* The author wants to thank Thomas Goll for fruitful discussions.

Nevertheless, martingale methods play a very important role in incomplete models. In utility maximization problems, conditional expectations still lead to the optimal wealth process if the right equivalent martingale measure is chosen. Intuitively, this measure leads to an in some sense *least favourable market completion)* (cf. [20], [21], [35], [36], [49], [8], [59], [52], [10], [66]).

As regards derivative pricing and hedging, we are facing the severe conceptual problem that unique arbitrage-free contingent claim values and perfect hedging strategies do not exist. Derivative pricing in incomplete models basically means choosing one of the set of equivalent martingale measures in some economically or mathematically motivated fashion, e.g. based on hedging arguments (cf. [67], [74]), on utility or equilibrium-type considerations (cf. [12], [47], [26], [63], [3], [45]), or on distance minimization (cf. [50], [55], [56], [6], [31], [32], [27], [30]). Similarly, various suggestions have been made to replace the perfect hedge in complete models with a reasonable counterpart in incomplete settings, e.g. based on hedging error minimization (cf. [23], [24], [69], [70], [71], [73]), superhedging (cf. [19]), or utility maximization (cf. [22], [41], [43], [10], [13]).

Although the problems and even more so their solutions look quite different on first glance, they are intimately related. The links between portfolio optimization, derivative pricing, and hedging start to become folklore among experts. Still, the results are scattered in the literature and have been derived in quite different settings. As far as derivative pricing is concerned, some precise definitions are missing altogether. This makes it hard for the non-specialist in the field to get an intuitive picture.

The aim of this paper is threefold. By focusing on one important result, we want to provide the novice with an idea how martingale methods enter portfolio optimization in incomplete models. Secondly, we give a precise definition of *neutral derivative prices* which occur if traders maximize their expected utility and if supply and demand of derivatives are balanced. Neutral prices are closely related to seemingly different concepts in the literature. Thirdly, we introduce the notion of *consistent derivative pricing* which is a way of constructing market models that are consistent with initially observed contingent claim quotations. Previous approaches in this direction include *inversion of the yield curve* (cf. [5]) and [2], [1], [42], [41], [29]. Since this paper aims at concepts rather than mathematical generality, we stick to a finite market model.

The paper is organized as follows: The role of martingale methods in portfolio optimization is reviewed in the subsequent section. Sections 3 and 4 deal with neutral and consistent derivative pricing, respectively.

2 Optimal Trading

2.1 General Notation

Our general mathematical framework for a finite frictionless market model is as follows. We work with a filtered probability space $(\Omega, \mathcal{F}, (\mathcal{F}_t)_{t\in\{0,1,\dots,T\}}, P)$, where Ω and the time set $\{0, 1, \ldots, T\}$ are finite. We consider traded securities $0, \ldots, d$ whose prices are expressed in terms of multiples of the *numeraire* security 0. Put differently, securities $0, \ldots, d$ are modelled by their adapted, $\mathbb{R}^{d+1}$-valued discounted *price process* $S = (S^0, \ldots, S^d)$, where $S^0 = 1$. In order to avoid lengthy notation, we let $\mathcal{F} = \mathfrak{P}(\Omega)$, $\mathcal{F}_0 = \{\varnothing, \Omega\}$, and $P(\{\omega\}) > 0$ for any $\omega \in \Omega$. Occasionally, we identify S with the $\mathbb{R}^d$-valued process $(S^1, \ldots, S^d)$.

We consider an investor (hereafter called "you") who disposes of an initial *endowment* $e \in \mathbb{R}$. Your preferences are expressed in terms of a proper, upper semi-continuous, concave *utility function* $u : \mathbb{R} \to [-\infty, \infty)$ which is differentiable on its effective domain (i.e. on $\{x \in \mathbb{R} : u(x) > -\infty\}$; in fact, functions like $x \mapsto \sqrt{x}$, which are not differentiable on the boundary of their effective domain, fit in our framework as well). *Trading strategies* are modelled by $\mathbb{R}^{d+1}$-valued, predictable stochastic processes $\varphi = (\varphi^0, \ldots, \varphi^d)$, where φ^i_t denotes the number of shares of security i in your portfolio at time t.

We write $\Delta X_t := X_t - X_{t-1}$ for any stochastic process X. Moreover, the notation $H \cdot X$ is used to denote the stochastic integral of H relative to X, which is just a sum in our simple setting, i.e. $H \cdot X_t = \sum_{s=1}^t H_s \Delta X_s$. The transposed of a vector x is written as $x^\top$.

A trading strategy φ is called *self-financing* if $\Delta\varphi_t^\top S_{t-1} = 0$ for $t = 1, \ldots, T$. An equivalent definition is $\varphi^\top S = \varphi_0^\top S_0 + \varphi^\top \cdot S$, which make sense in continuous-time as well. For any predictable process $(\varphi^1_t, \ldots, \varphi^d_t)_{t \in \{0,\ldots,T\}}$ and any $x \in \mathbb{R}$, there exists a unique predictable process $(\varphi^0_t)_{t \in \{0,\ldots,T\}}$ such that $\varphi := (\varphi^0, \ldots, \varphi^d)$ is self-financing and $\varphi_0^\top S_0 = x$ (cf. [54], Proposition 1.1.3). If your trading strategy φ is self-financing with $\varphi_0 = 0$, then the value of your portfolio is described by the discounted *wealth process* $V(\varphi)$ defined as $V_t(\varphi) := e + \varphi^\top \cdot S_t$ for $t = 0, \ldots, T$.

Trading constraints are given in terms of subsets of the set of all trading strategies. In this paper, we consider the following cases:

Definition 2.1. Let A denote an affine subspace of $\mathbb{R}^d$ and $K := \{\psi \in \mathbb{R}^d : \alpha_j^\top \psi \leq 0 \text{ for } j = 1, \ldots, p \text{ and } \alpha_j^\top \psi = 0 \text{ for } j = p+1, \ldots, q\}$ a polyhedral cone, where $\alpha_1, \ldots, \alpha_q \in \mathbb{R}^d$. By $\mathfrak{S}(A, K)$ we denote the set of all self-financing trading strategies φ with $\varphi_0 = 0$ that satisfy the following condition: There exists an $a \in A$ such that for any $(\omega, t) \in \Omega \times \{1, \ldots, T\}$ we have $(\varphi^1_t, \ldots, \varphi^d_t)(\omega) - a \in K$.

Convention. In the following, A always denotes an affine subspace and K a polyhedral cone as in the previous definition. Moreover, let $A' := A - a$ for some $a \in A$.

Typical choices for the constraint sets (A,K) are $(\{0\}, \mathbb{R}^d)$ (no constraints), $(\{0\}, (\mathbb{R}_+)^d)$ (no short sales), $(\{0, \ldots, 0, \xi\}, \mathbb{R}^{d-1} \times \{0\})$ (fixed position ξ in security d), $(\{0, \ldots, 0\} \times \mathbb{R}, \mathbb{R}^{d-1} \times \{0\})$ (Security d can only be traded once, namely at time 0, and then has to be held till time T.).

Definition 2.2. A strategy $\varphi \in \mathfrak{S}(A,K)$ is called *(A,K)-arbitrage* if $\varphi^\top \cdot S_T \geq 0$ P-almost surely and $P(\varphi^\top \cdot S_T > 0) > 0$. A $(\{0\},\mathbb{R}^d)$-arbitrage is simply called *arbitrage*. We say that *the market allows no (A,K)-arbitrage* (resp. *arbitrage*) if there is no such strategy.

Definition 2.3.

1. We say that $P^\star : \mathcal{F} \to \mathbb{R}$ is a *signed probability measure* with P-density $\frac{dP^\star}{dP}$ if $\frac{dP^\star}{dP}$ is a real-valued random variable with $E(\frac{dP^\star}{dP}) = 1$ and $P^\star(C) = \int_C \frac{dP^\star}{dP} dP$ for any $C \in \mathcal{F}$. We write $E_{P^\star}(X|\mathcal{F}_t) := E(X\frac{dP^\star}{dP}|\mathcal{F}_t)/E(\frac{dP^\star}{dP}|\mathcal{F}_t)$ for any real-valued random variable X if $E(\frac{dP^\star}{dP}|\mathcal{F}_t) \neq 0$.
2. Let $P^\star$ be a signed probability measure with density $\frac{dP^\star}{dP}$. An adapted process X is called *$P^\star$-(super)martingale* if the process $(E(\frac{dP^\star}{dP}|\mathcal{F}_t)X_t)_{t\in\{0,\dots,T\}}$ is a P-(super)martingale.
3. A signed probability measure $P^\star$ with P-density $\frac{dP^\star}{dP}$ is called *signed (A,K)-seperating measure* if $E(\frac{dP^\star}{dP}(\varphi^\top \cdot S_T)) \leq 0$ for any $\varphi \in \mathfrak{S}(A,K)$.
4. A probability measure $P^\star \sim P$ is called *equivalent (A,K)-seperating measure* if it is a signed (A,K)-seperating measure.
5. Any equivalent (resp. signed) $(\{0\},\mathbb{R}^d)$-seperating measure is simply called *equivalent martingale measure (EMM)* (resp. *signed martingale measure (SMM)*).

Remarks.

1. A probability measure $P^\star \sim P$ is an EMM if and only if S is a $P^\star$-martingale. A similar statement holds for SMM's.
2. The name *seperating measure* is taken from [39]. These measures play an important role in the work of Delbaen & Schachermayer [14], [16] on the fundamental theorem of asset pricing (cf. also Lemma 2.4 below). Note that $P^\star$ is a $(\{0\},K)$-seperating measure if and only if $\varphi^\top \cdot S$ is a $P^\star$-supermartingale for any $\varphi \in \mathfrak{S}(\{0\},K)$.

The following lemma is a simple form of the *fundamental theorem of asset pricing* which has been studied extensively (cf. [33], [34], [53], [76], [11], [64], [40], [62], [14], [15], [16], [68], [38]). In its definitive form in [16], arbitrage has to be replaced with *no free lunch with vanishing risk (NFLVR)* and EMM with *sigma-martingale measure*. For arbitrary closed convex cones, cf. in particular [57], Theorems 4.1 and 4.2.

Lemma 2.4 (Fundamental theorem of asset pricing). *The market does not allow arbitrage (resp. (A',K)-arbitrage) if and only if there exists an EMM (resp. equivalent (A',K)-seperating measure).*

PROOF. The general assertion follows almost literally as Theorem 1.2.7 in [54] for the unconstrained case if we set $\mathcal{V} := \{\varphi^\top \cdot S_T \in \mathbb{R}^\Omega : \varphi \in \mathfrak{S}(A',K)\}$ and show that the following two propositions hold.

1. $\mathcal{V}$ is a closed convex cone.
2. If $C \subset \mathbb{R}^\Omega$ is a closed convex cone and $B \subset \mathbb{R}^\Omega$ a compact convex set with $C \cap B = \varnothing$, then there exists some $\varrho \in \mathbb{R}^\Omega$ such that $\varrho^\top c \leq 0$ for any $c \in C$ and $\varrho^\top b > 0$ for any $b \in B$.

They can be proved as follows.

1. Here and in later proofs, we denote by $F_{t,1}, \dots, F_{t,m_t}$ the partition of Ω that generates $\mathcal{F}_t$ (for $t = 0, \dots, T$). Since a mapping is $\mathcal{F}_t$-measurable if and only if it is constant on the sets $F_{t,i}$, $i = 1, \dots, m_t$, and since φ is uniquely determined by $\varphi^1, \dots, \varphi^d$, we can identify the set of all self-financing strategies φ satisfying $\varphi_0 = 0$ with $\mathbb{R}^n := \mathbb{R}^{m_0 d} \times \dots \times \mathbb{R}^{m_{T-1} d}$. As K and A' are given by linear equality and inequality constraints, it follows that $\mathfrak{S}(A', K)$ corresponds to a polyhedral cone in $\mathbb{R}^n$. Therefore, there exist $a_1, \dots, a_k \in \mathbb{R}^n$ such that $\mathfrak{S}(A', K) = \{\sum_{i=1}^k \beta_i a_i \in \mathbb{R}^n : \beta_1, \dots, \beta_k \geq 0\}$ (cf. [61], Theorem 3.52). On the other hand, the mapping $L : \mathbb{R}^n \to \mathbb{R}^\Omega$: $\varphi \mapsto \varphi^\top \cdot S_T$ is linear. Hence, $\mathcal{V} = L(\mathfrak{S}(A', K)) = \{\sum_{i=1}^k \alpha_i L(a_i) \in \mathbb{R}^\Omega : \alpha_1, \dots, \alpha_k \geq 0\}$ is also a polyhedral cone and in particular closed and convex.
2. By [61], Theorem 2.39, there exist $\varrho \in \mathbb{R}^\Omega$, $r \in \mathbb{R}$ such that $\varrho^\top c < r$ for any $c \in C$ and $\varrho^\top b > r$ for any $b \in B$. Since $\varrho^\top 0 = 0$, it follows that $r \geq 0$. It remains to be shown that $\varrho^\top c \leq 0$ for any $c \in C$. Otherwise, there exists some $c \in C$ with $\varrho^\top c > 0$. This implies $\varrho^\top \widetilde{c} > r$ for a positive multiple $\widetilde{c} \in C$, which yields a contradiction. □

2.2 Utility of terminal wealth

As an investor with given endowment e, you may want to choose your trading strategy in some optimal way. Classical and intuitive suggestions are to maximize the expected utility from terminal wealth or consumption. A further related concept is introduced in Subsection 2.3 below. Due to limited space, we do not discuss the consumption case in this paper (except for Remark 7 below).

Definition 2.5. We say that $\overline{\varphi} \in \mathfrak{S}(A, K)$ is *optimal for terminal wealth under the constraints* (A, K) if it maximizes $E(u(V_T(\varphi))) = E(u(e + \varphi^\top \cdot S_T))$ over all $\varphi \in \mathfrak{S}(A, K)$. For $K = \mathbb{R}^d$ (no constraints) we call $\overline{\varphi}$ simply *optimal for terminal wealth.*

For the rest of this subsection, we make the following weak technical

Assumption. There exists a strategy $\chi \in \mathfrak{S}(A, K)$ such that $V_T(\chi) = e + \chi^\top \cdot S_T$ has only values in the interior of the effective domain of u.

The following result characterizes optimal strategies in terms of EMM's or, more generally, signed seperating measures. In analogy to the fundamental theorem of asset pricing and to underline its importance, we daringly call it a *fundamental theorem of utility maximization*. It is by no means new. Especially the inclusion 1⇒2 can be traced way back to Arrow and Debreu (cf. [17], Chapter 1c and the references therein). Moreover, it plays a key role in many papers that apply martingale or duality methods to utility maximization ([58], [48], [7], [20], [21], [35], [36], [49], [8], [59], [52], [10], [66]). Although easier to prove, the inclusion 2⇒1 is less often explicitly stated. Versions can be found in [20], [49], Theorem 9.3, [28], [41], [44], [30]. The extent to which the equivalence holds in more general settings depends sensitively on the chosen set of trading strategies and probability measures (cf. [65] for an overview).

Lemma 2.6. *Let φ be a trading strategy in $\mathfrak{S}(A,K)$ such that $E(u'(e+\varphi^\top\cdot S_T))>0$. Write φ as $\varphi=a+\psi$ with $a\in A$, $\psi\in\mathfrak{S}(\{0\},K)$. Then we have equivalence between:*

1. *φ is optimal for terminal wealth under the constraints (A,K).*
2. *There exists a signed (A',K)-seperating measure $P^\star$ and a number κ such that*
   (a) $\kappa\frac{dP^\star}{dP}=u'(e+\varphi^\top\cdot S_T)$,
   (b) $\psi^\top\cdot S$ is a $P^\star$-martingale.
   (Of course, we have $\kappa=E(u'(e+\varphi^\top\cdot S_T))$ in this case.)

Proof. 1⇒2: Let $a_0\in A$. Since A is an affine subspace of $\mathbb{R}^d$, there exist $\beta_1,\dots,\beta_r\in\mathbb{R}^d$ such that $A=\{a\in\mathbb{R}^d:\beta_i^\top(a-a_0)=0\text{ for }i=1,\dots,r\}$. As in the proof of Lemma 2.4, we denote by $F_{t,1},\dots,F_{t,m_t}$ the partition of Ω that generates $\mathcal{F}_t$, $t=0,\dots,T$. We use the notation $\psi_t(l):=(\psi_t^1(\omega),\dots,\psi_t^d(\omega))$ for $t=1,\dots,T$, $l=1,\dots,m_{t-1}$, $\omega\in F_{t-1,l}$. Similarly, we set $c(l):=c(\omega)$ for $l=1,\dots,m_T$, $\omega\in F_{T,l}$ and any real-valued random variable c. Therefore, any triple (a,ψ,c) consisting of $a\in\mathbb{R}^d$, a strategy $\psi\in\mathfrak{S}(\{0\},\mathbb{R}^d)$, and a real-valued random variable c can be identified with an element of $\mathbb{R}^n:=(\mathbb{R}^d\times(\mathbb{R}^{m_0d}\times\dots\times\mathbb{R}^{m_{T-1}d})\times\mathbb{R}^{m_T})$ and vice versa, namely with $(a,\psi_1(1),\dots,\psi_T(m_{T-1}),c(1),\dots,c(m_T))$. Using this identification, we can define mappings $f:\mathbb{R}^n\to\mathbb{R}\cup\{\infty\}$, $g_{j,t,l}:\mathbb{R}^n\to\mathbb{R}$ (for $j=1,\dots,q$, $t=1,\dots,T,l=1,\dots,m_{t-1}$), $h_i:\mathbb{R}^n\to\mathbb{R}$ (for $i=1,\dots,r$), and $k_\omega:\mathbb{R}^n\to\mathbb{R}$ (for any $\omega\in\Omega$) by

$$
\begin{aligned}
f(a,\psi,c)&:=E(-u(c)),\\
g_{j,t,l}(a,\psi,c)&:=\alpha_j^\top\psi_t(l),\\
h_i(a,\psi,c)&:=\beta_i^\top(a-a_0),\\
k_\omega(a,\psi,c)&:=c(\omega)-e-(a+\psi)^\top\cdot S_T(\omega).
\end{aligned}
$$

Note that all these are convex mappings on $\mathbb{R}^n$.

With this notion, $a+\psi$ is optimal for terminal wealth under the constraints (A,K) if and only if $(a,\psi,e+(a+\psi)^\top\cdot S_T)$ minimizes f subject to the constraints $g_{j,t,l}\le 0$ (for $j=1,\dots,p$, $t=1,\dots,T$, $l=1,\dots,m_{t-1}$), $g_{j,t,l}=0$ (for $j=p+1,\dots,q$, $t=1,\dots,T$, $l=1,\dots,m_{t-1}$), $h_i=0$ (for $i=1,\dots,r$),

$k_\omega = 0$ (for any $\omega \in \Omega$). From [60], Theorems 28.2 and 28.3 it follows that $a + \psi$ is optimal for terminal wealth under the constraints (A, K) if and only if there exists a real-valued random variable λ and real numbers $\mu_{j,t,l}$ (for $j = 1, \dots, q$, $t = 1, \dots, T$, $l = 1, \dots, m_{t-1}$), ν_i (for $i = 1, \dots, r$) such that we have for $c := e + (a + \psi)^\top \cdot S_T$:

1. $\mu_{j,t,l} \geq 0$ and $\mu_{j,t,l}\alpha_j^\top \psi_t(l) = 0$ for $j = 1, \dots, p$, $t = 1, \dots, T$, $l = 1, \dots, m_{t-1}$,
2. $0 = \nabla f(a, \psi, c) + \sum_{j=1}^q \sum_{t=1}^T \sum_{l=1}^{m_{t-1}} \mu_{j,t,l} \nabla g_{j,t,l}(a, \psi, c)$ $+ \sum_{i=1}^r \nu_i \nabla h_i(a, \psi, c) + \sum_{\omega \in \Omega} \lambda(\omega) \nabla k_\omega(a, \psi, c)$,

where ∇ denotes the gradient of a mapping $\mathbb{R}^n \to \mathbb{R}$. Note that $\lambda, \mu_{j,t,l}, \nu_i$ can be chosen independently of (a, ψ, c). Replacing $\lambda(\omega)$ with $\varrho(\omega) := \lambda(\omega)/P(\{\omega\})$ as well as $\mu_{j,t,l}$ with $\widetilde{\mu}_t^j(\omega) := \mu_{j,t,l}/P(F_{t-1,l})$ (for $\omega \in F_{t-1,l}$) and straightforward calculations yield the following equivalence: $a + \psi$ is optimal for terminal wealth under the constraints (A, K) if and only if there exists a real-valued random variable ϱ, a vector $\nu = (\nu_1, \dots, \nu_r) \in \mathbb{R}^r$, and a predictable $\mathbb{R}^q$-valued process $\widetilde{\mu}$ such that we have for $c := e + (a + \psi)^\top \cdot S_T$:

1. $\widetilde{\mu}_t^j \geq 0$ and $\widetilde{\mu}_t^j \alpha_j^\top (\psi_t^1, \dots, \psi_t^d) = 0$ for $j = 1, \dots, p$, $t = 1, \dots, T$,
2. (a) $0 = -u'(c) + \varrho$,
   (b) $0 = \sum_{j=1}^q \widetilde{\mu}_t^j \alpha_j - E(\varrho \Delta S_t | \mathcal{F}_{t-1})$ for $t = 1, \dots, T$.
   (c) $0 = \sum_{i=1}^r \nu_i \beta_i - E(\varrho(S_T - S_0))$.

Suppose now that $\varphi = a + \psi$ is optimal for terminal wealth under the constraints (A, K). Let $\kappa := E(\varrho) = E(u'(e + \varphi^\top \cdot S_T))$ and $\frac{dP^\star}{dP} := \frac{\varrho}{\kappa}$. Then Statement 2 in Lemma 2.6 holds.

2⇒1: Similarly as in the first part of the proof, one shows that a random variable c minimizes $E(-u(c))$ under the constraint $E(\frac{dP^\star}{dP}(c - e - a^T(S_T - S_0))) \leq 0$ if there exists a non-negative real number κ such that we have

1. $E(\frac{dP^\star}{dP}(c - e - a^\top(S_T - S_0))) \leq 0$,
2. $\kappa E(\frac{dP^\star}{dP}(c - e - a^\top(S_T - S_0))) = 0$,
3. $0 = -u'(c) + \kappa \frac{dP^\star}{dP}$.

It follows that $c := e + \varphi^\top \cdot S_T$ minimizes $E(-u(c))$ under the constraint $E(\frac{dP^\star}{dP}(c - e - a^\top(S_T - S_0))) \leq 0$. But note that $\widetilde{c} := e + \widetilde{\varphi}^\top \cdot S_T = (\widetilde{\varphi} - a)^\top \cdot S_T + e + a^\top(S_T - S_0)$ satisfies the constraint $E(\frac{dP^\star}{dP}(\widetilde{c} - e - a^\top(S_T - S_0))) \leq 0$ for any $\widetilde{\varphi} \in \mathfrak{S}(A, K)$ because $P^\star$ is a $\mathfrak{S}(A', K)$-seperating measure. Therefore $E(u(e + \varphi^\top \cdot S_T)) \geq E(u(e + \widetilde{\varphi}^\top \cdot S_T))$ for any $\widetilde{\varphi} \in \mathfrak{S}(A, K)$, which implies that φ is optimal for terminal wealth under the constraints (A, K). □

Corollary 2.7 (Fundamental theorem of utility maximization). *Let u be strictly increasing on its effective domain. Moreover, let φ be a self-financing trading strategy. Then we have equivalence between:*

1. φ is optimal for terminal wealth.

2. *There exists a number* $\kappa \in (0,\infty)$ *such that* $\frac{dP^\star}{dP} = \frac{1}{\kappa}u'(e+\varphi^\top \cdot S_T)$ *for some equivalent martingale measure* $P^\star$. *(Of course, we have* $\kappa = E(u'(e+\varphi^\top \cdot S_T))$ *in this case.)*

Definition 2.8. We call the measure $P^\star$ in Lemma 2.6 (A,K)*-dual measure for terminal wealth.*

Remarks.

1. If there is more than one optimal solution φ, then $P^\star$ in Lemma 2.6 can be chosen independently of the particular solution (cf. the respective proof). Therefore, the (A,K)-dual measure for terminal wealth is unique. If u is strictly concave, then any two optimal solutions $\varphi, \widetilde{\varphi}$ have the same wealth process $e+\varphi^\top \cdot S$.
2. If u is strictly increasing on its effective domain, then $P^\star$ in Lemma 2.6 is in fact an equivalent (A',K)-seperating measure.
3. Observe that $\dot{\varphi}^0$ does not affect the stochastic integral $\varphi^\top \cdot S$ in Definition 2.5. You only need to consider $(\varphi^1_t,\dots,\varphi^d_t)_{t\in\{1,\dots,T\}}$ without any self-financiability constraint and choose φ^0 such that φ is indeed self-financing.
4. If K is a subspace of $\mathbb{R}^d$, then it suffices to assume that $E(u'(e+\varphi^\top \cdot S_T)) \neq 0$ (instead of >0) in Lemma 2.6. Moreover, Statement 2b follows from the fact that $P^\star$ is a (A',K)-seperating measure.
5. By choosing the constraint sets (A,K) appropriately, Definition 2.5 also allows to address *hedging problems*. Suppose you have a fixed number ξ of shares of security d and you want to hedge the resulting risk by trading in securities $0,\dots,d-1$. A reasonable suggestion is to choose an optimal strategy in $\mathfrak{S}(\{0,\dots,0,\xi\},\mathbb{R}^{d-1}\times\{0\})$. This utility-based approach to hedging has been taken e.g. in [71], [73], [22], [13]. If the utility function is not differentiable, one has to replace the derivative u' in Lemma 2.6 with the subdifferential of u (cf. [41]).
6. If the "right" $P^\star$ is known and if u is strictly concave, then the optimal payoff is obtained as $e+\varphi^\top \cdot S_T = (u')^{-1}(\kappa\frac{dP^\star}{dP})$. This is the case in complete models and it gives rise to the *martingale method* in portfolio optimization (cf. [51] for an introduction). In general, it is not evident how to find an EMM whose density is related in this way to the terminal payoff of some trading strategy. Therefore, the fundamental theorem of utility maximization does not lead to explicit solutions in general. Nevertheless, it is very useful in order to verify that some given candidate is indeed optimal (cf. [28], [44]).
7. In utility maximization problems with consumption one usually tries to maximize $E(\sum_{t=0}^T u_t(c_t))$ over all pairs of self-financing trading strategies φ and consumption plans c satisfying the financiability constraint $\sum_{t=0}^T c_t \leq e+\varphi^\top \cdot S_T$. Here, the utility function may depend on $t \in \{0,\dots,T\}$. The corresponding version of Corollary 2.7 then reads as follows: (φ,c) is optimal if and only if there exists a number $\kappa \in (0,\infty)$ and an equivalent martingale measure $P^\star$ such that $E(\frac{dP^\star}{dP}|\mathcal{F}_t) = \frac{1}{\kappa}u_t'(c_t)$ for $t \in \{0,\dots,T\}$ (cf. [41]).

Lemma 2.9. *Suppose that u is increasing and for any $\varepsilon > 0$ there exist $x_1 > x_2$ with $\frac{u'(x_1)}{u'(x_2)} < \varepsilon$. If there is no (A', K)-arbitrage, then an optimal strategy φ for terminal wealth under the constraints (A, K) exists.*

PROOF. *Step 1:* Similarly as in the proof of Lemma 2.4, we identify any trading strategy $\varphi \in \mathfrak{S}(\{0\}, \mathbb{R}^d)$ with an element of $\mathbb{R}^n := (\mathbb{R}^{m_0 d} \times \cdots \times \mathbb{R}^{m_{T-1} d})$ via $\varphi_t(l) := (\varphi_t^1(\omega), \ldots, \varphi_t^d(\omega))$ for $t = 1, \ldots, T$, $l = 1, \ldots, m_{T-1}$, $\omega \in F_{t-1,l}$. In this sense $\mathfrak{S}(A, K)$ corresponds to a polyhedral subset C of $\mathbb{R}^n$ (cf. [60], Section 19). Define the convex function $f : \mathbb{R}^n \to \mathbb{R}$, $\varphi \mapsto -E(u(e + \varphi^\top \cdot S_T))$. Then φ is optimal for terminal wealth under the constraints (A, K) if it minimizes f on C. By [60], Theorem 27.3 f attains its minimum on C if f is constant on the common directions of recession of f and C.

Step 2: Let φ be a direction of recession of f and C. Then φ corresponds to a strategy in $\mathfrak{S}(A', K)$. If $\varphi^\top \cdot S_T \geq 0$, then $\varphi^\top \cdot S_T = 0$ because there is no (A', K)-arbitrage. In this case f is constant in direction φ. Hence we may assume $P(\varphi^\top \cdot S_T < 0) > 0$. Note that u must be finite on $\mathbb{R}$: Otherwise $f(\lambda\varphi) = -E(u(e + \lambda\varphi^\top \cdot S_T)) = \infty$ for $\lambda \in \mathbb{R}_+$ large enough, which implies that φ cannot be a direction of recession of f.

Step 3: W.l.o.g. $e = 0$, $u(0) = 0$. Let $\varepsilon > 0$ and choose $x_2 < 0 < x_1$ such that $\frac{u'(x_1)}{u'(x_2)} < \varepsilon$. Note that $\frac{u(\lambda x)}{\lambda} \leq u'(x_1)x + O(\frac{1}{\lambda})$ for $x > 0$ and $\lambda \to \infty$. Similarly, $\frac{u(\lambda x)}{\lambda} \leq u'(x_2)x + O(\frac{1}{\lambda})$ for $x < 0$ and $\lambda \to \infty$. Therefore, $\limsup_{\lambda\to\infty} \frac{1}{\lambda} E(u(\lambda(\varphi^\top \cdot S_T)^+)) \leq \varepsilon u'(x_2) E((\varphi^\top \cdot S_T)^+)$ and $\limsup_{\lambda\to\infty} \frac{1}{\lambda} E(u(-\lambda(\varphi^\top \cdot S_T)^-)) \leq u'(x_2) E(-(\varphi^\top \cdot S_T)^-)$. For small ε we have $\varepsilon E((\varphi^\top \cdot S_T)^+) + E(-(\varphi^\top \cdot S_T)^-) < 0$, which implies $-f(\lambda\varphi) = E(u(\lambda\varphi^\top \cdot S_T)) \to -\infty$ for $\lambda \to \infty$. Therefore φ cannot be a direction of recession of f. □

For more general versions of this lemma cf. [65].

2.3 Local Utility

Especially in hedging applications it may be very hard to compute optimal portfolios explicitly because one has to deal with constraints resp. random endowment. A way out is to work instead with the simpler concept of local utility maximization. The notion of *local utility* has been introduced in [42], [43], but the concept of step-by-step optimization goes at least back to [23] (cf. Remark 4 at the end of this section). If you are a local utility maximizer, you try to do your best on a, say, day-by-day basis. You do not care about past profits and losses or your exact current wealth. Mathematically, this means that optimal portfolios can be computed seperately for any one-period step. This makes it much easier to obtain explicit solutions than in the terminal wealth case. The concept is related to maximization of consumption, but here financial gains are consumed immediately (cf. the remark following Definition 2.10). For a continuous-time extension of this approach cf. [43]. In this subsection, we assume throughout that $A = \{a\}$ for some $a \in \mathbb{R}^d$.

Definition 2.10. We say that a strategy $\overline{\varphi} \in \mathfrak{S}(\{a\},K)$ is *locally optimal under the constraints* $(\{a\},K)$ if it maximizes $E(\sum_{t=1}^T u(\Delta V_t(\varphi))) = E(\sum_{t=1}^T u(\varphi_t^\top \Delta S_t))$ over all $\varphi \in \mathfrak{S}(\{a\},K)$. For $K = \mathbb{R}^d$ (no constraints) we simply call $\overline{\varphi}$ *locally optimal.*

Remark. Observe that Definition 2.10 corresponds to an optimal investment with consumption problem for $e = 0$, $c_0 = 0$ (cf. Remark 7 above) if we require in addition that $\sum_{s=1}^t c_s = \varphi^\top \cdot S_t$ for *any* $t \in \{1,\dots,T\}$ (not only for $t = T$).

As in the previous subsection, we make a weak technical

Assumption. There exists a strategy $\chi \in \mathfrak{S}(\{a\},K)$ such that $\chi_t^\top \Delta S_t$, $t = 1,\dots,T$ has only values in the interior of the effective domain of u.

In order to state the analogue of Lemma 2.6, we need the following

Definition 2.11. We say that a signed probability measure $P^\star$ with P-density $\frac{dP^\star}{dP}$ has *logarithm process* N if N is an adapted process with $N_0 = 0$ such that $E(\frac{dP^\star}{dP}|\mathcal{F}_t) = \prod_{s=1}^t (1 + \Delta N_s)$ for $t = 1,\dots,T$.

Remarks.

1. The name is motivated by the fact that $\frac{dP^\star}{dP} = \mathcal{E}(N)_T$, where $\mathcal{E}(N) = \prod_{t=1}^{\cdot}(1 + \Delta N_t)$ denotes the stochastic exponential of N. For details on stochastic logarithms in continuous-time cf. [46].
2. If $E(\frac{dP^\star}{dP}|\mathcal{F}_t) \neq 0$ P-almost surely for $t = 0,\dots,T$, then the logarithm process of $P^\star$ is unique and $E(1 + \Delta N_t|\mathcal{F}_{t-1}) = 1$ for $t = 1,\dots,T$.

Theorem 2.12. *Let φ be a trading strategy in $\mathfrak{S}(\{a\},K)$ such that $u'(\varphi_t^\top \Delta S_t) > 0$ P-almost surely for $t = 1,\dots,T$. Then we have equivalence between*

1. *φ is locally optimal under the constraints $(\{a\},K)$.*
2. *There exists a signed $(\{0\},K)$-seperating measure $P^\star$ with logarithm process N and a predictable process κ such that*
   (a) $\kappa_t(1 + \Delta N_t) = u'(\varphi_t^\top \Delta S_t)$ for $t = 1,\dots,T$,
   (b) $(\varphi - a)^\top \cdot S$ is a $P^\star$-martingale.
   (Obviously, we have $\kappa_t = E(u'(\varphi_t^\top \Delta S_t)|\mathcal{F}_{t-1})$ in this case.)

PROOF. 1⇒2: As in the proof of Lemma 2.6, we denote by $F_{t,1},\dots,F_{t,m_t}$ the partition of Ω that generates $\mathcal{F}_t$, $t = 0,\dots,T$ and we use the notation $\varphi_t(l) := (\varphi_t^1(\omega),\dots,\varphi_t^d(\omega))$ for $t = 1,\dots,T$, $l = 1,\dots,m_{t-1}$, $\omega \in F_{t-1,l}$. Similarly, we write $c_t(l) := c_t(\omega)$ for $t = 1,\dots,T$, $l = 1,\dots,m_t$, $\omega \in F_{t,l}$, and any adapted real-valued process c. Therefore, any pair (φ,c) consisting of a strategy $\varphi \in \mathfrak{S}(\{a\},\mathbb{R}^d)$ and an adapted real-valued process c can be identified with an element of $\mathbb{R}^n := ((\mathbb{R}^{m_0 d} \times \cdots \times \mathbb{R}^{m_{T-1} d}) \times (\mathbb{R}^{m_1} \times \cdots \times \mathbb{R}^{m_T}))$ and vice versa, namely with $(\varphi_1(1),\dots,\varphi_T(m_{T-1}),c_1(1),\dots,c_T(m_T))$. Using this identification, we can define mappings $f : \mathbb{R}^n \to \mathbb{R} \cup \{\infty\}$, $g_{j,t,l} : \mathbb{R}^n \to \mathbb{R}$ (for $j = 1, \dots, q$, $t = 1, \dots, T$, $l = 1, \dots, m_{t-1}$), $k_{t,l} : \mathbb{R}^n \to \mathbb{R}$ (for $t = 1, \dots, T$, $l = 1, \dots, m_t$) by

$$\begin{aligned} f(\varphi, c) &:= E\Big(- \sum_{t=1}^{T} u(c_t) \Big), \\ g_{j,t,l}(\varphi, c) &:= \alpha_j^\top (\varphi_t(l) - a), \\ k_{t,l}(\varphi, c) &:= c_t(l) - \varphi_t(\omega)^\top \Delta S_t(\omega) \end{aligned}$$

for an arbitrary $\omega \in F_{t,l}$. Note that all these are convex mappings on $\mathbb{R}^n$.

With this notion, φ is locally optimal under the constraints $(\{a\}, K)$ if and only if $(\varphi, (\varphi_t^\top \Delta S_t)_{t \in \{1,\dots,T\}})$ minimizes f subject to the constraints $g_{j,t,l} \leq 0$ (for $j = 1, \dots, p$, $t = 1, \dots, T$, $l = 1, \dots, m_{t-1}$), $g_{j,t,l} = 0$ (for $j = p+1, \dots, q$, $t = 1, \dots, T$, $l = 1, \dots, m_{t-1}$), $k_{t,l} = 0$ (for $t = 1, \dots, T$, $l = 1, \dots, m_t$). From [60], Theorems 28.2 and 28.3 it follows that $\varphi \in \mathfrak{S}(\{a\}, K)$ is locally optimal under the constraints $(\{a\}, K)$ if and only if there exist real numbers $\lambda_{t,l}$ (for $t = 1, \dots, T$, $l = 1, \dots, m_t$) and real numbers $\mu_{j,t,l}$ (for $j = 1, \dots, q$, $t = 1, \dots, T$, $l = 1, \dots, m_{t-1}$) such that we have for $c := \varphi^\top \Delta S$:

1. $\mu_{j,t,l} \geq 0$ and $\mu_{j,t,l} \alpha_j^\top \varphi_t(l) = 0$ for $j = 1, \dots, p$, $t = 1, \dots, T$, $l = 1, \dots, m_{t-1}$,
2. $0 = \nabla f(\varphi, c) + \sum_{j=1}^{q} \sum_{t=1}^{T} \sum_{l=1}^{m_{t-1}} \mu_{j,t,l} \nabla g_{j,t,l}(\varphi, c)$
$+ \sum_{t=1}^{T} \sum_{l=1}^{m_t} \lambda_{t,l} \nabla k_{t,l}(\varphi, c)$,

where ∇ denotes the gradient of a mapping $\mathbb{R}^n \to \mathbb{R}$. Note that $\lambda_{t,l}, \mu_{j,t,l}$ can be chosen independently of (φ, c). Replacing $\lambda_{t,l}$ with $\varrho_t(\omega) := \lambda_{t,l}/P(F_{t,l})$ (for $\omega \in F_{t,l}$) as well as $\mu_{j,t,l}$ with $\widetilde{\mu}_t^j(\omega) := \mu_{j,t,l}/P(F_{t-1,l})$ (for $\omega \in F_{t-1,l}$) and straightforward calculations yield the following equivalence: φ is locally optimal for terminal wealth under the constraints $(\{a\}, K)$ if and only if there exists an adapted real-valued process ϱ and a predictable $\mathbb{R}^q$-valued process $\widetilde{\mu}$ such that we have for $c := \varphi^\top \Delta S$:

1. $\widetilde{\mu}_t^j \geq 0$ and $\widetilde{\mu}_t^j \alpha_j^\top (\varphi_t^1, \dots, \varphi_t^d) = 0$ for $j = 1, \dots, p$, $t = 1, \dots, T$,
2. (a) $0 = -u'(c_t) + \varrho_t$ for $t = 1, \dots, T$,
   (b) $0 = \sum_{j=1}^{q} \widetilde{\mu}_t^j \alpha_j - E(\varrho_t \Delta S_t | \mathcal{F}_{t-1})$ for $t = 1, \dots, T$.

Suppose now that φ is locally optimal under the constraints $(\{a\}, K)$. Define $\kappa_t := E(\varrho_t | \mathcal{F}_{t-1}) = E(u'(\varphi_t^\top \Delta S_t) | \mathcal{F}_{t-1})$ and $\Delta N_t := \frac{\varrho_t}{\kappa_t} - 1$ for $t = 1, \dots, T$. Finally, set $N := \sum_{t=1}^{\cdot} \Delta N_t$ and $\frac{dP^\star}{dP} := \mathcal{E}(N)_T$. Then Statement 2 in Theorem 2.12 holds.

2⇒1: Similarly as in the first part of the proof, one shows that an adapted process c minimizes $E(-\sum_{t=1}^{T} u(c_t))$ under the constraints $E((1 + \Delta N_t)(c_t - a^\top \Delta S_t) | \mathcal{F}_{t-1}) \leq 0$ (for $t = 1, \dots, T$) if there exists a non-negative predictable process κ such that we have

1. $E((1 + \Delta N_t)(c_t - a^\top \Delta S_t) | \mathcal{F}_{t-1}) \leq 0$,

2. $\kappa_t E((1+\Delta N_t)(c_t - a^\top \Delta S_t)|\mathcal{F}_{t-1}) = 0$,
3. $0 = -u'(c_t) + \kappa_t(1+\Delta N_t)$

for $t = 1,\dots,T$. It follows that $c := \varphi^\top \Delta S$ minimizes $E(-\sum_{t=1}^T u(c_t))$ under the constraints $E((1+\Delta N_t)(c_t - a^\top \Delta S_t)|\mathcal{F}_{t-1}) \le 0$ for $t = 1,\dots,T$. But note that $\widetilde{c} := \widetilde{\varphi}^\top \Delta S$ satisfies the constraints $E((1+\Delta N_t)(\widetilde{c}_t - a^\top \Delta S_t)|\mathcal{F}_{t-1}) \le 0$, $t = 1,\dots,T$ for any $\widetilde{\varphi} \in \mathfrak{S}(\{a\},K)$ because $P^\star$ is a $\mathfrak{S}(\{0\},K)$-seperating measure. Therefore $E(\sum_{t=1}^T u(\varphi_t^\top \Delta S_t)) \ge E(\sum_{t=1}^T u(\widetilde{\varphi}_t^\top \Delta S_t))$ for any $\widetilde{\varphi} \in \mathfrak{S}(\{a\},K)$, which implies that φ is locally optimal under the constraints $(\{a\},K)$. □

Definition 2.13. We call the measure $P^\star$ in the previous theorem $(\{a\},K)$-*dual measure for local utility.*

Remarks.

1. If K is a subspace of $\mathbb{R}^d$, then it suffices to assume that $u'(\varphi_t^\top \Delta S_t) \neq 0$ and $E(u'(\varphi_t^\top \Delta S_t)|\mathcal{F}_{t-1}) \neq 0$ P-almost surely for $t = 1,\dots,T$ (instead of $u'(\varphi_t^\top \Delta S_t) > 0$). Moreover, Statement 2b follows from the fact that $P^\star$ is a $(\{0\},K)$-seperating measure.
2. Remarks 1–3 from Subsection 2.2 hold accordingly.
3. In the local utility optimization problem, φ can be chosen independently for $t = 1,\dots,T$. Therefore, local utility maximization practically means maximization of $E(u(\varphi_t^\top \Delta S_t))$ over all $\mathcal{F}_{t-1}$-measurable, $(a+K)$-valued random vectors φ_t seperately for any $t \in \{1,\dots,T\}$, i.e. maximization of the utility of discounted one-period gains (cf. also Lemma 2.15 below).
4. The approach of [23] to option hedging in incomplete markets is based on one-period hedging error minimization. More specifically, they propose the trading strategy ψ appearing in Definition 3.9 below (in the case $n = 1$) as an optimal hedge. A comparison with Definition 2.10 and the previous remark shows that this ψ is locally optimal for the utility function $u(x) := -x^2$, constraint set $\mathfrak{S}(\{0,\dots,0,1\},\mathbb{R}^m \times \{0\})$ and S^{m+1} as security $m+1$.
5. As in the terminal wealth case, hedging problems can be addressed by considering appropriate constraint sets (cf. [43]).

Lemma 2.14. *Suppose that u is increasing and for any $\varepsilon > 0$ there exist $x_1 > x_2$ with $\frac{u'(x_1)}{u'(x_2)} < \varepsilon$. If there is no $(\{0\},K)$-arbitrage, then a locally optimal strategy φ under the constraints $(\{a\},K)$ exists.*

PROOF. Note that φ maximizes $\varphi \mapsto E(\sum_{t=1}^T u(\varphi_t^\top \Delta S_t))$ if and only if φ_t maximizes $\varphi_t \mapsto E(u(\varphi_t^\top \Delta S_t)) =: -f_t(\varphi_t)$ seperately for $t = 1,\dots,T$. The statement follows along the same lines as the proof of Lemma 2.9. □

Lemma 2.15. *Suppose that $K = \mathbb{R}^k \times \{0\} \subset \mathbb{R}^d$ for some $k \in \{0,1,\dots,d\}$. Then $\varphi \in \mathfrak{S}(\{a\},K)$ is locally optimal under the constraints $(\{a\},K)$ if and only if $E(u'(\varphi_t^\top \Delta S_t)\, \Delta S_t^i|\mathcal{F}_{t-1}) = 0$ for $i = 1,\dots,k$ and $t = 1,\dots,T$.*

PROOF. Similarly as in the proof of Lemma 2.9, we identify any trading strategy $\varphi \in \mathfrak{S}(\{0\}, \mathbb{R}^d)$ with an element of $\mathbb{R}^n := (\mathbb{R}^{m_0 d} \times \cdots \times \mathbb{R}^{m_{T-1} d})$ via $\varphi_t(l) := (\varphi_t^1(\omega), \ldots, \varphi_t^d(\omega))$ for $t = 1, \ldots, T$, $l = 1, \ldots, m_{T-1}$, $\omega \in F_{t-1,l}$. Define $f_{t,l}(\varphi_t(l)) := -E(u(\varphi_t^\top \Delta S_t) | \mathcal{F}_{t-1})(\omega)$ for $t = 1, \ldots, T$, $l = 1, \ldots, m_{t-1}$, $\omega \in F_{t-1,l}$. Then $\varphi \in \mathfrak{S}(\{a\}, K)$ is locally optimal under the constraints $(\{a\}, K)$ if and only if $\varphi_t(l)$ minimizes $f_{t,l} : \mathbb{R}^d \to \mathbb{R}$ subject to the constraint $\varphi_t(l) \in \{a\} + K$ seperately for $t = 1, \ldots, T$. This is the case if and only if $0 = \nabla^i f_{t,l}(\varphi_t(l)) = -E(u'(\varphi_t^\top \Delta S_t) \Delta S_t^i | \mathcal{F}_{t-1})(\omega)$ for $i = 1, \ldots, k$ and $t = 1, \ldots, T$, $l = 1, \ldots, m_{t-1}$, where $\omega \in F_{t-1,l}$. □

3 Neutral Derivative Pricing

In this section we turn to derivative pricing. More exactly, we propose a way to extend a market model for the underlyings to a model for both underlyings and derivatives. Later, we will see that other valuation approaches as e.g. distance minimization and L^2-approximation pricing often lead to the same prices.

Derivative pricing in complete models is well understood. Since contingent claims are redundant securities, it suffices to assume absence of arbitrage in order to obtain unique prices. The situation is less obvious in incomplete markets. Even in many models of practical importance arbitrage arguments yield only trivial bounds on contingent claim prices (cf. [18], [25], [9]). Unique values can only be derived under stronger assumptions.

Sometimes it is suggested to apply some kind of equilibrium asset pricing based on production and the preference structure of agents in the economy. As an example let us consider a market consisting of a number of identical investors (or a representative agent as a stand-in) who all maximize their expected utility of terminal wealth in the sense of Definition 2.5. Moreover, assume that they all dispose of a random aggregate endowment e_T up to time T. From the market clearing condition and from standard variation arguments it follows that the discounted price process of any traded security in this economy must be a martingale with respect to some probability measure $P^\star$ whose density $\frac{dP^\star}{dP}$ is a multiple of $u'(e_T)$ where u denotes the utility function of the investors. Hence we can compute prices for any asset whose terminal payoff at time T is given. Of course, this reasoning can and has been modified, generalized, and refined in many ways. We only want to stress two important features: On the one hand, not only the preferences but also the random endowment of the representative agent has to be given exogenously as input. On the other hand, this is not a genuine derivative pricing approach. By contrast, we obtain prices for *any* security with given payoff at time T.

Note that we run into a consistency problem if we want to apply this concept naively to derivative pricing. Since the underlying price processes that are given in the first place are not incorporated in this approach, they may turn out not to be $P^\star$-martingales. Put differently, we do not recover the right price even for trivial contingent claims whose payoff at maturity equals the underlying. A way out is to choose the exogenous endowment e_T such that $P^\star$ is in fact a martingale measure for the given underlyings. In view of the fundamental theorem of asset pricing we obtain a price system that is consistent with absence of arbitrage. Note that we certainly have to consider random endowment in this case. Otherwise $P^\star = P$ which is generally no EMM.

A problem with this approach is its arbitrariness. Since the endowment is not observable, one may be tempted to just guess some e_T such that $\frac{u'(e_T)}{E(u'(e_T))}$ is the density of an EMM. However, by letting $e_T := (u')^{-1}(\frac{dP^\star}{dP})$ one can obtain *any* equivalent martingale measure $P^\star$ in this fashion. This means that any derivative price in the no-arbitrage interval can be recovered by choosing e_T appropriately. Therefore the use of this approach to derivative pricing is limited unless one has good reason to assume a specific kind of endowment.

We want to suggest an alternative that mediates between derivative pricing in complete models and general equilibrium asset pricing. The idea is to apply the equilibrium arguments only to the secondary market. We take the underlying price processes as given ignoring the market mechanisms they result from. On the other hand, we assume that only some identical investors trade in derivatives as well. (This assumption will be relaxed in the next section.) Suppose that these investors are expected utility maximizers as in Definition 2.5 with fixed non-random endowment $e \in \mathbb{R}$. Since any derivative that is bought has to be sold and since the derivative traders behave identically, we end up with the following market clearing condition:

> Derivative prices should be such that the optimal portfolio of the representative investor contains no contingent claim.

Such prices are termed *neutral* in Definition 3.1 below. Intuitively, they are stable in the sense that they do not lead to unmatched supply or demand of derivatives. Note that we no longer need to consider a random endowment e_T to reproduce underlying prices because the clearing condition applies only to contingent claims. Of course, neutral price processes still depend on the preferences and the initial endowment of the derivative traders. However, the degree of arbitrariness is greatly reduced.

To an economist, this valuation principle may sound quite natural or even familiar. Nevertheless, we can produce almost no reference where such an approach has been taken. In fact, the first to suggest this kind of valuation seems to be [12] (cf. also [47]), but the corresponding price system has been considered before by [35], [36] and [49] in the context of portfolio optimization. Davis' suggestion for a reasonable derivative price is such that among all strategies that buy an infinitesimal number of contingent claims and hold it till maturity, a portfolio containing no derivative is optimal. Although he does not claim that this remains true if we compare among all portfolios trading *arbitrarily* with the contingent claim, this follows in an Itô-process setting from duality results by [36] and [49]. The first to notice the duality between portfolio optimization and *least favourable market completion* seem to be [35], [36].

The general setting is as in the previous section. We distinguish two kinds of securities: *underlyings* $0, \dots, m$ and *derivatives* $m+1, \dots, m+n$. The underlyings are given in terms of their discounted price process $S = (S^0, \dots, S^m)$. At this stage, the only information on the derivatives is their discounted terminal payoffs $R^{m+1}, \dots, R^{m+n}$ at time T, which are supposed to be $\mathcal{F}_T$-measurable random variables. We call adapted processes $S^{m+1}, \dots, S^{m+n}$ *derivative price processes* if $S_T^{m+i} = R^{m+i}$ for $i = 1, \dots, n$.

3.1 Terminal Wealth

Fix a utility function u and an initial endowment $e \in \mathbb{R}$ such that e belongs to the interior of the effective domain of u. For the following, we assume that there exists an optimal strategy for terminal wealth φ in the underlyings' market $S^0, \dots, S^m$ such that $E(u'(e + \varphi^\top \cdot S_T)|\mathcal{F}_t) \neq 0$ P-almost surely for $t = 0, \dots, T$ (cf. Lemma 2.9). For reasons that will become clear below, we call the unique $(\{0\}, \mathbb{R}^m)$-dual measure for terminal wealth $P^\star$ *neutral pricing measure for terminal wealth.* Recall that $P^\star$ is an EMM if u is strictly increasing on its effective domain.

Definition 3.1. We call derivative price processes $S^{m+1}, \dots, S^{m+n}$ *neutral derivative price processes for terminal wealth* if there exists a strategy $\overline{\varphi}$ in the extended market $S^0, \dots, S^{m+n}$ which is optimal for terminal wealth (without constraints, i.e. for $(\{0\}, \mathbb{R}^{m+n})$) and satisfies $\overline{\varphi}^{m+1} = \dots = \overline{\varphi}^{m+n} = 0$.

A slightly more general definition that avoids claiming the existence of an optimal strategy could be based on the property that the maximal expected utility in the market $S^0, \dots, S^{m+n}$ is not higher than in the market $S^0, \dots, S^m$. However, for simplicity we stick to the above version.

Theorem 3.2. *There exist unique neutral derivative price processes. They are given by* $S_t^{m+i} = E_{P^\star}(R^{m+i}|\mathcal{F}_t)$ *for* $t = 0, \dots, T$, $i = 1, \dots, n$.

PROOF. Let $S_t^{m+i} = E_{P^\star}(R^{m+i}|\mathcal{F}_t)$ for $t = 0, \dots, T$, $i = 1, \dots, n$ and define a trading strategy $\overline{\varphi} \in \mathfrak{S}(\{0\}, \mathbb{R}^{m+n})$ in the extended market by $\overline{\varphi} := (\varphi, 0)$. Lemma 2.6 yields that $\overline{\varphi}$ is optimal for terminal wealth, which in turn implies that $S^{m+1}, \dots, S^{m+n}$ are neutral derivative price processes.

Conversely, let $S^{m+1}, \dots, S^{m+n}$ be neutral derivative price processes and $\overline{\varphi} \in \mathfrak{S}(\{0\}, \mathbb{R}^{m+n})$ a corresponding optimal strategy for terminal wealth. Obviously, $\psi := (\overline{\varphi}^0, \dots, \overline{\varphi}^m) \in \mathfrak{S}(\{0\}, \mathbb{R}^m)$ is an optimal strategy for terminal wealth in the market $S^0, \dots, S^m$. Since φ is an optimal strategy in this market as well, we have $u'(\varphi^\top \cdot S_T) = u'(\psi^\top \cdot S_T)$ and hence $u'((\varphi, 0)^\top \cdot (S^0, \dots, S^{m+n})) = u'(\overline{\varphi}^\top \cdot (S^0, \dots, S^{m+n}))$. In particular, $(\varphi, 0) \in \mathfrak{S}(\{0\}, \mathbb{R}^{m+n})$ is an optimal strategy in the market $S^0, \dots, S^{m+n}$. From

Lemma 2.6 it follows that $S^{m+1}, \dots, S^{m+n}$ are $P^\star$-martingales, which yields the uniqueness. □

Note that $P^\star$ is an EMM for the extended market $S^0, \dots, S^{m+n}$ if u is strictly increasing on its effective domain. In particular, the corresponding price system allows no arbitrage.

If an EMM Q is interpreted as a pricing rule, then it induces a market completion: In the completed market any $\mathcal{F}_T$-measurable claim is traded at time 0 at a price $E_Q(X)$. Your optimal expected utility in this completed market obviously equals

$$\begin{aligned} &\sup\{E(u(X)) : X \ \ \mathcal{F}_T\text{-measurable random variable with } E_Q(X) = e\} \\ &= \sup\{E(u(e + X - E_Q(X))) : X \ \ \mathcal{F}_T\text{-measurable random variable}\}. \quad (3.1) \end{aligned}$$

Investing in the underlyings does not increase your utility because you can buy the terminal payoff $\varphi^\top \cdot S_T$ of any strategy φ as a contingent claim at time 0. The *minimax martingale measure* $P^\star$ is defined as the measure which minimizes the expression (3.1) (cf. [36]). It has been studied in the context of derivative pricing by [3]. It corresponds to the market completion that is *least favourable* to you as an investor. The following result shows that this measure leads to neutral derivative prices.

Lemma 3.3 (Least favourable market completion). *If u is strictly increasing on its effective domain, then the neutral pricing measure $P^\star$ minimizes*

$$\sup\{E(u(e + X - E_Q(X))) : X \ \ \mathcal{F}_T\text{-measurable random variable}\}$$

over all equivalent martingale measures Q.

PROOF. Note that $\sup\{E(u(e + X - E_Q(X))) : X \ \mathcal{F}_T\text{-measurable random variable}\} = \sup\{E(u(c)) : c \ \mathcal{F}_T\text{-measurable random variable with } E_Q(c-e) = 0\}$. Similarly as in the second part of the proof of Lemma 2.6 it is shown that for $Q = P^\star$ the supremum is attained in $c = e + \varphi^\top \cdot S_T$, where φ denotes the optimal strategy for terminal wealth. Since $E_Q(\varphi^\top \cdot S_T) = 0$ for any EMM Q, the claim follows. □

Some authors suggest to choose an EMM $P^\star$ as a pricing rule that is closest in some sense to the original probability measure P. In particular, the relative entropy and the Hellinger distance or more generally L^p-distances have been considered (cf. [50], [55], [56], [6], [31], [32], [27], [30]). One may critisize that distance minimization is a purely mathematically motivated criterion which is hard to interpret economically. However, [3] observed that for some standard "distances", the minimizing measures are in fact minimax measures for HARA-type utility functions (cf. also [30] in this context). In view of the previous lemma, we conclude that they lead to neutral price processes as well.

Lemma 3.4 (Distance minimization). *1. If $u(x) = -a\exp(-bx) + c$ for some $a, b > 0$, $c \in \mathbb{R}$, and any $x \in \mathbb{R}$, then the neutral pricing measure minimizes $E_Q(\log \frac{dQ}{dP}))$ over all equivalent martingale measures Q, i.e. the entropy of Q relative to P.*

2. If $u(x) = a\log(x+b) + c$ for some $a > 0$, $b, c \in \mathbb{R}$ and any $x \in (-b, \infty)$, then the neutral pricing measure minimizes $-E(\log \frac{dQ}{dP})) = E(\log \frac{dP}{dQ}))$ over all equivalent martingale measures Q, i.e. the entropy of P relative to Q.

3. If $u(x) = -a(x+b)^p + c$ for some $p \in (-\infty, 0)$, $a > 0$, $b, c \in \mathbb{R}$ and any $x \in (-b, \infty)$, then the neutral pricing measure minimizes $-E((\frac{dQ}{dP})^{\frac{p}{p-1}})$ over all equivalent martingale measures Q.

4. If $u(x) = a(x+b)^p + c$ for some $p \in (0,1)$, $a > 0$, $b, c \in \mathbb{R}$ and any $x \in [-b, \infty)$, then the neutral pricing measure minimizes $E((\frac{dQ}{dP})^{\frac{p}{p-1}})$ over all equivalent martingale measures Q.

5. If $u(x) = -a|x+b|^p + c$ for some $p \in (1, \infty)$, $a > 0$, $b, c \in \mathbb{R}$ and any $x \in \mathbb{R}$, then the neutral pricing measure minimizes $E(|\frac{dQ}{dP}|^{\frac{p}{p-1}})$ over all signed martingale measures Q.

PROOF. *1.* W.l.o.g. $a = 1$, $c = 0$. Set $f(x) := -u(e+x)$ and let Q be an EMM. Note that $\sup\{E(u(e+X-E_Q(X))) : X \text{ random variable}\} = -\inf\{E(f(X)) : X \text{ random variable with } E_Q(X) \leq 0\}$. By [60], Theorem 28.2, this equals

$$
\begin{aligned}
& - \sup_{\kappa\in(0,\infty)} \inf\{E(f(X)) + \kappa E_Q(X) : X \text{ random variable}\} \\
&= \inf_{\kappa\in(0,\infty)} E\Big(- \inf_{x\in\mathbb{R}} \Big(f(x) + \kappa\frac{dQ}{dP}x\Big)\Big) \\
&= \inf_{\kappa\in(0,\infty)} E\Big(f^*\Big(- \kappa\frac{dQ}{dP}\Big)\Big),
\end{aligned}
$$

where f^* denotes the convex conjugate of f (cf. [60], Section 12). For $f(x) = \exp(e+x)$ we have $f^*(x) = |\frac{x}{b}|(\log|\frac{x}{b}| - 1 + eb)$ for $x < 0$. Therefore $\inf_{\kappa\in(0,\infty)} E(f^*(-\kappa \frac{dQ}{dP})) = \inf_{\kappa\in(0,\infty)} \kappa(E_Q(\log(\frac{dQ}{dP}) + \log(\kappa) - 1 + eb)$. Note that this expression gets minimal in Q if and only if $E_Q(\log(\frac{dQ}{dP}))$ gets minimal. In view of Lemma 3.3, the claim follows.

Statements 2–5 are shown similarly. For Statement 5 observe that Lemma 3.3 holds also for non-increasing u if *EMM* is replaced with *SMM*. □

Mean-variance hedging tries to approximate contingent claims as close as possible in an L^2-sense by the payoff of some portfolio with initial investment $y \in \mathbb{R}$ and trading strategy $\psi \in \mathfrak{S}(\{0\}, \mathbb{R}^m)$. Mathematically speaking, the claim is projected onto the space of replicable payoffs. This problem has been studied extensively, cf. [75] for an overview. The optimal initial investment $y \in \mathbb{R}$ has been suggested as a pricing rule e.g. by [67], [74].

Definition 3.5. (Global L^2-approximation pricing) We call $S_0^{m+1}, \ldots, S_0^{m+n} \in \mathbb{R}$ *global L^2-approximation derivative prices* if, for $i = 1, \ldots, n$, there exists a strategy $\psi \in \mathfrak{S}(\{0\}, \mathbb{R}^m)$ such that

$$E\Big((R^{m+i} - S_0^{m+i} - \psi^\top \cdot S_T)^2\Big) \leq E\Big((R^{m+i} - \widetilde{y} - \widetilde{\psi}^\top \cdot S_T)^2\Big)$$

for any $\widetilde{y} \in \mathbb{R}$ and any strategy $\widetilde{\psi} \in \mathfrak{S}(\{0\}, \mathbb{R}^m)$.

It is well-known that the global L^2-approximation price can be obtained as expectation of the payoff under the variance-optimal signed martingale measure, which in turn corresponds to the case $p = 2$ in Statement 5 of Lemma 3.4. Therefore, it can also be interpreted as a neutral price. However, since neutral prices may lead to arbitrage in this case, one should use this valuation rule with care.

Lemma 3.6. *If $u(x) = -a(x+b)^2+c$ for some $a > 0$, $b, c \in \mathbb{R}$, then the global L^2-approximation prices are the initial values of the neutral price processes for terminal wealth.*

PROOF. Fix $i \in \{1, \ldots, n\}$. Observe that $S_0^{m+i} + \psi^\top \cdot S_T$ in the previous definition is the L^2-projection of R^{m+i} on the vector space $\{y + \varphi^\top \cdot S_T : y \in \mathbb{R}, \varphi \in \mathfrak{S}(\{0\}, \mathbb{R}^m)\}$. Hence, we have $E((R^{m+i} - S_0^{m+i} - \psi^\top \cdot S_T)(y + \varphi^\top \cdot S_T)) = 0$ for any $y \in \mathbb{R}$, $\varphi = \mathfrak{S}(\{0\}, \mathbb{R}^m)$. Since $u'(x) = -2ax - 2ab$, we have in particular $E((R^{m+i} - S_0^{m+i} - \psi^\top \cdot S_T)\frac{dP^\star}{dP}) = 0$ for the neutral pricing measure for terminal wealth (cf. Lemma 2.6). This in turn implies that $S_0^{m+i} = E_{P^\star}(R^{m+i} - \psi^\top \cdot S_T) = E_{P^\star}(R^{m+i})$. □

3.2 Local Utility

In this subsection, we consider the local utility-based version of neutral pricing. Fix a utility function u such that 0 belongs to the interior of the effective domain of u. Similarly as in the last subsection, we assume that there exists a locally optimal strategy φ in the underlyings' market $S^0, \ldots, S^m$ such that $u'(\varphi_t^\top \Delta S_t) \neq 0$ and $E(u'(\varphi_t^\top \Delta S_t)|\mathcal{F}_{t-1}) \neq 0$ P-almost surely for $t = 1, \ldots, T$ (cf. Lemma 2.14). We call the unique $(\{0\}, \mathbb{R}^m)$-dual measure for local utility $P^\star$ *neutral pricing measure for local utility.* Recall that $P^\star$ is an equivalent martingale measure if u is strictly increasing on its effective domain.

Definition 3.7. We call derivative price processes $S^{m+1}, \ldots, S^{m+n}$ *neutral derivative price processes for local utility* if there exists a strategy $\overline{\varphi}$ in the extended market $S^0, \ldots, S^{m+n}$ which is locally optimal (without constraints, i.e. for $(\{0\}, \mathbb{R}^{m+n})$) and satisfies $\overline{\varphi}^{m+1} = \cdots = \overline{\varphi}^{m+n} = 0$.

Theorem 3.8. *There exist unique neutral derivative price processes. They are given by $S_t^{m+i} = E_{P^\star}(R^{m+i}|\mathcal{F}_t)$ for $t = 0, \ldots, T$, $i = 1, \ldots, n$.*

PROOF. This is shown in the same way as Theorem 3.2. □

Note that $P^\star$ is an EMM for the extended market $S^0, \dots, S^{m+n}$ if u is strictly increasing on its effective domain. In particular, the corresponding price system allows no arbitrage.

A local L^2-approximation procedure has been suggested by [23] for contingent claim hedging (cf. the subsequent definition). The processes S^{m+1}, $\dots, S^{m+n}$ appearing in that procedure are obtained via conditional expectation relative to the (signed) *minimal martingale measure* in the sense of [24], [72]. Lemma 3.10 shows that they can be interpreted in terms of neutral pricing for local utility. The minimal martingale measure has been proposed for derivative pricing e.g. in [37].

Definition 3.9. (Local L^2-approximation pricing) Let $S^{m+1}, \dots, S^{m+n}$ be derivative price processes. We call $S^{m+1}, \dots, S^{m+n}$ *local L^2-approximation price processes* if, for $i = 1, \dots, n$, there exists a strategy $\psi \in \mathfrak{S}(\{0\}, \mathbb{R}^m)$ such that the following condition holds: For $t = 1, \dots, T$ we have

$$E\left(\left(S_t^{m+i} - S_{t-1}^{m+i} - \psi_t^\top \Delta S_t\right)^2 \middle| \mathcal{F}_{t-1}\right) \leq E\left(\left(S_t^{m+i} - Y - \xi^\top \Delta S_t\right)^2 \middle| \mathcal{F}_{t-1}\right)$$

for all $\mathcal{F}_{t-1}$-measurable, $\mathbb{R} \times \mathbb{R}^m$-valued random variables (Y, ξ).

Lemma 3.10. *If $u(x) = -a(x+b)^2 + c$ for some $a > 0$, $b, c \in \mathbb{R}$, then the local L^2-approximation price processes are the neutral price processes for local utility. In other words, the neutral pricing measure for local utility is the minimal martingale measure in the sense of [24], [72].*

PROOF. Fix $i \in \{1, \dots, n\}$ and $t \in \{1, \dots, T\}$. Observe that $S_{t-1}^{m+i} + \psi_t^\top \Delta S_t$ is the L^2-projection of S_t^{m+i} on the vector space $\{y + \xi^\top \Delta S_t : y, \xi\ \mathcal{F}_{t-1}$-measurable random variables with values in $\mathbb{R}$ resp. $\mathbb{R}^m\}$. Hence, we have $E((S_t^{m+i} - S_{t-1}^{m+i} - \psi_t^\top \Delta S_t)(y + \xi^\top \Delta S_t)|\mathcal{F}_{t-1}) = 0$ for any such y, ξ. In particular, we have $E((S_t^{m+i} - S_{t-1}^{m+i} - \psi_t^\top \Delta S_t)(1 + \Delta N_t)|\mathcal{F}_{t-1}) = 0$, where N denotes the logarithm process of the neutral pricing measure $P^\star$ for local utility (cf. Theorem 2.12). This implies $E_{P^\star}(S_t^{m+i} - S_{t-1}^{m+i}|\mathcal{F}_{t-1}) = E_{P^\star}(S_t^{m+i} - S_{t-1}^{m+i} - \psi_t^\top \Delta S_t|\mathcal{F}_{t-1}) = 0$. Consequently S^{m+i} is a $P^\star$-martingale, which proves the claim. □

In general, neutral prices for terminal wealth and for local utility differ even if the same utility function is chosen. Indeed, a main motivation to introduce local utility maximization was that hedging strategies and neutral pricing measures can be computed more easily (cf. Lemma 2.15). Even so, neutral prices for terminal wealth and for local utility do in fact coincide for logarithmic utility:

Lemma 3.11. *If $u(x) = a\log(x+b)+c$ for some $a > 0$, $b \in (-\min(1, e), \infty)$, $c \in \mathbb{R}$, then the neutral pricing measures for terminal wealth and for local utility coincide. Moreover, this measure does not depend on a, b, c and the initial endowment e.*

PROOF. W.l.o.g. $c = 0$, $a = 1$. Suppose that $\varphi \in \mathfrak{S}(\{0\}, \mathbb{R}^m)$ is locally optimal and let $P^\star$ be the neutral pricing measure for local utility with logarithm process N. Moreover, let $\kappa_t := E(u'(\varphi_t^\top \Delta S_t)|\mathcal{F}_{t-1})$ for $t = 1, \dots, T$. Then $1 + \Delta N_t = (\kappa_t(b + \varphi_t^\top \Delta S_t))^{-1}$ and hence $1 = E(\kappa_t(b + \varphi_t^\top \Delta S_t)(1 + \Delta N_t)|\mathcal{F}_{t-1}) = \kappa_t b$. Therefore $\frac{dP^\star}{dP} = \prod_{t=1}^T (1 + \frac{1}{b}\varphi^\top \Delta S_t)^{-1}$. Define $\psi \in \mathfrak{S}(\{0\}, \mathbb{R}^m)$ recursively by $\psi_t := \varphi_t \frac{b + e + \psi^\top \cdot S_{t-1}}{b}$ for $t = 1, \dots, T$. A simple calculation yields that $\frac{dP^\star}{dP} = (e + b)(b + e + \psi^\top \cdot S_T)^{-1} = (e + b)u'(e + \psi^\top \cdot S_T)$. In view of Lemma 2.6, this implies that $P^\star$ is the neutral pricing measure for terminal wealth.

From Lemma 3.4 it follows that the neutral pricing measure for terminal wealth does not depend on a, b, c, e. Hence the second statement holds as well. □

4 Consistent Derivative Pricing

In practise, valuation rules are used for different purposes. If you are an issuer of a contingent claim that is not yet traded, the approaches in the previous section provide some guidance as to what a reasonable price could be. The situation is different for a risk manager who must assess the riskyness of a company's portfolio. For her, the current derivative quotations are observable in the market. She is interested in the distribution of *future* asset price changes, especially for short time horizons. Of course, the pricing measures from the previous section can be used for this purpose as well. However, typically neutral derivative prices do not match observed quotations exactly even at time 0. So one may doubt whether they provide a good description of future price movements. It would be more satisfactory to use a valuation rule that incorporates the initially observed market quotations. A solution of this kind is presented in this section.

In the derivation of neutral prices we assumed that all participants in the derivative market are identical expected utility maximizers. Let us now consider the more realistic situation that some traders (from now on called *other investors*) behave differently. They may e.g. take a derivative position for insurance purposes. How are market prices affected if the aggregate demand for derivatives by these other investors does not sum up to 0? Since the utility maximizers have to take the counterposition, market prices must be such that the number of derivatives in their optimal portfolio exactly offsets the demand from the other traders. This demand is reflected in the initially observed market quotations. Intuitively, a positive demand by the other investors should lead to market prices which are higher than the neutral value in order to prompt utility maximizers to sell contingent claims.

We make the simplifying assumption that the aggregate demand from the other investors is deterministic and constant. Put differently, we relax the market clearing condition from the previous section as follows:

> Derivative prices should be such that the optimal portfolio of the representative utility maximizer contains a constant number of any contingent claim.

Such prices are called *constant demand price processes* in Definition 4.2 below. Of course, we cannot hope any more for unique derivative prices in this more general setting. But we can hope for unique constant demand price processes that match initially observed market quotations.

The issue of consistent derivative pricing is barely touched in the literature. Maybe closest in spirit is the concept of *inverting the yield curve* in interest rate theory (cf. [5]). As in the approach below a pricing measure is chosen from a parametric set of EMM's in such a way that theoretical and observed initial prices match. The Heath-Jarrow-Morton approach in interest rate theory avoids the consistency problem elegantly by treating bonds as underlyings rather than derivatives on the short rate. For a discussion of [2], [1] cf. the end of this section. Recently, [41] and [29] addressed this issue. In [45] a local utility-based variant of consistent pricing is discussed.

The general setting is as in the previous section. As before, we assume that derivative securities $m+1, \dots, m+n$ are given in terms of their discounted $\mathcal{F}_T$-measurable payoff $R^{m+1}, \dots, R^{m+n}$. Moreover, suppose that their initial prices $\pi^{m+1}, \dots, \pi^{m+n} \in \mathbb{R}$ are given. We call adapted processes $S^{m+1}, \dots,$ S^{m+n} *consistent derivative price processes* if $S_T^{m+i} = R^{m+i}$ and $S_0^{m+i} = \pi^{m+i}$ for $i = 1, \dots, n$. Fix an initial endowment $e \in \mathbb{R}$ and a utility function u such that e belongs to the interior of the effective domain of u.

Observe that *any* choice of consistent price processes leads to arbitrage if the initial prices are unreasonably chosen. This situation occurs if the market $S^0, \dots, S^{m+n}$ allows $(\{0\} \times \mathbb{R}^n, \mathbb{R}^m \times \{0\})$-arbitrage, i.e. the derivatives have to be traded only at times 0 and T in order to obtain a riskless gain. By Lemma 2.4 there is no such arbitrage if and only if threre exists an equivalent $(\{0\} \times \mathbb{R}^n, \mathbb{R}^m \times \{0\})$-seperating measure, or put differently, a consistent equivalent martingale measure in the sense of the following

Definition 4.1. We call an EMM $P^\star$ *consistent equivalent martingale measure (CEMM)* if $E_{P^\star}(R^{m+i}) = \pi^{m+i}$ for $i = 1, \dots, n$. *Consistent signed martingale measures (CSMM)* are defined accordingly.

Note that the terminal wealth of strategies $\varphi \in \mathfrak{S}(\{0\} \times \mathbb{R}^n, \mathbb{R}^m \times \{0\})$ in the extended market $S^0, \dots, S^{m+n}$ does not depend on the particular choice of consistent derivative price processes $S^{m+1}, \dots, S^{m+n}$. Therefore, we may assume that there exists an optimal strategy φ for terminal wealth under the constraints $(\{0\} \times \mathbb{R}^n, \mathbb{R}^m \times \{0\})$ without fixing $S^{m+1}, \dots, S^{m+n}$ in the first place. Moreover, we suppose that $E(u'(e + \varphi^\top \cdot S_T)|\mathcal{F}_t) \neq 0$ P-almost surely for $t = 0, \dots, T$. By Lemma 2.9, these assumptions follow from the absence of arbitrage for decent utility functions. We call the corresponding $(\{0\} \times \mathbb{R}^n, \mathbb{R}^m \times \{0\})$-dual measure $P^\star$ for terminal wealth the *least favourable consistent pricing measure*. Recall that $P^\star$ is an EMM for $S^0, \dots, S^m$ if u is strictly increasing on its effective domain.

Definition 4.2. We call consistent price processes $S^{m+1}, \dots, S^{m+n}$ *constant demand derivative price processes* if there exists a strategy $\overline{\varphi}$ in the extended market $S^0, \dots, S^{m+n}$ which is optimal for terminal wealth (without constraints, i.e. for $(\{0\}, \mathbb{R}^{m+n})$) and satisfies $\overline{\varphi}^{m+1} = a^{m+1}, \dots, \overline{\varphi}^{m+n} = a^{m+n}$ for some constants $a^{m+1}, \dots, a^{m+n} \in \mathbb{R}$.

Theorem 4.3. *There exist unique constant demand derivative price processes. They are given by* $S_t^{m+i} = E_{P^\star}(R^{m+i}|\mathcal{F}_t)$ *for* $t = 0, \dots, T$, $i = 1, \dots, n$.

PROOF. This is shown similarly as Theorem 3.2 if the neutral pricing measure is replaced with the least favourable consistent pricing measure. □

The name least favourable consistent pricing measure is motivated by the following analogue of Lemma 3.3:

Lemma 4.4 (Least favourable market completion). *If u is strictly increasing on its effective domain, then the least favourable consistent pricing measure $P^\star$ minimizes*

$$\sup\{E(u(e + X - E_Q(X))) : X \quad \mathcal{F}_T\text{-measurable random variable}\}$$

over all consistent equivalent martingale measures Q.

PROOF. This is shown in the same way as Lemma 3.3. □

The following result shows that Lemma 3.4 also has a counterpart for consistent pricing. It is shown exactly in the same way. Distance minimizing CEMM's are considered in [29].

Lemma 4.5 (Distance minimization). *Lemma 3.4 still holds if* neutral pricing measure *is replaced with* least favourable consistent pricing measure *and* EMM *(resp.* SMM*) with* CEMM *(resp.* CSMM*).*

[2], [1] suggest to calibrate a market model to initially observed derivative quotations $\pi^{m+1}, \dots, \pi^{m+n}$ by minimizing the entropy or pseudo entropy of the pricing measure relative to P. In view of the previous lemma this seems to be closely related to constant demand derivative pricing. However, instead of focusing on equivalent martingale measures, their minimization extends over a larger class of probability measures Q that reproduce the observed initial prices. In these papers, either the equivalence to P (cf. [2]) or the martingale property of S (cf. [1]) is dropped. This allows for easier numerical computations but applied to our setting, the resulting price system is typically not arbitrage-free.

References

1. Avellaneda, M. (1998). The minimum-entropy algorithm and related methods for calibrating asset-pricing models. *Documenta Mathematica Extra Volume ICM III*, 545–563.

2. Avellaneda, M., C. Friedman, R. Holmes, and D. Samperi Calibrating volatility surfaces via relative-entropy minimization. *Applied Mathematical Finance 4*, 37–64.
3. Bellini, F. and M. Frittelli (2000). On the existence of minimax martingale measures. Technical Report 14/2000, Università degli Studi di Milano - Bicocca.
4. Bismut, J. (1975). Growth and optimal intertemporal allocations of risks. *Journal of Economic Theory 10*, 239–287.
5. Björk, T. (1997). Interest rate theory. In W. Runggaldier (Ed.), *Financial Mathematics*, Volume 1656 of *Lecture Notes in Mathematics*, pp. 53–122. Berlin: Springer.
6. Chan, T. (1999). Pricing contingent claims on stocks driven by Lévy processes. *The Annals of Applied Probability 9*, 504–528.
7. Cox, J. and C.-F. Huang (1989). Optimal consumption and portfolio policies when asset prices follow a diffusion process. *Journal of Economic Theory 49*, 33–83.
8. Cvitanić, J. and I. Karatzas (1992). Convex duality in constrained portfolio optimization. *The Annals of Applied Probability 2*, 767–818.
9. Cvitanić, J., H. Pham, and N. Touzi (1999). Super-replication in stochastic volatility models under portfolio constraints. *Journal of Applied Probability 36*, 523–545.
10. Cvitanić, J., W. Schachermayer, and H. Wang (2001). Utility maximization in incomplete markets with random endowment. *Finance & Stochastics 5*, 259–272.
11. Dalang, R., A. Morton, and W. Willinger (1990). Equivalent martingale measures and no-arbitrage in stochastic security market models. *Stochastics and Stochastics Reports 29*, 185–202.
12. Davis, M. (1997). Option pricing in incomplete markets. In M. Dempster and S. Pliska (Eds.), *Mathematics of Derivative Securities*,, pp. 216–226. Cambridge: Cambridge University Press.
13. Delbaen, F., P. Grandits, T. Rheinländer, D. Samperi, M. Schweizer, and C. Stricker (2000). Exponential hedging and entropic penalties. Preprint Technische Universität Berlin.
14. Delbaen, F. and W. Schachermayer (1994). A general version of the fundamental theorem of asset pricing. *Mathematische Annalen 300*, 463–520.
15. Delbaen, F. and W. Schachermayer (1995). The existence of absolutely continuous local martingale measures. *The Annals of Applied Probability 5*, 926–945.
16. Delbaen, F. and W. Schachermayer (1998). The fundamental theorem of asset pricing for unbounded stochastic processes. *Mathematische Annalen 312*, 215–250.
17. Duffie, D. (1992). *Dynamic Asset Pricing Theory*. Princeton: Princeton University Press.
18. Eberlein, E. and J. Jacod (1997). On the range of option prices. *Finance & Stochastics 1*, 131–140.
19. El Karoui, N. and M. Quenez (1995). Dynamic programming and pricing of contingent claims in an incomplete market. *SIAM Journal on Control and Optimization 33*, 29–66.
20. Foldes, L. (1990). Conditions for optimality in the infinite-horizon portfolio-cum-saving problem with semimartinagle investments. *Stochastics and Stochastics Reports 29*, 133–170.

21. Foldes, L. (1992). Existence and uniqueness of an optimum in the infinite-horizon portfolio-cum-saving problem with semimartinagle investments. *Stochastics and Stochastics Reports 41*, 241–267.
22. Föllmer, H. and P. Leukert (2000). Efficient hedging: cost versus shortfall risk. *Finance & Stochastics 4*, 117–146.
23. Föllmer, H. and M. Schweizer (1989). Hedging by sequential regression: an introduction to the mathematics of option trading. *ASTIN Bulletin 18*, 147–160.
24. Föllmer, H. and M. Schweizer (1991). Hedging of contingent claims under incomplete information. In M. H. A. Davis and R. J. Elliott (Eds.), *Applied Stochastic Analysis*, Volume 5 of *Stochastics Monographs*, pp. 389–414. London: Gordon & Breach.
25. Frey, R. and C. Sin (1999). Bounds on european option prices under stochastic volatility. *Mathematical Finance 9(2)*, 97–116.
26. Frittelli, M. (1998). Introduction to a theory of value coherent with the no-arbitrage principle. Technical Report, Università degli Studi di Milano
27. Frittelli, M. (2000). The minimal entropy martingale measure and the valuation problem in incomplete markets. *Mathematical Finance 10(1)*, 39–52.
28. Goll, T. and J. Kallsen (2000). Optimal portfolios for logarithmic utility. *Stochastic Processes and their Applications 89*, 31–48.
29. Goll, T. and L. Rüschendorf (2000). Minimal distance martingale measures and optimal portfolios under moment constraints. Preprint.
30. Goll, T. and L. Rüschendorf (2001). Minimax and minimal distance martingale measures and their relationship to portfolio optimization. *Finance & Stochastics, 5*, 557–581.
31. Grandits, P. (1999a). On martingale measures for stochastic processes with independent increments. *Theory of Probability and its Applications 44*, 39–50.
32. Grandits, P. (1999b). The *p*-optimal martingale measure and its asymptotic relation with the minimal-entropy martingale measure. *Bernoulli 5*, 225–247.
33. Harrison, M. and D. Kreps (1979). Martingales and arbitrage in multiperiod securities markets. *Journal of Economic Theory 20*, 381–408.
34. Harrison, M. and S. Pliska (1981). Martingales and stochastic integrals in the theory of continuous trading. *Stochastic Processes and their Applications 11*, 215–260.
35. He, H. and N. Pearson (1991a). Consumption and portfolio policies with incomplete markets and short-sale constraints: The finite-dimensional case. *Mathematical Finance 1(3)*, 1–10.
36. He, H. and N. Pearson (1991b). Consumption and portfolio policies with incomplete markets and short-sale constraints: The infinite-dimensional case. *Journal of Economic Theory 54*, 259–304.
37. Hofmann, N., E. Platen, and M. Schweizer (1992). Option pricing under incompleteness and stochastic volatility. *Mathematical Finance 2(3)*, 153–187.
38. Jacod, J. and A. Shiryaev (1998). Local martingales and the fundamental asset pricing theorems in the discrete-time case. *Finance & Stochastics 2*, 259–273.
39. Kabanov, Yu. (1997). On the FTAP of Kreps-Delbaen-Schachermayer. In *Statistics and control of stochastic processes (Moscow, 1995/1996)*, pp. 191–203. River Edge, NJ: World Scientific.
40. Kabanov, Yu. and D. Kramkov (1994). No arbitrage and equivalent martingale measures: An elementary proof of the Harrison-Pliska theorem. *Theory of Probability and its Applications 39*, 523–527.

41. Kallsen, J. (1998a). Duality links between portfolio optimization and derivative pricing. Technical Report 40/1998, Mathematische Fakultät Universität Freiburg i. Br.
42. Kallsen, J. (1998b). *Semimartingale Modelling in Finance*. Dissertation Universität Freiburg i. Br.
43. Kallsen, J. (1999). A utility maximization approach to hedging in incomplete markets. *Mathematical Methods of Operations Research 50*, 321–338.
44. Kallsen, J. (2000). Optimal portfolios for exponential Lévy processes. *Mathematical Methods of Operations Research 51*, 357–374.
45. Kallsen, J. (2001). Derivative pricing based on local utility maximization. *Finance & Stochastics*, forthcoming.
46. Kallsen, J. and A. Shiryaev (2000). The cumulant process and Esscher's change of measure. Preprint.
47. Karatzas, I. and S. Kou (1996). On the pricing of contingent claims under constraints. *The Annals of Applied Probability 6*, 321–369.
48. Karatzas, I., J. Lehoczky, and S. Shreve (1987). Optimal portfolio and consumption decisions for a small investor on a finite horizon. *SIAM Journal on Control and Optimization 25*, 1557–1586.
49. Karatzas, I., J. Lehoczky, S. Shreve, and G. Xu (1991). Martingale and duality methods for utility maximization in an incomplete market. *SIAM Journal on Control and Optimization 29*, 702–730.
50. Keller, U. (1997). *Realistic Modelling of Financial Derivatives*. Dissertation Universität Freiburg i. Br.
51. Korn, R. (1997). *Optimal Portfolios: Stochastic Models for Optimal Investment and Risk Management in Continuous Time*. Singapore: World Scientific.
52. Kramkov, D. and W. Schachermayer (1999). The asymptotic elasticity of utility functions and optimal investment in incomplete markets. *The Annals of Applied Probability 9*, 904–950.
53. Kreps, D. (1981). Arbitrage and equilibrium in economics with infinitely many commodities. *Journal of Mathematical Economics 8*, 15–35.
54. Lamberton, D. and B. Lapeyre (1996). *Stochastic Calculus Applied to Finance*. London: Chapman & Hall.
55. Miyahara, Y. (1996). Canonical martingale measures of incomplete assets markets. In S. Watanabe, M. Fukushima, Yu. Prohorov, and A. Shiryaev (Eds.), *Probability Theory and Mathematical Statistics: Proc. Seventh Japan-Russia Symposium Tokyo 1995*, pp. 343–352. Singapore: World Scientific.
56. Miyahara, Y. (1999). Minimal entropy martingale measures of jump type price processes in incomplete assets markets. *Asia-Pacific Financial Markets 6*, 97–113.
57. Pham, H. and N. Touzi (1999). The fundamental theorem of asset pricing with cone constraints. *Journal of Mathematical Economics 31*, 265–279.
58. Pliska, S. (1986). A stochastic calculus model of continuous trading: Optimal portfolios. *Mathematics of Operations Research 11*, 371–384.
59. Pliska, S. (1997). *Introduction to Mathematical Finance*. Malden, MA: Blackwell.
60. Rockafellar, T. (1970). *Convex Analysis*. Princeton: Princeton University Press.
61. Rockafellar, T. and R. Wets (1998). *Variational Analysis*. Berlin: Springer.
62. Rogers, C. (1994). Equivalent martingale measures and no-arbitrage. *Stochastics and Stochastics Reports 51*, 41–49.

63. Rouge, R. and N. El Karoui (2001). Pricing via utility maximization and entropy. *Mathematical Finance*, forthcoming.
64. Schachermayer, W. (1992). A Hilbert space proof of the fundamental theorem of asset pricing in finite discrete time. *Insurance: Mathematics and Economics 11*, 249–257.
65. Schachermayer, W. (2001a). Optimal Investment in Incomplete Financial Markets. In H. Geman, D. Madan, S. Pliska, and T. Vorst (Eds.), *Mathematical Finance – Bachelier Congress 2000*, Berlin: Springer.
66. Schachermayer, W. (2001b). Optimal investment in incomplete markets when wealth may become negative. *The Annals of Applied Probability*, forthcoming.
67. Schäl, M. (1994). On quadratic cost criteria for options hedging. *Mathematics of Operations Research 19*, 121–131.
68. Schürger, K. (1996). On the existence of equivalent τ-measures in finite discrete time. *Stochastic Processes and their Applications 61*, 109–128.
69. Schweizer, M. (1991). Option hedging for semimartingales. *Stochastic Processes and their Applications 37*, 339–363.
70. Schweizer, M. (1992). Mean-variance hedging for general claims. *The Annals of Applied Probability 2*, 171–179.
71. Schweizer, M. (1994). Approximating random variables by stochastic integrals. *The Annals of Probability 22*, 1536–1575.
72. Schweizer, M. (1995a). On the minimal martingale measure and the Föllmer-Schweizer decomposition. *Stochastic Analysis and Applications 13*, 573–599.
73. Schweizer, M. (1995b). Variance-optimal hedging in discrete time. *Mathematics of Operations Research 20*, 1–32.
74. Schweizer, M. (1996). Approximation pricing and the variance-optimal martingale measure. *The Annals of Probability 24*, 206–236.
75. Schweizer, M. (1999). A guided tour through quadratic hedging approaches. Preprint.
76. Willinger, W. and M. Taqqu (1987). The analysis of finite security markets using martingales. *Advances in Applied Probability 19*, 1–25.

Pricing Credit Derivatives in Credit Classes Frameworks

Franck Moraux, Patrick Navatte

Université de Rennes 1-IGR and CREREG-Axe Finance. 11, rue Jean Macé 35000 Rennes, France

1 Introduction

Many credit management systems, based on different underlying frameworks, are now available to measure and control default and credit risks[1]. Homogeneous credit classes and associated transition matrix may thus be constructed within many different frameworks. For illustration, the KMV Corporation provides a transition matrix within a structural approach *à la* Black-Sholes-Merton (Crouhy-Galai-Mark [5]). Independentely from the underlying framework, a methodology based on credit classes may therefore be used to price any claim contingent on credit events among which the default.

Altman and Kao [1] have early suggested the Markov chain methodology to model a credit process based on credit classes. Skinner [21], studying a trinomial tree with both a reflecting boundary and a absorbing one for valuing corporate bonds, has in fact explored a special Markov Chain. However Jarrow-Lando-Turnbull [10] (JLT hereafter) have offered the first complete arbitrage-free framework to price risky bonds by identifying credit classes to the rating agencies' ones[2]. Other contributors are Kijima [13], Kijima-Komoribayashi [14] (KIK hereafter) and Arvanitis-Gregory-Laurent [3] (AGL hereafter).

This paper focuses on the ability of the Markov Chain methodology for pricing credit derivatives in credit classes framework, matching any observable risky term structures when the generator matrix is held constant or piecewised constant (as in JLT, KIK, AGL). Chosen credit derivatives include those already priced in other approaches by Das [4] and Pierides [19] (in the structural setting), and by Longstaff-Schwartz [17], Schönbucher [20] and Duffie [6] (in the pure intensity framework). KIK [14], who concentrate on the European spread put option, are thus systematically extended.

The rest of this paper is organized as follow. Section 1 recalls the general framework. Section 2 derives an useful conditional probability density

[1] See the *Journal of Banking and Finance* special issues on credit risk modelling, Lando [15] and Cathcart-El Jahel [7] for surveys on the *structural* or *reduced form* theoretical backgrounds.

[2] Overlapping w.r.t. the credit spreads, lack of reaction are the most documented problems of the rating process - see among others Weinstein [23] or Hand-Holthausen-Leftwich [8].

function. Section 3 reviews many credit derivatives valuable in the chosen setting. Section 4 considers the calibration procedures. Section 5 is devoted to a numerical examples and a final section concludes.

2 The General Framework

Following JLT [10], let it be a frictionless continuous economy with a finite horizon $[0, T_H]$, whose uncertainty is represented by $(\Omega, \mathcal{F}, (\mathcal{F}_t)_{0 \le t \le T_H}, \mathcal{P})$ and where riskless and risky term structures of interest rates are available. One then assumes that *first* there exists a unique equivalent martingale measure denoted $\tilde{\mathcal{P}}$ and that *second* under this measure, the interest rate and the credit risks are non correlated. The credit process $(\nu_t)_t$ is supposed to be correctly described by a continuous Markov chain defined on the finite set of credit status denoted $\mathcal{S} = (S_i)_{i=1,K+1}$. K depends on the number of classes retained in the discriminating procedure and while the first state S_1 describes the surest state, S_{K+1} stands for bond in default. Moreover $\mathcal{S}^* = (S_i)_{i=1,K}$. The one period transition matrix associated with ν is written $\mathcal{Q} = (q_{ij})_{i,j=1,K+1}$, $q_{ij} = P[\nu_{t+1} = j/\nu_t = i], \forall i, j \in \mathcal{S}$ is therefore the probability of going from a state i to an other j, $1 - q_{j,K+1}$ the survival probability. Since the $(K+1)^{th}$ state is an absorbing one, $q_{K+1,j} = P[\nu_{t+1} = j/\nu_t = K+1] = \delta_{j=K+1}$ with δ the Dirac measure.

Following JLT [10], KIK [14], AGL [3], the credit process is assumed to be both Markovian and time homogeneous by period. Standard results then tell us that tractable Markov chains may be specified in terms of a constant generator matrix, $\Lambda = (\lambda_{i,j})_{i,j=\mathcal{S}\times\mathcal{S}}$, any $t-$transition matrix being given by $\mathcal{Q}_t = e^{t\Lambda} = \sum_{k\ge 0} \frac{1}{k!}(t\Lambda)^k$. Historical transition matrix $\mathcal{Q}$ cannot however be used for risk neutral pricing without correction. And the risk neutral transition matrix (RNTM hereafter) $\tilde{\mathcal{Q}}$ must verify properties to be a transition matrix and generate *a priori* correct credit spreads. Expressed in term of the risk neutral generator matrix (RNGM hereafter), $\tilde{\Lambda}$, these follow. First, $\forall j : \tilde{\lambda}_{K+1,j} = 0$ warranties that the default state is an absorbing state. Second, $\forall i \in$ cal S, 0x2200j0x2208$\mathcal{S}, i \neq j, \tilde{\lambda}_{i,j} \ge 0$ ensures that the transition probabilities cannot be negative. Third, $\sum_{kAAA,,D} \tilde{\lambda}_{i,k} = -\tilde{\lambda}_{i,i}$ assures a total of transition probabilities equals to one. Finally, $\sum_{j \ge k} \tilde{\lambda}_{i,j} \le \sum_{j \ge k} \tilde{\lambda}_{i+1,j}$ implies that *lower credit classes are riskier* (JLT [10], lemma 2.). Additional conditions related to empirical monotonicities may be found in Kijima [13].

Assuming that bondholders face a constant writedown ω at maturity in case of an early default, the price of a risky bond is:

$$p(t, T; \Theta) = p(t, T)(\delta + \omega \tilde{\mathcal{P}}_t[\tau > T]) \tag{2.1}$$

where $p(t, T)$ is the price of a default-free zero-coupon bond paying surely a dollar at T and $\tau = \inf[t : \nu_t = K+1]$ the default time. The credit status Θ includes both the credit class and the seniority, the recovery rate is

$\delta = 1-\omega^3$. Since the risky yield Y is such that $p(t,T;\Theta) = e^{-Y(t,T;\Theta)(T-t)}$, the yield spread over the Treasury one (or credit spread) is defined by $h(t,T;\Theta) = -\frac{1}{T-t}\ln(\delta+\omega\tilde{\mathcal{P}}_t[\tau > T])$. By analogy with the interest rate theory, the instantaneous forward credit spread (or local spread), $h(t,T;\Theta) = \frac{1}{T-t}\int_t^T s(t,s)\,ds$, can be computed as

$$s(t,T) = -\frac{\partial}{\partial T}\ln(\delta + \omega\tilde{\mathcal{P}}_t[\tau > T]). \tag{2.2}$$

Several remarks may be done. First, the survival probability, *i.e.* $\frac{1}{\omega}(e^{-h(t,T;\Theta)(T-t)} - \delta)$ implies, among many things, that the survival probability may be interpreted as the price of a bond with yield h if (and only if) its recovery rate is null. Second, the credit margin is infinite if (and only if) $q_{t,T}(j, K+1) = 1$ and $\delta = 0$ (roughly speaking when one loses everything surely). Third, the instantaneous forward credit spread is seen to be closely related to the conditional default time density under the martingale measure.

3 The Conditional Default Time Density

Since Lando [15] and Schönbucher [20], local credit spreads are known to be critical for credit derivatives valuation. JLT [10] have already computed them as the difference of the risky and risk free instantaneous forward rates. Equation 2.2, nevertheless, allows one to closely relate them to the risk neutral probability density function of the default time conditional on survival (until then).

Proposition 3.1. *The risk neutral default time conditional density (given Λ and no early default), in a credit class framework and $\forall i \in \mathcal{S}^*$, is :*

$$f_\tau^t(s) = \sum_{k=1}^{K} \tilde{q}_{i,k}(t,s)\mu_k(s)\lambda_{k,K+1}. \tag{3.3}$$

where $\mu_k(s)$ denotes the $s-$risk premium.

Proof. Following JLT [10], the transition probability is known to solve the following forward equation $\frac{\partial\tilde{\mathcal{Q}}(t,T)}{\partial T} = \tilde{\mathcal{Q}}(t,T)\tilde{\Lambda}(T)$ where $\tilde{\Lambda}(T)$ denotes the $T-$risk neutral generator. This equation defines :

$$\left(\sum_{k=1}^{K+1} \tilde{q}_{i,k}(t,T)\mu_k(T)\lambda_{k,j}\right)_{i,j}$$

And therefore: $\frac{\partial\tilde{Q}_t^i[\tau<T]}{\partial T} = \sum_{k=1}^{K} \tilde{q}_{i,k}(t,T)\mu_k(T)\lambda_{k,K+1}$.

[3] A constant recovery hypothesis may be regarded as costly, especially for speculative grade bond, but it allows one to neglect the recovery risk and the required risk premium. One refers to Altman-Kishore [2] for a detailed study on recovery rates.

The risk neutral default time density function at time T is then the sum of attainable instantaneous risk neutral probabilities of default at that time. This result has two major interests. First, it extends the Poisson model with constant intensity ($\lambda e^{-\lambda(T-t)}$) to the case where there are many credit states. Second, any instantaneous forward credit spread may then be explicitly derived in this framework (in particular the spot credit spread $s(t)$). Since $\lim_{T\to t}\tilde{q}_{i,k}(t,T)=\delta_{i,k}$ and $\lim_{T\to t}\mu(T)=\mu(t)$, Eq.(2) and proposition yield to: $\lim_{T\to t}s(t,T)=s(t,t)=s(t)=(1-\delta)\lambda_{i,K}$. The spot credit spread may therefore be different from zero as soon as λ_{iK} is not null as confirmed by JLT [10].

4 Credit Derivatives Contracts

The Markov Chain methodology is expected to succeed in pricing credit derivatives hedging pure default, credit event and credit class events. Let's denote Φ the payoff function. The objective transition matrix $\mathcal{Q}$ may be used without loss of generality because risk premia are easily introduced.

4.1 Credit Derivatives and Default Events

First will be considered the European and American default digital put options. These contracts are important because, as demonstrated by Schönbucher [20], the pricing of many others reduces to them.

– An European default digital put option, dig_e, delivers a unit at maturity if the default occurs during the life of the contract. It is the simplest credit derivative that concerns the risky bond price and a default event. Its payoff is given at any date t by $\Phi_t(dig_e(0,T,\Theta))=1_{\{\tau\leq T\}\cap\{t=T\}}(t)$ and thanks to the independence assumption, its price is simply given by :

$$dig_e(0,T,\Theta)=E^{\tilde{\mathcal{P}}}[\beta_T^{-1}\Phi_T(dig_e(0,T,\Theta))]=p(0,T)q_{i,K+1}(0,T)$$

Proof. Independence and lemma 1 of JLT [10][4].

– An American default digital put option, dig_a, is comparable to the previous contract except that it can be exercised as soon as the default occurs. Its payoff is given by $\Phi_t(dig_a(0,T,\Theta))=1_{\{\tau\leq T\}\cap\{t=\tau\}}(t)$ at any date t and it therefore needs to be discounted only from τ :

$$dig_a(0,T,\Theta)=E^{\tilde{\mathcal{P}}}[\beta_\tau^{-1}\Phi_\tau(dig_a(0,T,\Theta))]=\sum_{k\in\mathcal{S}^*}\lambda_{k,K+1}\int_0^T p(0,s)q_{i,k}(0,s)\,ds$$

4 Following results are indebted to lemma 1 of JLT [10] which states that, since the default is an absorbing state, the transition matrix between t and T is suitable to provide the survival probability and hence the default probability in T.

Proof. Thanks to the iterated expectations principle,

$$E^{\tilde{\mathcal{P}}}[\beta_\tau^{-1} 1_{\{\tau \le T\} \cap \{t=\tau\}}] = E^{\tilde{\mathcal{P}}}[E^{\tilde{\mathcal{P}}}[\beta_\tau^{-1} 1_{\{\tau \le T\} \cap \{t=\tau\}} | \Lambda]] = E^{\tilde{\mathcal{P}}}[\int_0^T \beta_s^{-1} f_\tau^0(s)\, ds].$$

One has $diga_a(0,T,\Theta) = \int_0^T p(0,s) E^{\tilde{\mathcal{P}}}[f_\tau^0(s)]\, ds$ and the result then follows from the deterministic behavior of the generator matrix.

Let's verify the two states case. Schönbucher [20] has proved that $diga_a(0,T,\Theta) = \int_0^T p(0,t) S(t) h(0,t)\, dt$ where $h(0,t)$ denotes the forward spread associated to a zero recovery bond and $S(t)$ the survival probability. Letting $i = k = K = 0$ in the previous equation gives $\int_0^T p(0,s) q_{0,0}(0,s) \lambda_{0,1}\, ds$ and recalling that Eq.(2) provides the forward spread $(\frac{q_{0,0}\lambda_{0,1}}{q_{0,0}})$ completes the argument.

4.2 Credit Derivatives and Credit Events

Credit derivatives hedging against a credit event exploit the fact that any deterioration (*resp.* improvement) of the quality of a credit exposure implies a reduction (*resp.* a growth) in its market price value or equivalently an increase (*resp.* a decrease) in its market credit spread value.

Denoting t the horizon contract and T the maturity of the underlying credit exposure ($T > t$), let's review the different pay-offs. The fall in credit quality is accompanied by a rise of the credit margin required by investors. This event may be hedged by an European call on the credit spread (Longstaff-Schwartz [17]), or by an European put on the price (Das [4]). Denoting C_h, P_p respectively the spread call and bond put, respective payoffs are given by:

$$\Phi(C_h(t,T;\Theta)) = \max[h(t,T;\Theta) - K, 0]$$
$$\Phi(P_p(t,T;\Theta)) = \max[Kp(t,T) - p(t,T;\Theta), 0].$$

Symmetrically, the rise in the credit quality can be hedged either by a European put on the credit spread (KIK [14]) or by an European call on the risky bond price whose payoffs are respectively :

$$\Phi(P_h(t,T;\Theta)) = \max[K - h(t,T;\Theta), 0],$$
$$\Phi(C_p(t,T;\Theta)) = \max[p(t,T;\Theta) - Kp(t,T), 0].$$

Following standard arbitrage arguments, these contracts are priced by discounting the payoffs. The market risk and the credit risk being not correlated each other, option prices written on credit margin are

$$H_h(t',t,T,\Theta) = E_{t'}^{\tilde{\mathcal{P}}}[\beta_{t'} \beta_t^{-1} \Phi(H_h(t,T,\Theta))] = p(t',t) E_{t'}^{\tilde{\mathcal{P}}}[\Phi(H_h(t,T,\Theta))]$$

where H stands for C, P ((call, put) and then nowadays values are

$$p(0,t) E_0^{\tilde{\mathcal{P}}}[\Phi(H_h(t,T,\Theta))] = p(0,t) \sum_{j=1}^{K+1} q_{0,t}(i,j) \Phi(H_h(t,T,\Theta_j)) \qquad (4.4)$$

where $q_{0,t}(i,j) = (e^{t\Lambda})_{ij}$. Option prices involving risky bonds are given by:

$$\begin{aligned} H_p(t',t,T,\Theta) &= E_{t'}^{\tilde{\mathcal{P}}}[\beta_{t'}\beta_t^{-1}\Phi(H_p(t,T,\Theta))] \\ &= E_{t'}^{\tilde{\mathcal{P}}}[\beta_{t'}\beta_t^{-1}p(t,T)]E_{t'}^{\tilde{\mathcal{P}}}[\max[(-1)^{1_{H=C}}(K-\delta-\omega\tilde{\mathcal{P}}_t(\tau>T)),0]] \\ &= p(t',T)E_{t'}^{\tilde{\mathcal{P}}}[\max[(-1)^{1_{H=C}}(K-\delta-\omega\tilde{\mathcal{P}}_t(\tau>T)),0]]. \end{aligned}$$

In particular, nowadays values are:

$$p(0,T)\sum_{j=1}^{K+1} q_{0,t}(i,j)\max[(-1)^{1_{H=C}}(K-\delta-\omega Q^S_{t,T}(j)),0]$$

where $q_{0,t}(i,j) = (e^{t\Lambda})_{ij}$ and $Q^S_{t,T}(j) = 1 - q_{t,T}(j,K+1)$.

Let's point out that these credit derivatives do not involve the same information regarding the riskless term structure. Furthermore, strike prices appear critical whereas they would have been endogenously computed in other framework (Das [4] and Pierides [19]).

4.3 Credit Derivatives and Credit Class Events

Credit derivatives hedging against a credit class change tend to be more important as the credit portfolio management becomes a risk classes management[5]. These contingent contracts include both digital contracts delivering a unit in case of a change in credit class and credit options whose strike prices are fixed at a suitable level. These latter contracts may however be shown not to be valuable in our framework because they need border credit spread (between credit classes) unattainable in a constant generator framework.

An European credit class digital put option, dig_{ecc}, delivers a unit at maturity if the change of credit class occurs during the life of the contract. Its payoff is given at any date t by $\Phi_t(dig_{ecc}(0,T,\Theta)) = 1_{\{\tau_{cc}\le T\}\cap\{t=T\}}(t)$ where τ_{cc} denotes the time the credit status changes. Its American version, dig_{acc}, pays the unit as soon as the event occurs; its payoff is therefore $\Phi_t(dig_{acc}(0,T,\Theta)) = 1_{\{\tau_{acc}\le T\}\cap\{t=\tau_{acc}\}}(t)$. Once the proper transition matrix defined, these contracts are simple generalisations of previous default digitals.

For ease of presentation, let's consider the special case of a credit class digital option paying one unit if the creditworthiness changes to the BB one or less. The relevant transition matrix for pricing is defined on $\mathbf{S} = (S_i)_{i=1,BB}$ and considers the BB-rated class as an absorbing state. More precisely, one needs the squared matrix $\mathbf{Q}$ defined by

$$\mathbf{Q} = (\mathbf{q}_{ij})_{i,j=1,BB}$$

with $\mathbf{q}_{i,j} = q_{i,j}$ for any $i,j < BB$, $\mathbf{q}_{i,BB} = \sum_{k\ge BB} q_{i,k}$ for all $i < BB$ and $\mathbf{q}_{BB,j} = \delta_{BB}(j)$.

[5] *e.g.* speculative classes are often non desired.

Then, thanks to the independence assumption, the respective prices are:

$$
\begin{aligned}
dig_{ecc}(0,T,\Theta) &= E^{\tilde{\mathcal{P}}}[\beta_T^{-1}\Phi_T(dig_{ecc}(0,T,\Theta))] = p(0,T)\mathbf{q}_{0,T}(i,BB) \\
dig_{acc}(0,T,\Theta) &= E^{\tilde{\mathcal{P}}}[\beta_\tau^{-1}\Phi_\tau(dig_{acc}(0,T,\Theta))] \\
&= \sum_{k\in\mathbf{S}} \lambda_{k,BB} \int_0^T p(0,s))\mathbf{q}_{i,k}(0,s)\,ds
\end{aligned}
$$

Formal proofs are similar to the default digitals ones. Other credit class digitals are valued along the same lines by associating the proper transition matrix.

5 Implementation

Any objective transition matrix must be calibrated before risk neutral pricing. JLT [10], KIK [14] and AGL [3] have proposed different procedures to estimate credit risk premia.

5.1 A Comparison of Calibration Procedures

Apart minor differences, all of them have suggested risk premia as solutions of a squared errors minimization between observed and theoretical prices[6]. Let's denote $\Lambda = (\lambda_{i,j})_{i,j\in\mathcal{S}\times\mathcal{S}}$ the historical generator, $\tilde{\Lambda} = (\tilde{\lambda}_{i,j})_{i,j\in\mathcal{S}\times\mathcal{S}}$ the risk neutral generator matrix or RNGM, I the identity matrix, e the square zero matrix whose last column is the unity and $\hat{p}$ the observed bond prices. The recursive relation $\tilde{\mathcal{Q}}_{0,t+1} = \tilde{\mathcal{Q}}_{0,t}\tilde{\mathcal{Q}}_{t,t+1} = \tilde{\mathcal{Q}}_{0,t}e^{\tilde{\Lambda}(t)}$ must hold for any horizon and, as a result, two distinct assumptions concerning the RNTM over a period are possible.

First, following AGL [3], a unique RNGM may be considered. Any rated bond prices available on the market are used and the generator matrix is designed to match simultaneously, any risky term structures and therefore any horizon. Formally, the RNGM $\tilde{\Lambda}$ is solution of the following least square optimization:

$$
\tilde{\Lambda} = \arg\min_{\bar{\Lambda}} \left[\sum_{i,j\in\mathcal{S}} (\beta_{i,j})^{-1}[\bar{\lambda}_{i,j} - \lambda_{i,j}]^2 + \sum_{i\in\mathcal{S}}\sum_{n=1}^{N_i} [\hat{p}_n(0,T;\Theta_n) - p(t,T;\bar{\Lambda})]^2 \right]
$$

subject to conditions warrantying a well-defined risk neutral generator recalled in the first section. β stands for some subjective weights.

Second, along the lines of JLT [10] and KIK [14], the RNGM may be assumed as a deterministic function of the historical generator and the horizon

[6] To warranty some consistent value, JLT [10] have constrained solutions to positive ones, AGL [3] have prevented them to be far from the historical one and KIK [14] have bounded them.

considered - $\tilde{\Lambda}(t) = U(t)\Lambda$. Both recursive procedures identify $U(t)$ to $\Pi(t) = (\pi_{i,j}(t))_{i,j \in S\times S}$ the diagonal matrix that contains the K (transition matrix) risk premia and the unity. Both also assume proportional adjustments between the one-period RNTM and the historical one, as exposed in Table 1. Risk premia are supposed *invariant* either for any attainable class from the credit class considered (JLT [10]) or for any attainable class except the default class (KIK [14]). As demonstrated by KIK [14], arbitrage conditions must be also verified. It is noteworthy that they have excluded the JLT [10] 'unconstrained' procedure in most empirical expriments conducted[7]. Other operational details are given in appendix, we now turn to the numerical examples.

Table 1: Horizon-dependent Risk Premia for the Transition Matrix
Recall that $\Pi(t) = (\pi_{i,j}(t))_{i,j \in S\times S}$, $\Pi(0) = I$. Conditions are $\forall i \in S^*$

	Alternative Procedures	
	JLT [10]	KIK [14]
Proportional Adjustment	$[\tilde{Q}_{t,t+1} - I] = \Pi(t)[Q - I]$	$[\tilde{Q}_{t,t+1} - e] = \Pi(t)[Q - e]$
Conditions	$\pi_i(t) \in]0, 1/(1 - q_{i,i})[$	$\pi_i(t) \in]0, 1/(1 - q_{i,K+1})[$

5.2 Numerical Illustrations of Credit Derivatives Valuation

An investor who fears the credit quality to fall may consider many different credit derivatives among which calls on credit spread, puts on the risky bond price with a strike price sensible to interest rates and default digital put options.

In the following, observable term structures of interest rates, plotted in Figure 1, are used. Based on data reported in KIK [14] (16/05/97), they are interpolated with a standard "modified Nielsen-Siegel's" procedure ([18]). This is consistent with Helwege-Turner [9] who find that risky bonds (better than B) have upward-sloping credit yield curves[8]. For calibration, the *S&P*'s average one-period transition matrix[9] has been chosen. Finally, subjective weights in the AGL [3] procedure have been chosen to match "at the best" the observed credit risky term structure.

Figure 2 first compares American digital put option prices to their European counterparts. While the European digital put option has been shown

[7] JLT's proportional adjustment may imply a division by zero avoided by imposing a striict postive value for any $q_{j,K+1}$ even $j = 1$. In the KIK procedure, the "first computed" risk premia are well defined *iif* the write down faced by the bondholder is such that $\omega > 1 - \hat{p}(0,T;\Theta_i)/\hat{p}(0,T)$. This restriction is avoided by assuming below that the recovery rate is null.

[8] For the riskiest exposure, a strictly downward sloping term structure is nevertheless defined as $a + be^{-ct}$ with a,b some strict positive values. It allows the AGL calibration procedure. See the discussion below.

[9] available in *Standard & Poor's Special Report* ([22], Table 8) or from the authors.

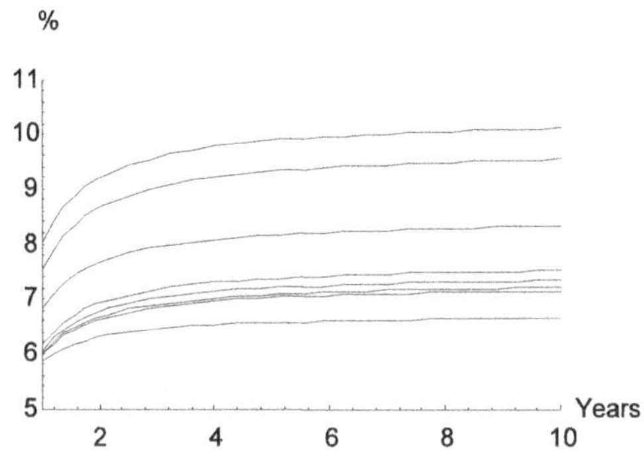

Fig. 1. Term structures of interest rates (from the risk free to the CC risky one). Term structurs of interest rates are based on observable data reported in KIK (1998) and interpolated with the Nielsen-Siegel (1987) modified procedure

proportional to the default probability, the American digital put option may be computed as

$$dig_a(0,T,\Theta) = \sum_{j=1}^{T/\Delta t} \sum_{k \in S^*} p(0, j\Delta t)\tilde{q}_{i,k}(0, j\Delta t)\tilde{\lambda}_{k,K+1}(j\Delta t)\Delta t$$

where T is the expiration date. The supervision frequency, Δt, of the American digital put option is chosen daily and $p(0, j\Delta t) = e^{-NS(j\Delta t)}$ where NS stands for the modified Nielsen-Siegel function. Let's precise that $\tilde{\lambda}_{k,K+1}(j\Delta t) = \tilde{\lambda}_{k,K+1}$ for AGL [3] and $\tilde{\lambda}_{k,K+1}(j\Delta t) \approx \overline{q}_{k,K+1} \frac{\ln \overline{q}_{k,k}}{\overline{q}_{k,k}-1}$ for KIK [14] and JLT [10] where $\overline{q}$ is the calibrated risk neutral matrix over the suitable "one year"-period As expected, the price of an American digital is always greater than its European equivalent and riskier is the exposure, more costly is the insurance contract. This graph also illustrates the pricing difference induced by the risk neutral calibration procedure.

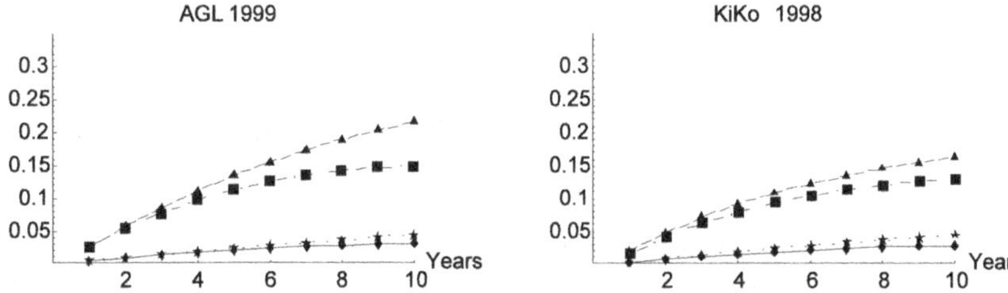

Fig. 2. European vs. American default digital put option prices for two underlying exposures rated AA/B

Figure 3 finally plots credit spread call options and risky bonds put options hedging credit exposures (rated from AAA to B and maturing 10 years later) to be secured at a BBB-rated level. It then appears that these two insurance contracts behave quite similarly suggesting a parity relation. This latter has already been pointed by Pierides [19][10].

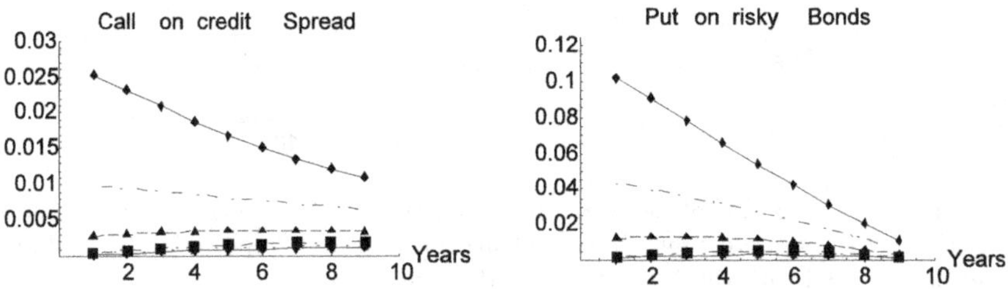

Fig. 3. Credit derivatives hedging against a fall in the credit quality of a 10-maturing bond below a BBB level. Underlying exposures are rated from AAA to B

5.3 The Constant Generator Case

In fact, it can be shown that a single constant risk neutral generator matrix *á la* AGL [3] cannot, by construction, handle with any shapes of risky term structures. This claim may be illustrated by Figure 4 that provides the credit spreads implied by the constant RNGM after risk neutral calibration on (strictly increasing) term structures of Figure 1. The CCC-rated credit spread is seen to be constant w.r.t. the maturity. This property of constant risk neutral generator matrix may be formally proved as follows.

Proposition 5.1. *A Markov Chain process with a constant generator (well defined w.r.t. credit risk) cannot generate, by construction, any strictly increasing tem structure of credit spread for the last credit class before default. Moreover, the lowest possible (flat) CCC-term structure of implied credit spreads is directly given by* $-\ln(\tilde{q}_{CCC,CCC}) = -\tilde{\lambda}_{CCC,CCC} = \tilde{\lambda}_{CCC,D}$.

Proof. First, let's recall that the CCC-rated class is, in our setting, the last class before default. Let's also point out that if $\mathcal{Q} = (\mathbf{q}_{ij})_{i,j=1,K+1}$ is such that $q_{CCC,j} > 0$ for any $j < CCC$ then the probability of being upgraded (at the next instant) is not null and as a result the credit margin is decreasing with respect to the horizon. $q_{CCC,j}$ are therefore null $\forall j < CCC$ and it then remains $q_{CCC,CCC} \neq 0$ and $q_{CCC,D} \neq 0$. Recalling furthermore that the default state is absorbing, one has $\tilde{\mathcal{P}}_t[\tau > T] = q^{T-t}_{CCC,CCC}$. As a result,

[10] and may considered in the present framework as well (available upon request).

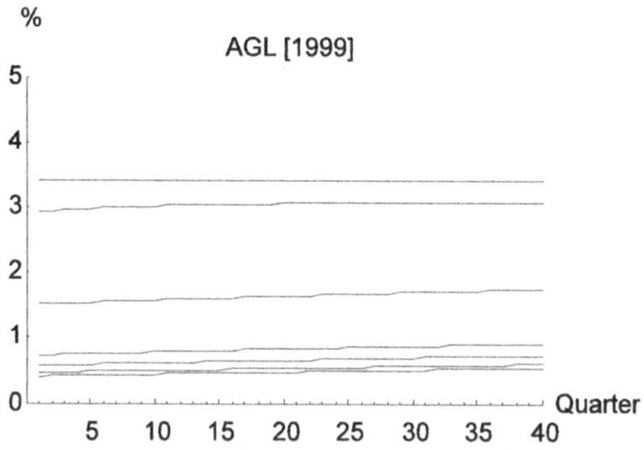

Fig. 4. Term structures of credit spreads implied by a single subjective generator matrix. Term structures are based on observable data plotted in Figure 1.

the term structure of implied credit spreads given by $-\frac{1}{T}\ln\tilde{\mathcal{P}}_0[\tau > T]$ is $-\ln q_{CCC,CCC}$. At last, equalities are obtained using (eq. 28, JLT (1997)) where $\tilde{q}_{i,K+1} = \tilde{\lambda}_{i,K+1}(\frac{\tilde{q}_{i,i}-1}{\ln\tilde{q}_{i,i}})$. Indeed, since $\tilde{q}_{i,i} = 1 - \tilde{q}_{i,K+1}$ for CCC, one has $\tilde{q}_{i,i} = e^{\tilde{\lambda}_{i,i}}$.

The risk neutral calibration suggested in AGL [3] therefore fails to match risky term structures whose CCC one is strictly increasing as that reported in KIK [14]. This is not the case of KIK [14]'s recursive procedure.

6 Conclusion

This paper has questioned the ability of the Markov Chain methodology with a constant or piecewised constant generator matrix for valuing credit derivatives, matching the observable risky term structures and generating implied credit spreads. This setting has been shown capable to price many credit derivatives. However since it generates *finite* and *deterministic* credit spreads and cannot handle with correlation between exposures, nor spread risk credit derivatives nor those involving a basket of credit exposures can be considered. Moreover it has been demonstrated that the constant risk neutral generator matrix cannot match any shapes of risky term structures. In particular, no strictly increasing CCC-term structure of credit spread, as the one reported in Kijima-Komoribayashi [14], can be implied by the framework.

7 Appendix

7.1 Operational Notes

Effective estimation procedures depend on the available information. Let's assume available a set of available risky zero-coupon bond prices

$(\hat{p}(0,T,\Theta_i))_{i\in\mathcal{S},T}$ as numerous as the number of credit class and maturities. Since the theoretical pricing equation holds $\forall T > 0, \forall i \in \mathcal{S}$ one has $\hat{p}(0,T;\Theta_i) = p(0,T)(\delta + \omega\tilde{\mathcal{P}}_i[\tau > T])$ or equivalently

$$q_{i,K+1}(T) = a + b\hat{p}(0,T;\Theta_i), \forall i \in \mathcal{S},$$

with $a = 1 + \delta\omega^{-1}$, $b = -[\omega p(0,T)]^{-1}$. Since, on the other hand, $\forall T$:

$$q_{i,K+1}(T) = \sum_{j\in\mathcal{S}} a(i,j)\pi_j(T), \forall i \in \mathcal{S}$$

with $a(i,j) = \tilde{q}_{ij}(T-1)q_{j,K+1}$, one has a system of $(K+1)$ equations with $(K+1)$ unknowns. At each horizon there exists a unique set of risk premia if and only if first the previous $(\tilde{q}_{i,j}(T-1))_{i,j\in\mathcal{S}}$ (and therefore $(\pi_j(T-1))_{j\in\mathcal{S}}$) have been previously calculated and second the resulting $(a(i,j))_{i,j\in\mathcal{S}\times\mathcal{S}}$ matrix is inversible. In this case, however, nothing warranties that this one-to-one relation is costless (in particular w.r.t. the stripping procedure). More general recursive least square minimization procedure subject to conditions ensuring suitable properties appears then reasonable to provide risk premia.

Finally, to compute the RNGM $\tilde{\Lambda}$ from the RNTM $\tilde{Q}$, a couple of numerical approximations are available, A first one, presented in JLT [10] (Eq. 28), may be applied to $I + \Pi(t)(Q - I)$. It holds as soon as 'more than one transition per year' is a low probability event (see Appendix A, 517-520.). A second one is related to an analytical approximation suggested in JLT [10] and exploits both $\tilde{Q}_{t,t+1} \approx I + \tilde{\Lambda}(t)$ and that U is identified to Π.

7.2 Default Swaps Valuation

As already claimed, digital default options allow one to price many other credit derivatives among which the default swaps. Default swaps are insurance contracts paying, at the default time and against an annuity premium, an amount covering the incured loss ([6]). Two default swaps with constant fee may be considered.

– The default digital swap is an exchange contract of a regular fee (paid until default) against a unit amount paid at default. Pricing this contract is equivalent to valuing the fee noted x. Along Schönbucher's lines [20], the constant and continuous fee rate is given by:

$$x\int_0^T [1 - q_{i,K+1}(0,t)]\,dt = \sum_{k\in\mathcal{S}^*} \lambda_{k,K+1}\int_0^T p(0,s)q_{i,k}(0,s)\,ds$$

– The default swap, dsw, is an exchange contract of a regular fee (paid until default) against an amount paid at default equaling the resulting lost value. Schönbucher [20] and Duffie [6] have demonstrated that it is duplicated by a portfolio containing a long position in a default free floating rate note and a short position in default risky one, this latter being obtained with the help of a digital default put option. The price of a default swap is given by:

$$dsw = frn(0,T) + c \sum_{k \in \mathcal{S}^*} \lambda_{k,K+1} \int_0^T p(0,s) q_{i,k}(0,s)\, ds - 1$$

where frn denotes the price of a par default free floating rate note. These formulae are valid for any model of term structure of interest rates.

Acknowledgements. We thank participants at the BFS Congress - Paris 2000 for comments and stimulating discussion, in particular T. Bielecki, H. Geman, JN. Hugonnier, M. Jeanblanc, JL. Prigent, O. Renault, O. Scaillet, P. Schönbucher and L. Schloegl. We are especially indebted to P. Schönbucher for suggesting digital contracts contingent to a change in credit class. S. Aboura, M. Bellalah, F. Quittard-Pinon are also acknowledged.

References

1. E. Altman, D. Kao : The Implications of Corporate Bond Ratings Drift", Financial Analyst Journal, May-June, (1992), 64-75.
2. E. Altman, V. Kishore : Almost Everything You Wanted to Know About Recoveries on Defaulted Bonds. Financial Analyst Journal Nov (1996) 57–64
3. A. Arvanitis, J. Gregory, J.P. Laurent : Building Models for Credit Spreads. Journal of Derivatives Spring (1999) 27–43
4. S. Das : Credit Risk Derivatives. Journal of Derivatives **2** (1995) 7–23
5. M. Crouhy, D. Galai, R. Mark : A Comparative Analysis of Current Credit Risk Models. Journal of Banking & Finance **24**(1/2) (2000) 59–117
6. D. Duffie : Credit Swap Valuation. Financial Analyst Journal Jan (1999) 73–87
7. L. Cathcart, L. El Jahel : Valuation of Defaultable Bonds. Journal of Fixed Incomes (1998) 65–78
8. J. Hand, R. Holthausen, R. Leftwich : The Effect of a Bond Rating Announcements on Bond and Stock Price. Journal of Finance **47** (1992) 733–750
9. J. Helwege, C. Turner : The Slope of the Credit Yield Curve for Speculative-Grade Issuers. Journal of Finance **54**(5), (1999) 1869–1884
10. R. Jarrow R., D. Lando, S. Turnbull : A Markov Model for the Term Structure of Credit Risk Spread. Review of Financial Studies **10** (1997) 481–523
11. R. Jarrow, S. Turnbull : Pricing Derivatives on Financial Securities subject to Credit Risk, Journal of Finance **50** (1995) 53–86
12. *Journal of Banking & Finance.* Special issues **22**(10/11), **24**(1/2)
13. M. Kijima : Monotonicities in Markov Chain Model for Valuing Corporate Bonds subject to Credit Risk. Mathematical Finance **8** (1998) 229–247
14. M. Kijima, K. Komoribayashi : A Markov Chain Model for Valuing Credit Risk Derivatives. Journal of Derivatives Fall (1998) 97–108.
15. D. Lando : Modeling Bonds and Derivatives with Default Risk, in M. Dempster and S. Pliska (eds.) *Mathematics of Derivatives Securities* Cambridge University Press (1997a) 369–393
16. D. Lando : On Cox processes and credit risky bonds. Review of Derivatives Research **2**(2/3) (1997b) 99–120

17. F. Longstaff, E. Schwartz : Valuing Credit Derivatives. Journal of Fixed Income **5** (1995) 25–32
18. C. Nielsen, A. Siegel : Parsimonious modeling of Yield Curves. Journal of Business **60**(4) (1987) 473-489
19. Y. Pierides : The Pricing of Credit Risk Derivatives. Journal of Economic Dynamics and Control **21** 1997 1579–1611.
20. P. Schönbucher : Pricing Credit Risk Derivatives. Working paper University of Bonn 1998
21. F. Skinner : A trinomial model of bonds with default risk. Financial Analyst Journal **50**(2) (1994) 73–78
22. *Standard & Poor's Special Reports.* 1998
23. M. Weinstein : The effect of a rating change announcement on bond price. Journal of Financial Economics **5** (1977) 329–350

An Autoregressive Conditional Binomial Option Pricing Model

Jean-Luc Prigent[1], Olivier Renault[2], and Olivier Scaillet[3]

[1] THEMA, Université de Cergy Pontoise, 33 bd du Port, 95011 Cergy-Pontoise, France
[2] Financial Markets Group, London School of Economics, Houghton Street, London WC2A 2AE, UK
[3] IRES and IAG, Université Catholique de Louvain, 3 place Montesquieu, 1348 Louvain-la-Neuve, Belgium

1 Introduction

On financial markets, option traders typically readjust their hedging portfolios when the underlying stock price has moved by a given percentage. This trading rule implies that rebalancing occurs at random times and that the Black and Scholes [4] model which relies on the assumption of continuous rebalancing is no longer appropriate. Although this continuity assumption is clearly unrealistic for transaction cost reasons, because prices are quoted in ticks, or because of the mere impossibility of continuous trading, it has received comparatively less academic attention than other assumptions such as constant volatility.

As a discrete approximation to the continuous time model of Black and Scholes, the binomial tree of Cox, Ross and Rubinstein [12] is widely employed by practitioners. Thanks to its elegant simplicity it is used as an introductory example to pricing theory in most finance textbooks. The simple structure of this model unfortunately yields its major drawback. Observed price variations do not follow i.i.d. binomial variables and the model cannot cope well with empirical data.

The purpose of this paper is to introduce an easily implementable pricing methodology which captures some of the salient features of observed market data and trading behaviour while enabling to derive option prices and deltas consistent with the trading rule described above. To this aim, we propose a model of discrete trading where market participants rebalance their position at random times triggered by variations in the underlying stock price by a given percentage a. In essence we introduce a binomial tree with random time spacings and probabilities in which the size of price changes is kept fixed. The structure of the traditional binomial pricing model is modified in order to improve its empirical fit and relax some of its unrealistic assumptions.

First we relax the stringent constraint of a fixed time interval between two successive price variations. The random intervals between two arrival times are called durations. Their conditional expectations depend on past durations in an autoregressive way. Such modelling for high frequency transaction data has been proposed by Engle and Russell ([18], [19]), and has already proved successful in examining empirical predictions of microstructure theory (see O'Hara [36] for a thorough exposition of microstrucure theory) on how the frequency of transactions (clustering) should carry information about the state of the market (see also Engle [17]).

Second we allow for a similar past dependence in the probabilities of up-moves and down-moves. We therefore acknowledge the possibility that the direction of previous price jumps may influence forthcoming variations in the stock price. Such a relationship can occur if traders have a herding behaviour (positive relationship) or if a deviation from the fundamental value of the stock tends to be corrected by market participants (negative relationship). The autoregressive specification for up-move probabilities mainly follows the framework of Cox ([10],[11]) and its recent extension by Russell and Engle [44]. It consists of a pure time series version of the specification adopted by Hausman, Lo and MacKinlay [27] for analysing transaction stock prices. This joint dynamic modelling of the price transition probabilities and the arrival times of the transactions is rich and flexible enough to capture the historical behaviour of price change data.

The structure of the paper is the following. We first review in section 2 some basic concepts about marked point processes (MPP) which embody our binomial tree specification. We outline our framework and introduce the main notations and mathematical tools used in the modelling. Dynamic specifications for the arrival times and the jumps (namely the ACD and ACB models) are discussed in some detail. These models will be implemented in the empirical application. Section 3 provides an introduction to the minimal martingale measure (MMM), before giving the option pricing formula based on it. The minimal martingale measure is the main building block of our pricing strategy. Derivative asset prices are obtained by taking discounted expectations of future payoffs with respect to this measure. The option pricing formulae rely on pricing tools derived in Prigent, Renault and Scaillet [39]. In Section 4, an empirical application is provided using IBM intraday transaction data on the NYSE (New York Stock Exchange). Model parameters are estimated and used as input to compute European call option prices. These prices are compared with Black-Scholes prices based on an historical volatility estimate from daily closing prices. The model is able to capture the shape of the volatility smile usually observed on stock option markets. Section 5 concludes.

2 Framework

We first start by reviewing some basic facts about marked point processes (MPP). The initials JS stand for the book of Jacod and Shiryaev [30], which gathers major contributions to the theory of MPPs.

Let us consider an increasing sequence of non overlapping random times T_j, $j = 1, \ldots$. To each of them, we associate a random variable Z_j, called a mark and defined on the same probability space: $(\Omega, \mathcal{F}, P)$. Each (T_j, Z_j) is said to be a marked point and the sequence $((T_j, Z_j))_j$ of marked points is referred to as a marked point process (see Figure 1).

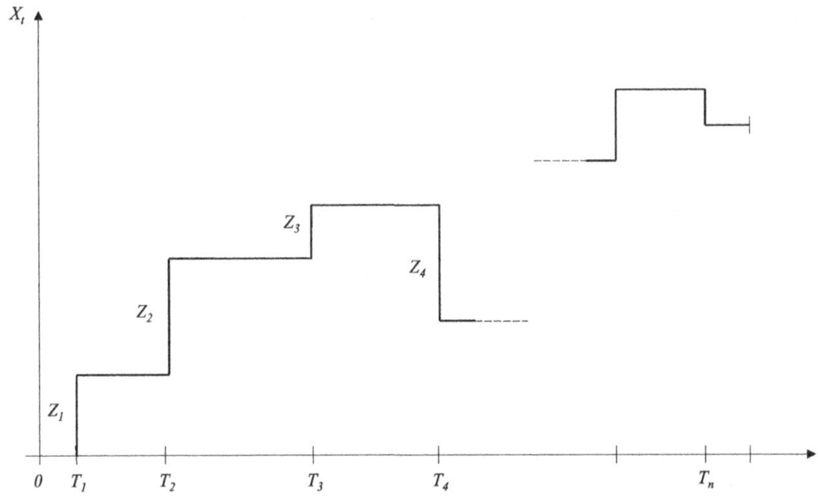

Fig. 1. A typical realisation of a marked point process

There are therefore three elements which caracterize a MPP. The law of arrival times of the jumps, the size (or amplitude) of the marks and their "direction" (either up or down). In our framework, we restrict the mark size to be constant and equal to a and denote the mark space by $E = \{a, -a\}$, therefore leaving two elements to be specified.

We assume that the logarithm of the stock price S_t follows such a MPP so that:

$$S_t = S_0 e^{X_t}, \tag{2.1}$$

where the random variable:

$$X_t = \sum_{j:T_j \leq t} Z_j, \tag{2.2}$$

corresponds to the sum of jumps taken over the random times T_j of their arrivals. The process X is a purely discontinuous process with jumps $Z_j = \Delta X_{T_j}$. The logarithmic variations of the stock price: $\Delta \log S_{T_j} = \Delta X_{T_j}$ thus either take the values a or $-a$.

The process X can be written as the sum of jumps over time and over the mark space (see JS p. 69-72):

$$X_t = \int_0^t \int_E x\mu(dt, dx)$$

The random measure $\mu(dt, dx)$ is the counting measure associated to the marked point process. This measure records the number of jumps occuring in the time interval dt and whose size falls in the interval dx (here $dx = a$ or $-a$). Introducing the predictable measure ν, called the compensator, with the property that $\mu - \nu$ is a local martingale measure, we can write:

$$X_t = \int_0^t \int_E x\nu(dt, dx) + \int_0^t \int_E x(\mu - \nu)(dt, dx). \tag{2.3}$$

Although this rewritting may look only technical, it enables us to see the crucial role of the compensator in our pricing model. We will be concerned with the expected returns on the stock price or equivalently with the expected value of X. By definition of the compensator, we know that the second term in (2.3) is a martingale, only leaving the first term to specify and estimate from the data.

Recall that two elements were left to fully specify our marked point process: the law of arrival times and the probability of an up-move (the jump size a being held constant). The compensator can be disintegrated (JS p. 67) so that these two elements are clearly identified:

$$\nu(dt, dx) = d\Lambda_t K(t, dx),$$

where Λ is a predictable integrable increasing process and K is a transition kernel.

The process Λ represents the intensity of the arrival times of jumps, and when $d\Lambda_t = \lambda_t dt$, the process λ is called the directing intensity and corresponds to a conditional hazard function. The transition kernel $K(t, dx)$ is given by: $\sum_j 1\!\!1_{\{T_j < t \leq T_{j+1}\}} P(Z_{j+1} \in dx | \mathcal{F}_{T_j}, T_{j+1})$ i.e. the conditional probability of an up-jump (if $x = a$) or a down-jump (if $x = -a$) given the current information set $\mathcal{F}_{T_j}$ (made of current and past realisations of the MPP) and the fact that there is a jump at time T_{j+1}.

Note that the standard Poisson process corresponds to the constant directing intensity case: $d\Lambda_t = \lambda dt$. Its jumps are always 1, so all the probability is concentrated on the point $dx = 1$ which implies that the compensator can be written:

$$\nu(dt, dx) = \lambda dt \epsilon_1(dx),$$

where $\epsilon_1(dx)$ denotes the Dirac measure at point 1 with property $\epsilon_1(dx) = 1$ for $dx = 1$ and $\epsilon_1(dx) = 0$ otherwise.

Let us summarize what we have obtained so far. We have a model where the log of the stock price follows a marked point process with constant jump size a. We have shown the usefulness of the compensator which enables us to calculate expected price variations. This measure can be broken down into two parts, one of which is the intensity of jump arrival times and the second which determines the up-move probability. The model is fully specified up to the choice of these two terms which we will now discuss.

In Prigent, Renault and Scaillet [39], we studied two specifications for the compensator. We considered the case where the true stock price is an unobservable geometric Brownian motion whose values were only known when its logarithm crossed boundaries spaced by a. Although intuitive from a continuous time finance point of view, this specification was not appropriate for empirical purposes because of the cumbersome and restrictive form of the kernel $K(t, dx)$. A marked Poisson model (i.e. a Poisson process whose jumps follow i.i.d. binomial variables) was then proposed. Both parameters (the directing intensity λ and the up-move probability p) were easily estimated on IBM transaction data and prices were derived for European call options on IBM stock for values of a between 3% and 5%. The marked Poisson specification was in the same spirit as the stochastic volatility model proposed by Bossaerts, Ghysels and Gourieroux [6], and based on a time deformed binomial model.

However, we noticed that the hypothesis of exponentially distributed intertrade durations was rejected for small values of a ($< 3\%$). This was due to the phenomenon of overdispersion of durations (the standard deviation of the durations exceeds their mean) in intraday data, a well documented fact in the microstructure literature. Besides, a constant up-move probability appears not to be a satisfactory assumption as already mentioned because of a possible herding behaviour of traders or, on the contrary, because of a tendancy to revert towards the fundamental value of the stock.

We will now tackle these two issues by turning to other possible specifications of the compensator which will be easily estimable and testable from market data. These specifications will also allow to derive expressions for option prices. Building on the econometrics of high frequency data (Engle [17]), we have chosen to use autoregressive conditional specifications for the conditional distributions of both arrival times and marks.

2.1 Conditional Distribution of Durations

The conditional distribution of arrival times is specified according to an ACD(m,q) model proposed by Engle and Russell ([18], [19]). The Autoregressive Conditional Duration (ACD) class of models consists of assuming that the durations $d_{j+1} = T_{j+1} - T_j$ are such that:

$$d_{j+1} = \psi_{j+1}\xi_{j+1},$$

where ξ_j are positive i.i.d. variables and the conditional expectation: $\psi_j = E\left[d_j|\mathcal{F}_{T_{j-1}}\right]$ is:

$$\psi_j = \omega + \sum_{k=1}^{m} \alpha_k d_{j-k} + \sum_{k=1}^{q} \beta_k \psi_{j-k}.$$

These models are analogous to ARCH and GARCH models (Engle [16], Bollerslev [5]) and share many of the same properties. For example, some constraints must be satisfied by the parameters, specifically:

$$\omega > 0, \alpha_k \geq 0 \text{ and } \beta_k \geq 0\,,^1 \tag{2.4}$$

in order to ensure the positivity of the durations, and

$$\sum\nolimits_{k=1}^{m} \alpha_k + \sum\nolimits_{k=1}^{q} \beta_k < 1, \tag{2.5}$$

to guarantee model stationarity and allow to use Maximum Likelihood (see Engle and Russell [19], Carrasco and Chen [8]). The conditional hazard function of an ACD model for t in the random time interval $]]T_j, T_{j+1}]]$ is given by:

$$\lambda_t = \psi_{j+1}^{-1} \lambda_0 \left(\frac{t - T_j}{\psi_{j+1}} \right),$$

where λ_0 is the baseline hazard of ξ (the ratio of the density and survival functions of ξ). Two choices are usually adopted for the distribution of ξ, either the exponential or the Weibull, which give respectively:

$$\lambda_t = \psi_{j+1}^{-1},$$

or:

$$\lambda_t = \left(\psi_{j+1}^{-1} \Gamma(1 + 1/\gamma)\right)^{\gamma} (t - T_j)^{\gamma - 1} \gamma, \tag{2.6}$$

where Γ is the gamma function and γ is the second Weibull parameter. The first Weibull parameter must be equal to $\Gamma(1 + 1/\gamma)$ in order to ensure that ξ has mean 1. When $\gamma = 1$, the Weibull distribution coincides with the exponential distribution.

In trying to fit the above ACD model to IBM transaction data, we found that the constraints on the parameters ((2.4) and (2.5)) were not always satisfied. We thus propose to use the Log-ACD model proposed by Bauwens and Giot [3] (see also Russell and Engle [44]) which is the analogue of the Log-GARCH model of Geweke [23] applied to durations. It is specified as:

$$d_{j+1} = \exp\left(\psi_{j+1}\right) \xi_{j+1},$$

[1] The positivity of the parameters is a sufficient but not necessary condition to ensure that ψ_i be positive. These constraints can be weakened as shown by Nelson and Cao (1992).

with a conditional expectation $\exp(\psi_j)$ satisfying:

$$\psi_j = \omega + \sum_{k=1}^{m} \alpha_k \ln(d_{j-k}) + \sum_{k=1}^{q} \beta_k \psi_{j-k}.$$

and its conditional hazard function in the random time interval $]]T_j, T_{j+1}]]$ is given by:

$$\lambda_t = \exp(-\psi_{j+1}),$$

for the exponential case and

$$\lambda_t = (\exp(-\psi_{j+1})\,\Gamma(1+1/\gamma))^{\gamma}\,(t - T_j)^{\gamma-1}\,\gamma,$$

for the Weibull case.

2.2 Conditional Distribution of Marks

Concerning the conditional distribution of marks we use the extension of the logistic linear model of Cox ([10],[11]) given by Russell and Engle [44]. We define:

$$Y_j = \begin{cases} 1 & \text{if} \quad Z_j = +a, \\ 0 & \text{if} \quad Z_j = -a. \end{cases}$$

The probability $\pi_j = P[Y_j = 1|\mathcal{F}_{T_{j-1}}]$, resp. $1 - \pi_j$, gives the conditional probability of an up-move, resp. a down-move. In a logistic linear model, it satisfies:

$$l(\pi_j) = \bar{\omega} + \sum_{k=1}^{\bar{m}} \bar{\alpha}_k Y_{j-k}$$

with $l(\pi) = \log(\pi/(1-\pi))$. The logistic transformation l ensures the interpretation of π as a probability. Russell and Engle [44] propose to extend this specification by incorporating lagged values of the conditional probabilities themselves:

$$l(\pi_j) = \bar{\omega} + \sum_{k=1}^{\bar{m}} \bar{\alpha}_k Y_{j-k} + \sum_{k=1}^{\bar{p}} \bar{\beta}_k \pi_{j-k} + \sum_{k=1}^{\bar{r}} \bar{\kappa}_k l(\pi_{j-k}).$$

This model has an Autoregressive Conditional Binomial (ACB) structure and is a binomial version of the Autoregressive Conditional Multinomial (ACM) model of Russell and Engle [44]. We refer to it as an ACB($\bar{m}$,$\bar{p}$,$\bar{r}$) model. The transition kernel of the MPP is taken for $t \in]]T_j, T_{j+1}]]$ equal to:

$$K(t, dx) = \begin{cases} \pi_{j+1} & \text{for} \quad dx = +a, \\ 1 - \pi_{j+1} & \text{for} \quad dx = -a. \end{cases}$$

Both autoregressive conditional models, ACD for the arrival times and ACB for the marks, are thus the building blocks of our specification for the stock price dynamics. This is summarized by the next assumption.

Assumption 2.1. ***(model specification)***
The compensator $\nu(dt, dx)$ on $\mathbb{R}_+ \times \{a, -a\}$ satisfies:

$$\nu(dt, dx) = \lambda_t dt K(t, dx),$$

where for $t \in]]T_j, T_{j+1}]]$, the directing intensity λ_t is given by the conditional hazard function of an ACD(m,q) model and the transition kernel $K(t, dx)$ corresponds to an ACB($\bar{m}$,$\bar{q}$,$\bar{r}$) model.

Now that the setting is described we may turn to the next step: the choice of an equivalent martingale measure and the derivation of an option pricing formula.

3 Option Pricing and the Minimal Martingale Measure

From Harrison and Kreps [25] and Harrison and Pliska [26], we know that in order to preclude arbitrage in a market, there must exist an equivalent martingale measure (EMM) under which discounted asset prices are martingales. However, this measure needs not be unique unless markets are complete.

The presence of jumps in our framework implies that the market is incomplete. We therefore need a criterion by which to choose among all EMM, one measure under which to calculate option prices as expectations of discounted future payoffs. In this paper we have chosen the minimal martingale measure (MMM) initially proposed by Föllmer and Schweizer [21]. Let us first recall some results about pricing in incomplete markets before motivating our choice of measure. We consider for simplicity the case of a market with one risky asset (the stock with price S_t) and one riskless asset whose growth rate is set to zero.

The standard approach to pricing by arbitrage consists of finding a portfolio (i.e. an investment policy of α in the riskless asset and β in the stock) which replicates the payoffs H of the option we want to price. Let $V_t = \alpha_t + \beta_t S_t$ denote the value of our hedging portfolio. The cost process of following a trading strategy from time 0 to time t is $C_t = V_t - V_0 - \int_0^t \beta_u dS_u$

Recall that a trading strategy is said to be self-financing if the cost is 0. When markets are incomplete, no self-financing strategy will provide a perfect hedge ($V_T = H$ a.s.) for the option, or conversely, if we adopt a strategy which replicates the payoff of the contingent claim perfectly, it will in general not be self-financing. The natural way forward is to try to minimize some definition of the remaining risk or equivalently the cost associated with the replicating strategy.

One possibility is to choose a strategy (α^*, β^*) which minimizes the total risk under the historical measure P as defined by $R_t = E^P\left[(C_T - C_t)^2 \middle| \mathcal{F}_t\right]$. This corresponds to the choice of pricing under the variance optimal measure (for this measure see e.g. Föllmer and Sondermann [20], Bouleau and Lamberton [7], Duffie and Richardson [14], Schweizer ([46], [48]), Gouriéroux, Laurent and Pham [24], Laurent and Scaillet [31]). This strategy is mean self-financing, i.e., on average, its cost is zero. However the associated measure has two main drawbacks. First, an optimal (i.e. minimizing the total risk) strategy does not always exist. Second, the variance optimal measure does not have an analytic form in general and is therefore unpractical.

Another possibility is to adopt another (also mean self-financing) strategy which minimizes the local risk in the sense of Schweizer [45]. Instead of considering the total variation of the cost between dates t and T, it consists in minimizing all the variations of the costs over successive "small" periods between t and T (see Frey [22]).

This policy corresponds to the choice of the minimal martingale measure which we will be using in this paper. This measure is characterized by the fact that it sets to zero all risk premia on sources of risk orthogonal to the martingale part of the underlying's price process. An example, borrowed from Hofmann, Platen and Schweizer [28], will help understand this statement. One of the most famous models of option pricing with stochastic volatility has been proposed by Hull and White [29]. It is particularly convenient to model volatility smiles (see Renault and Touzi [41]). The authors assume that the square of the volatility follows a geometric Brownian motion which is uncorrelated to the Brownian motion driving the price process. The market is incomplete because there are two sources of risk and only one risky asset to trade with. However Hull and White [29] argue that if one assumes that the CAPM holds and that volatility risk is diversifiable, this source of uncertainty should not bear a risk premium. They then proceed to derive their option price. This price coincides with that given by the MMM which precisely assigns a zero value to the market price of risks orthogonal to the martingale part (here the Brownian motion) of the price.

The MMM has several appealing features which motivate its use in practical applications. First, there always exists an explicit form for the Radon-Nikodym derivative enabling to switch from the historical probability measure to the minimal measure. This makes it a computationally convenient tool as will become apparent later on. Then, recent work on the topic shows that it induces good convergence properties (Runggaldier and Schweizer [43], Prigent [38], Mercurio and Vorst [34], Lesne, Prigent and Scaillet [32]). Furthermore, in our framework (see Prigent, Renault and Scaillet [39]), jump boundedness ensure that the MMM is a probability measure (i.e. is always positive) and therefore that the value of the trading strategy is an actual no-arbitrage price.

Finally this measure can also be linked to other possible choices of measures. For example Schweizer [47] shows that in some cases the expectation of the final payoff under the minimal measure is equal to the value of the variance optimal hedging strategy (α^*, β^*) described above. When the mean-variance trade-off (i.e. the market price of risk) of the price process is deterministic, the MMM is the closest of all EMM to the historical measure P, as measured by the relative entropy criterion (Föllmer and Schweizer [21]). Concerning existence and uniqueness of the minimal measure, we refer to Ansel and Stricker ([1], [2]).

We now turn to the derivation of the option pricing formula under the minimal measure. We take as discount factor (or numéraire) a savings account whose growth rate r_{T_j} on the random time interval: $]]T_{j-1}, T_j]]$ satisfies:

$$r_{T_j} = e^{\rho(T_j - T_{j-1})} - 1, \tag{3.7}$$

with $\rho > 0$.

The discounted stock price is then equal to:

$$\tilde{S}_t = S_t / \prod_{j:T_j \le t} (1 + r_{T_j}).$$

Let us introduce the discounted excess return process δ such that for $t \in]]T_j, T_{j+1}]]$:

$$\delta(t,x) = \delta(T_j, Z_j) = \frac{e^{Z_j} - (1 + r_{T_j})}{1 + r_{T_j}} = e^{Z_j - \rho(T_j - T_{j-1})} - 1. \tag{3.8}$$

The minimal measure is characterized by its Radon-Nicodym derivative w.r.t. P. It takes the following form in our framework (Prigent, Renault and Scaillet [39]).

Proposition 3.1. ***(minimal probability measure)***
Under Assumption 1, the minimal martingale measure $\hat{P}$ is a probability measure characterized by its density process $\hat{\eta}$ relative to P:

$$\hat{\eta}_t = \prod_{j:0<T_j \le t} (1 + \hat{h}_j(Z_j)) \exp\left(- \int_0^t \int_E \hat{H}(s,x) K(s,dx) \lambda_s ds \right). \tag{3.9}$$

where for $t \in]]T_j, T_{j+1}]]$, $\hat{H}(t,x) = \hat{h}_j(x)$ with:

$$\hat{h}_j(x) = - \frac{\delta(T_j, a)\pi_{j+1} + \delta(T_j, -a)(1 - \pi_{j+1})}{\delta^2(T_j, a)\pi_{j+1} + \delta^2(T_j, -a)(1 - \pi_{j+1})} \delta(T_j, x).$$

$\hat{H}(t,x)$ takes the interpretation of a jump risk premium process. Once the Radon-Nicodym derivative $\hat{\eta}_t$ is computed, it is straightforward to derive the price $C_t = C(t, S_t)$ of a contingent claim with final payoff $C(T, S_T)$. For a European call option with maturity T and strike price $\overline{K}$, the final payoff is $(S_T - \overline{K})_+ = \max(0, S_T - \overline{K})$. By taking its expectation under $\hat{P}$ after an adequate discounting, we get:

$$C(t, S_t) = E^{\hat{P}} \left[(S_T - \overline{K})_+ \prod_{j:t<T_j \le T} (1 + r_{T_j})^{-1} | \mathcal{F}_t \right], \tag{3.10}$$

which leads to:

Proposition 3.2. *(minimal option price)*
Under Assumption 1, the call price given by the minimal martingale measure is:

$$C(t,S_t) = E^P\left[(S_T-\overline{K})_+\frac{\hat{\eta}_T}{\hat{\eta}_t}\prod_{j:t<T_j\leq T}(1+r_{T_j})^{-1}|\mathcal{F}_t\right]. \tag{3.11}$$

If the exponential or the Weibull distribution underlies the ACD model, we have:

$$\frac{\hat{\eta}_T}{\hat{\eta}_t} = \prod_{j:t<T_j\leq T}\left(1-\frac{\delta(T_j,a)\pi_{j+1}+\delta(T_j,-a)(1-\pi_{j+1})}{\delta^2(T_j,a)\pi_{j+1}+\delta^2(T_j,-a)(1-\pi_{j+1})}\delta(T_j,Z_j)\right) \tag{3.12}$$
$$\exp\left(-\frac{(\delta(T_j,a)\pi_{j+1}+\delta(T_j,-a)(1-\pi_{j+1}))^2}{\delta^2(T_j,a)\pi_{j+1}+\delta^2(T_j,-a)(1-\pi_{j+1})}\log G(\min(T_{j+1},T)-T_j)\right),$$

where $G(u)$ is the survival function (i.e. the probability of not jumping over the period u). $-\log G(\min(T_{j+1},T)-T_j)$ is equal to $(\min(T_{j+1},T)-T_j)/\psi_{j+1}$ for the exponential ACD, to $(\min(T_{j+1},T)-T_j)/\exp(\psi_{j+1})$ for the exponential Log-ACD. For Weibull distributed noise $-\log G(\min(T_{j+1},T)-T_j)$ is equal to $(\Gamma(1+1/\gamma)(\min(T_{j+1},T)-T_j)/\psi_{j+1})^\gamma$ for the ACD model and to $(\Gamma(1+1/\gamma)(\min(T_{j+1},T)-T_j)/\exp(\psi_{j+1}))^\gamma$ for the Log-ACD model.

Expectation (3.11) can in principle be valuated by Monte-Carlo integration. Indeed S_T, $\hat{\eta}_T/\hat{\eta}_t$, and $(r_{T_j})_j$ can be computed from simulated paths of the MPP $((T_j,Z_j))_j$ once the parameters of the ACD and ACB models have been estimated. However, since $\hat{\eta}_T/\ \hat{\eta}_t$ is made of a product of terms, it will not be accurately estimated through simulations. Therefore it is wiser to use expression (3.10) and work directly with the dynamics of S_t under $\hat{P}$. The process of S_t under $\hat{P}$ can be derived using relationships between the directing intensities and transition kernels under P and $\hat{P}$ (see appendix). These relationships come from a direct application of Girsanov theorem for jumps (JS p.157).

4 An Empirical Illustration on IBM Trades

This section illustrates the empirical application of our marked point model to the pricing of European call options.[2] The parameters of the ACD and ACB models are estimated from intraday transaction data. The data were extracted from the Trades and Quotes Database (TAQ Database) released by the NYSE and span the period beginning on Thursday January 2th 1997 and ending on Wednesday September 30th 1997 (9 months). We report results obtained on trades of the IBM stock, which is one of the most liquid stocks in this market and is the support of actively traded options.

[2] Gauss programs developed for this section are available on request.

The observations are the trades recorded every second from market opening (9:30:00) to market closure (16:00:00). The trades dataset consists of 498, 692 transactions. We removed all trades which took place outside market opening hours and were left with 486, 506 data points. We also adjusted our series for a 2:1 stock split which occurred on 28th May 1997 before market opening.

We now proceed to estimate the intensity and the kernel, namely the ACD and ACB models. Given Assumption 1, the likelihood function is separable as in Russell and Engle [44] and we can estimate the parameters of the two models separately.

4.1 Estimation of the Log-ACD Model

We first consider the dynamics of durations between trades. As mentioned above, we have chosen a Logarithmic Autoregressive Conditional Duration specification. Recall that in a Log-ACD model durations d_j are assumed to follow:

$$d_{j+1} = \exp(\psi_{j+1})\, \xi_{j+1},$$

where ξ_j are i.i.d. variables (typically exponential for the Log-EACD model and Weibull for the Log-WACD model) and $\psi_j = \omega + \sum_{k=1}^{m} \alpha_k \ln(d_{j-k}) + \sum_{k=1}^{q} \beta_k \psi_{j-k}$.

Parameter estimates are obtained by maximum likelihood for $m = 1, 2$ and $q = 1, 2$ and both specifications of the distribution of ξ_j (exponential or Weibull).

The most successful specification is the Log-WACD(1,1) whose parameter estimates are given in Table 1. Parametrizations with more lags are rejected by a likelihood ratio test at the 5% level and the hypothesis of exponentially distributed ξ_j is strongly rejected (γ is ranging from 0.40 to 0.48 for different jump sizes and is statistically different from unity).

These results are in line with what was previously documented in tick-by-tick transaction studies: durations exhibit clustering, i.e. a short time between

Table 1. Parameter estimates of the Log-WACD(1,1) model

a	0.5%	1%	2%
ω	0.6111*	0.8607*	2.3173*
α	0.1844*	0.1701*	0.2293*
β	0.7740*	0.7705*	0.5912*
γ	0.4806*	0.4180*	0.3998*
$LB(40)$	21.0283**	13.4486**	9.5661**
$BI-1$	3.2917	2.8791	0.3300
$BI-2$	1.7112	2.3959	1.0351

* significant at 1% level, ** significant at 5% level.

two price variations tends to be followed by another short interval. This is suggested by Figure 2 and is confirmed by the positivity of $\widehat{\alpha}$ and $\widehat{\beta}$. Estimates of ω, α and β are all significantly different from zero at the 1% confidence level in the Log-WACD(1,1) model.

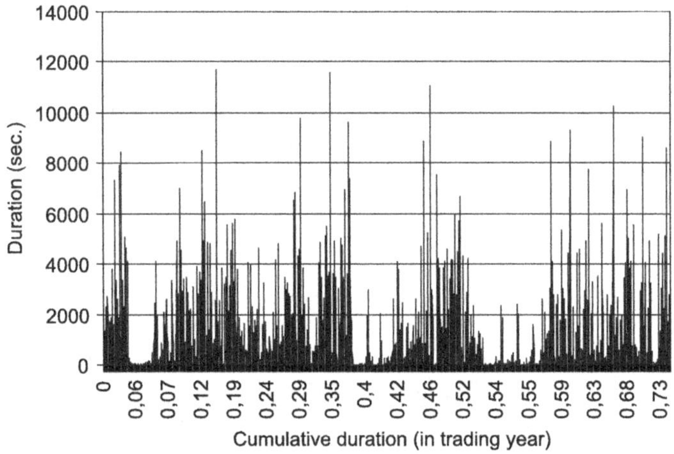

Fig. 2. Durations between two 0.5% jumps

Under our hypothesis, the residuals $\widehat{\xi}_j$ should be i.i.d. Weibull. We start by testing the absence of autocorrelation. $LB(40)$ reported in Table 1 denotes the Ljung-Box test with 40 lags. The hypothesis of zero autocorrelation in the residuals cannot be rejected at the 5% level for $a = 0.5\%$, $a = 1\%$ and $a = 2\%$.

We now want to test for Weibull-distributed residuals. Recall that if ξ_j is Weibull with parameter γ, then $(\xi_j)^\gamma$ follows an exponential distribution. This is the latter hypothesis which we test using Bartlett identities tests for the exponential distribution introduced by Chesher, Dhaene, Gourieroux and Scaillet [9].

The Bartlett Identity test of order 1 considers the equality between mean and standard deviation (overdispersion test on residuals), while the Bartlett Identity test of order 2 examines a restriction on the first three moments. Taking this second restriction into account helps gaining power against alternative specifications. Results of these tests (BI-1 and BI-2) are provided in the last two rows of Table 1. All values are below their critical $\chi^2(1)$ distribution at the 5% level $\left(\chi^2_{0.05}(1) = 3.841\right)$, so we cannot reject the hypothesis of Weibull distributed residuals.

4.2 Estimation of the ACB Model

We now turn to the estimation of the process governing up-moves and down-moves from trades data. We have adopted the specification of Autoregressive Conditional Binomial jumps where the logistic transformation $l(.)$ of the up-move probability π_j is given by:

$$l(\pi_j) = \bar{\omega} + \sum_{k=1}^{\bar{m}} \bar{\alpha}_k Y_{j-k} + \sum_{k=1}^{\bar{p}} \bar{\beta}_k \pi_{j-k} + \sum_{k=1}^{\bar{r}} \bar{\kappa}_k l(\pi_{j-k}),$$

and $Y_j = 1$ if the jump at the period j is an up-move and 0 otherwise.

Again, we find that a simple specification with one lag in all parameters was successful in capturing the dynamics of the kernel. Table 2 reports the parameter estimates obtained for this $ACB(1,1,1)$.

Table 2. Parameter estimates of the $ACB(1,1,1)$ model

a	0.5%	1%	2%
$\bar{\omega}$	0.3675*	0.5018*	0.3608*
$\bar{\alpha}$	−0.6420*	−0.8899*	−0.8043*
$\bar{\beta}$	0.0634*	0.1582*	0.4863**
$\bar{\kappa}$	−0.7790*	−0.6979*	−0.2827*
# jumps	3715	1065	222

* significant at 1% level, ** significant at 10% level.

All parameters but one are significant at the 1% level. The dynamics are similar for the various values of a, with the past jump and the logistic transform entering negatively in the recursion and the past probability entering positively.

We show in Figure 3 the relationship between the successive up-move probabilities π_j and π_{j-1} setting alternatively $Y_{j-1} = 0$ (down-move) or $Y_{j-1} = 1$ (up-move), for $a = 0.5\%$. We can observe that π_j is a decreasing function of π_{j-1} in both cases. It means that a high probability of an up-move will be followed by a lower probability at the next arrival time. This observation is thus not in favour of a herding behaviour of market participants. This decrease is less pronounced when a down-move ($Y_{j-1} = 0$) has occurred, introducing an asymmetry in the response of the probability levels. Such an effect reinforces the mean reversion type of behaviour of the stock price since the up-move probability is comparatively higher when the stock has gone down ($Y_{j-1} = 0$) than when it has gone up ($Y_{j-1} = 1$). This remark also applies for other jump sizes.

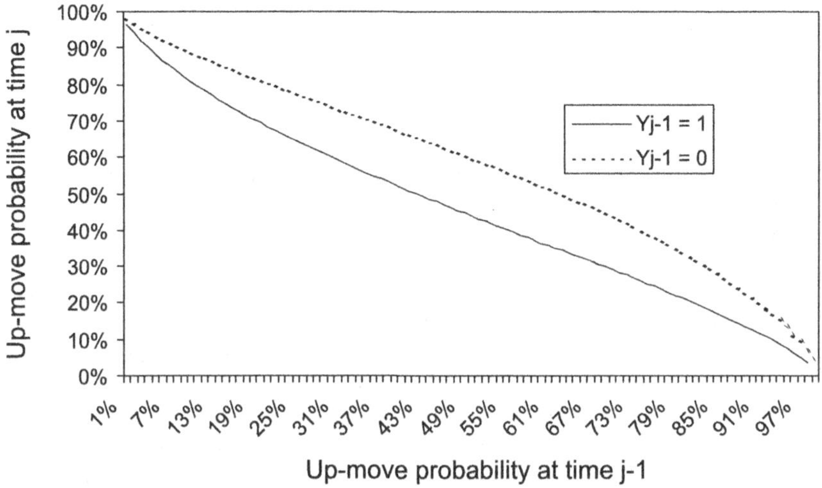

Fig. 3. Up-move probability π_j as a function of past probability π_{j-1}

Figure 4 plots the jump risk premium $\widehat{h}_j(x)$ for $a = 0.5\%$ over the whole observation period. We can clearly see a clustering phenomenon in the risk premium. Large changes in the risk premium tend indeed to be followed by other large changes of either sign, and small changes tend to be followed by small changes. This type of clustering corresponds to the well known Mandelbrot [33] observation on price changes (also valid for the IBM stock).

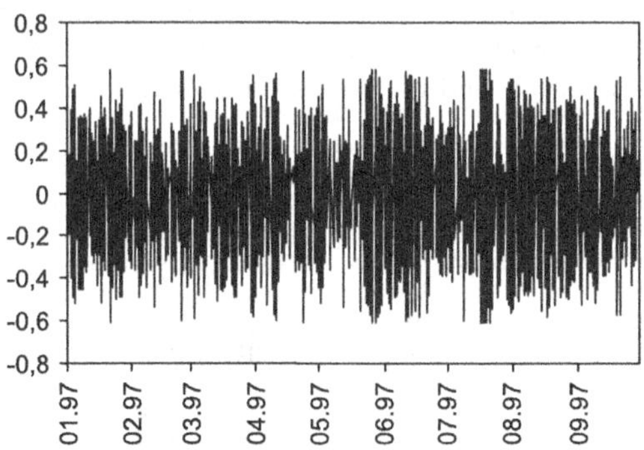

Fig. 4. Evolution of the risk premium $\widehat{h}_j\ (x)$

Trade prices used so far are appealing because they correspond to prices at which real transactions take place while quotes are only indicative values at which market markers are willing to trade. However, our database does not enable us to identify when a trade corresponds to a purchase of stocks from the market-maker or to a sale to the market-maker. Trade prices are thus potentially influenced by the bid-ask bounce (a trade at bid price followed by a trade at offer price leading to an observed price change although the mid price remained unchanged). This phenomenon has been reported to generate spurious autocorrelation (see Roll [42]) and could also lead to an over-estimation of the number of jumps for small a. Typical values for the bid-ask spread are one or two ticks (\$0.125 or \$0.25) as reported in the NYSE fact book which represents about 0.125% or 0.25% (the IBM stock price oscillated around \$100 in our sample period). We do not believe that the bid-ask bounce should substantially affect the dynamics of arrival times of the jumps but it could bias the estimation of the ACB specification for $a = 0.5\%$ as considered in this paper. Especially the negative sign of $\overline{\alpha}$ may partially be attributable to it. However we have seen that the negative sign of $\overline{\alpha}$ persists even for large values of a where the bid-ask bounce surely is not at work. Furthermore the same methodology has been applied to mid-prices (i.e. the mean of bid and ask prices) which are free of any bid-ask bounce effect and has lead to similar results. We prefer to proceed further with trades rather than quotes because we believe it is more appropriate to work with actual transaction prices, especially for option pricing and hedging purposes.

4.3 Option Pricing with the ACB Model

We are now able to price options with the underlying Log-WACD and ACB models and formulae (3.11) and (3.12). We carry out Monte Carlo simulations with 100,000 replications and use S_T as control variate device, knowing that $S_t = E^{\hat{P}}\left[S_T \prod_{j:t<T_j\leq T}(1+r_{T_j})^{-1}|\mathcal{F}_t\right]$, to reduce the variance of Monte Carlo estimates.

Table 3. Option prices: ACB vs Black-Scholes

$a \setminus K$	90	95	100	105	110
0.5%	12.2633	8.5415	5.5638	3.4205	2.0076
1%	12.8271	9.3088	6.4706	4.3594	2.8766
2%	13.1964	9.8069	7.0430	4.9518	3.4201
BS	13.5285	10.2305	7.5030	5.3400	3.6926

Option prices based on the estimated Log-WACD(1,1) and ACB(1,1,1) models are compared with Black-Scholes prices in Table 3. Black-Scholes prices are computed using historical volatility of daily IBM stock returns taken over a 3 month period (from 2nd January to 27th March 1997). This 3 month length corresponds to market standards. Over this period the annualised volatility was 34.67%. For comparison, calculating the volatility over the whole sample (9 months) would yield an annualised volatility of 31.98%. All ACB prices are below their Black-Scholes equivalent. However this needs not be the case for all specifications. We have indeed tested the model on other data and other values of a and we have found that the Black-Scholes price cannot be seen as an upper bound on the values of the ACB model. Finally, prices generated by the ACB model can be inverted using the Black-Scholes formula in order to derive implied volatilities. Implied volatilities for various strike prices and values of a are displayed in Table 4. We find that the ACB model exhibits a volatility smile for all a which is asymetric with respect to the at-the-money level $\overline{K}$. The implied volatility ratio is here defined as the implied volatility for a given strike price $\overline{K}$ divided by the implied volatility for $\overline{K} = 100$, while moneyness is $\left(\overline{K} - S_0\right)/S_0$. The asymetry (smirk) with respect to $\overline{K}$ is usually observed on stock option markets (for recent evidence see Dumas, Fleming and Whaley [15] or Peña, Rubio and Serna [37]). Such a smile is present for all option maturities (ranging from one month $T = 1/12$ to one year $T = 1$) and is steeper for short dated options as shown on Figure 5. This result is in accordance with the observed fact reported in Das and Sundaram [13], that the smile is deepest for short maturities in most markets and flattens out as the time to maturity increases.

Table 4. Implied volatilities in the ACB model

$a \setminus \overline{K}$	90	95	100	105	110
0.5%	25.44%	25.04%	24.82%	24.90%	25.16%
1%	29.77%	29.49%	29.43%	29.70%	30.18%
2%	32.40%	32.30%	32.33%	32.71%	33.19%

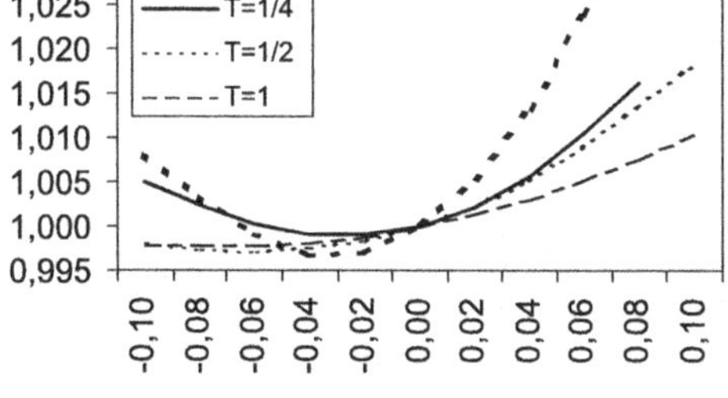

Fig. 5. Implied volatility smile for various maturities

5 Conclusion

In this paper, we have derived option-pricing formulae grounded in microstructure econometric modelling. Our results are very general and can be applied to various price dynamics which can be described as pure jump processes. This kind of process can in particular arise if traders rebalance their portfolio whenever the underlying price process has changed by a given percentage. We believe that this model is new to the literature not only because we propose new option pricing formulae for discontinuous processes but also because our approach to contingent claim pricing is essentially data-oriented (see Renault [40] for related empirical option modelling issues). Most models since Black and Scholes [4] have proceeded to derive their pricing formulae by first assuming a process and then pricing options based on this *a priori*. In this paper we first let the dynamics of the price process be estimated by econometric techniques and then derive option pricing formulae based on the estimated dynamics.

6 Appendix

In this appendix, we derive the dynamics of the price process under $\widehat{P}$. This allows to use expression (3.10) directly. We know that the compensator under the MMM is linked to the compensator under the historical measure (see Girsanov theorem for jumps in JS p.157) by the following relation:

$$\widehat{\nu}\left(dt, dx\right) = \nu\left(dt, dx\right)\left(\widehat{h}\left(t, x\right) + 1\right),$$

where $\nu\left(dt, dx\right) = d\Lambda_t K\left(t, dx\right)$. Besides the compensator $\widehat{\nu}\left(dt, dx\right)$ under $\widehat{P}$ can also be disintegrated into a kernel part $\widehat{K}\left(t, dx\right)$ and an intensity component $d\widehat{\Lambda}_t$. Hence after identification, we deduce:

$$d\widehat{\Lambda}_t = \left(\int_E \left(\widehat{h}\left(t, x\right) + 1\right) K\left(t, dx\right)\right) d\Lambda_t, \tag{6.13}$$

$$\widehat{K}\left(t, dx\right) = \frac{K\left(t, dx\right)\left(\widehat{h}\left(t, x\right) + 1\right)}{\int_E \left(\widehat{h}\left(t, x\right) + 1\right) K\left(t, dx\right)}. \tag{6.14}$$

The normalisation factor $\int_E \left(\widehat{h}\left(t, x\right) + 1\right) K\left(t, dx\right)$ in (6.14) comes from the condition on the transition kernel to integrate to 1.

Thus, by using the directing intensity and transition kernel of the ACD and ACB models estimated under P, we can immediately deduce the dynamics of the stock price under $\widehat{P}$ thanks to (6.13) and (6.14).

In the specific case of Log-WACD distributed durations with ACB marks under P, we obtain under $\widehat{P}$:

$$\widehat{h}(T_j, x) = -\delta(T_{j+1}, x) \frac{\pi_{j+1}\delta(T_{j+1}, a) + (1 - \pi_{j+1})\delta(T_{j+1}, -a)}{(\pi_{j+1}\delta^2(T_{j+1}, a) + (1 - \pi_{j+1})\delta^2(T_{j+1}, -a))},$$

$$\widehat{K}(t, a) = \widehat{p}_j(a) = \frac{\pi_{j+1}\left(\widehat{h}(T_j, a) + 1\right)}{I(T_j, a)},$$

with:

$$I(T_j, a) = 1 - \frac{(\pi_{j+1}\delta(T_{j+1}, a) + (1 - \pi_{j+1})\delta(T_{j+1}, -a))^2}{(\pi_{j+1}\delta^2(T_{j+1}, a) + (1 - \pi_{j+1})\delta^2(T_{j+1}, -a))}.$$

We know that the conditional intensity of the Log-WACD model under the historical measure is $\lambda_t = (\exp(-\psi_{j+1})\Gamma(1 + 1/\gamma))^\gamma (t - T_j)^{\gamma - 1}\gamma$, corresponding to the process

$$d_{j+1} = \exp(\psi_{j+1})\xi_{j+1}.$$

Under the MMM, we have $\widehat{\lambda}_t = I(T_j, a)\lambda_t$, thus corresponding to the modified process

$$d_{j+1} = \frac{\exp(\psi_{j+1})}{I(T_j, a)^{1/\gamma}}\xi_{j+1}.$$

This enables us to simulate durations directly under the MMM.

References

1. Ansel J.P. and C. Stricker (1992) "Lois de martingale, densités et décomposition de Föllmer-Schweizer", *Annales de l'Institut Henri Poincaré*, **28**, 375-392.
2. Ansel J.P. and C. Stricker (1993) "Unicité et existence de la loi minimale", *Lecture notes in math. Sem. Prob. XXVII.*, **1557**, 22-29, Springer-Verlag, Berlin.
3. Bauwens L. and P. Giot (2000) "The logarithmic ACD model: an application to the bid-ask quote process of two NYSE stocks, Annales d'Economie et Statistique, **60**, 117-149.
4. Black F. and M. Scholes (1973) "The pricing of options and corporate liabilities", *Journal of Political Economy*, **81**, 637-654.
5. Bollerslev T. (1986) "Generalized Autoregressive Conditional Heteroskedasticity", *Journal of Econometrics*, **31**, 307-327.
6. Bossaerts P., E. Ghysels and C. Gouriéroux (1996) "Arbitrage-based pricing when volatility is stochastic", DP #9641, CREST.
7. Bouleau N. and D. Lamberton (1989) "Residual risks and hedging strategies in Markovian markets", *Stochastic Processes and their Applications*, **33**, 131-150.
8. Carrasco M. and X. Chen (1999) "β-mixing and moment properties of various GARCH, stochastic volatility and ACD models", DP CREST.

9. Chesher A., Dhaene G., Gouriéroux C. and O. Scaillet (1999), "Bartlett Identities Tests", DP CREST.
10. Cox D. (1970) *The analysis of binary data*, Chapman and Hall, London.
11. Cox D. (1981) "Statistical analysis of time series: some recent developments", *Scandinavian Journal of Statistics*, **8**, 93-115.
12. Cox J., S. Ross and M. Rubinstein (1979) "Option pricing: a simplified approach", *Journal of Financial Economics*, **7**, 229-264.
13. Das S.R. and R. Sundaram (1999) "Of smiles and smirks: a term structure approach", *Journal of Financial and Quantitative Analysis*, **34**, 211-239.
14. Duffie D. and H. Richardson (1991) "Mean-variance hedging in continuous time", *Annals of Applied Probability*, **1**, 1-15.
15. Dumas B., J. Fleming and R. Whaley (1998) "Implied volatility functions: empirical tests", *Journal of Finance*, **53**, 2059-2106.
16. Engle R. (1982) "Autoregressive Conditional Heteroskedasticity with estimates of the variance of United Kingdom inflation", *Econometrica*, **50**, 987-1008.
17. Engle R. (2000) "The Econometrics of Ultra-High Frequency Data.", *Econometrica*, **68**, 1-22.
18. Engle R. and J. Russell (1997) "Forecasting the frequency of changes in quoted foreign exchange prices with the autoregressive conditional duration model", *Journal of Empirical Finance*, **4**, 187-212.
19. Engle R. and J. Russell (1998) "Autoregressive conditional duration: a new model for irregularly-spaced transaction data", *Econometrica*, **66**, 1127-1162.
20. Föllmer H. and D. Sondermann (1986) "Hedging of non-redundant contingent claims", in *Contributions to Mathematical Economics in Honor of Gérard Debreu*, eds. W. Hildenbrand and A. Mas-Colell, North-Holland, 205-223.
21. Föllmer H. and M. Schweizer (1991) "Hedging of contingent claims under incomplete information", in *Applied Stochastic Analysis, Stochastics Monographs*, eds. M. H. A. Davis and R. J. Elliott, Gordon and Breach, **5**, London/New York, 389-414.
22. Frey R. (1997) "Derivative asset analysis in models with level-dependent and stochastic volatility", *CWI Quarterly*, **10**, 1-34.
23. Geweke J. (1986) "Modelling the persistence of conditional variances: a comment", *Econometric Reviews*, **5**, 57-61.
24. Gouriéroux C., J.P. Laurent and H. Pham (1998) "Mean-Variance Hedging and Numéraire", *Mathematical Finance*, **8**, 179-200.
25. Harrison J. and D. Kreps (1979) "Martingale and arbitrage in multiperiod securities markets", *Journal of Economic Theory*, **20**, 348-408.
26. Harrison J. and S. Pliska (1981) "Martingales and stochastic integrals in the theory of continuous trading", *Stochastic Processes and their Applications*, **11**, 215-260.
27. Hausman J. A. Lo and C. MacKinlay (1992) "An ordered probit analysis of transaction stock prices", *Journal of Financial Economics*, **31**, 319-379.
28. Hofmann N., E. Platen and M. Schweizer (1992) "Option pricing under incompleteness and stochastic volatility", *Mathematical Finance*, **2**, 153-187.
29. Hull J. and A. White (1987) "The pricing of options on assets with stochastic volatilities", *Journal of Finance*, **42**, 281-300.
30. Jacod J. and A. Shiryaev (1987) *Limit theorems for stochastic processes*, Springer-Verlag, Berlin.
31. Laurent J.P. and O Scaillet (1998) "Variance optimal cap pricing models", DP CREST.

32. Lesne J.P., J.L. Prigent and O. Scaillet (2000) "Convergence of discrete time option pricing formulas under stochastic interest rates", *Finance and Stochastics,* **4**, 81-93.
33. Mandelbrot B. (1963) "The variation of certain speculative prices", *Journal of Business*, **36**, 394-419.
34. Mercurio F. and T. Vorst (1996) "Option pricing with hedging at fixed trading dates", *Applied Mathematical Finance*, **3**, 135-158.
35. Nelson D. and C. Cao (1992) "Inequality Constraints in the Univariate GARCH Model", *Journal of Business & Economic Statistics*, **10**, 229-235.
36. O'Hara (1995) *Market microstucture theory*, Blackwell, Cambridge MA.
37. Peña I., G. Rubio and G. Serna (1999), "Why do we smile ? On the determinants of the implied volatility function", *Journal of Banking and Finance,* **23**, 1151-1179.
38. Prigent J.L. (1995) "Incomplete markets: Convergence of options values under the minimal martingale measure", *Advances in Applied Probability*, **31**, 1058-1077.
39. Prigent J.L., O. Renault and O. Scaillet (1999) "Option pricing with discrete rebalancing", DP Université Catholique de Louvain.
40. Renault E. (1997) "Econometric models of option pricing errors", *Advances in economics and econometrics*, eds. D. Kreps and K. Wallis, Cambridge University Press.
41. Renault E. and N. Touzi (1996) "Option hedging and implied volatilities in a stochastic volatility model", *Mathematical Finance*, **6**, 279-302.
42. Roll R. (1984) "A simple implicit measure of the effective bid-ask spread in an efficient market", *Journal of Finance*, **39**, 1127-1140.
43. Runggaldier W. J. and M. Schweizer (1995) "Convergence of option values under incompleteness" in *Seminar on stochastic analysis, random fields and applications*, eds. E. Bolthausen, M. Dozzi and F. Russo, Birkhäuser, 365-384.
44. Russell J. and R. Engle (1998), "Econometric analysis of discrete-valued irregularly-spaced financial transaction data using a new autoregressive conditional multinomial model", CRSP working paper #470.
45. Schweizer M. (1991) "Option hedging for semimartingales", *Stochastic Processes and their Applications*, **37**, 339-363.
46. Schweizer M. (1992) "Mean-variance hedging for general claims", *Annals of Applied Probability*, **2**, 171-179.
47. Schweizer M. (1993) "Variance-optimal hedging in discrete time", *Mathematics of Operations Research*, **20**, 1-32.
48. Schweizer M. (1994) "Approximation pricing and the variance-optimal martingale measure", *Annals of Probability*, **24**, 206-236.

Markov Chains and the Potential Approach to Modelling Interest Rates and Exchange Rates

L.C.G. Rogers and F.A. Yousaf*

Department of Mathematical Sciences, University of Bath, Bath, BA2 7AY, UK

1 Introduction

Within the mathematical finance literature, there have been several distinct classes of interest-rate model. The first historically was the family of *spot-rate models*, where one proposes a model for the evolution of the spot rate of interest under the pricing measure, and then attempts to find expressions for the prices of derivatives; the models of Vasicek [16], Cox, Ingersoll & Ross [7], Black, Derman & Toy [3] and Black & Karasinski [4] are well-known examples of this type. Next came the *whole-yield models*, starting with Ho & Lee [10] in a discrete setting, and then in the continuous setting Babbs [1] and Heath, Jarrow & Morton [9]. Lately, there has been much interest in so-called *market models*, whose chief characteristic is the choice of some suitable numéraire process, relative to which the prices of various derivatives have some particularly tractable form; see Miltersen, Sandmann & Sondermann [12] and Brace, Gatarek & Musiela [5] for examples of such models. These three classes of models have been developed extensively; a thorough survey would be outside the aims of this paper, but we refer the reader to the excellent recent monograph of Musiela & Rutkowski [11] for more details and references.

In amongst these, with elements in common but seemingly little noticed by the mathematical finance community at large, there was another approach, advocated by Constantinides [6] and by Rogers [14], named the *potential approach.* The key element of this approach is to view the state-price density process as the modelling primitive, and to express the prices of derivatives directly in terms of this. From one point of view, this method is based on the choice of a numéraire process, rather as in the market models, but the emphasis is very different; in the market model approach, the numéraire is taken to be something very concrete, closely related to some particular derivative of interest, and possibly to be chosen differently when dealing with another range of derivatives, whereas in the potential approach, the numéraire is something very abstract, and is viewed as something quite universal, to be used for pricing *every* interest-rate derivative. This leads to models which are typically harder to calibrate (and ease of calibration was a major reason for the

* Work partly supported by EPSRC Research Studentship 97003358

development of market models), but the reward is a consistent interest-rate modelling system. As Rogers [14] emphasises, this consistency extends across many different currencies very simply; valuing cross-currency derivatives is only a little more difficult than valuing single-currency products.

To date, there has been very little work on fitting potential models to data (the paper of Rogers & Zane [15] appears to be the only study so far), and this paper is another contribution in that direction. Earlier references concentrated exclusively on the situation where the underlying Markov process was a diffusion, but in this paper we shall focus exclusively on the case where the underlying Markov process is a *finite Markov chain.* There are advantages and disadvantages to this modelling choice, which we shall discuss at length later. But for now, notice one clear advantage which comes when we are trying to price a very general derivative. European-style derivative prices are computed as an average over the statespace, so for a Markov chain, this is just a finite sum. Pricing an American-style derivative is just an optimal-stopping problem for a finite Markov chain, and provided the number of states of the chain is not too big, this will be a very simple numerical exercise. In fact, the number of states used in our calibrations was of the order of tens, so these pricing calculations are always going to be extremely fast, in contrast to many other methods.

The plan of the paper is as follows. In Section 2, we shall briefly summarise the main ideas of the potential approach, as a way of setting up our notation, and pointing out the special forms that some of the pricing expressions take in the Markov chain situation. Section 3 describes the dataset used, and discusses various issues to do with the calibration. In Section 4, we present and discuss the results of the calibration, and finally in Section 5 we draw conclusions.

2 The Potential Approach

We begin by recalling the main elements of the potential approach, as set forth in Rogers [14], and making more explicit the forms they take when the underlying Markov process is a finite-statespace chain. Arbitrage-pricing theory gives the time-t price of a contingent claim Y payable at time $T > t$ to be

$$Y_t = E\left[\exp\left(-\int_t^T r_s ds\right) Y|\mathcal{F}_t\right], \tag{2.1}$$

where $(r_t)_{t\geq 0}$ is the *spot rate of interest* process. The probability P used for the expectation is some fixed risk-neutral measure. By taking some equivalent reference measure $\tilde{P}$, we can express this price in terms of an expectation with respect to $\tilde{P}$ as

$$Y_t = \tilde{E}_t[\zeta_T Y]/\zeta_t, \tag{2.2}$$

where the *state-price density process* ζ is defined by

$$\zeta_t \equiv \exp\left(-\int_0^t r_s ds\right) \cdot \left.\frac{dP}{d\tilde{P}}\right|_{\mathcal{F}_t} \equiv \exp\left(-\int_0^t r_s ds\right) \cdot Z_t. \tag{2.3}$$

Assuming $r \geq 0$ (which we always shall), the process ζ_t is a positive supermartingale, and for *any* positive supermartingale ζ, (2.2) determines an arbitrage-free pricing system. The potential approach therefore seeks to model the state-price density process ζ with respect to the reference probability $\tilde{P}$, and computes prices using the characterisation (2.2).

One very natural way to build positive supermartingales is to take some Markov process $(X_t)_{t\geq 0}$ with resolvent $(R_\lambda)_{\lambda>0}$, fix some $\alpha > 0$, and some positive function g on the statespace of X and make an interest-rate model by setting

$$\zeta_t = e^{-\alpha t} R_\alpha g(X_t). \tag{2.4}$$

A particularly attractive feature of this modelling approach is that the spot rate process r can be expressed very simply as

$$r_t = \frac{g(X_t)}{R_\alpha g(X_t)} . \tag{2.5}$$

See Rogers [14], p.161 for the derivation.

In the context of a finite Markov chain X with finite statespace I and infinitesimal generator (or Q-matrix) Q, the resolvent has the simple expression

$$R_\lambda = (\lambda - Q)^{-1},$$

when we regard the transition semigroup $(P(t))_{t\geq 0}$ as a semigroup of matrices acting on the vector space $\mathbb{R}^I$, expressible in terms of Q as $P(t) = \exp(tQ)$. Thus, for example, the time-0 price of a zero-coupon bond delivering a riskless \$1 at time T is just

$$P(0,T) = \exp(-\alpha T)P(T)(\alpha - Q)^{-1}\mathbf{1}/R_\alpha g, \tag{2.6}$$

regarded as a function on I.

A further feature of the potential approach is the ease with which yield curves in several countries can be modelled. Indeed, we can introduce another country j without introducing any further sources of randomness, simply by taking a new positive function g^j and positive real α^j and defining the state-price density process ζ^j for country j by

$$\zeta_t^j = R_{\alpha^j} g^j(X_t).$$

As Rogers [14] shows, if Y_t^{ij} is the time-t price in currency i of one unit of currency j, then we have in general that

$$\zeta_t^i Y_t^{ij} / \zeta_t^j \equiv N_t^{ij} \tag{2.7}$$

is a $\tilde{P}$-martingale orthogonal to all the $\tilde{P}$-martingales of the form $\zeta_t^j S_t^j$, where S_t^j is a traded asset, valued in currency j. A special case of this (which we shall focus on exclusively below) is when the martingale N^{ij} is constant.

3 Discussion of the Data and Calibration Methodology

The data which is used in this study is daily yield curve data covering the period from 2nd January 1992 to 1st March 1996[1]

For each day we have values of the yield of bonds with maturity 1 month, 3 months, 6 months, 1 year, 2 years, 5 years, 7 years and 10 years. We shall use daily yield curve data for three currencies; these are sterling (GBP), the US dollar (USD) and the German Mark (DEM).

We also have daily exchange rate data between these three currencies, obtained from the United States Federal Reserve Data Exchange[2].

As a preliminary data-cleaning, any dates that were not common to every set were removed from all sets. This included public holidays and other days where one or more of the three markets was closed. In total we have 1029 days of data. Surface plots of the yield curve for each country, together with graphs of the exchange rates are shown in Fig. 1.

It is worth pointing out that the period under consideration in this study represented a turbulent time in the world markets. The years of 1992 and 1993 saw both the US and UK economies in the middle of deep recessions. Indeed, 1992 was a year of huge turmoil for the UK economy; it saw the surprise re-election of the Conservative party for a third consecutive term of office, and this was followed a few months later by the embarrassing débacle of 16th September 1992 - "Black Wednesday" - in which the UK was embarrassingly forced out of the ERM, losing 4 billion GBP trying to stop the pound devaluing. On this day, the UK government announced a 5% rise in the base rate taking the rate to 15% in a desperate attempt to stop the pound's value sliding. The turmoil in the UK economy at this point was partly attributed (by many analysts) to the strength and dominance of the German Mark. In fact it can be seen that the German economy had a strong influence on most of the other major European economies at this time.

Conversely, 1994 to 1996 saw a weakening of the German dominance and a recovery in the UK and US economies. These countries slowly came out of their long recession and this is reflected in the shape of the yield curve and exchange rate over this period. We have therefore chosen quite a varied and turbulent period for the calibration exercise.

We shall attempt to fit the data using a potential model based on an underlying Markov chain. In any of the calibration exercises, the first step is to fix the number N of states of the chain. This done, there are in total N^2 free parameters to be estimated: $N^2 - N$ off-diagonal entries of the Q-matrix Q, $N-1$ entries[3] of g, and the one value α. However, to make the problem somewhat easier, we restricted the fitting to *reversible* chains, where for some

[1] We are grateful to Dr Simon Babbs for supplying the GBP and DEM data. The USD data was taken from the website http://www.stls.frb.org/fred/index.html

[2] See http://www.federalreserve.gov/releases/H10/hist

[3] One degree of freedom represents a redundant scaling of g.

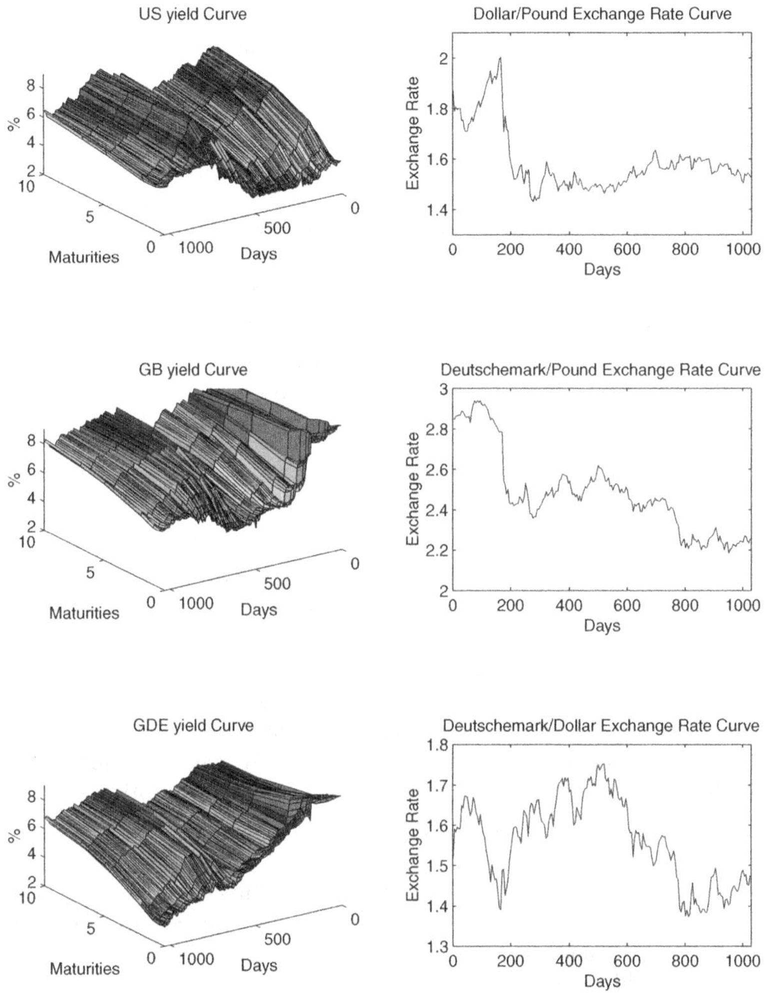

Fig. 1. Yield and Exchange Rate Curves

vector m of positive entries

$$m_i q_{ij} = m_j q_{ji} \quad \text{for all } i, j.$$

Thus the flux matrix $A \equiv (m_i q_{ij})_{i,j=1,\dots,N}$ is symmetric with zero row sums. In choosing the reversible Q-matrix, we therefore have the choice of the $N(N-1)/2$ above-diagonal entries of A, and of $N-1$ of the entries of m; the diagonal entries of A are then determined by the zero-row-sum condition, and the last entry of m is fixed by the fact that the entries of m have to sum to 1. We therefore have in total $(N^2 + 3N - 2)/2$ free parameters to estimate. By restricting to reversible Markov chains, we have thus reduced the number of parameters by about half, but the principal reason for making this restriction is that by so doing we guarantee that all the eigenvalues of the Q-matrix are *real*, thereby avoiding the need to program with complex variables throughout.

Nevertheless, it is clear that our modelling assumptions involve a large number of parameters; in our examples, we took N in the range 10 to 25, so that the number of parameters to be estimated was of the order of hundreds. Using daily yield curve data for one week, the number of parameters is far in excess of the the number of data-points. Conventional statistical wisdom would frown on such a model, for a variety of reasons:

If you have more parameters than data points, you will be able to fit the data perfectly. This is clearly false. If, for example, you wish to model real-valued observations $y_1, \dots, y_n$ taken at increasing times $t_1, \dots, t_n$ as

$$y_i = \sum_{j=1}^{J} \alpha_j \exp(-\beta_j t_i) + \varepsilon_i$$

for non-negative parameters α_j, β_j, then however large you take J you will be unable to fit the y_i if they are not decreasing. The same is true for our application; we are trying to fit a model with very strong structural properties, and there is no guarantee that we will be able to get a perfect fit (in fact, we don't). We *need* a highly-structured model because we do not simply wish to be able to fit yield curve data, we have to be able to price general derivatives; if we had simply done a principal-components analysis of yield curve data, we would have been unable to begin to value an American swaption.

Some of the parameters will be indeterminate. There are examples (such as a two-way analysis of variance) where this does indeed happen, but this can arise even when there are far more data points than parameters. Our estimation procedure looks for the minimum of a real-valued function of many variables, and there is no reason based on the number of data points why this minimum should not be unique.

The estimates of many of the parameters will be subject to large error. Though there is no general reason why this *must* happen, we do observe this. But if we find that a particular parameter cannot be estimated with high precision, this is because it has relatively little influence on the model values for the observables, so it really does not *matter* what value it takes! What matters is how well the fitted model fits the data.

In summary, we regard such conventional statistical wisdom in this case rather as the split infinitive (see Fowler [8], p 579, who distinguishes the meanings of 'to just have heard' and 'to have just heard'); we know the objections, and shall not hesitate to completely ignore them. Our methods will be justified by the quality of the fit that they achieve, and by the stability of the estimates we come up with. It should be remarked that the finance industry routinely works with models with time-dependent coefficients, in which the parameter space is every bit as large as those we shall be dealing with here, and in which problems of parameter stability are very hard to deal with in a satisfactory manner.

To introduce the estimation methods we shall use, we now explain carefully the modelling assumptions in use. Our model is parametrised by a vector[4] θ. The underlying Markov chain X takes values in a finite set I, and on day n we have a vector y_n of observations[5]. If the model were correct, the value of this observation vector y_n would be $Y(X_n, \theta)$, but we suppose that the observed values are the true values plus some independent Gaussian noise. We adopt a Bayesian standpoint, and suppose that the initial law of X is given by $\pi = (\pi_i)_{i=1}^N$, and the initial law of θ is given by density $f_0(\theta)$; conceptually, θ is unchanging with time, even though our knowledge of it varies[6].

We shall use the notation $\mathbf{z}_n \equiv (z_0, \dots, z_n)$ in what follows to reduce the acreage of formulae. Based on the assumptions above, and ignoring irrelevant constants, the likelihood Λ_n of $(\mathbf{X}_n, \mathbf{y}_n, \theta)$ is

$$\begin{aligned} \Lambda_n &\equiv \Lambda_n(\mathbf{X}_n, \mathbf{y}_n, \theta) \\ &= f_0(\theta)\, \pi_{X_0} \prod_{j=1}^{n} p_{X_{j-1}X_j}(s_j; \theta) \exp[-b(y_j - Y(X_j; \theta))] \end{aligned} \tag{3.1}$$

where $p_{ij}(s; \theta) = P_\theta(X_s = j | X_0 = i)$, and $b(z) \equiv \frac{1}{2} z \cdot V^{-1} z$, where V is the covariance matrix of the Gaussian errors. We have also used the notation $s_j = t_j - t_{j-1}$ for the time between the $(j-1)$th and jth observations. We shall be more interested in the posterior distribution of (X_n, θ) given $\mathbf{y}_n$, so we introduce the notation

$$L_n(x, \mathbf{y}_n, \theta) = \sum_{\mathbf{X}_n : X_n = x} \Lambda_n(\mathbf{X}_n, \mathbf{y}_n, \theta), \tag{3.2}$$

and notice that directly from the definitions

$$L_n(x, \mathbf{y}_n, \theta) = \sum_{\xi} L_{n-1}(\xi, \mathbf{y}_{n-1}, \theta) p_{\xi x}(s_n; \theta) \exp[-b(y_n - Y(x; \theta))]. \tag{3.3}$$

It is clear that for the Markov chain model in mind this expression will be far too complicated to allow exact analysis, and we shall have to make simplifying assumptions in order to make progress. Here are the simplifications which we used.

---

[4] We can think of this as the above-diagonal entries of A, the first $N-1$ entries of m, the first $N-1$ entries of g, and the value of α stacked into a single vector if we wish.

[5] The observations happen to be the yields of the different maturities, though this is irrelevant for the present discussion.

[6] We shall later consider what happens if we modify this assumption.

Day-by-day calibration. In this case, we simply ignore all the 'earlier' information in (3.3) and, given the observations y_n on day n, we just compute

$$\min_{\theta} b(y_n - Y(x;\theta)), \tag{3.4}$$

where in the minimisation we make the arbitrary convention that x is some distinguished state (say, the first) in the statespace. The labelling of the states of the chain is clearly irrelevant under this simplifying assumption. This particular method can be expected to be simple to implement, but cannot be expected to be very stable. Nevertheless, it should furnish a lower bound for the fitting error; if the results of fitting under this assumption are disappointing, then the results will be disappointing under more realistic assumptions.

Rigid calibration. In this approach, we take some initial period of K days data, and then try to fit the model using an approximation to the likelihood (3.1). This calibration is more honest than the day-by-day fit, in that it requires the parameters to be the same for all days. The simplification used is based on the observation that the underlying state of the Markov chain does not change very frequently, so we replace the true likelihood (3.2) - which involves a sum over all possible paths of the chain during the K days of the calibration period - by the single term corresponding to a path which remains at its initial state throughout the calibration period. This is a reasonable thing to do when the length of the calibration period is up to a few tens of days, during which period a change of underlying state is comparatively unlikely. Since the particular state is not important, we may as well assume that it is the first one labelled, 1, say.[7] The true calibration, involving a sum over all possible paths of the underlying chain during the K days, would be far too slow. So the calibration is achieved by minimising the expression

$$-\log f_0(\theta) + \sum_{j=1}^{K} [b(y_j - Y(1;\theta)) + q_1 s_j], \tag{3.5}$$

where $-q_1$ is the diagonal entry in the first position of the Q-matrix Q.

Having found our calibrated values θ^*, we can then check the model out-of-sample by taking the days after the calibration period and trying to fit the yield curves by allowing only changes in the (posterior) distribution of X.

Conditional-independence (CI) calibration. In this case, we imagine the situation where there has been a large amount of observed data, and we postulate that

$$L_n(x, \mathbf{y}_n, \theta) = \pi_n(x, \mathbf{y}_n)\, l_n(\theta, \mathbf{y}_n). \tag{3.6}$$

The motivation for this is that we have seen so much data that we have a pretty good idea what the values of the parameters must be; the values of

[7] An extension of this fitting would be to allow the chain just one jump during the K days.

θ will largely be determined by the long-run historical average behaviour of the system. On the other hand, the posterior distribution of X_n will be more influenced by recent history, because of the ergodicity of the Markov chain, and so some approximate conditional independence is reasonable; recent history tells us all we can know of X_n, distant history tells us all we can know of θ. We shall further assume that

$$l_n(\theta, \mathbf{y}_n) \propto \exp(-\frac{1}{2}(\theta - \hat{\theta}_n) \cdot S_n(\theta - \hat{\theta}_n)) \tag{3.7}$$

for some positive-definite symmetric matrix S_n. If we think that we have nearly identified the true value of θ, then such a quadratic approximation to the likelihood is quite natural.

The values $\hat{\theta}_n$, S_n, and $\pi_n(\cdot, \mathbf{y}_n)$ are computed recursively, using the assumed form (3.6) of the likelihood. Supposing that we know already $\hat{\theta}_{n-1}$, S_{n-1}, and $\pi_{n-1}(\cdot, \mathbf{y}_{n-1})$, returning to (3.3) and using (3.6) we see that

$$\begin{aligned} L_n(x, \mathbf{y}_n, \theta) &= \sum_\xi \pi_{n-1}(\xi, \mathbf{y}_{n-1})\, l_{n-1}(\theta, \mathbf{y}_{n-1}) p_{\xi x}(s_n; \theta) \exp[-b(y_n - Y(x;\theta))] \\ &\propto \sum_\xi \pi_{n-1}(\xi, \mathbf{y}_{n-1})\, p_{\xi x}(s_n; \theta) \exp[-b(y_n - Y(x;\theta))] \\ &\qquad .\exp\big[-\frac{1}{2}(\theta - \hat{\theta}_{n-1}) \cdot S_{n-1}(\theta - \hat{\theta}_{n-1}))\big] \end{aligned} \tag{3.8}$$

We now sum this expression over x, and numerically pick θ to maximise; the maximising value is our new estimate $\hat{\theta}_n$ of θ. By computing the second derivative matrix with respect to θ at $\hat{\theta}_n$ we find the value of S_n,[8] and finally we compute π_n by

$$\pi_n(x, \mathbf{y}_n) \propto \sum_\xi \pi_{n-1}(\xi, \mathbf{y}_{n-1}) p_{\xi x}(s_n; \hat{\theta}_n) \exp[-b(y_n - Y(x; \hat{\theta}_n))].$$

Properly speaking, the posterior distribution π_n for X_n should be obtained by integrating the likelihood (3.8) with respect to θ, but we approximate this by assuming that the posterior distribution for θ can be replaced by the point mass at $\hat{\theta}_n$, to avoid the need to integrate over a large number of dimensions.

Random walk (RW) calibration. This method is very similar to the previous method, which can be seen as a special case. The theoretical justification is explained in Appendix A in more detail, and is based on the Kalman filter. The idea is that we shall now allow the value of θ to change from day to day according to a random walk. If the variance of the steps of the random walk is zero, then we arrive at the CI method, but if we allow the variance of the random walk step to be a fixed multiple of the posterior covariance of θ,

[8] In practice, we compute only the diagonal terms of S_n

then we obtain

$$\sum_{\xi} \pi_{n-1}(\xi, \mathbf{y}_{n-1})\, p_{\xi x}(s_n;\theta) \exp[-b(y_n - Y(x;\theta)) \\ -\frac{\beta}{2}(\theta - \hat{\theta}_{n-1}) \cdot S_{n-1}(\theta - \hat{\theta}_{n-1}))], \tag{3.9}$$

where $\beta \in (0,1)$ is fixed. The closer β is to 1, the closer we are to the CI fit.

In the CI calibration, we expect that the matrices S_n will be growing approximately linearly with n, by analogy with the situation where we attempt to estimate the mean of a Gaussian distribution using a sequence of noisy observations of the mean; when we have seen $n-1$ observations, the nth receives weight $1/n$ in the estimation. The same thing happens with our CI calibration, so the most recent observations get relatively little weight in relation to the average over earlier times. On the other hand, we do not believe that there is *no* change in the interest-rate environment, and by introducing the parameter β, we allow the new day's observations to have the same importance in the estimation as yesterday's new observations did yesterday; the analogy is with the estimation of an underlying random walk process based on noisy observations of that process.

The last three approaches to calibration are (quasi-)Bayesian and produce estimates of the posterior distribution π_n of the underlying Markov chain X_n at time n, as well as point (ML) estimates $\hat{\theta}_n$ of the parameter θ. Thus to price a derivative on day n, we shall use the expression

$$\sum_{x} \pi_n(x, \mathbf{y}_n) F(x, \hat{\theta}_n), \tag{3.10}$$

where $F(x,\theta)$ is the price which the Markov chain potential model would produce if the starting state were x and the true parameter value were θ. This would apply, for example, to the pricing of zero-coupon bonds; so, in particular, we end up with a continuum of possible yield curves at any given time, even though the model with known θ could only produce one yield curve for each possible state of the Markov chain.

4 Numerical Results

The heart of the calibration procedure is a minimisation routine, and for this we chose the NAG routine E04JYF. Of several which we investigated, this one seemed to do the best job. Our first fitting attempt was a day-by-day calibration; we do not of course believe in this approach, but if the results of this fit were poor, then it would be impossible that a more realistic fitting procedure will produce anything other than poor results. For purposes of comparision, we split the dataset into 19 overlapping blocks of 100 days, and computed summary statistics, which we present in Table 1. The data was

Table 1. Results of fits for the day-by-day calibration using the 19 sample periods of 100 days. Note that these results were obtained using a 15-state data imposed underlying Markov chain. The column marked 'mean' above refers to the average basis points (bp) error between model and observed values per day. The standard deviation is that of the daily basis point error in the period. Q1 and Q3 denote the first and third quartiles. Min, Median, Max again refer to the basis point error per day. BRC is used to denote the number of Bank of England base rate changes

	Calendar Period	Day numbers	BRC	Statistics of day-by-day calibration, all values in bp						
				Mean	Std. Dev.	Min	Q1	Median	Q3	Max
1	17th Feb 1992 - 15th July 1992	30-129	1	18.538	6.453	10.076	13.796	16.565	21.385	35.305
2	5th May 1992 - 25th Sept 1992	80-179	2	19.015	8.334	9.395	13.768	16.543	21.672	69.703
3	16th July 1992 - 7th Dec 1992	130-229	3	18.196	26.629	0.001	6.939	14.672	21.268	226.958
4	28th Sept 1992 - 18th Feb 1993	180-279	3	10.238	6.147	1.70	5.773	9.44	13.117	26.423
5	8th Dec 1992 - 7th May 1993	230-329	1	9.677	4.77	0.975	6.107	9.443	12.852	21.472
6	19th Feb 1993- 19th July 1993	280-379	0	6.678	4.613	0.001	2.94	6.422	8.992	23.569
7	10th May 1993 - 28th Sept 1993	330-429	0	5.261	3.382	0.128	2.874	4.5	7.631	15.704
8	20th July 1993 - 8th Dec 1993	380-479	1	5.397	3.488	0.055	2.692	4.463	8.144	15.510
9	29th Sept 1993 - 23rd Feb 1994	430-529	2	6.310	3.949	0.12	3.358	5.181	9.322	16.944
10	9th Dec 1993 - 11th May 1994	480-579	1	7.248	4.738	0.039	4.099	6.833	9.466	31.231
11	24th Feb 1994 - 22 July 1994	530-629	0	9.649	6.118	0.164	5.39	8.263	13.457	30.662
12	12th May 1994 - 4th Oct 1994	580-679	1	10.005	5.32	0.443	6.337	9.482	13.508	27.236
13	25th July 1994 - 16th Dec 1994	630-729	2	7.02	4.188	0.558	3.646	6.328	10.086	18.910
14	5th Oct 1994 - 6th Mar 1995	680-779	2	8.402	4.03	1.036	5.273	8.463	11.09	21.629
15	19th Dec 1994 - 23rd May 1995	730-829	1	8.437	2.608	0.77	6.666	8.243	10.081	15.573
16	7th Mar 1995 - 3rd Aug 1995	780-879	0	7.846	3.370	0.795	5.78	8.132	9.687	17.092
17	24th May 1995 - 16th Oct 1995	830-929	0	4.524	3.286	0.081	2.152	3.733	5.952	16.660
18	4th Aug 1995 - 28th Dec 1995	880-979	1	2.586	1.844	0.001	1.058	2.231	3.576	8.969
19	17th Oct 1995 - 8th Mar 1996	930-1029	3	5.503	3.436	0.001	2.387	4.956	8.552	14.726

GBP data, and we used 15 states in the Markov chain. Here we took the covariance matrix V, in (3.4), to be the identity matrix.

Perhaps the most interesting figures in this table are in the *Median* column. These present the median values of the sum of absolute errors in basis points for each day's fit. This sum consists of 8 terms, one for each maturity, so the basis-point error per maturity is $1/8$ of the figure given in the *Median* column. The worst values are in the turbulent months of 1992, when the median error per maturity is 2bp, but for most of the periods under study, the error is 1bp or even a lot less. Even looking at the upper quartile, we find that only in three of the 19 periods did the error exceed 2bp per maturity. For more detailed analysis, we chose to use an 11 state Markov chain and focus on period 14 which contains two base rate changes occuring on 7th December 1994 (day 43) and 2nd February 1995 (day 79).

Table 2. Summary statistics for the day-by-day calibration using an 11-state Markov chain on period 14 GBP data

Day-by-day calibration statistics (all values are in basis points)							
Data	Mean	Std. Dev.	Min	Q1	Median	Q3	Max
Period 14	7.585	3.588	0.693	4.842	7.302	10.400	16.619

The plots in Fig. 2 refer to this period and show the stability of the parameters g_i and α, as well as the contributions of different maturities to the total residual error. We normalised the g values to sum to one, so as to remove the degree of indeterminacy and all maturities were weighted equally. The parameters exhibit no particular stability, which is not a surprise, but what is encouraging about these fits is that the errors are *small*; the median fit per maturity is consistently below 2 bp, and the upper quartile is below 4 bp, often a lot less. A model that is fitting yields to within a basis point is good enough to trade off, and we are here getting close to that degree of precision, without any particular effort, and with relatively few states.

The next fitting exercise we carried out was the rigid calibration, which one would expect to be quite poor in comparison with the day-by-day fit, and indeed it was. Working again with the GBP data, and taking a chain with 11 states, we used five consecutive days of data to calibrate the model, and then stepped ahead through the next 100 days (period 14) computing the fit each day. So at the end of the five-day calibration period, we have found a value θ^* for the parameter θ, and for subsequent days we hold this value fixed, but use the data to update the posterior distribution for X_n by the recipe

$$\pi_n(x, \mathbf{y}_n) \propto \sum_{\xi} \pi_{n-1}(\xi, \mathbf{y}_{n-1}) p_{\xi x}(s_n; \theta^*) \exp\left[-b(y_n - Y(x, \theta^*))\right].$$

Diagnostic plots for day-by-day calibration

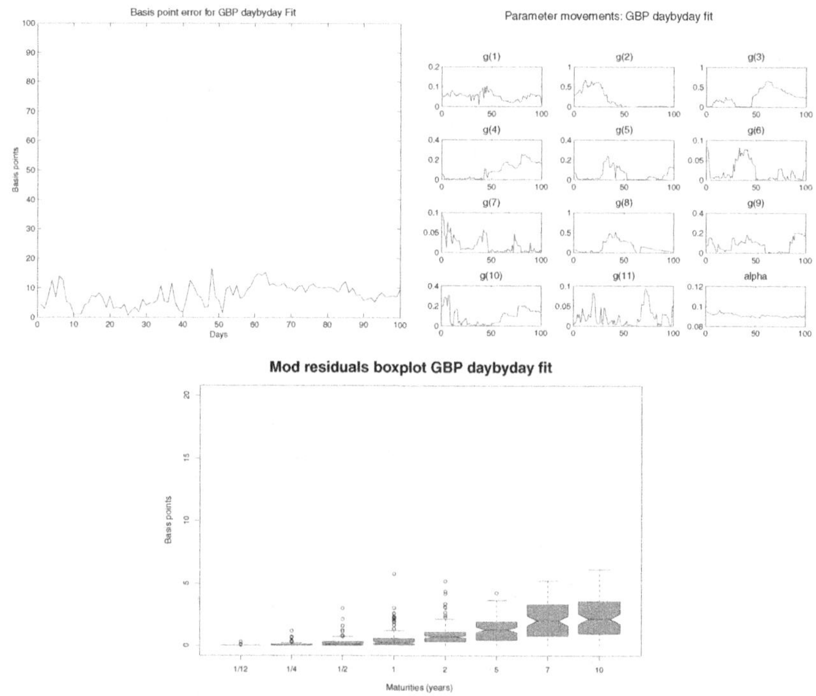

Fig. 2. Diagnostic plots for the day-by-day calibration using period 14 GBP data and an 11-state Markov chain. The basis point error plot (top left) shows the total error, given in basis points, between the market and model yield curves for each fitted day. The evolution of the parameters $\mathbf{g}$ and α over the whole fitting period are given in the top right plot. Finally we give a series of boxplots showing the mean and quartiles of the mod residuals for each maturity

The bond prices were then computed following (3.10). It is inconceivable that in practice one would fit a model to just five days' data and then run with that unaltered for the next 100 days, and the results of this fitting procedure, presented in Table 3 and in Figures 3 and 4, show why. These Figures and Table show also the results of variants of the rigid calibration, where we recalibrate the model every J days, using again the latest $K = 5$ days of data.

The panels in Figures 3 and 4, correspond to $J = 100$, $J = 10$ and $J = 1$. For the case $J = 100$, we see from Table 3 that the median error in bp per maturity is of the order of 35, which is really quite useless. Notice that Fig. 3 shows how the quality of fit deteriorates as we get further into the 100-day period, as one would expect. The fits for the case $J = 1$ are a lot better, but even here the median error is three times the worst that occurred in Table 3, amounting to around 6bp per maturity.

This calibration is poor not only because of the rigidity imposed by the assumptions, but also because we have trained the model on just 5 consecutive days' data. Since this tiny calibration set cannot possibly represent the variety of yield curves that might arise, it is not surprising that as time rolls forward we encounter days where the yield curve is far from the possibilities of the 5 day calibration period, and so the fit is very poor. A better recipe might be to take the last 5 Mondays for our calibration set. The problem with this is that the assumption that the underlying state has not changed in this time becomes untenable, and we would have to evaluate a sum over all possible paths of the chain during this calibration period, and this would be slow and clumsy. We do need to have more influence of past data in our calibration method, but the obvious way to do this is via some recursive approach, and this was what we tried next.

The next fitting exercise was an implementation of the CI/RW fitting strategy (3.8) and (3.9); since the CI fitting is the special case $\beta = 1$ of the RW fitting, it makes sense to consider them all together. We started with S_0 equal to the identity, $\hat{\theta}_0$ equal to zero, and the prior distribution for X to be uniform over the 11 states. The data used was period 14 of the GBP data. Table 4 shows summary statistics for the fits. Taking $\beta = 1$, we obtained a median error of just over 5bp per maturity, already better than even the one-day-ahead form of the rigid fit, and with $\beta = 0.2$ - allowing a random step with 4 times the posterior covariance - we obtained a median error of 2.5 bp per maturity, with the upper quartile at a little over 3 bp per maturity. Figures 5 and 6 display various results of the fitting procedure: notice the quite impressive stability of the g_i for the $\beta = 0.2$ case (compare with Fig. 2). This justifies empirically the (at first sight) low value of β; although we have

Table 3. Summary statistics for the rigid calibration procedure. The results are for fits over a 100 day period using different re-calibration intervals. All results are for an 11-state chain using GBP data

Rigid calibration statistics (all values are in basis points) 5 day calibration period							
Re-calibrate After	Mean	Std. Dev.	Min	Q1	Median	Q3	Max
100 days	257.09	101.87	42.24	175.79	270.46	349.67	416.87
50 days	140.04	68.08	34.50	90.40	124.44	180.32	313.13
25 days	105.78	54.04	21.64	63.58	95.44	136.31	246.78
10 days	94.12	53.45	19.73	49.82	82.93	128.65	232.33
5 days	76.17	45.65	19.71	41.97	63.46	99.50	200.14
2 days	60.11	38.63	7.34	30.51	48.62	74.47	192.79
1 day	55.34	33.99	7.22	29.46	47.81	66.60	162.49

Daily basis point error plots for the rigid calibration

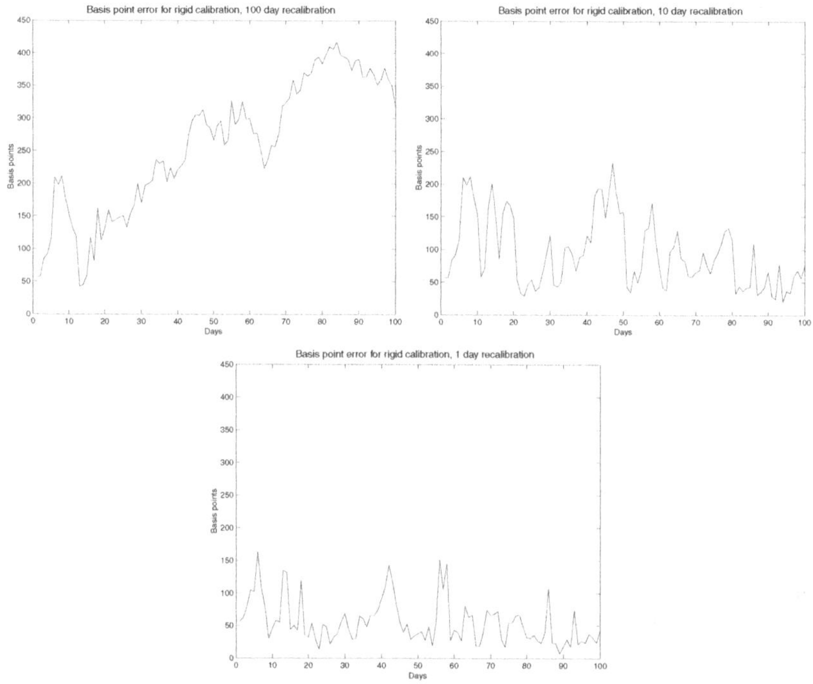

Fig. 3. 'Basis point error' plots showing the cumulative error in basis points for each day over the 100 day fitting period, recalibrating after 100 days, 10 days, and 1 day

in principle allowed the random walk a lot of freedom to move, it turns out in practice that it is not moving very much.

It appears therefore that the CI/RW fitting methodology represents a good compromise between the unstable but close fitting day-by-day approach, and the very stable but poorly-fitting rigid approach. Moving on to the simultaneous fitting of yield curves in more than one country, and the exchange rate(s), we concentrated on the CI/RW calibration approach. The first fitting exercise we carried out was using USD and GBP data from the period 5th October 1994 to 6th March 1995, with 11 states in the Markov chain; we report summary statistics for these in Table 5, with various diagnostics displayed in Figures 7 and 8. The fit was noticeably poorer than the single-country fit, as one would expect; for $\beta = 0.2$ we found a median fitting error of 3.5-4.5 bp per maturity. We then moved on to fit three currencies, USD, GBP and DEM, summarising the results in Table 6, with diagnostics displayed in Fig. 9. The inclusion of Germany worsens the fit of the US and UK very slightly, but with $\beta = 0.2$ we are still finding median errors of 3.5-4.5 bp per maturity.

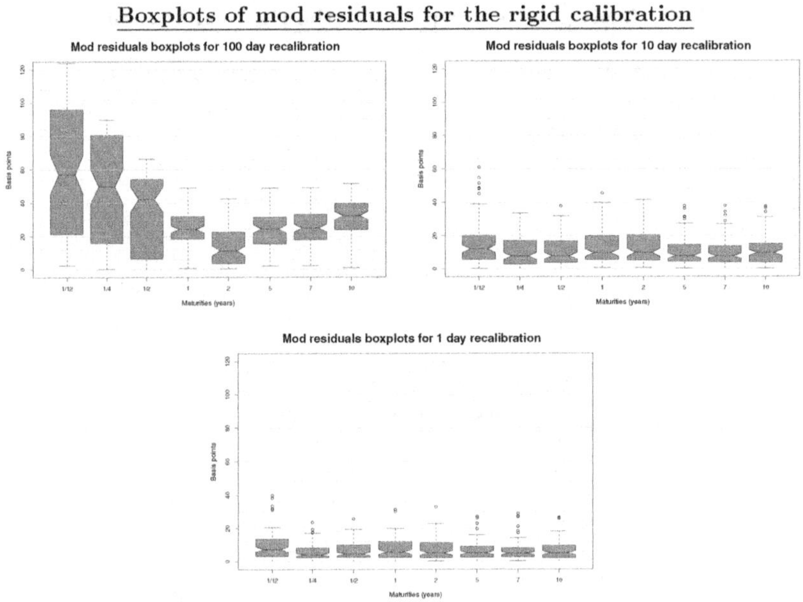

Fig. 4. Three boxplots of the mod residuals for each maturity for the 100 day, 10 day and 1 day recalibration

The final fitting study we carried out was to include exchange rate data. Once again, we took USD and GBP data from 5th October 1994 to 6th March 1995, with 11 states in the Markov chain; we report summary statistics for these in Table 7, with various diagnostics displayed in Figures 10, 11, 12 and 13 By including the exchange rate in the calculation, we worsen the fit of the yield curves by about 1 bp per maturity at $\beta = 0.2$. The fit of the exchange rate is very good, mostly within about 0.5 bp. We tried to trade off the quality of the fit of the exchange rate and the fit of the yield curves, by attaching more weight to poorly fitting yields, but it seemed impossible to improve the fit of the yield curves very much by this. Rogers & Zane [15] found a similar behaviour. In view of the fact that we were fitting the exchange rate much better than the yield curves, it seems that the assumption made at (2.7) that the martingale N^{ij} is constant is relatively harmless; taking something more general would give greater flexibility to fit the exchange rate, but that is not where we appear to need the flexibility.

Table 4. This table contains summary statistics relating to the one country (GBP) CI/RW fits. Note that the case $\beta = 1.0$ corresponds to the CI calibration. These are daily fits on period 14 for varying values of the RW parameter β

CI/RW GBP calibration statistics (all values are in basis points)							
β	Mean	Std. Dev.	Min	Q1	Median	Q3	Max
1.0 (CI)	47.936	22.253	8.441	30.155	41.755	65.786	107.383
0.8	34.995	14.942	8.432	23.944	34.272	45.493	78.581
0.6	26.398	11.243	7.734	18.004	25.72	34.134	62.514
0.4	23.957	10.172	5.751	15.95	23.216	30.261	53.292
0.2	21.346	9.48	5.462	13.35	20.529	26.704	45.744
0.1	20.339	9.31	4.917	13.062	18.942	26.088	45.509

Table 5. Summary statistics for the two country CI/RW fits, period 14 of GBP and USD. This table gives breakdowns for the GBP and USD fits individually

CI/RW calibration statistics for two country fit
(all values are in basis points)

USD FIT							
β	Mean	Std. Dev.	Min	Q1	Median	Q3	Max
1.0 (CI)	93.082	36.671	28.606	60.205	87.108	127.584	168.642
0.8	59.183	19.086	16.899	45.524	55.416	71.192	123.935
0.6	47.947	13.160	18.402	38.282	47.458	54.798	91.380
0.4	43.583	11.195	16.978	36.625	42.664	50.295	74.498
0.2	38.153	11.564	15.619	30.589	37.258	45.501	73.653
0.1	36.775	11.586	15.335	29.052	35.939	44.924	73.549

GBP FIT							
β	Mean	Std. Dev.	Min	Q1	Median	Q3	Max
1.0 (CI)	62.968	26.815	9.552	41.592	62.022	81.679	128.058
0.8	37.816	13.280	9.292	30.283	37.566	46.950	76.318
0.6	33.551	12.923	8.444	24.170	33.806	41.459	76.651
0.4	31.264	12.261	12.330	22.219	28.115	38.620	74.929
0.2	29.184	12.738	6.061	19.819	27.085	38.287	73.427
0.1	28.338	12.902	7.086	19.283	26.448	37.447	72.702

Diagnostic plots for GBP CI calibration

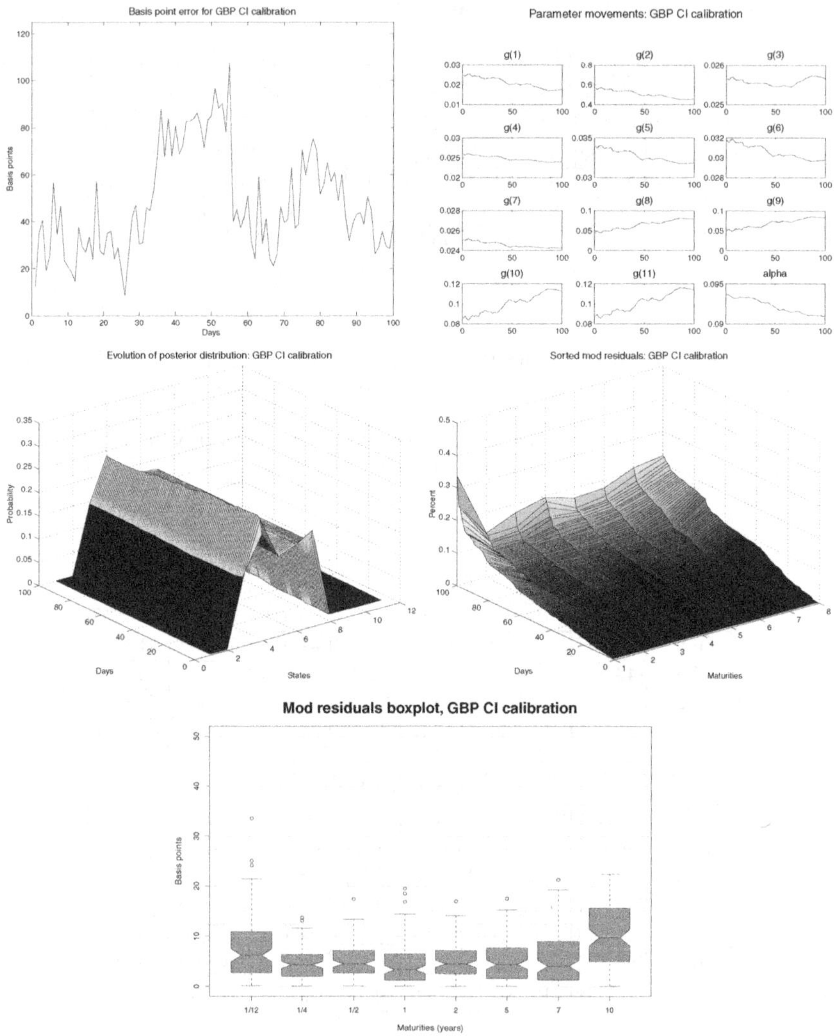

Fig. 5. These plots relate to the one country (GBP) CI calibration for period 14. The 'Parameter change' plot shows how the **g** vector and α scalar change over the 100 day fitting period. We give a surface plot which shows the evolution of the posterior distribution over the 100 day fit. We also show the characteristics of the residuals in the 'Sorted mod residual' and the boxplots

Diagnostic plots for GBP RW calibration ($\beta = 0.2$)

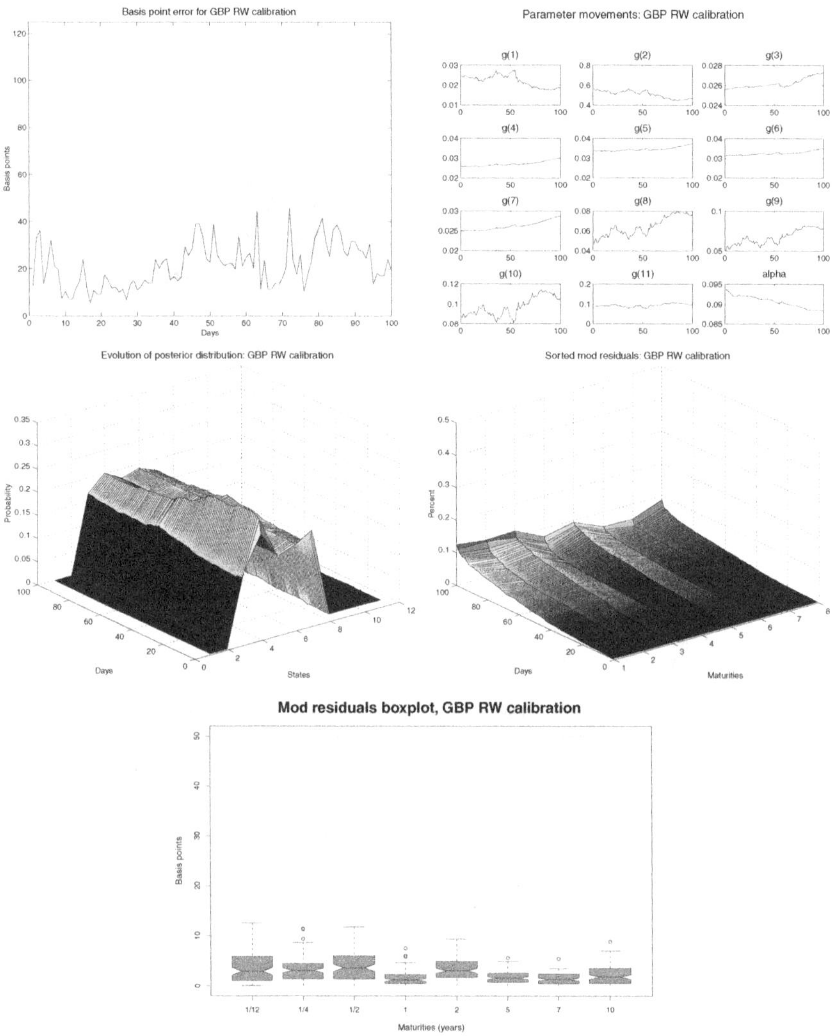

Fig. 6. These plots relate to the one country CI calibration described in Case A. The 'Basis point error' plot shows the cumulative error in basis points for each day over the 100 day fitting period. The 'Parameter change' plot shows how the **g** vector and α scalar change over the 100 day fitting period. We give a surface plot which shows the evolution of the posterior distribution over the 100 day fit. We also show the characteristics of the residuals in the 'Sorted mod residual' and the boxplots

Diagnostic plots for two country (USD & GBP) CI calibration

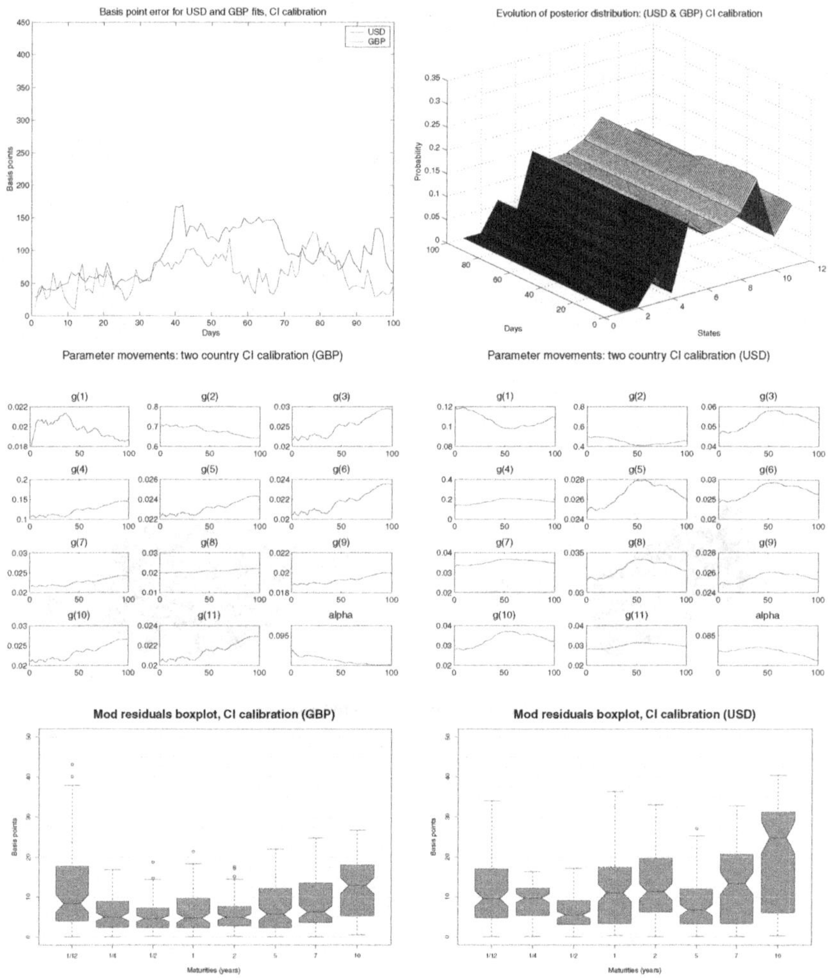

Fig. 7. These plots refer to the two country fits for the CI calibration, USD and GBP, period 14. In this figure we show the basis point error plots (top left) for both the USD and the GBP. The worst fit is the USD

Diagnostic plots for two country (USD & GBP) RW calibration ($\beta = 0.2$)

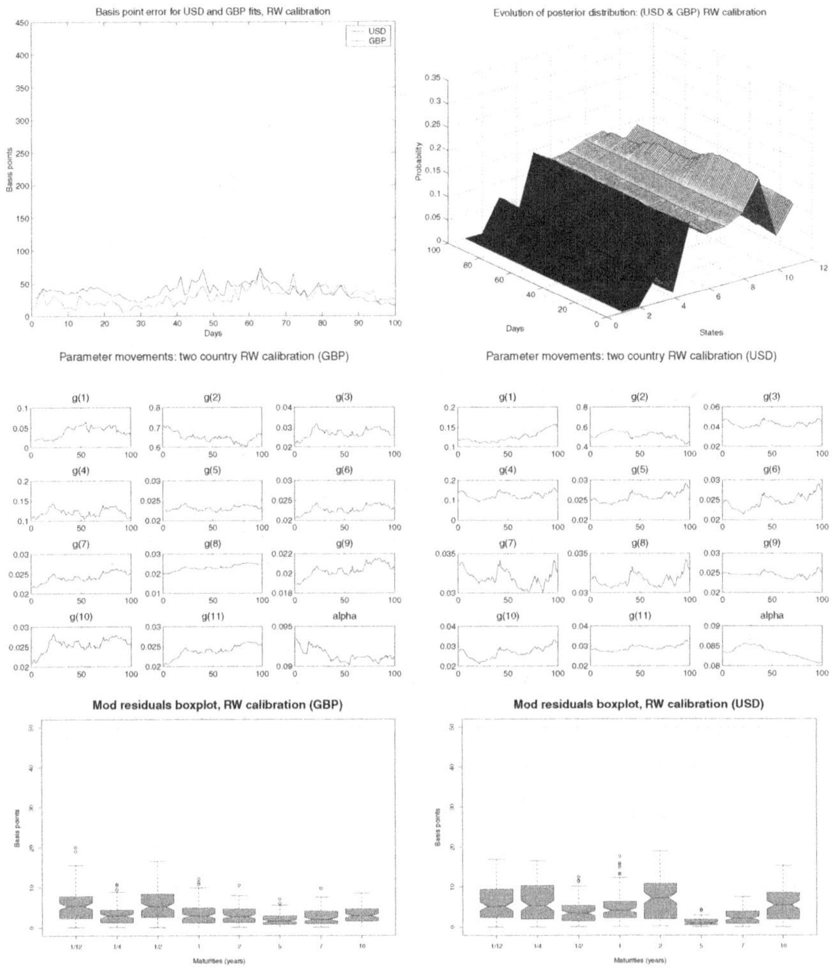

Fig. 8. These plots refer to the two country fits for the RW calibration, USD and GBP, period 14. In this figure we show the basis point error plots (top left) for both the USD and the GBP. The worst fit is the USD

Table 6. Summary statistics for the 100 day, three country CI/RW fits on period 14. This table gives breakdowns for USD, GBP and DEM fits

CI/RW calibration statistics for the three country fit
(all values are in basis points)

USD FIT							
β	Mean	Std. Dev.	Min	Q1	Median	Q3	Max
1.0 (CI)	115.678	32.023	48.308	100.543	115.006	136.413	211.331
0.8	86.260	22.713	34.222	66.194	86.097	106.046	130.637
0.6	67.065	20.279	14.589	52.759	63.658	83.099	113.933
0.4	50.983	14.494	23.901	41.129	50.661	57.949	89.969
0.2	39.511	10.414	21.605	30.209	39.640	47.340	67.369
0.1	35.153	10.444	17.487	27.883	33.912	42.666	66.016

GBP FIT							
β	Mean	Std. Dev.	Min	Q1	Median	Q3	Max
1.0 (CI)	84.958	29.853	28.229	60.984	81.629	107.327	149.090
0.8	45.772	13.036	12.617	36.644	46.403	52.012	82.245
0.6	39.031	13.808	12.529	31.140	37.188	46.512	87.466
0.4	35.006	12.430	13.078	27.435	33.425	40.875	76.725
0.2	31.140	10.532	12.523	24.456	29.408	37.529	70.204
0.1	29.123	10.458	7.467	22.943	27.858	35.956	67.502

DEM FIT							
β	Mean	Std. Dev.	Min	Q1	Median	Q3	Max
1.0 (CI)	66.798	21.886	32.260	50.378	62.351	79.501	132.468
0.8	49.857	10.955	30.902	41.828	47.704	56.204	86.001
0.6	45.070	9.925	28.391	38.174	43.194	50.179	74.346
0.4	40.557	8.014	25.880	35.359	40.172	45.085	72.847
0.2	37.050	6.360	23.124	32.613	37.530	41.188	59.138
0.1	35.640	5.704	23.380	32.013	34.881	39.556	55.702

Diagnostic plots for three country (USD, GBP & DEM) RW calibration ($\beta = 0.2$)

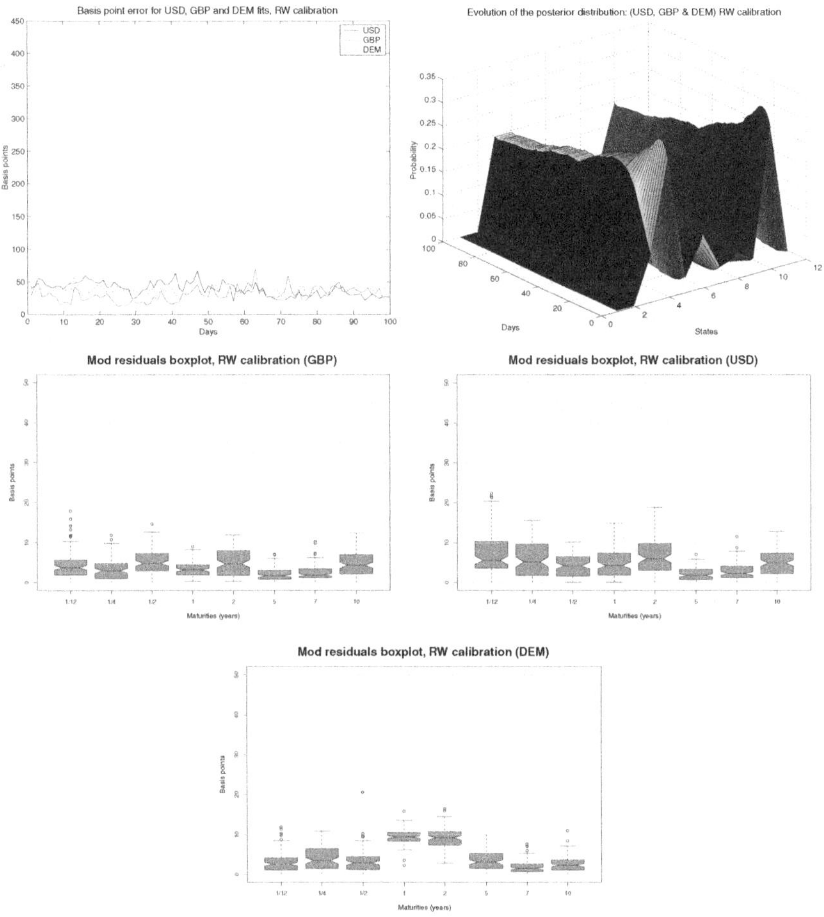

Fig. 9. These plots are for a three country fit using the RW calibration method with $\beta = 0.2$. The first plot shows the basis point error for each of the three countries. The worst fit is achieved by the USD and the the best fit (red line) is the GBP. The second plot (top right) shows the evolution in the posterior distribution during the fitting process. The boxplots are of the mod residuals for each maturity

Table 7. Summary statistics for the two country and exchange rate CI/RW fits over 100 days (period 14). This table has breakdowns for the USD and GBP fitting errors

CI/RW calibration statistics for two country and exchange rate fit (all values are in basis points)

USD FIT							
β	Mean	Std. Dev.	Min	Q1	Median	Q3	Max
1.0 (CI)	146.245	54.529	38.552	101.450	158.334	190.084	310.810
0.8	82.139	29.645	35.550	60.063	78.209	105.595	144.969
0.6	72.696	29.291	30.794	51.712	62.452	98.970	138.709
0.4	58.471	20.738	21.819	43.719	52.419	71.981	115.074
0.2	49.979	16.727	19.442	37.926	47.491	60.373	96.033
0.1	42.505	12.762	5.670	34.353	42.572	48.860	82.313

GBP FIT							
β	Mean	Std. Dev.	Min	Q1	Median	Q3	Max
1.0 (CI)	103.685	41.776	16.543	72.962	101.316	129.850	211.00
0.8	48.495	17.366	10.098	39.003	47.939	58.941	86.957
0.6	42.928	14.936	13.318	32.262	40.831	53.962	83.303
0.4	32.557	12.410	10.285	22.825	32.933	40.758	74.011
0.2	35.910	12.144	13.076	27.088	34.235	43.552	79.165
0.1	30.402	11.581	10.049	21.721	29.326	36.612	73.550

Diagnostic plots for two-country (USD & GBP) CI calibration with exchange rates

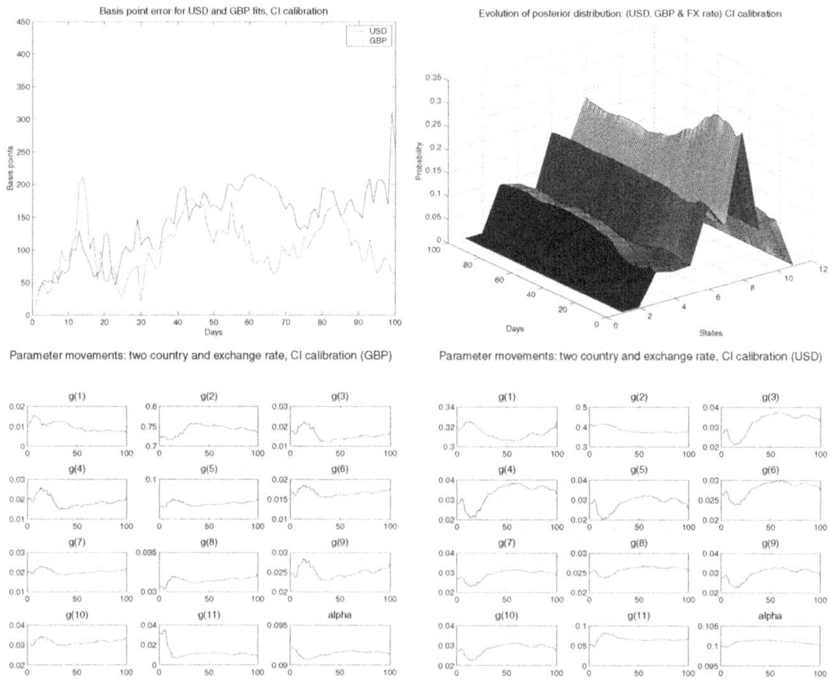

Fig. 10. These plots refer to the two country and exchange rate fits for the RW calibration, period 14, over a 100 day period using USD and GBP data. In this figure we show the basis point error plots (top left) for both the USD and the GBP, the worst fit is the USD

Diagnostic plots for two-country (USD & GBP) CI calibration with exchange rates

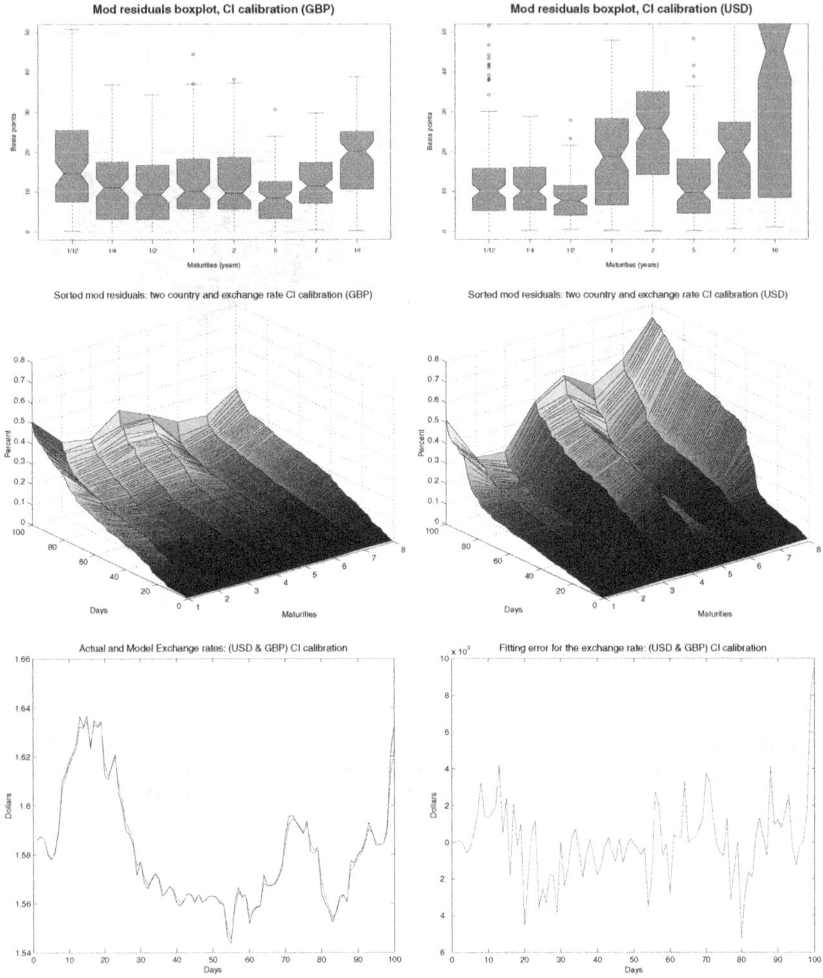

Fig. 11. These plots refer to the two country and exchange rate fits for the CI calibration of the USD and GBP, period 14 data. The penultimate plot in this figure shows the observed data and the fitted curve for the exchange rates (there are two curves in this picture). The final plot is of the fitting error in the exchange rate

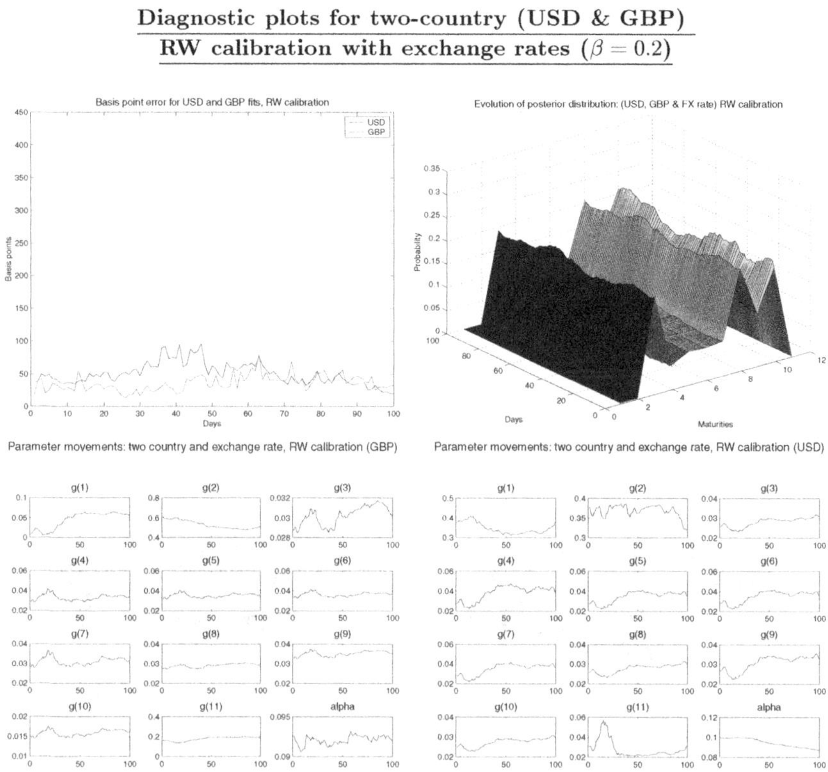

Fig. 12. These plots refer to the two country and exchange rate fits for the CI calibration, period 14, over a 100 day period using USD and GBP data. In this figure we show the basis point error plots (top left) for both the USD and the GBP, the worst fit is the USD

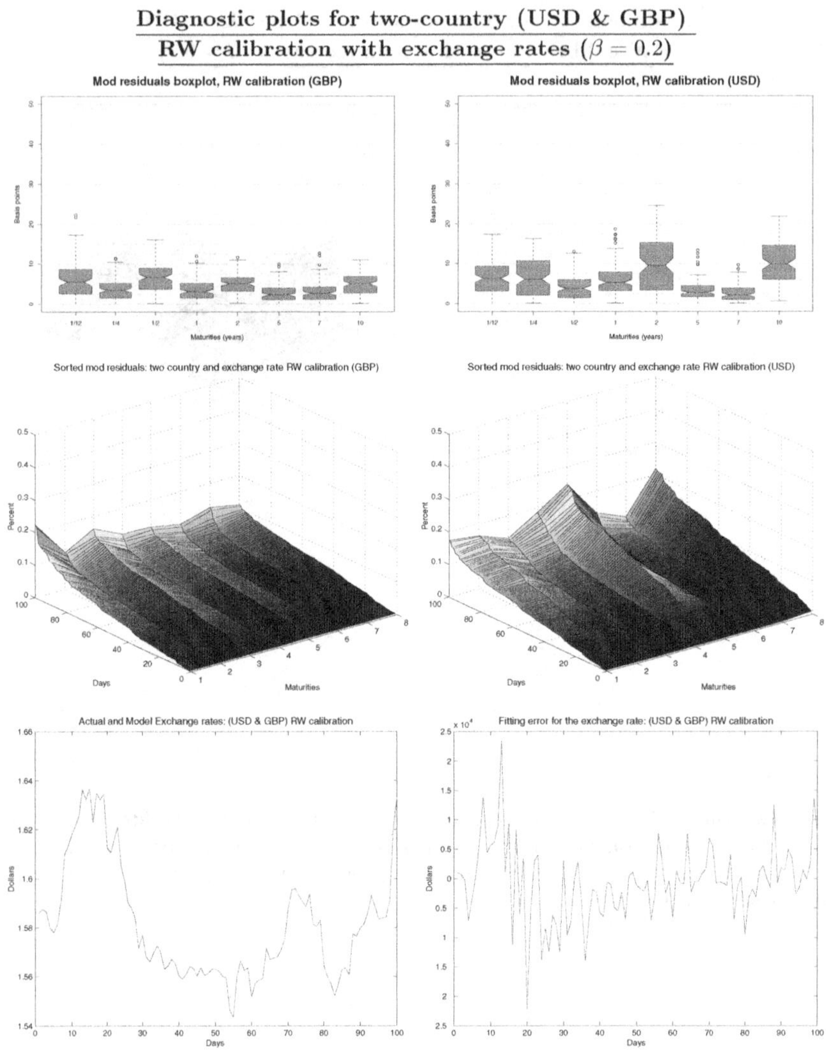

Fig. 13. These plots refer to the two country and exchange rate fits for the RW calibration of the USD and GBP, period 14 data. The penultimate plot in this figure shows the observed data and the fitted curve for the exchange rates (there are two curves in this picture). The final plot is of the fitting error in the exchange rate

5 Conclusions

In this study, we have carried out a number of calibration exercises for potential models of interest rates based on an underlying Markov chain. At a theoretical level, such models offer persuasive advantages:

- the approach generates a model to account for all derivatives;
- pricing of a European-style derivative is simply a sum over a (typically small) finite number of states, and pricing of an American-style derivative is an optimal stopping problem for a Markov chain with (typically few) states;
- adding a new country can be done without complicating the underlying Markov process;
- exchange rates are modelled within the same modelling framework as interest rates.

What we have done here is by way of a pilot study, to investigate the feasibility of this approach. Most of the fitting runs were done using only 11 states of the Markov chain, and we were insisting on fitting a time-homogeneous model, both very stringent requirements which would undoubtedly be abandoned in practice. If we allowed a different model to be fitted each day, we were able to come up with fits of the yield curve in one country with median errors of the order of 1bp per maturity; sometimes more, sometimes less. At the other end of the scale, by calibrating to 5 days' data and then using the calibrated model to fit the next day, we were coming up with median errors of the order of 6 bp per maturity, which is too high to be much use. By taking a fitting methodology in between these two extremes, we were able to produce one-country fits with median errors of around 2.5 bp per maturity, with good parameter stability.

Incorporating more than one country inevitably worsened the fit; when we fitted USD and GBP data, we came up with median errors of the order of 3.5-4.5 bp per maturity, and including DEM as well increased the errors very slightly. However, including the exchange rate in the USD/GBP fitting exercise worsened the median fit by about 1 bp per maturity, which would lead to quite significant mispricing.

Given the restrictions to time-homomogeneous chains with no more than 11 states, the fits we have come up with are very encouraging. There are obvious extensions which could be carried out, and some will be the subject of a later study. For example, we could simply increase the number of states. Since the calibration procedure was quite lengthy on the machine[9] available to us (of the order of 200 CPU minutes to fit a single country, of the order of 300 CPU minutes to fit two countries), we preferred to investigate a larger number of relatively small problems, rather than try a few huge fits. Another obvious place where the modelling could be extended would be by dropping the reversibility criterion; this requires code which can cope with complex

[9] A Sun Ultra E3500 with 400 MHz UltraSPARC II processor

eigenvalues, but the principles are the same. Another extension would be to allow time-dependent Markov processes. In some sense, this is completely trivial; as Rogers [13] remarked (p 101), if we take a time-homogeneous model and apply a deterministic time-change, we can exactly fit any given initial yield curve! However, this is really far too easy, and we need to be aware of the model changing completely in a day, always a problem with time-inhomogeneous models.

The relative success of such a primitive probabilistic model is either to be expected, or something quite remarkable, depending on your point of view. On the grounds that there are many parameters, it might be thought to be expected; but our earlier comments show that large parameter spaces are not in themselves guarantors of a close fit. To be able to fit more than one yield curve reasonably closely using nothing more sophisticated than an 11 state chain does seem to us to be remarkable. In Fig. 14, we present a plot of 1-month LIBOR and Bank of England band one stop rate. The agreement is evident, and the conclusion unavoidable: if we were able to model the band one rate, we would already have a good model for 1-month LIBOR! Now the band one rate is a jump process, taking relatively few values. It is not fanciful to imagine that *this* could be well modelled by a Markov chain with a small number of states. Indeed, looking at Fig. 14, the interpretation of 1-month LIBOR as a noisy observation of the band one rate seems quite natural, and the very interesting paper of Babbs & Webber [2] uses elements of this interpretation in its modelling. In short, focusing on the volatilities of various yields and rates may actually be concentrating on the *noise* in the system, and overlooking the *signal!*

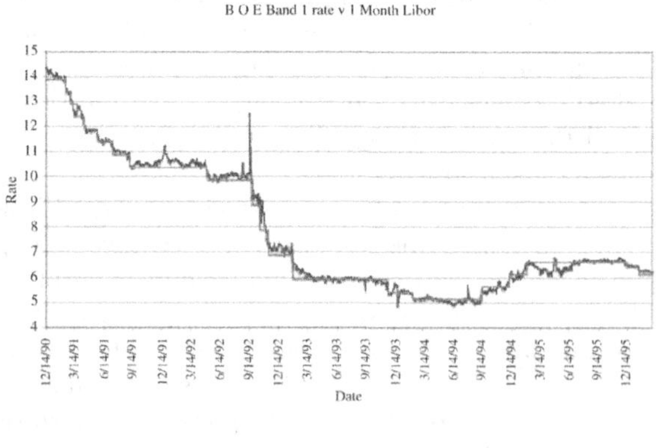

Fig. 14. Base Rate against 1 Month LIBOR

Appendix A

For ease of reference, we summarise here the Kalman filter argument which we used as the basis of the fitting procedures of the earlier parts of the paper. To begin with, suppose we have a pair of discrete-time vector processes θ and Y evolving according to the dynamic linear model

$$\theta_n = \theta_{n-1} + \varepsilon_n, \tag{A.1}$$

$$Y_n = C\theta_n + \eta_n, \tag{A.2}$$

where the ε are independent $N(0, Q)$ and the η are independent $N(0, R)$ random variables. If $\mathcal{Y}_n$ denotes the σ-field generated by $\{Y_k : k \leq n\}$, and if we have that conditional on $\mathcal{Y}_n$ the law of θ_n is $N(\hat{\theta}_n, V_n)$, then

$$\left(\begin{matrix}\theta_{n+1}\\ Y_{n+1}\end{matrix}\middle| \mathcal{Y}_n\right) \sim N\left(\begin{pmatrix}\hat{\theta}_n\\ C\hat{\theta}_n\end{pmatrix}, \begin{pmatrix}Q+V_n & (Q+V_n)C^T\\ C(Q+V_n) & R+C(Q+V_n)C^T\end{pmatrix}\right), \tag{A.3}$$

and likewise

$$\left(\begin{matrix}\theta_{n+1}\\ Y_{n+1} - C\theta_{n+1}\end{matrix}\middle| \mathcal{Y}_n\right) \sim N\left(\begin{pmatrix}\hat{\theta}_n\\ 0\end{pmatrix}, \begin{pmatrix}Q+V_n & 0\\ 0 & R\end{pmatrix}\right). \tag{A.4}$$

It is an easy though tedious exercise to confirm from (A.3) that the law of θ_{n+1} given $\mathcal{Y}_{n+1}$ is $N(\hat{\theta}_{n+1}, V_{n+1})$, where $\hat{\theta}_{n+1}$ is the value of θ maximising the joint density of the distribution (A.3) or equivalently (A.4):

$$\exp\left[-\frac{1}{2}(\theta - \hat{\theta}_n)^T(Q+V_n)^{-1}(\theta - \hat{\theta}_n) - \frac{1}{2}(y - C\theta)^T R^{-1}(y - C\theta) \right], \tag{A.5}$$

and $-V_{n+1}^{-1}$ is the second derivative of the log-likelihood with respect to θ. The actual estimation problem we face has non-linear dynamics, but we shall suppose that a local linear approximation is adequate, so we replace (A.2) with

$$Y_n = Y(x_n, \theta_n) + \eta_n,$$

giving the analogue of (A.5) to be

$$\exp\left[-\frac{1}{2}(\theta - \hat{\theta}_n)^T(Q+V_n)^{-1}(\theta - \hat{\theta}_n) - \frac{1}{2}(y - Y(x_n,\theta))^T R^{-1}(y - Y(x_n,\theta)) \right]. \tag{A.6}$$

If we have $Q = 0$, the full CI fitting assumption, we find that (A.6) reduces to the exponential terms in (3.8), and if we take $Q = (\beta^{-1} - 1)V_n$, we obtain exactly the exponential terms in (3.9).

References

1. S. H. Babbs: *A family of Itô process models for the term structure of interest rates*, University of Warwick, preprint 90/24, (1990).
2. S. H. Babbs and N. Webber: *Term Structure Modelling Under Alternative Official Regimes*, Mathematics of Derivative Securities, Hardback ISBN 0521-5842-48 (1997), 394–422.
3. F. Black, E. Derman and W. Toy: *A One-factor Model of Interest Rates and its Application to Treasury Bond Options*, Financial Analysts Journal, **46** (1990), 33–39.
4. F. Black and P. Karasinski: *Bond and Option Pricing when Short Rates are lognormal*, Journal of Financial Analysts, **47** (1991), 52–59.
5. A. Brace and M. Musiela: *A Multifactor Gauss Markov Implementation of Heath, Jarrow and Morton*, Mathematical Finance, **4** (1994), 259–283.
6. G. M. Constantinides: *A Theory of the Nominal Term Structure of Interest Rates*, Review of Financial Studies, **5** (1992), 531–552.
7. J. C. Cox, J. E. Ingersoll and S. A. Ross: *A Theory of the Term Structure of Interest Rates*, Econometrica, **53** (1985), 385–407.
8. H. W. Fowler: *A Dictionary of Modern English Usage*, 2nd Edition, Oxford University Press (revised by E. Gowers), (1968).
9. D. C. Heath, R. A. Jarrow, & A. Morton: *Bond Pricing and the Term Structure of Interest Rates: A Discrete Time Approximation*, Journal of Financial Quantitative Analysis, **25** (1990), 419–440.
10. T. S. Y. Ho and S. B. Lee: *Term Structure Movements and Pricing Interest Rate Contingent Claims*, Journal of Finance, **41** (1986), 1011–1029.
11. M. Musiela and M. Rutkowski: *Martingale Methods in Financial Modelling*, Springer-Verlag Berlin, ISBN 354061477X (1997).
12. K. R. Miltersen, K Sandmann and D. Sondermann: *Closed Form Solutions for Term Structure Derivatives with Log-normal Interest Rates*, Journal of Finance, **52** (1997), 409–430.
13. L. C. G. Rogers: *Which Model for the Term-Structure of Interest Rates Should One Use?*, Mathematical Finance, **65** (1995), 93–116.
14. L. C. G. Rogers: *The Potential Approach to the Term-Structure of Interest Rates and Foreign Exchange Rates*, Mathematical Finance, **7** (1997), 157–176.
15. L. C. G. Rogers and O. Zane: *Fitting Potential Models to Interest Rate and Foreign Exchange Data*, Vasicek and Beyond (book), ISBN 1899-3325-02 (1996), 327–342.
16. O. Vasicek: *An Equilibrium Characterisation of the Term Structure*, Journal of Financial Economics, **5** (1977), 177–188.

Theory and Calibration of HJM with Shape Factors

Andrea Roncoroni[1] and Paolo Guiotto[2]

[1] ESSEC Graduate Business School & Université Paris Dauphine, France*
[2] Università degli Studi di Padova, Italy**

Abstract. We construct arbitrage free dynamics for the term structure of interest rates driven by infinitely many factors, each one representing a basic shape for the instantaneous forward rate curve in a given market. The consistency between a finite-dimensional space of polynomials where the curve is day-to-day recovered and the proposed evolution equation is investigated. The main result is the developement of a historical-implicit hybrid calibration procedure for our infinite-dimensional shape factor model. In this context, we also derive a pricing formula for caplets.

Key words

Term Structure of Interest Rates, Yield Curve Estimation, Principal Components Analysis, Calibration

JEL Classification: E43

1 Introduction

The benchmark framework for building arbitrage free dynamics for the term structure of interest rates has been set up by Heath, Jarrow and Morton in [16] (HJM in what follows). The basic ingredients are an initially observed yield curve in terms of instantaneous forward rates and a volatility function defining the local covariance process for each of the modeled rates. One of their main results is a drift restriction for the risk-neutral term structure dynamics in terms of the volatility function, which is known to be independent of the underlying equivalent measure and thus amenable of direct historical

* Finance Department, ESSEC Graduate Business School, Av. Bernard Hirsch BP 105, 95021 Cergy-Pontoise, France; CEREG, Université Paris Dauphine, Pl. du Maréchal de Lattre-de-Tassigny, 75775 Paris, France; Tel. +33134433239; Fax. +33134433001; *e-mail: roncoroni@essec.fr*

** Dipartimento di Matematica Pura e Applicata, Università di Padova, via Belzoni 7, 35131 Padova, Italy; Tel.+390498275987; Fax. +390498758596; *e-mail: parsifal@galileo.math.unipd.it*

estimation, at least in principle. Several issues were left open to subsequent research. In particular, we investigate the following ones.

The original HJM model considers a continuum of stochastic differential equations, one for each instantaneous forward rate. Yet, a finite, though arbitrarily large, number of random forces, each one represented by a Brownian motion, is allowed too. The absence of arbitrage opportunities across discount bonds driven by n random noises requires as many as n linear constraints between excesses of instantaneous rates of bond prices returns over the risk-free short rate and local volatilities. In practice, the number of actually traded bonds is so huge with respect to the number of random noises that the n-dimensional linear constraint above mentioned never holds true. One may wonder whether there is any extension of HJM to infinitely many noises so to relax such a constraint.

Secondly, cross-sectional estimation procedure is usually employed to determine an initial continuous yield curve from a discrete set of observed points. This curve is required to belong to a suitable family of smooth functions and constitutes the initial condition for the system of infinitely many stochastic differential equations (or, as in our model, the single Hilbert space-valued stochastic differential equation) describing the random evolution of the term structure dynamics. At each time the cross-sectional estimation procedure is repeated. Thus it seems necessary to require that any such system of equations have support in the class of functions the yield curve is periodically recovered in. If this space is big enough, any arbitrage free dynamics is consistent. Otherwise, one may consider quasi-arbitrage-free dynamics as we will do in the case of polynomial families for the instantaneous forward rate curve.

As for the third issue we investigate, for a model to be of some practical use, one should use a calibration procedure embedding most available data into the model. This data consists of both standard (*e.g.* coupon-bearing bonds and swaps) and derivative (*e.g.* options and futures) assets. Therefore pricing formulae for these claims ought to be provided in either closed or numerically computable form. More significantly, a challenge is represented by the determination of a procedure for integrating knowledge about parameters inferred from historical data and parameters implied from observed option prices through risk-neutral formulae. This is a kind of hybrid calibration which seems to have had no evidence in current interest rate research yet.

We hereby construct a class of dynamic interest rate models in infinite dimensional spaces and show how they are particularly suitable to tackle all of the above mentioned issues. A short review of what has already been done in this direction may enlighten the nature and the use of our results.

To begin with factor modelling, we note that state-of-the-art interest rate models usually begin with exogenously imposed dynamics for a finite dimensional vector-valued process $\mathbf{x}(t)$, whose components are called factors and a suitable functional of which defines the interest rates of reference. For example, early arbitrage-free models assume $\mathbf{x}$ to be the short rate itself. More generally, under suitable, yet weak assumptions, any finite multi-factor term structure model can assume any finite set of rates as driving factors (see Geman, El Karoui and Lacoste [11]). We propose a dynamic interest rate model of HJM type where each of possibly *infinitely many factors* is no more a rate, but the weight assigned to a *whole basic shape* the yield curve may assume in a given market, this allowing for a free choice of the functions reconstructing the yield curve according to well known classical term structure estimation procedures. This is shown to be compatible with the construction of an arbitrage-free bond market by imposing a drift restriction of HJM kind to the yield curve movement, here expressed as a single diffusion process taking values in a functional space. To this aim, we make use of the theory of stochastic equations in Hilbert spaces as developed by Da Prato and Zabczyk in [6]: the geometric properties of these spaces will turn out to be of key importance for the calibration procedure of our model. Douady [7] also exploits the idea of identifying factors with basis functions generating a Hilbert space and develops a similar analysis in a yield curve framework. However he does not enter into the issues of derivative pricing and implicit calibration of the risk-free dynamics through option prices. From the purely theoretical point of view, the work by Goldys and Musiela [13] is the closest to ours: they obtained a similar condition for absence of arbitrage opportunities using a different method than ours. The importance of shapes in arbitrage-free modelling has been firstly underlined in the early nineties by Brown and Shaefer in [4].

The consistency issue can be summarized in the following terms. Given a particular class of functions used to fit observed points into a unique continuous current yield curve, one ideally wishes for an arbitrage-free model to prevent the resulting term structure from going out of it. Most classes presented in the literature on classical yield curve estimation turn out to be unstable under the best known arbitrage-free term structure dynamics (see Björk and Christiansen [2]). To our knowledge the first who stressed the economic significance of this inconsistency were Corielli and Petrone [5]. The selection of the functional class where a yield curve is supposed to live in is the well known issue of cross-sectional estimation (see Anderson *et al.* [1]). A systematic study of the consistency problem has been carried out in the excellent doctorate thesis by Filipovič [8] to which we refer the reader who is interested in deepening his or her own knowledge on the subject from a purely mathematical point of view. In our model, the user may choose any reasonable family of interpolating functions. We determine the closest consistent approximate dynamics to the perfectly arbitrage-free one, the latter being necessarily an element of a larger space. An explicit formula delivers an upper bound for this error.

As far as practical implementation of the model is concerned, Brace and Musiela [3] developed an implicit non-parametric procedure for calibrating an infinite dimensional HJM model where the forward rates dynamics are driven by a *finite* dimensional Brownian motion (see also Musiela [18]). We use a two-stage procedure for calibrating our model. First, we compute the historical volatility starting from observations on static (*i.e.*, whose arbitrage-free price is volatility independent) yield curve products (*e.g.* coupon bonds and swaps). Then, we integrate an implicit method, which is closely related to that of Brace and Musiela, into the previously obtained historical data. An optimization algorithm is provided to ensure the minimization of the inconsistency between historical and implicit calibrations. In practice we build a class of historical estimation procedures coupled with conjugate implicit calibration procedures from option prices. Then we choose the pair of historical estimation procedure and its conjugate implicit one giving rise to the closest calibrated parameters, this all over the set of historical estimation procedures.

The outline of this paper is as follows. In section 2 we set up notation and basic definitions. Section 3 delivers the arbitrage-free interest rate dynamics with infinitely many factors. Next, section 4 deals with a truncated version of the arbitrage-free dynamics which is consistent with a certain class of polynomials; a method for controlling the resulting error is also presented. In section 5 we provide an algorithm for performing calibration to market prices of both volatility independent and dependent financial products: pricing by the forward-risk-adjusted method developed by Geman [10], Geman, El Karoui and Rochet [12] and Jamshidian [17] is part of this program. Section 6 draws conclusions.

2 Notation and Definitions

Let H be a space of functions; we sometimes write $x(\cdot)$ instead of x to underline the function nature of x; both $y(t)$ and y_t may represent either a t-indexed stochastic process or its t-th random variable; for example, $r_t(\cdot)$, or much simpler r_t, denotes the element of H representing the time t yield curve (as defined below): the number $r_t(x)$ is the value of r_t at point x. $\langle x., y.\rangle_s$ denotes the covariance process between processes x_t, y_t and should not be confused with $\langle x, y\rangle_H$, which is an inner product in H. *Span* $\left(\mathbf{h}^{(1)}, ..., \mathbf{h}^{(n)}\right)$ is the linear space spanned by vectors $\mathbf{h}^{(i)}, i = 1, ..., n$. The i-th component of a vector $\mathbf{h}$ will be indifferently denoted by either h_i, $\mathbf{h}_i$ or $(\mathbf{h})_i$.

Let $P^T(t)$ be the time t price for a zero coupon bond maturing 1 Euro at a future date T, $r_t(x)$ (or $f^{t+x}(t)$) be the time t instantaneous forward rate expiring in x years ahead from t, $r(s)$ be the time s short rate and $\beta(t) := \exp\left(\int_0^t r(u)\,du\right)$ be the rolled over money market account starting at time 0. Since: $P_s(T-s) := P^T(s) = P^t(s)\exp\left(-\int_t^T f^u(s)\,du\right)$ the following definition makes sense.

For $t \geq 0$, the time t yield curve, or term structure of interest rates, is either of the following objects: $\left\{f^T(t), T \geq t\right\}$, $\left\{P^T(t), T \geq t\right\}$. They may be expressed in the time-to-maturity parametrization:

$$\{r_t(x), x \geq 0\}\ ;\ \{P_t(x), x \geq 0\}.$$

The former is referred to as the yield curve 'in terms of forward rates', whereas the latter is said to be 'in terms of zero coupon bonds'. Note that x will always represent time to maturity.

3 Infinite Dimensional HJM Model

3.1 The Musiela Equation

In the original HJM formulation, yield curves are functions defined on varying domains. For example, the time t_1 forward curve is $\left\{f^T\left(t_1\right), t_1 \leq T \leq T^*\right\}$, namely a function defined on $\left[t_1, T^*\right]$, while the time t_2 forward curve is defined on $\left[t_2, T^*\right]$. In order to have a space of yield curves defined on a unique interval, Musiela proposed a new parametrization where time t yield curve is indexed by the time-to-maturity instead of the maturity itself.

Assuming that the HJM forward curve $f(t, T)$ is C^1 in T, a formal differentiation of the forward stochastic differential equation [16, eq.(4)] with respect to t gives:

$$d_t r_t(x)=\frac{\partial}{\partial x} r_t(x) d t+\left(\alpha(t, x) d t+\sum_{i=1}^n \sigma_i(t, x) d W_i(t)\right),$$

where the coefficients notation has been adapted to the time to maturity representation: for instance, $\alpha(t, x)$ replaces the symbol $\alpha(t, t+x)$ in the HJH notation.

If the HJM drift restriction [16, eq.(18)] holds true, then:

$$d_t r_t(x)=\left(\frac{\partial}{\partial x} r_t(x)+\sum_{i=1}^n \sigma_i(t, x) \int_0^x \sigma_i(t, v) d v\right) d t+\sum_{i=1}^n \sigma_i(t, x) d W_i(t). \tag{1}$$

This is the so-called Musiela equation.

3.2 Model Ingredients

The whole time-t yield curve $\left\{r_t(x), 0 \leq x \leq X^*\right\}$ is represented by an element in an *a priori* chosen space of functions H, endowed with inner product $\langle\cdot, \cdot\rangle_H$ and spanned by an orthonormal complete system $\left\{f_j\right\}$: H is a separable Hilbert space. Here X^* is the greatest time to maturity (e.g. 30 years). Without loss of generality X^* is assumed to be equal to 1. Let $[0, T]$ be the time horizon upon consideration. The *idea* is to move the entire term structure of interest rates as seen as a single point in H in order to be able to reconstruct it by a series expansion in terms of a basis $\left\{f_j\right\}$ acting as a set of interpolating or approximating functions. Basically H will be $L^2\left(\left[0, X^*\right]\right)$, the set of square integrable function on $\left[0, T^*\right]$ with respect to the Lebesgue measure. For instance, let X^* be the greatest time to maturity and suppose we wish to assign diversified weights to rates along the spectrum $[0, X^*]$; as a norm in H one may use the one induced by an inner product of form:

$$\langle r_t(\cdot), s_t(\cdot)\rangle_H \equiv \left\{\int_0^{X^*} (r_t(x)\, s_t(x))^2\, p(x)\, dx\right\}^{1/2},$$

where p is a suitable regular weight function across time to maturities. For instance, $p(x) := 1/x$ corresponds to giving more importance to short maturity rates against long term ones.

We consider an evolution equation of form:

$$d_t r(t,x) = \alpha(t,x)\, dt + dW_Q(t,x), \tag{2}$$

where, differently from Musiela [18], we take a Brownian motion $W_Q(t,x)$ in both of the variables t and x. In order for this equation to have a meaning, some care is due. A complete treatment can be found in the monograph by Da Prato and Zabczyk [6]; however we will briefly sketch an outline of the main tools we need.

Let $\{W_k(t)\}_{t\geq 0}, k \geq 1$ be a countable collection of independent standard one dimensional Brownian motions and set:

$$W(t) := \sum_k W_k(t)\, f_k. \tag{3}$$

It is well known that this series is a.e. divergent in H, so that $W(t)$, known as cylindrical Brownian motion, is not properly defined.

In order to cope with this, the customary adjustment is to set:

$$W_Q(t) := \sum_k \sqrt{\lambda_k(t)} W_k(t)\, f_k \tag{4}$$

under the condition:

$$\sum_k \lambda_k(t) < \infty.$$

Formally $W_Q(t) = \sqrt{Q(t)} W(t)$, where $\sqrt{Q(t)}$ is the diagonal operator defined by:

$$\sqrt{Q(t)} h := \sum_k \sqrt{\lambda_k(t)}\, \langle h, f_k\rangle_H\, f_k.$$

We refer to $W_Q(t)$ as a $Q(t)$-nuclear Brownian motion.

The main tool of stochastic analysis is the stochastic integral, which may be defined by:

$$\left(\int_0^t dW_Q(s)\right)(\cdot) := \sum_k \int_0^t \sqrt{\lambda_k(s)} dW_k(s)\, f_k(\cdot).$$

This series makes sense for any $t \leq T$, provided that:

$$\sum_k \int_0^T \lambda_k(s)\,ds < \infty. \tag{5}$$

We are ready to give a meaning to equation (2). An H-valued process r_t satisfies (2) if:

$$r_t = r_0 + \int_0^t \alpha(s)\,ds + \int_0^t dW_Q(s).$$

3.3 General Theorem for Arbitrage Free Term Structure Dynamics

Theorem 3.1. *Let $\{\lambda_k(t)\}$ satisfy (5), $\{r_t, 0 \leq t \leq T\}$ be an $H := L^2([0,1])$-valued predictable process representing the instantaneous forward rate in the time to maturity parametrization, with $r_t(\cdot) \in AC([0,1])$, namely the class of absolutely continuous functions defined on $[0,1]$. Let $W_Q(t)$ be a $Q(t)$-nuclear Brownian motion with respect to the augmentation of its natural filtration. If $\{\alpha(s), 0 \leq s \leq T\}$ is an H-valued predictable process such that:*

$$r_t = r_0 + \int_0^t \alpha(s)\,ds + \int_0^t dW_Q(s) \tag{6}$$

holds and there are no arbitrage opportunities in the sense of Harrison and Pliska (see [15]), then:

$$\alpha(s,x) = \frac{\partial}{\partial x} r_s(x) + \sum_{j=1}^{\infty} \lambda_j(s) f_j(x) \int_0^x f_j(u)\,du. \tag{7}$$

Conversely, if $\{\alpha(s), 0 \leq s \leq T\}$ is defined by (7), then it is predictable, H-valued and $\{r_t, t \geq 0\}$ in (6) is the arbitrage-free forward rate dynamics.

Remark 3.2. We obtained arbitrage restrictions in a more general setting (see [14]); in the general case of instantaneous forward rate dynamics:

$$dr_t(x) = \alpha(t,x)\,dt + \tau(t,x)\,dW_Q(t,x),$$

the corresponding arbitrage-free drift turns out to be:

$$\alpha(s,x) = \frac{\partial}{\partial x} r_s(x) + \sum_{j=1}^{\infty} \lambda_j \tau(s,x) f_j(x) \int_0^x \tau(s,u) f_j(u)\,du.$$

In particular, if $W(t)$ is a one-dimensional Brownian motion, one obtains the Musiela drift restriction [18]:

$$\alpha(s,x) = \frac{\partial}{\partial x} r_s(x) + \tau(s,x) \int_0^x \tau(s,u)\,du.$$

Our present formulation allows for infinitely many factors, but the local covariance process of r_t is forced to assume a precise diagonal form: this is justified by the flexibility of the resulting calibration to be developed in the last section.

The economic significance of this results is twofold: 1) if one admits infinitely many sources of noise (and not just d like in HJM, where the Brownian motions are maturity independent), here represented by the sequence of coefficients in the series expansion of a given yield curve in a specific market, then it is still possible to obtain an arbitrage free model by imposing a drift restriction which is fully determined by the market volatility structure and, in addition, by the basis spanning the space of yield curves; 2) the above stated representation in series expansion describes any actual yield curve as a superposition of fundamental shapes f_i, each one carrying a quota of the total volatility embedded into the real term structure; this quota can be computed both theoretically and in practise by a simple calibration algorithm, as we will see in Sect. 5.

Proof. We rely on basic results in the theory of stochastic equations in infinite dimensions (see Da Prato and Zabczyk [6]). Because of the hypothesis on the filtration, showing that the discounted bond process:

$$\left\{\frac{P^T(t)}{\beta(t)}, 0 \le t \le T\right\}$$

is a martingale amounts to setting its drift to 0. Indeed, by Ito's product rule:

$$\frac{P^T(t, r_t(\cdot))}{\beta(t)} = P^T(0, r_0(\cdot)) + \int_0^t \left(-P^T(s, r_s(\cdot))\,\beta(s)^{-1} r_s(0)\right) ds + \int_0^t \beta(s)^{-1} d_s P^T(s, r_s(\cdot)), \tag{8}$$

because the money market account is a finite variation process. By Ito's formula for H-valued Ito processes, one computes the third term as:

$$\begin{aligned} P^T(t, r_t(\cdot)) = P^T(0, r_0(\cdot)) &+ \int_0^t \Big\{P_t^T(s, r_s(\cdot)) + \left\langle P_\phi^T(s, r_s(\cdot)), \alpha(s)\right\rangle_H \\ &+ \frac{1}{2}\mathrm{Tr}\left[P_{\phi\phi}^T(s, r_s(\cdot))\sqrt{Q(t)}\left(\sqrt{Q(t)}\right)^*\right]\Big\} ds \\ &+ \int_0^t \left\langle P_\phi^T(s, r_s(\cdot)), dW_Q(s)\right\rangle_H, \end{aligned} \tag{9}$$

where the subscripts denote differentiation in H and * indicates the corresponding adjoint operator. Easy computations lead to:

$$P_t^T(t, \phi)\,|_{(s, r_s(\cdot))} = P^T(s, r_s(\cdot))\, r_s(T - s), \tag{10}$$

$$P_{\phi}^{T}(t,\phi)\mid_{(s,r_s(\cdot))}=-P^{T}(s,r_s(\cdot))\,\chi_{[0,T-s)}(\cdot)\in H \tag{11}$$

and:

$$P_{\phi\phi}^{T}(t,\phi)\big|_{(s,r_s(\cdot))}=P^{T}(s,r_s)\left(\chi_{[0,T-s)}\otimes\chi_{[0,T-s)}\right)\in L(H). \tag{12}$$

Substituting (10), (11) and (12) into (9), the drift of (8) collapses to 0 if and only if:

$$\begin{aligned}0=&-r_s(0)+r_s(T-s)-\left\langle\chi_{[0,T-s)}(\cdot),\alpha(s)\right\rangle_H\\&+\frac{1}{2}\mathrm{Tr}\left[\left(\chi_{[0,T-s)}\otimes\chi_{[0,T-s)}\right)\sqrt{Q(t)}\left(\sqrt{Q(t)}\right)^*\right].\end{aligned} \tag{13}$$

(Note the financial meaning: if arbitrage is to be prevented, then the instantaneous expected return on the discounted bond price, namely the sum of the last three terms in (13), must equal the return from a deposit in the rolled over money market account, that is $r_s(0)$, and this has to hold all over the time horizon.)
Computing explicitly the trace:

$$\begin{aligned}&\mathrm{Tr}\left[\left(\chi_{[0,T-s)}\otimes\chi_{[0,T-s)}\right)\sqrt{Q(t)}\left(\sqrt{Q(t)}\right)^*\right]\\&=\sum_{j=1}^{\infty}\left\langle\left(\chi_{[0,T-s)}\otimes\chi_{[0,T-s)}\right)\sqrt{Q(t)}f_j,\sqrt{Q(t)}f_j\right\rangle_H\\&=\sum_{j=1}^{\infty}\left\langle\chi_{[0,T-s)},\sqrt{Q(t)}f_j\right\rangle_H^2\\&=\sum_{j=1}^{\infty}\lambda_j(s)\left(\int_0^{T-s}f_j(u)\,du\right)^2\end{aligned}$$

and setting $x:=T-s$, one comes up to:

$$0=-r_s(0)+r_s(x)-\int_0^x\alpha(s,u)\,du+\frac{1}{2}\sum_{j=1}^{\infty}\lambda_j(s)\left(\int_0^x f_j(u)\,du\right)^2, \tag{14}$$

for every $x\geq 0$ and $s\geq 0$. By taking partial derivatives with respect to x in (14), we end up with (7). *Q.E.D.*

Remark 3.3. Given a basis $\{g_i\}$ in H, it is likely that the driving Brownian motion is not in the form (4), as it often comes in calibration. Starting with a noise:

$$W(t)=\sum b_i(t)\,g_i,$$

where $b_i(t) = \langle W(t), g_i \rangle$ are mutually stochastically dependent Brownian motions, an infinite dimensional principal components analysis provides us with a basis $\{f_i\}$, whose elements are the eigenvectors of the linear operator $Q(t)$ defined by:

$$\langle Q(t) f_i, f_j \rangle_H := E\left[\langle W(t), f_i \rangle_H \cdot \langle W(t), f_j \rangle_H\right].$$

There is also a collection of numbers $\{\lambda_i(t)\}$, namely the corresponding eigenvalues, such that (3) holds true. The set $\{f_i\}$ is a rotated Cartesian system such that $W_i(t) = \frac{\langle W(t), f_i \rangle_H}{\lambda_i(t)}$ are independent standard Brownian motions (see Roncoroni [20]). The reducibility to form (3) is an assumption equivalent to supposing that the basis $\{f_i\}$ is time invariant, the variable t entering only through the $\lambda_i(t)$.

4 Cross-Sectional Estimation Consistency with Arbitrage-Free Dynamics

We derived the following arbitrage free evolution for the instantaneous forward curve:

$$dr_t = \left(\frac{\partial}{\partial x} r_t + \sum \lambda_k(t) g_k\right) dt + \sum \sqrt{\lambda_k(t)} f_k dW_k(t),$$

where $\{f_k\}$ is an orthonormal complete system and $g_k(x) := \int_0^x f_k(y)\, dy f_k(x)$. We wonder whether a finite dimensional linear subspace generated by a selection from the basis functions f_k is actually stable or not under the above dynamics. If the initial term structure r_0 belongs to, say, $H_k := Span\{f_1, ..., f_k\}$, then r_t may well be out of H_k; otherwise, the dynamics r_t is said to be consistent with the class H_k.

We construct quasi arbitrage free dynamics which are consistent with a suitably tailored class of polynomials and then derive a precise estimate of the error in terms of mean square distance (*i.e.* variance) from the arbitrage free solution.

First, let us write the solution in the classical semigroup notation:

$$r_t = e^{t\mathcal{D}} r_0 + \sum_k \int_0^t \lambda_k(s) e^{(t-s)\mathcal{D}} g_k ds + \sum_k \int_0^t \sqrt{\lambda_k(s)} e^{(t-s)\mathcal{D}} f_k dW_k(t), \tag{15}$$

where $\{e^{t\mathcal{D}}\}_{t \geq 0}$ is the semigroup of translations defined by: $\left(e^{t\mathcal{D}}\alpha\right)(x) = \alpha(t+x)$. We want $\{f_k\}$ such that:

$$e^{t\mathcal{D}} H_k \subset H_k, \qquad \forall k.$$

For any system $\{f_k\}$ of polynomials such that $\deg f_k = k$, this condition is clearly fulfilled. Moreover g_k is a polynomial of degree at most equal to $2k+1$, so $e^{t\mathcal{D}} g_k$ is in H_{2k+1}. Fix $N \geq 1$ and set:

$$r_t^N = e^{t\mathcal{D}} r_0 + \sum_{k=1}^{\left[\frac{N}{2}\right]-1} \int_0^t \lambda_k(s) e^{(t-s)\mathcal{D}} g_k ds + \sum_{k=1}^{N} \int_0^t \sqrt{\lambda_k(s)} e^{(t-s)\mathcal{D}} f_k dW_k(t). \tag{16}$$

These dynamics are consistent with H_N: if $r_0 \in H_N$, then $e^{t\mathcal{D}} r_0$, $e^{(t-s)} g_k$ $\left(0 \leq k \leq \left[\frac{N}{2}\right] - 1\right)$ and $e^{(t-s)\mathcal{D}} f_k$ $(0 \leq k \leq N)$ belong to H_N, so that r_t^N still lives there.

We quantify the approximation bias in terms of mean square error, namely the H-distance between r_t^N and r_t:

$$\left\|r_t - r_t^N\right\|_H^2 \leq 2 \left\{ \left[\sum_{k=\left[\frac{N}{2}\right]}^{\infty} \int_0^t \lambda_k(s) \left\| e^{(t-s)\mathcal{D}} g_k \right\| ds \right]^2 + \left\| \sum_{k=N+1}^{\infty} \int_0^t \sqrt{\lambda_k(s)} e^{(t-s)\mathcal{D}} f_k dW_k(s) \right\|_H^2 \right\}.$$

By the Hölder inequality: $\left\|e^{(t-s)\mathcal{D}} g_k\right\|_H \leq \|g_k\|_H \leq \|f_k\|_H \leq 1$.
Furthermore:

$$\begin{aligned}
&E \left\| \sum_{k=N+1}^{\infty} \int_0^t \sqrt{\lambda_k(s)} e^{(t-s)\mathcal{D}} f_k dW_k \right\|_H^2 \\
&= \sum_{k=N+1}^{\infty} E \left[\int_0^1 \left(\int_0^t \sqrt{\lambda_k(s)} e^{(t-s)\mathcal{D}} f_k(\cdot)\, dW_k(s) \right)^2 (x)\, dx \right] \\
&= \sum_{k=N+1}^{\infty} \int_0^1 \int_0^t \lambda_k(s) \left| e^{(t-s)\mathcal{D}} f_k(\cdot) \right|^2 (x)\, ds dx \\
&\leq \sum_{k>N} \int_0^t \lambda_k(s)\, ds,
\end{aligned}$$

because of the inequality: $\left\|e^{(t-s)\mathcal{D}} f_k\right\|_H^2 \leq 1$.
Therefore, the bound:

$$E\left[\left\|r_t - r_t^N\right\|^2\right] \leq 2 \left\{ \left[\sum_{k \geq \left[\frac{N}{2}\right]} \int_0^t \lambda_k(s)\, ds \right]^2 + \sum_{k>N} \int_0^t \lambda_k(s)\, ds \right\}$$

holds true.
We can state the following:

Theorem 4.1. *If r_t is the arbitrage free term structure dynamics* (15) *and $\widehat{r}_t^N$ is its projection onto H_N, then the arbitrage bias of $\widehat{r}_t^N$ with respect to r_t in terms of mean square error admits the following bound:*

$$E\left[\left\|r_t - \widehat{r}_t^N\right\|^2\right] \leq 2\left\{\left[\sum_{k\geq\left[\frac{N}{2}\right]} \int_0^t \lambda_k(s)\, ds\right]^2 + \sum_{k>N} \int_0^t \lambda_k(s)\, ds\right\},$$

for each $t \leq T$. Since $\sum \int_0^t \lambda_k(s)\, ds < \infty$, one can make this error arbitrarily small.

This result states the stability of a polynomial class of fundamental shapes of the yield curve with respect to dynamics which are arbitrage free up to an error which explicitly depends on the market volatility and is directly controllable by suitably enlarging the starting space.

Of course the degree at which this enlargement is enforced depends upon the behavior of the series $\left\{\int_0^T \lambda_k(s)\, ds\right\}$, what is a structural property of the market volatility. Practically one selects a suitably wide, yet finite (say M-sized) class of polynomials, performs calibration, identifies the λ_k's up to the M-th and then assumes the hypothesis that the remaining λ_k's (associated to a supposed set of functions completing the finite dimensional basis into an orthonormal complete system) are so rapidly decreasing that the sum of the series converges to a number ζ such that $\left|\zeta - \sum_{k=1}^M \int_0^T \lambda_k(s)\, ds\right| << \varepsilon$. Thus $N(\varepsilon) \leq M$.

5 Hybrid Calibration

A model specification corresponding to actually observed prices may result from two procedures: by a slight modification of the Brace-Musiela non parametric method, one can implicitly fit prices of volatility dependent derivatives; by a direct matching to coupon bond and swap prices, one can historically infer an average realized volatility. These two methods will be linked by a minimization problem.

5.1 Historical Calibration

If one wishes to implement even a simplified (*e.g.* with constant covariance) infinite dimensional evolution law for the yield curve of the above kind, countably many coefficients should be provided: this is practically unfeasible. Therefore, instead of considering r_t as an element of a Hilbert space H spanned by an infinite orthonormal complete system $\{\psi_1, ...\}$[1], one is forced to restrict the

[1] In general, this is not the system $\{f_k\}$ of section 3, but an arbitrary orthonormal complete system; detecting $\{f_k\}$ from $\{\psi_k\}$ and market data is the very aim of the calibration procedure to follow.

analysis to a suitable *finite* dimensional subspace $H_n := Span\{\psi_1, ..., \psi_n\}$, such that enough information is available to obtain reliable data matching. On the choice of such a space, see Anderson *et al.* [1].

Step 1. Fix $n \geq 1$ and choose H_n as above. The yield curve r_t in H is identified with its orthogonal projection $\widehat{r}_t^n$ onto H_n: yield curves in the original space H are thus indistinguishable up to such a projection. The classical yield curve estimation technologies make use of polynomials or, more generally, splines: these must be orthonormalized by a Gram-Schmidt procedure.

Step 2. If $\psi_1, ..., \psi_n$ are orthonormal and a yield curve $\widehat{r}_t^n$ is a point in H_n of form $a_1(t)\psi_1 + ... + a_n(t)\psi_n$, the map I_n:

$$\widehat{r}_t^n \overset{I_n}{\to} (a_1(t), ..., a_n(t)) := \mathbf{a}_t \tag{17}$$

is an isomorphism from H_n onto $\Re^n$ preserving inner products. This means that the evolution law for $\widehat{r}_t^n$ in H_n induces corresponding dynamics for the vector $\mathbf{a}_t$ of coefficients in $\Re^n$ such that covariances are preserved in both of the spaces. The main consequence is that a principal components analysis may be performed in $\Re^n$.

We can therefore consider a stochastic evolution law for $\widehat{r}_t^n$ as represented by a corresponding equation in $\Re^n$ for the associated vector of coefficients:

$$d\mathbf{a}(t) = \mathbf{b}dt + \Sigma dW(t).$$

The coefficient $\mathbf{b}$ is immaterial for the estimation of the local covariance process of $\mathbf{a}(t)$ (and thus $\widehat{r}_t^n$): actually any drift would give rise to an equivalent probability measure over the path space and the volatility coefficient would remain unchanged because of a well-known property of continuous diffusions; the drift in the equation for $\widehat{r}_t^n$ is given by drift restriction (7).

Step 3. In order to estimate Σ, one needs explicit formulae for securities whose prices do not depend on the term structure of volatilities, namely coupon bonds and swaps. For the bond case, one may use:

$$P^{\mathbf{C},\mathbf{T}}(t) = \sum_{j=1}^{n} C_j P^{T_j}(t) + NP^{T_n}(t), \tag{18}$$

with $\mathbf{C} := (C_1, ..., C_n)$ denoting the coupon sequence, $\mathbf{T} := (T_1, ..., T_n)$ the bond tenor and N its face value. For a fixed-leg-payer swap with tenor $\mathbf{T}$ and fixed-leg rate R, the value is:

$$S(t) = P^{T_0}(t) - P^{T_n}(t) - \sum_{i=1}^{n} R\Delta T_i P^{T_i}(t). \tag{19}$$

Step 4. Observed prices are taken from market data and corresponding vectors $\mathbf{a}(t)$ are implicitly derived by inverting the following relation:

$$\mathbf{a}(t) \to \widehat{r}_t^n \to P^{\cdot}(t) \to V(t)(\mathbf{a}(t)),$$

where the first arrow stems from (17), the second from $P^T(t) = e^{\int_0^{T-t} \widehat{r}_t^n(s)ds}$, the third either from (18) or (19) according to the case and $V(t)(\mathbf{a}(t))$ denotes the time t security price as a functional of the yield curve corresponding to the vector of coefficients $\mathbf{a}(t)$.

The idea is to find the n-tuple $\mathbf{a}(t)$ selecting the best possible approximation $V(\mathbf{a}(t))$ to observed prices. Suppose we observe time t prices of products $i = 1...,n$. Let $V_1(t)(\mathbf{a}), ..., V_n(t)(\mathbf{a})$ denote their *theoretical* values corresponding to a given vector of coefficients $\mathbf{a}$. If $V_i^-(t)$ is the time t *observed* price of the i-th security, we wish to solve:

$$\min_{\mathbf{a}\in\Re^n} \left\{ \sum_{i=1}^{n} w_i(t) \left(V_i(t)(\mathbf{a}) - V_i^-(t) \right)^2 \right\}, \tag{20}$$

where the time t discrepancies between the i-th security approximating price and its observed value can be judiciously weighted by the user according to his or her own preferences expressed by $\{w_i\}$. For instance, if calibration is to produce a yield curve for pricing derivatives whose value is strongly linked to the prices of some of the products used in calibrating the model, then heavier weights ought to be imposed to price discrepancies relative to these latter. Later we will show how the selection of these weights is connected to the degree of consistency between the present historical calibration and the Brace-Musiela implicit procedure.

Step 5. After repeating this procedure at dates $t_0, t_1, ..., t_n$, one comes up to a sequence $\mathbf{a}(t_0), ..., \mathbf{a}(t_n)$. We obtained a sample movement of the yield curve along time in the form of a time-varying coefficients vector. Assuming these observations stem from a diffusion process and that time lags $t_i - t_{i-1}$, $i = 1, ..., n$ are sufficiently small to let the drift be well approximated by a constant, we may identify the corresponding covariance as follows: first, define $d\mathbf{a}(t_i) := \mathbf{a}(t_i) - \mathbf{a}(t_{i-1})$; second, estimate the drift by the normalized sample mean:

$$\mathbf{b} := \frac{1}{n} \sum_{i=1}^{n} \frac{d\mathbf{a}(t_i)}{t_i - t_{i-1}};$$

third, center and normalize all the $d\mathbf{a}(t_i)$ by setting:

$$\widetilde{d\mathbf{a}}(t_i) := \frac{d\mathbf{a}(t_i) - \mathbf{b}[t_i - t_{i-1}]}{\sqrt{t_i - t_{i-1}}};$$

finally, put:

$$A \equiv \begin{pmatrix} {}^t\widetilde{d\mathbf{a}}(t_1) \\ {}^t\widetilde{d\mathbf{a}}(t_2) \\ \dots \\ {}^t\widetilde{d\mathbf{a}}(t_n) \end{pmatrix} \equiv \left(A_1 \quad | \, A_2 \quad | \dots | \quad A_n \right).$$

The estimate of the realized average covariance matrix over $[t_1, t_n]$ is given by:

$$\Sigma \equiv {}^t A' A = \left(\mathrm{Cov}\left(\widetilde{d\mathbf{a}}_i, \widetilde{d\mathbf{a}}_k \right) \right)_{0 \leq i,k \leq n}.$$

The isometry above discussed guarantees that Σ is indeed the sampled average covariance operator for the evolution law of $\widehat{r}_t^n$ in H_n, on which one may perform a principal components analysis. If $t_0 := 0$ is kept as present time (*i.e.* $\mathcal{F}_{t_0}$ is the trivial σ-algebra), then:

$$\langle \widehat{r}_\cdot^n \rangle_{t_n} = \Sigma t_n.$$

Step 6. We are now ready to perform a principal components analysis of the stochastic movement for $\widehat{r}_t^n$ in H_n, so to further cut the number of factors and get the wanted precision for that dynamics. The basic result of principal components analysis states that the axes making a multidimensional Gaussian random movement componentwise uncorrelated are exactly the normalized eigenvectors $\mathbf{v}_i$ of Σ (see Rebonato [19, ch.3] and [20]). After decreasingly reordering the eigenvalues such that $\lambda_1 > ... > \lambda_n > 0$, let S be the matrix columnwise collecting the corresponding eigenvectors and $\Lambda := diag(\lambda_1, ..., \lambda_n)$ (positivity of λ_i follows from S being symmetric). If $\Sigma = S^T \Lambda S$ and $S = \left(\mathbf{v}^{(1)} \mid \mathbf{v}^{(k)} \mid ... \mid \mathbf{v}^{(k)} \right)$, then formally:

$$\begin{aligned} d\left[S\mathbf{a}\right](t) &= \left[S\mathbf{b}\right] dt + \sum_{k=1}^{n} \left\langle d\mathbf{a}(t), \mathbf{v}^{(k)} \right\rangle_{\Re^n} \mathbf{v}^{(k)} \\ &= \left[S\mathbf{b}\right] dt + \sum_{k=1}^{n} \left(\sqrt{\lambda_k} dW_k(t) \right) \mathbf{v}^{(k)}. \end{aligned}$$

By the above mentioned isometry between H_n and $\Re^n$, the matrix S induces an orthonormal transformation $\mathcal{S}$ in H_n. In particular, the starting basis $\{\psi_k\}_{k=1}^n$ of H_n is transformed into $\{f_k\}_{k=1}^n$ according to the rule:

$$f_k = I_n^{-1} \mathbf{v}^{(k)}.$$

Schematically:

$$\begin{array}{ccc} H_n & & \Re^n \\ \left(\{\langle r_t, \psi_k \rangle_H\}_{1\leq k \leq n}, \{\psi_k\} \right) & \xrightarrow{I_n} & \left(\mathbf{a}(t), \{\mathbf{e}^{(k)}\}_{k=1}^n \right) \\ \downarrow \mathcal{S} & & \downarrow S \\ \left(\{\langle r_t, f_k \rangle_H\}_{1\leq k \leq n}, \{f_k\} \right) & \xleftarrow{I_n^{-1}} & \left([S\mathbf{a}](t), \{\mathbf{v}^{(k)}\}_{k=1}^n \right) \end{array}$$

Thus:

$$d\widehat{r}_t^n(x) = \text{drift} + \sum_{k=1}^{n} \sqrt{\lambda_k} f_k(x)\, dW_k(t),$$

where f_i is the i-th eigenmode of $\widehat{r}_t^n$, that is $f_k(x) = \sum_{i=1}^n (\mathbf{v}^{(k)})_i \psi_i(x)$. An easy computation based on mutually independence among the $W_k(t)$ gives:

$$\begin{aligned}\langle \widehat{r}_\cdot^n(x), \widehat{r}_\cdot^n(y)\rangle_{t_n} &= \int_{t_0}^{t_n} \mathrm{Cov}\left(\sum_{k=1}^n \sqrt{\lambda_k} f_k(x)\, dW_k(t), \sum_{j=1}^n \sqrt{\lambda_j} f_j(y)\, dW_j(t)\right) \\ &= \sum_{k=1}^n \lambda_k f_k(x) f_k(y)\, t_n\end{aligned}$$

for the $\{t_0, ..., t_n\}$-averaged local covariance between times-to-maturity x and y. In particular:

$$\langle \widehat{r}_\cdot^n(x)\rangle_{t_n} = \sum_{k=1}^n \lambda_k f_k(x)^2\, t_n.$$

The instantaneous covariance between the time t rates for times-to-maturity x and y can be computed adding up all of the products between the corresponding "eigenrates" f_k, properly weighted by the relative eigenvalues, times dt, namely the local variance of both $d\mathbf{a}(t)$ and $\widehat{r}_t^n$ on their proper axes:

$$\begin{aligned}\langle\langle \widehat{r}_\cdot^n(\cdot), f_k(\cdot)\rangle_H\rangle_{t_n} &= \mathrm{Var}\left[\left\langle d\mathbf{a}(t), \mathbf{v}^{(k)}\right\rangle_{\Re^n}\right]_{t_n} \\ &= \lambda_k t_n.\end{aligned}$$

Fixing a level M of total variance one wishes for the model to embed, a truncation of the sum to the C-th element defined by:

$$m \equiv \inf\left\{k \geq 1 : \sum_{i=1}^k \lambda_i \geq C\right\}$$

gives our final outcome as:

$$d\widehat{r}_t^m(x) = \text{drift} + \sum_{k=1}^m \sqrt{\lambda_k} f_k(x)\, dW_k(t).$$

Everything has been computed from the real world: the λ_k come from the sampled matrix A and the g_k are determined via the eigenvectors for A and the given orthonormal complete system. As for the drift, one may use any truncation of the drift restriction derived in the previous section up to the m-th term; we suggest to take the order $\left[\frac{m}{2}\right] - 1$ for the dynamics is consistent according to the result of the previous section. Finally the arbitrage-approximated, polynomial-stable, historically calibrated instantaneous forward rate dynamics are given by:

$$r_t^m = e^{t\mathcal{D}} r_0 + \sum_{k=1}^{\left[\frac{m}{2}\right]-1} \lambda_k \int_0^t e_k^{(t-s)\mathcal{D}} g_k(s)\, ds + \sum_{k=1}^m \sqrt{\lambda_k} \int_0^t e^{(t-s)\mathcal{D}} f_k dW_k(t).$$

In the next section we will propose a method to correct this historical calibration procedure so to match implied volatilities as much as possible.

5.2 Historical-Implicit Hybrid Calibration

The idea is to reconstruct functions $\lambda_i(t)$ by inversion of explicit derivative formulae as the ones derived by Brace and Musiela [3]. The basic ingredient is the local covariance process of the forward bond yield. In our model, expression (4) and the time-s diffusion term of $P^T(s)$:

$$-\int_0^t P^T(s, r_s(\cdot)) \left\langle \chi_{[0,T-s)}(\cdot), dW(s,\cdot)\right\rangle_H$$

lead to the diffusion term of $\log P^T(t)$ as:

$$\begin{aligned} & -\int_0^t \left\langle \chi_{[0,T-s)}(\cdot), \sum_j \sqrt{\lambda_j(s)} f_j(\cdot)\, dW_j(s) \right\rangle_H \\ & = -\int_0^t \sum_j \lambda_j(s) \int_0^{T-s} f_j(u)\, du dW_j(s). \end{aligned}$$

From this, one gets to:

$$\begin{aligned} \left\langle \lg \frac{P^{T_1}(\cdot)}{P^{T_0}(\cdot)} \right\rangle_t &= \sum_j \int_0^t \lambda_j(s) \left| \int_{T_0-s}^{T_1-s} f_j(x)\, dx \right|^2 ds \qquad (21) \\ &: = \sum_j \int_0^t \lambda_j(s)\, g_j^{T_0,T_1}(s)\, ds, \end{aligned}$$

that is: for each component j, one cumulates the instantaneous variances $\lambda_j(s)\, g_j^{T_0,T_1}(s)$ over the interval $[0,t]$. This can be plugged into any of the option formulae by Brace and Musiela [3]. For instance, the time t price of a caplet spanning $[T_0, T_1]$ is:

$$Caplet^{T_0,T_1}(t) = (1+\delta K)\left(\frac{P^{T_0}(t,r_t)}{1+\delta K} F(-a_1) - P^{T_1}(t,r_t)\, F(-a_2) \right) \qquad (22)$$

with:

$$a_{1-2} := \frac{\lg\left(\frac{P^{T_1}(t,r_t)(1+\delta K)}{P^T(t,r_t)}\right)}{\nu(t,T_0)} \pm \frac{1}{2}\nu(t,T_0)$$

and:

$$\nu^2(t,T_0) := \left\langle \lg\left(\frac{P^{T_1}(\cdot)}{P^{T_0}(\cdot)} \right) \right\rangle_{T_0} = \sum_j \int_t^{T_0} \lambda_j(s)\, g_j^{T_0,T_1}(s)\, ds$$

denoting the cumulated volatility for the forward bond P^{T_0,T_1} returns over the interval $[t, T_0]$ as seen from time t.

A trivial mimic of the Brace-Musiela calibration procedure as developed in [3] is not possible: there the covariance operator is assumed to have a form $\tau(t,x) = {}^t(\tau_1(x), \tau_2(t+x)\chi_{[0,M]})$ and the model is taken to be driven by two Brownian motions; here we allow for arbitrarily many noises and the covariance is assumed to have diagonal form with respect to a suitable basis in H_n; this latter may be recovered from market data by performing the historical calibration developed in the last paragraph: such a procedure provides us with both $\{f_k\}_{k=1}^n$ and $\{\lambda_k\}_{k=1}^n$, which we collect in a vector $\lambda(\mathbf{w})$ explicitly dependent on the vector of weights $\mathbf{w} :={}^t (w_1, ..., w_n)$ employed in (20) and here set to a constant for simplicity.

We assume that still at time t_n, $\{f_k\}_{k=1}^n$ is the orthonormal complete system diagonalizing the local covariance operator of the yield curve dynamics. The implicit calibration is set up at time t_n. Fix some caplet maturities $T_1, ..., T_n$ relative to market observable prices; at time t_n, observe caplet prices $C_1, ..., C_n$ spanning time intervals $[T_i, T_{i+}]$, for $i = 1, ..., n$; then, implicitly recover the approximated local covariances as:

$$\nu^2(t_n, T_i) \simeq \sum_{k=1}^n \lambda_k \int_{t_n}^{T_i} g_k^{T_i, T_{i+}}(s)\, ds := \sum_{k=1}^n \lambda_k H_k(t_n, T_i),$$

leading to a system:

$$\begin{bmatrix} H_1(t_n, T_1), \ ..., \ H_n(t_n, T_1) \\ \cdots \\ H_1(t_n, T_n), \ ..., \ H_n(t_n, T_n) \end{bmatrix} \cdot \begin{bmatrix} \lambda_1 \\ \cdots \\ \lambda_n \end{bmatrix} = \begin{bmatrix} \nu^2(t_n, T_1) \\ \cdots \\ \nu^2(t_n, T_n) \end{bmatrix},$$

or:

$$H \cdot \lambda_{imp.}(\mathbf{w}) = \nu^2, \tag{23}$$

where the matrix H is known and the vector ν^2 stems from market prices via (22). The model is thus implicitly calibrated.

For both the implicit and explicit methods to be consistent, the solution $\lambda_{imp.}(\mathbf{w})$ is required to match the historically sampled vector $\lambda(\mathbf{w})$ of squared eigenvalues; because market prices are not perfectly consistent one another, this is not likely to be the case. Yet, we propose a trivial method to make the two calibrations be as consistent as possible. Indeed the best match is accomplished by solving:

$$\min_{\mathbf{w}} \|\lambda(\mathbf{w}) - \lambda_{imp.}(\mathbf{w})\|,$$

for some suitable norm depending on the relative importance of the basic yield curve shapes f_j in the pricing.

By supposing that the eigenvalues increments are time homogeneous, one may recover a time dependent volatility as follows. Perform the above implicit procedure over $t_0 \leq t_1 < \ldots < t_n < T_1$ and get to $\{\lambda_{imp.}(\mathbf{w})(t_0), \ldots,$

$\lambda_{imp.}(\mathbf{w})(t_n)\}$; on $[t_0, t_n]$, interpolate these points into a smooth vector-valued function $\lambda_{imp.}(\mathbf{w})(t)$; then a better consistency comes from:

$$\min_{\mathbf{w}} \left\{ \left\| \int_{t_0}^{t_n} |\lambda_{imp.}(\mathbf{w})(u) - \lambda(\mathbf{w})| \, du \right\| \right\}.$$

Let $\mathbf{w}'$ be a solution; define the sample function of increments as:

$$\Delta\lambda_{imp.}(\mathbf{w}')(x) := \lambda_{imp.}(\mathbf{w}')(t_0 + x) - \lambda_{imp.}(\mathbf{w}')(t_0).$$

One may compute time t_n derivative prices by making use of (21) with:

$$\lambda_i(u) = \left(\lambda_{imp.}(\mathbf{w}')(t_n) + \Delta\lambda_{imp.}(\mathbf{w}')(u)\right)_i .$$

6 Conclusion

Arbitrage free term structure dynamics have been derived in terms of infinitely many factors, each one representing the weight assigned to a specified basic yield curve shape. The consistency problem of cross-sectional estimation and arbitrage free dynamics has been studied yielding n-degree polynomial stable, 'quasi arbitrage free' dynamics. It remains to study the consistency problem for other function spaces. Another open problem is to link the proposed concept of 'quasi arbitrage free' interest rate dynamics with a measure of approximation of arbitrage free prices. This is not trivial since derivative prices are non linear functionals of the term structure of interest rates, but we conjecture that some Lipschitz conditions on parameters ought to suffice for this purpose. An implementation algorithm has been developed based upon both historical and implicit procedures. These have been integrated into a unified framework: the idea was to select the historical calibration procedure giving the closest possible result to the conjugate implicit one. An empirical study of the performance of our model can be found in a paper by Galluccio and Roncoroni [9].

Acknowledgements. We would like to thank seminar partecipants at ESSEC, Université Paris Dauphine, Ecole Polytechnique, Université Paris Nord, Università degli Studi di Torino, Association Française de Finance, Université d'Aix-En-Provence, Carnegie Mellon University, Politecnico di Milano, XXII Convegno AMASES and Bocconi University in Milan for their helpful comments and discussion on an earlier version of this paper, in particular Francesco Corielli, Rama Cont, Farshid Jamshidian, Elisa Luciano, Dilip Madan, Giovanni Peccati, Francesco Russo, Wolfgang Runggaldier, Sandro Salsa and anonymous referees. Discussion with Hélyette Geman, Nicole El Karoui, Raphaël Douady and David Heath gave a substiantial improvement to the developement of our ideas on the subject. The first author thanks ESSEC for providing financial support. The second author acknowledges Hélyette Geman for the kind ospitality at CEREG and MURST for financial support. As for the content, the usual disclaimers apply.

References

1. Anderson, N., Breedon, F., Deacon, M., Derry, A., Murphy, G.: Estimating and Interpreting the Yield Curve, John Wiley & Sons, New York (1996)
2. Björk, T., Christensen, J.: Forward Rate Models and Invariant Manifolds, Working Paper., Stockholm School of Economics (1997)
3. Brace, A., Musiela, M.: A Multifactor Gauss-Markov Implementation of Heath, Jarrow and Morton, Mathematical Finance **3**, 259-283 (1994)
4. Brown, R., Schaefer, S.: Interest rate volatility and the shape of the term structure, IFA Working Paper 177, London Business School (1994)
5. Corielli, F., Petrone, S.: Polinomi di Bernstein e Modelli di Non Arbitraggio per la Curva dei Rendimenti, Proceedings of AMASES Conference, Roma (1997)
6. Da Prato, G., Zabczyk, J.: Stochastic Equations in Infinite Dimensions, Cambridge University Press, Cambridge (1992)
7. Douady, R.: Yield Curve Smoothing and Residual Variance of Fixed Income Positions, Working Paper, CIMS, New York (1997)
8. Filipović, D.: Consistency Problems for HJM Interest Rate Models, Thesis Dissertation, ETH, Zurich (2000)
9. Galluccio, S., Roncoroni, A.: Shape Hedging: A New Paradigm for Measuring Yield Curve Volatility and Selecting Immunization Strategies (forthcoming)
10. Geman, H.: The Importance of the Forward Neutral Probability in a Stochastic Approach of Interest Rates, Working Paper, ESSEC, Cergy Pontoise (1989)
11. Geman, H., El Karoui, N., Lacoste, V.: On the Role of State Variables in Interest Rates Models, Applied Stochastic Models in Business and Industry **16**, 197-217 (2000)
12. Geman, H., El Karoui, N., Rochet, J.C.: Changes of Numeraire, Changes of Probability Measures and Option Pricing, Journal of Applied Probability **32**, 443-458 (1995)
13. Goldys, B., Musiela, M.: Infinite Dimensional Diffusions, Kolmogorov equations and Interest Rate Models, Report No. S98-22 (dec.), Department of Statistics, University of New South Wales (1998)
14. Guiotto, P., Roncoroni, A.: Infinite Dimensional HJM Dynamics for the Term Structure of Interest Rates, D.R. 99006, ESSEC, Cergy-Pontoise (1999)
15. Harrison, R.J., Pliska, S.: Martingales and Stochastic Integrals in the Theory of Continuous Trading, Stochastic Processes and their Applications **11**, 215-260 (1981)
16. Heath, D., Jarrow, R., Morton, A.: Bond Pricing and the Term Structure of Interest Rates: A New Methodology, Econometrica **61**, 77-105 (1992)
17. Jamshidian, F.: Pricing of Contingent Claims in the One-Factor Term Structure Model, Working Paper, Merril Lynch, New York (1987); reprinted in: Vasicek and Beyond, Risk Books, London (1997)
18. Musiela, M.: Stochastic PDEs and Term Structure Models, Working Paper, University of New South Wales (1994)
19. Rebonato, R.: Interest Rate Option Models (2nd ed.), John Wiley & Sons, New York (1998)
20. Roncoroni, A.: Infinite Dimensional Principal Component Decomposition for Yield Curve Modeling: Theory and Implementation Techniques, Working Paper, Courant Institute, New York (1997)

Optimal Investment in Incomplete Financial Markets

Walter Schachermayer*

Department of Financial and Actuarial Mathematics
Vienna University of Technology
Wiedner Hauptstrasse 8-10 / 105, 1040 Vienna, Austria
e-mail: wschach@fam.tuwien.ac.at

Abstract. We give a review of classical and recent results on maximization of expected utility for an investor who has the possibility of trading in a financial market. Emphasis will be given to the duality theory related to this convex optimization problem.

For expository reasons we first consider the classical case where the underlying probability space Ω is finite. This setting has the advantage that the technical difficulties of the proofs are reduced to a minimum, which allows for a clearer insight into the basic ideas, in particular the crucial role played by the Legendre-transform. In this setting we state and prove an existence and uniqueness theorem for the optimal investment strategy, and its relation to the dual problem; the latter consists in finding an equivalent martingale measure optimal with respect to the conjugate of the utility function. We also discuss economic interpretations of these theorems.

We then pass to the general case of an arbitrage-free financial market modeled by an $\mathbb{R}^d$-valued semi-martingale. In this case some regularity conditions have to be imposed in order to obtain an existence result for the primal problem of finding the optimal investment, as well as for a proper duality theory. It turns out that one may give a *necessary and sufficient* condition, namely a mild condition on the asymptotic behavior of the utility function, its so-called *reasonable asymptotic elasticity*. This property allows for an economic interpretation motivating the term "reasonable". The remarkable fact is that this regularity condition only pertains to the behavior of the utility function, while we do not have to impose any regularity conditions on the stochastic process modeling the financial market (to be precise: of course, we have to require the arbitrage-freeness of this process in a proper sense; also we have to assume in one of the cases considered below that this process is locally bounded; but otherwise it may be an arbitrary $\mathbb{R}^d$-valued semi-martingale).

We state two general existence and duality results pertaining to the setting of optimizing expected utility of terminal consumption. We also survey some of the ramifications of these results allowing for intermediate consumption, state-dependent utility, random endowment, non-smooth utility functions and transaction costs.

* Support by the Austrian Science Foundation (FWF) under the Wittgenstein-Preis program Z36-MAT and grant SFB#010 and by the Austrian National Bank under grant 'Jubiläumsfondprojekt Number 8699' is gratefully acknowledged.

Key words

Optimal Portfolios, Incomplete Markets, Replicating Portfolios, No-arbitrage bounds, Utility Maximization, Asymptotic Elasticity of Utility Functions.
JEL classification: C 60, C 61, G 11, G 12, G 13

1 Introduction

A basic problem of mathematical finance is the problem of an economic agent, who invests in a financial market so as to maximize the expected utility of her terminal wealth. As we shall see in (16) below, this problem can be written in an abstract way as

$$\mathbb{E}\left[U\left(x+\int_0^T H_u dS_u\right)\right] \longrightarrow \max!, \tag{1}$$

where we optimize over all "admissible" trading strategies H. In the framework of a continuous-time model the problem was studied for the first time by R. Merton in two seminal papers [M 69] and [M 71] (see also [M 90] as well as [S 69] for a treatment of the discrete time case). Using the methods of stochastic optimal control Merton derived a non-linear partial differential equation (Bellman equation) for the value function of the optimization problem. He also produced the closed-form solution of this equation, when the utility function is a power function, the logarithm, or of the form $-e^{-\gamma x}$ for $\gamma > 0$.

The Bellman equation of stochastic programming is based on the assumption of Markov state processes. The modern approach to the problem of expected utility maximization, which permits us to avoid the assumption of Markovian asset prices, is based on duality characterizations of portfolios provided by the set of martingale measures. For the case of a *complete* financial market, where the set of martingale measures is a singleton, this "martingale" methodology was developed by Pliska [P 86], Cox and Huang [CH 89], [CH 91] and Karatzas, Lehoczky and Shreve [KLS 87]. It was shown that the marginal utility of the terminal wealth of the optimal portfolio is proportional to the density of the martingale measure; this key result naturally extends the classical Arrow-Debreu theory of an optimal investment derived in a one-step, finite probability space model.

Considerably more difficult is the case of incomplete financial models. It was studied in a discrete-time, finite probability space model by He and Pearson [HP 91], and, in a continuous-time diffusion model, by He and Pearson [HP 91a], and by Karatzas, Lehoczky, Shreve and Xu in their seminal paper [KLSX 91]. The central idea here is to solve a *dual* variational problem and then to find the solution of the original problem by convex duality, the latter step being similar as in the case of a complete model.

We now formally assemble the ingredients of the optimization problem.

We consider a model of a security market which consists of $d+1$ assets. We denote by $S = ((S_t^i)_{0\le t\le T})_{0\le i\le d}$ the price process of the d stocks and suppose that the price of the asset S^0, called the "bond" or "cash account", is constant, $S_t^0 \equiv 1$. The latter assumption does not restrict the generality of the model as we always may choose the bond as numéraire (c.f., [DS 95]). In other words, $((S_t^i)_{0\le t\le T})_{1\le i\le d}$, is an $\mathbb{R}^d$-valued semi-martingale modeling the discounted price process of d risky assets.

The process S is assumed to be a semimartingale based on and adapted to a filtered probability space $(\Omega, \mathcal{F}, (\mathcal{F}_t)_{0\le t\le T}, \mathbf{P})$ satisfying the usual conditions of saturatedness and right continuity. As usual in mathematical finance, we consider a finite horizon T, but we remark that our results can also be extended to the case of an infinite horizon.

In section 2 we shall consider the case of finite Ω, in which case the paths of S are constant except for jumps at a finite number of times. We then can write S as $(S_t)_{t=0}^T = (S_0, S_1, \ldots, S_T)$, for some $T \in \mathbb{N}$.

The assumption that the bond is constant is mainly chosen for notational convenience as it allows for a compact description of self-financing portfolios: a self-financing portfolio Π is defined as a pair (x, H), where the constant x is the initial value of the portfolio and $H = (H^i)_{1\le i\le d}$ is a predictable S-integrable process specifying the amount of each asset held in the portfolio. The value process $X = (X_t)_{0\le t\le T}$ of such a portfolio Π at time t is given by

$$X_t = X_0 + \int_0^t H_u dS_u, \qquad 0 \le t \le T, \tag{2}$$

where $X_0 = x$ and the integral refers to stochastic integration in $\mathbb{R}^d$.

In order to rule out doubling strategies and similar schemes generating arbitrage-profits (by going deeply into the red) we follow Harrison and Pliska ([HP 81], see also [DS 94]), calling a predictable, S-integrable process *admissible*, if there is a constant $C \in \mathbb{R}_+$ such that, almost surely, we have

$$(H \cdot S)_t := \int_0^t H_u dS_u \ge -C, \quad \text{for } 0 \le t \le T. \tag{3}$$

Let us illustrate these general concepts in the case of an $\mathbb{R}^d$-valued process $S = (S_t)_{t=0}^T$ in finite, discrete time adapted to the filtration $(\mathcal{F}_t)_{t=0}^T$. In this case each $\mathbb{R}^d$-valued process $(H_t)_{t=1}^T$, which is predictable (i.e. each H_t is $\mathcal{F}_{t-1}$-measurable), is S-integrable, and the stochastic integral reduces to a finite sum

$$(H \cdot S)_t = \int_0^t H_u dS_u \tag{4}$$

$$= \sum_{u=1}^t H_u \Delta S_u \tag{5}$$

$$= \sum_{u=1}^t H_u (S_u - S_{u-1}), \tag{6}$$

where $H_u \Delta S_u$ denotes the inner product of the vectors H_u and $\Delta S_u = S_u - S_{u-1}$ in $\mathbb{R}^d$. Of course, each such trading strategy H is admissible if the underlying probability space Ω is finite.

Passing again to the general setting of an $\mathbb{R}^d$-valued semi-martingale $S = (S_t)_{0 \leq t \leq T}$ we denote as in [KS 99] by $\mathcal{M}^e(S)$ (resp. $\mathcal{M}^a(S)$) the set of probability measures Q equivalent to $\mathbf{P}$ (resp. absolutely continuous with respect to $\mathbf{P}$) such that for each admissible integrand H, the process $H \cdot S$ is a local martingale under Q.

Throughout the paper we assume the following version of the no-arbitrage condition on S:

Assumption 1.1. *The set $\mathcal{M}^e(S)$ is not empty.*[1]

We note that in this paper we shall mainly be interested in the case when $\mathcal{M}^e(S)$ is not reduced to a singleton, i.e., the case of an *incomplete* financial market.

After having specified the process S modeling the financial market we now define the function $U(x)$ modeling the utility of an agent's wealth x at the terminal time T.

We make the classical assumptions that $U : \mathbb{R} \to \mathbb{R} \cup \{-\infty\}$ is *increasing on* $\mathbb{R}$, *continuous* on $\{U > -\infty\}$, *differentiable and strictly concave* on the interior of $\{U > -\infty\}$, and that marginal utility tends to zero when wealth tends to infinity, i.e.,

$$U'(\infty) := \lim_{x \to \infty} U'(x) = 0. \tag{7}$$

These assumptions make good sense economically and it is clear that the requirement (7) of marginal utility decreasing to zero, as x tends to infinity, is necessary, if one is aiming for a general existence theorem for optimal investment. Indeed, if $U'(\infty) > 0$, then even in the case of the Black-Scholes model the solution to the optimization problem (1) fails to exist.

[1] If follows from [DS 94] and [DS 98a] that Assumption 1.1 is equivalent to the condition of "no free lunch with vanishing risk". This property can also be equivalently characterised in terms of the existence of a measure $Q \sim \mathbf{P}$ such that the process S itself (rather than the integrals $H \cdot S$ for admissible integrands) is "something like a martingale". The precise notion in the general semi-martingale setting is that S is a sigma-martingale under Q (see [DS 98a]); in the case when S is locally bounded (resp. bounded) the term "sigma-martingale" may be replaced by the more familiar term "local martingale" (resp. "martingale").

Readers who are not too enthusiastic about the rather subtle distinctions between martingales, local martingales and sigma-martingales may find some relief by noting that, in the case of finite Ω, or, more generally, for bounded processes, these three notions coincide. Also note that in the general semi-martingale case, when S is locally bounded (resp. bounded), the set $\mathcal{M}^e(S)$ as defined above coincides with the set of equivalent measures $Q \sim \mathbf{P}$ such that S is a local martingale (resp. martingale) under Q (see [E 80] and [AS 94]).

As regards the behavior of the (marginal) utility at the other end of the wealth scale we shall distinguish throughout the paper two cases.

Case 1 (negative wealth not allowed): in this setting we assume that U satifies the conditions $U(x) = -\infty$, for $x < 0$, while $U(x) > -\infty$, for $x > 0$, and that

$$U'(0) := \lim_{x \searrow 0} U'(x) = \infty. \tag{8}$$

Case 2 (negative wealth allowed): in this case we assume that $U(x) > -\infty$, for all $x \in \mathbb{R}$, and that

$$U'(-\infty) := \lim_{x \searrow -\infty} U'(x) = \infty. \tag{9}$$

Typical examples for case 1 are

$$U(x) = \ln(x) \tag{10}$$

or

$$U(x) = \frac{x^\alpha}{\alpha}, \quad 0 < \alpha < 1, \tag{11}$$

whereas a typical example for case 2 is

$$U(x) = -e^{-\gamma x}, \quad \gamma > 0. \tag{12}$$

We again note that it is natural from economic considerations to require that the marginal utility tends to infinity when the wealth x tends to the infimum of its allowed values.

For later reference we summarize our assumptions on the utility function:

Assumption 1.2. *Throughout the paper the utility function $U : \mathbb{R} \to \mathbb{R} \cup \{-\infty\}$ is increasing on $\mathbb{R}$, continuous on $\{U > -\infty\}$, differentiable and strictly concave on the interior of $\{U > -\infty\}$, and satisfies*

$$U'(\infty) := \lim_{x \to \infty} U'(x) = 0. \tag{13}$$

Denoting by $\operatorname{dom}(U)$ *the interior of* $\{U > -\infty\}$, *we assume that we have one of the two following cases.*

Case 1: $\operatorname{dom}(U) =]0, \infty[$ *in which case U satisfies the condition*

$$U'(0) := \lim_{x \searrow 0} U'(x) = \infty. \tag{14}$$

Case 2: $\operatorname{dom}(U) = \mathbb{R}$ *in which case U satisfies*

$$U'(-\infty) := \lim_{x \searrow -\infty} U'(x) = \infty. \tag{15}$$

We now can give a precise meaning to the expression (1) at the beginning of this section. Define the value function

$$u(x) := \sup_{H \in \mathcal{H}} \mathbb{E}\left[U(x + (H \cdot S)_T)\right], \quad x \in \mathrm{dom}(U), \tag{16}$$

where H ranges through the admissible S-integrable trading strategies. To exclude trivial cases we shall assume throughout the paper that the value function u is not degenerate:

Assumption 1.3.

$$u(x) < \sup_{\xi} U(\xi), \quad \textit{for some} \quad x \in \mathrm{dom}(U). \tag{17}$$

One easily verifies that this assumption implies that

$$u(x) < \sup_{\xi} U(\xi), \quad \text{for all} \quad x \in \mathrm{dom}(U), \tag{18}$$

and that, in the case of finite Ω, Assumptions 1.1 and 1.2 already imply Assumption 1.3.

2 Utility Maximization on Finite Probability Spaces

In this section we consider an $\mathbb{R}^{d+1}$-valued process $(S_t)_{t=0}^T = (S_t^0, S_t^1, \ldots, S_t^d)_{t=0}^T$ with $S_t^0 \equiv 1$, based on and adapted to the *finite* filtered probability space $(\Omega, \mathcal{F}, (\mathcal{F}_t)_{t=0}^T, \mathbf{P})$, which we write as $\Omega = \{\omega_1, \ldots, \omega_N\}$. Without loss of generality we assume that $\mathcal{F}_0$ is trivial, that $\mathcal{F}_T = \mathcal{F}$ is the power set of Ω, and that $\mathbf{P}[\omega_n] > 0$, for all $1 \leq n \leq N$.

Assumption 1.1 is the existence a measure $Q \sim \mathbf{P}$, i.e., $Q[\omega_n] > 0$, for $1 \leq n \leq N$, such that S is a Q-martingale.

2.1 The Complete Case (Arrow–Debreu)

As a first case we analyze the situation of a financial market which is *complete*, i.e., the set $\mathcal{M}^e(S)$ of equivalent probability measures under which S is a martingale is reduced to a singleton $\{Q\}$. In this setting consider the *Arrow-Debreu assets* $\mathbf{1}_{\{\omega_n\}}$, which pay 1 unit of the numéraire at time T, when ω_n turns out to be the true state of the world, and 0 otherwise. In view of our normalization of the numéraire $S_t^0 \equiv 1$, we get for the price of the Arrow-Debreu assets at time $t = 0$ the relation

$$\mathbb{E}_Q\left[\mathbf{1}_{\{\omega_n\}}\right] = Q[\omega_n], \tag{19}$$

and each Arrow-Debreu asset $\mathbf{1}_{\{\omega_n\}}$ may be represented as $\mathbf{1}_{\{\omega_n\}} = Q[\omega_n] + (H \cdot S)_T$, for some predictable trading strategy $H \in \mathcal{H}$.

Hence, for fixed initial endowment $x \in \mathrm{dom}(U)$, the utility maximization problem (16) above may simply be written as

$$\mathbb{E}_{\mathbf{P}}\left[U(X_T)\right] = \sum_{n=1}^{N} p_n U(\xi_n) \to \max! \tag{20}$$

$$\mathbb{E}_Q[X_T] = \sum_{n=1}^{N} q_n \xi_n \quad \leq x. \tag{21}$$

To verify that (20) and (21) indeed are equivalent to the original problem (16) above (in the present finite, complete case), note that a random variable $X_T(\omega_n) = \xi_n$ can be dominated by a random variable of the form $x + (H \cdot S)_T = x + \sum_{t=1}^{T} H_t \Delta S_t$ iff $\mathbb{E}_Q[X_T] = \sum_{n=1}^{N} q_n \xi_n \leq x$. This basic relation has a particularly evident interpretation in the present setting, as q_n is simply the price of the Arrow-Debreu asset $\mathbf{1}_{\{\omega_n\}}$.

Let us fix some notation for the domain over which the problem (20) is optimized:

$$C(x) = \left\{X_T \in L^0(\Omega, \mathcal{F}_T, \mathbf{P}) : \mathbb{E}_Q[X_T] \leq x\right\}. \tag{22}$$

The notation $L^0(\Omega, \mathcal{F}_T, \mathbf{P})$ only serves to indicate that X_T is an $\mathcal{F}_T$-measurable random variable at this stage, as for finite Ω all the L^p-spaces coincide. But we have chosen the notation to be consistent with that of the general case below.

We have written ξ_n for $X_T(\omega_n)$ to stress that (20) simply is a concave maximization problem in $\mathbb{R}^N$ with one linear constraint. To solve it, we form the Lagrangian

$$L(\xi_1, \ldots, \xi_N, y) = \sum_{n=1}^{N} p_n U(\xi_n) - y\left(\sum_{n=1}^{N} q_n \xi_n - x\right) \tag{23}$$

$$= \sum_{n=1}^{N} p_n \left(U(\xi_n) - y \tfrac{q_n}{p_n} \xi_n\right) + yx. \tag{24}$$

We have used the letter $y \geq 0$ instead of the usual $\lambda \geq 0$ for the Lagrange multiplier; the reason is the dual relation between x and y which will become apparent in a moment.

Writing

$$\Phi(\xi_1, \ldots, \xi_N) = \inf_{y>0} L(\xi_1, \ldots, \xi_N, y), \quad \xi_n \in \mathrm{dom}(U), \tag{25}$$

and

$$\Psi(y) = \sup_{\xi_1, \ldots, \xi_N} L(\xi_1, \ldots, \xi_N, y), \quad y \geq 0, \tag{26}$$

it is straight forward to verify that we have

$$\sup_{\xi_1,\dots,\xi_N} \Phi(\xi_1,\dots,\xi_N) = \sup_{\substack{\xi_1,\dots,\xi_N \\ \sum_{n=1}^N q_n\xi_n \le x}} \sum_{n=1}^N p_n U(\xi_n) = u(x). \tag{27}$$

As regards the function $\Psi(y)$ we make the following pleasant observation which is the basic reason for the efficiency of the duality approach: using the form (24) of the Lagrangian and fixing $y > 0$, the optimization problem appearing in (26) splits into N independent optimization problems over $\mathbb{R}$

$$U(\xi_n) - y\tfrac{q_n}{p_n}\xi_n \mapsto \max!, \quad \xi_n \in \mathbb{R}. \tag{28}$$

In fact, these one-dimensional optimization problems are of a very convenient form: recall (see, e.g., [R 70], [ET 76] or [KLSX 91]) that, for a concave function $U : \mathbb{R} \to \mathbb{R} \cup \{-\infty\}$, the *conjugate function* V (which — up to the sign — is just the Legendre-transform) is defined by

$$V(\eta) = \sup_{\xi\in\mathbb{R}} \left[U(\xi) - \eta\xi\right], \quad \eta > 0. \tag{29}$$

The following facts are well known (and easily verified by one-dimensional calculus): if U satisfies Assumption 1.2, we have that V is finitely valued, differentiable, strictly convex on $]0,\infty[$, and satisfies

$$V'(0) := \lim_{y\searrow 0} V'(y) = -\infty, \qquad V(0) := \lim_{y\searrow 0} V(y) = U(\infty). \tag{30}$$

As regards the behavior of V at infinity, we have to distinguish between case 1 and case 2 in Assumption 1.2 above:

$$\text{case 1:} \quad \lim_{y\to\infty} V(y) = \lim_{x\to 0} U(x) \quad \text{and} \quad \lim_{y\to\infty} V'(y) = 0 \tag{31}$$

$$\text{case 2:} \quad \lim_{y\to\infty} V(y) = \infty \quad \text{and} \quad \lim_{y\to\infty} V'(y) = \infty \tag{32}$$

We also note that these properties of the conjugate function V are, in fact, equivalent to the properties of U listed in Assumption 1.2. We also have the inversion formula to (29)

$$U(\xi) = \inf_{\eta} \left[V(\eta) + \eta\xi\right], \quad \xi \in \mathrm{dom}(U) \tag{33}$$

and that $-V'(y)$, denoted by $I(y)$ for "inverse function" in [KLSX 91], is the inverse function of $U'(x)$; of course, U' has a good economic interpretation as the *marginal utility* of an economic agent modeled by the utility function U.

Here are some concrete examples of pairs of conjugate functions:

$$U(x) = \ln(x), \quad x > 0, \qquad V(y) = -\ln(y) - 1, \tag{34}$$

$$U(x) = \tfrac{x^\alpha}{\alpha}, \quad x > 0, \qquad V(y) = \tfrac{1-\alpha}{\alpha} y^{\frac{\alpha}{\alpha-1}}, \quad 0 < \alpha < 1, \tag{35}$$

$$U(x) = -\tfrac{e^{-\gamma x}}{\gamma}, \quad x \in \mathbb{R}, \quad V(y) = \tfrac{y}{\gamma}(\ln(y) - 1), \quad \gamma > 0. \tag{36}$$

We now apply these general facts about the Legendre transformation to calculate $\Psi(y)$. Using definition (29) of the conjugate function V and (24), formula (26) becomes

$$\Psi(y) = \sum_{n=1}^{N} p_n V\left(y\tfrac{q_n}{p_n}\right) + yx \tag{37}$$

$$= \mathbb{E}_{\mathbf{P}}\left[V\left(y\tfrac{dQ}{d\mathbf{P}}\right)\right] + yx. \tag{38}$$

Denoting by $v(y)$ the dual value function

$$v(y) := \mathbb{E}_{\mathbf{P}}\left[V\left(y\tfrac{dQ}{d\mathbf{P}}\right)\right] = \sum_{n=1}^{N} p_n V\left(y\tfrac{q_n}{p_n}\right), \quad y > 0, \tag{39}$$

the function v clearly has the same qualitative properties as the function V listed above. Hence by (30), (31), and (32) we find, for fixed $x \in \mathrm{dom}(U)$, a unique $\widehat{y} = \widehat{y}(x) > 0$ such that $v'(\widehat{y}(x)) = -x$, which therefore is the unique minimizer to the dual problem

$$\Psi(y) = \mathbb{E}_{\mathbf{P}}\left[V\left(y\tfrac{dQ}{d\mathbf{P}}\right)\right] + yx = \min! \tag{40}$$

Fixing the critical value $\widehat{y}(x)$ of the Lagrange multiplier, the concave function

$$(\xi_1, \dots, \xi_N) \mapsto L(\xi_1, \dots, \xi_N, \widehat{y}(x)) \tag{41}$$

defined in (24) assumes its unique maximum at the point $(\widehat{\xi}_1, \dots, \widehat{\xi}_N)$ satisfying

$$U'(\widehat{\xi}_n) = \widehat{y}(x)\tfrac{q_n}{p_n} \quad \text{or, equivalently,} \quad \widehat{\xi}_n = I\left(\widehat{y}(x)\tfrac{q_n}{p_n}\right), \tag{42}$$

so that we have

$$\inf_{y>0} \Psi(y) = \inf_{y>0} (v(y) + xy) \tag{43}$$

$$= v(\widehat{y}(x)) + x\widehat{y}(x) \tag{44}$$

$$= L(\widehat{\xi}_1, \dots, \widehat{\xi}_N, \widehat{y}(x)). \tag{45}$$

Note that $\widehat{\xi}_n$ are in $\mathrm{dom}(U)$, for $1 \le n \le N$, so that L is continuously differentiable at $(\widehat{\xi}_1, \dots, \widehat{\xi}_N, \widehat{y}(x))$, which implies that $\frac{\partial}{\partial y} L(\xi_1, \dots, \xi_N, y)|_{(\widehat{\xi}_1, \dots, \widehat{\xi}_N, \widehat{y}(x))}$ $= 0$; hence we infer from (23) and the fact that $\widehat{y}(x) > 0$ that the constraint (21) is binding, i.e.,

$$\sum_{n=1}^{N} q_n \widehat{\xi}_n = x, \tag{46}$$

and that

$$\sum_{n=1}^{N} p_n U(\widehat{\xi}_n) = L(\widehat{\xi}_1, \dots, \widehat{\xi}_N, \widehat{y}(x)). \tag{47}$$

In particular, we obtain that

$$u(x) = \sum_{n=1}^{N} p_n U(\widehat{\xi}_n). \tag{48}$$

Indeed, the inequality $u(x) \geq \sum_{n=1}^{N} p_n U(\widehat{\xi}_n)$ follows from (46) and (27), while the reverse inequality follows from (47) and the fact that for all $\xi_1, \dots, \xi_N$ verifying the constraint (21)

$$\sum_{n=1}^{N} p_n U(\xi_n) \leq L(\xi_1, \dots, \xi_N, \widehat{y}(x)) \leq L(\widehat{\xi}_1, \dots, \widehat{\xi}_N, \widehat{y}(x)). \tag{49}$$

We shall write $\widehat{X}_T(x) \in C(x)$ for the optimizer $\widehat{X}_T(x)(\omega_n) = \widehat{\xi}_n$, $n = 1, \dots, N$.

Combining (43), (47) and (48) we note that the value functions u and v are conjugate:

$$\inf_{y>0} (v(y) + xy) = v(\widehat{y}(x)) + x\widehat{y}(x) = u(x), \quad x \in \mathrm{dom}(U), \tag{50}$$

which, by the remarks after equations (32) and (39), implies that u inherits the properties of U listed in Assumption 1.2. The relation $v'(\widehat{y}(x)) = -x$ which was used to define $\widehat{y}(x)$, therefore translates into

$$u'(x) = \widehat{y}(x), \quad \text{for } x \in \mathrm{dom}(U). \tag{51}$$

Let us summarize what we have proved:

Theorem 2.1 (finite Ω, complete market). *Let the financial market $S = (S_t)_{t=0}^T$ be defined over the finite filtered probability space $(\Omega, \mathcal{F}, (\mathcal{F})_{t=0}^T, \mathbf{P})$ and satisfy $\mathcal{M}^e(S) = \{Q\}$, and let the utility function U satisfy Assumption 1.2.*

Denote by $u(x)$ and $v(y)$ the value functions

$$u(x) = \sup_{X_T \in C(x)} \mathbb{E}[U(X_T)], \qquad x \in \mathrm{dom}(U), \tag{52}$$

$$v(y) = \mathbb{E}\left[V\left(y\frac{dQ}{d\mathbf{P}}\right)\right], \qquad y > 0. \tag{53}$$

We then have:

(i) The value functions $u(x)$ and $v(y)$ are conjugate and u inherits the qualitative properties of U listed in Assumption 1.2.

(*ii*) *The optimizer* $\widehat{X}_T(x)$ *in* (52) *exists, is unique and satisfies*

$$\widehat{X}_T(x) = I(y\tfrac{dQ}{d\mathbf{P}}), \quad \textit{or, equivalently,} \quad y\tfrac{dQ}{d\mathbf{P}} = U'(\widehat{X}_T(x)), \tag{54}$$

where $x \in \operatorname{dom}(U)$ *and* $y > 0$ *are related via* $u'(x) = y$ *or, equivalently,* $x = -v'(y)$.

(*iii*) *The following formulae for* u' *and* v' *hold true:*

$$u'(x) = \mathbb{E}_{\mathbf{P}}[U'(\widehat{X}_T(x))], \qquad v'(y) = \mathbb{E}_Q\left[V'\left(y\tfrac{dQ}{d\mathbf{P}}\right)\right] \tag{55}$$

$$xu'(x) = \mathbb{E}_{\mathbf{P}}\left[\widehat{X}_T(x)U'(\widehat{X}_T(x))\right], \; yv'(y) = \mathbb{E}_{\mathbf{P}}\left[y\tfrac{dQ}{d\mathbf{P}}V'\left(y\tfrac{dQ}{d\mathbf{P}}\right)\right]. \tag{56}$$

Proof Items (i) and (ii) have been shown in the preceding discussion, hence we only have to show (iii). The formulae for $v'(y)$ in (55) and (56) immediately follow by differentiating the relation

$$v(y) = \mathbb{E}_{\mathbf{P}}\left[V\left(y\tfrac{dQ}{d\mathbf{P}}\right)\right] = \sum_{n=1}^{N} p_n V\left(y\tfrac{q_n}{p_n}\right). \tag{57}$$

Of course, the formula for v' in (56) is an obvious reformulation of the one in (55). But we write both of them to stress their symmetry with the formulae for $u'(x)$.

The formula for u' in (55) translates via the relations exhibited in (ii) into the identity

$$y = \mathbb{E}_{\mathbf{P}}\left[y\tfrac{dQ}{d\mathbf{P}}\right], \tag{58}$$

while the formula for $u'(x)$ in (56) translates into

$$v'(y)y = \mathbb{E}_{\mathbf{P}}\left[V'\left(y\tfrac{dQ}{d\mathbf{P}}\right)y\tfrac{dQ}{d\mathbf{P}}\right], \tag{59}$$

which we just have seen to hold true. ■

Remark 2.2. Firstly, let us recall the economic interpretation of (54)

$$U'\left(\widehat{X}_T(x)(\omega_n)\right) = y\frac{q_n}{p_n}, \quad n = 1, \ldots, N. \tag{60}$$

This equality means that, in every possible state of the world ω_n, the *marginal utility* $U'(\widehat{X}_T(x)(\omega_n))$ of the wealth of an optimally investing agent at time T is *proportional to the ratio of the price* q_n *of the corresponding Arrow-Debreu security* $\mathbf{1}_{\{\omega_n\}}$ *and the probability of its success* $p_n = \mathbf{P}[\omega_n]$. This basic relation was analyzed in the fundamental work of K. Arrow and G. Debreu and allows for a convincing economic interpretation: considering for a moment the situation where this proportionality relation fails to hold true, one immediately deduces from a marginal variation argument that the investment of the agent cannot be optimal. Hence for the optimal investment the proportionality must hold true. The above result also identifies the proportionality factor as $y = u'(x)$, where x is the initial endowment of the investor.

Theorem 2.1 indicates an easy way to solve the utility maximization at hand: calculate $v(y)$ by (53), which reduces to a simple one-dimensional computation; once we know $v(y)$, the theorem provides easy formulae to calculate all the other quantities of interest, e.g., $\widehat{X}_T(x)$, $u(x)$, $u'(x)$ etc.

Another message of the above theorem is that the value function $x \mapsto u(x)$ may be viewed as a utility function as well, sharing all the qualitative features of the original utility function U. This makes sense economically, as $u(x)$ denotes the expected utility at time T of an agent with initial endowment x, after having optimally invested in the financial market S.

Let us also give an economic interpretation of the formulae for $u'(x)$ in item (iii) along these lines: suppose the initial endowment x is varied to $x+h$, for some small real number h. The economic agent may use the additional endowment h to finance, in addition to the optimal pay-off function $\widehat{X}_T(x)$, h units of the cash account, thus ending up with the pay-off function $\widehat{X}_T(x)+h$ at time T. Comparing this investment strategy to the optimal one corresponding to the initial endowment $x+h$, which is $\widehat{X}_T(x+h)$, we obtain

$$\lim_{h\to 0} \frac{u(x+h)-u(x)}{h} = \lim_{h\to 0} \frac{\mathbb{E}[U(\widehat{X}_T(x+h)) - U(\widehat{X}_T(x))]}{h} \tag{61}$$

$$\geq \lim_{h\to 0} \frac{\mathbb{E}[U(\widehat{X}_T(x)+h) - U(\widehat{X}_T(x))]}{h} \tag{62}$$

$$= \mathbb{E}[U'(\widehat{X}_T(x))]. \tag{63}$$

Using the fact that u is differentiable, and that h may be positive as well as negative, we have found another proof of formula (55) for $u'(x)$; the economic interpretation of this proof is that the economic agent, who is optimally investing, is indifferent of first order towards a (small) additional investment into the cash account.

Playing the same game as above, but using the additional endowment $h \in \mathbb{R}$ to finance an additional investment into the optimal portfolio $\widehat{X}_T(x)$ (assuming, for simplicity, $x \neq 0$), we arrive at the pay-off function $\frac{x+h}{x}\widehat{X}_T(x)$. Comparing this investment with $\widehat{X}_T(x+h)$, an analogous calculation as in (61) leads to the formula for $u'(x)$ displayed in (56). The interpretation now is, that the optimally investing economic agent is indifferent of first order towards a marginal variation of the investment into the optimal portfolio.

It now becomes clear that formulae (55) and (56) for $u'(x)$ are just special cases of a more general principle: for each $f \in L^\infty(\Omega, \mathcal{F}, \mathbf{P})$ we have

$$\mathbb{E}_Q[f]u'(x) = \lim_{h\to 0} \frac{\mathbb{E}_{\mathbf{P}}[U(\widehat{X}_T(x)+hf) - U(\widehat{X}_T(x))]}{h}. \tag{64}$$

The proof of this formula again is along the lines of (61) and the interpretation is the following: by investing an additional endowment $h\mathbb{E}_Q[f]$ to finance the contingent claim hf, the increase in expected utility is of first order equal to $h\mathbb{E}_Q[f]u'(x)$; hence again the economic agent is of first order indifferent towards an additional investment into the contingent claim f.

2.2 The Incomplete Case

We now drop the assumption that the set $\mathcal{M}^e(S)$ of equivalent martingale measures is reduced to a singleton (but we still remain in the framework of a finite probability space Ω) and replace it by Assumption 1.1 requiring that $\mathcal{M}^e(S) \neq \emptyset$.

In this setting it follows from basic linear algebra that a random variable $X_T(\omega_n) = \xi_n$ may be dominated by a random variable of the form $x + (H \cdot S)_T$ iff $\mathbb{E}_Q[X_T] = \sum_{n=1}^N q_n \xi_n \leq x$, for each $Q = (q_1 \dots, q_N) \in \mathcal{M}^a(S)$ (or equivalently, for every $Q \in \mathcal{M}^e(S)$). This basic result is proved in [KQ 95], [J 92], [AS 94], [DS 94] and [DS 98a] in varying degrees of generality; in the present finite-dimensional case this fact is straightforward to prove, using elementary linear algebra (see, e.g, [S 00a]).

In order to reduce the infinitely many constraints, where Q runs through $\mathcal{M}^a(S)$, to a finite number, make the easy observation that $\mathcal{M}^a(S)$ is a bounded, closed, convex polytope in $\mathbb{R}^N$ and therefore the convex hull of its finitely many extreme points $\{Q^1, \dots, Q^M\}$. Indeed, $\mathcal{M}^a(S)$ is given by finitely many linear constraints. For $1 \leq m \leq M$, we identify Q^m with its probabilites $(q_1^m, \dots, q_N^m)$.

Fixing the initial endowment $x \in \text{dom}(U)$, we therefore may write the utility maximization problem (16) similarly as in (20) as a concave optimization problem over $\mathbb{R}^N$ with finitely many linear constraints:

$$(\mathbf{P_x}) \quad \mathbb{E}_\mathbf{P}\left[U(X_T)\right] = \sum_{n=1}^N p_n U(\xi_n) \to \max! \tag{65}$$

$$\mathbb{E}_{Q^m}[X_T] = \sum_{n=1}^N q_n^m \xi_n \quad \leq \quad x, \quad \text{for} \quad m = 1, \dots, M. \tag{66}$$

Writing again

$$C(x) = \left\{X_T \in L^0(\Omega, \mathcal{F}, \mathbf{P}) : \mathbb{E}[X_T] \leq x, \quad \text{for all } Q \in \mathcal{M}^a(S)\right\} \tag{67}$$

we define the value function

$$u(x) = \sup_{H \in \mathcal{H}} \mathbb{E}\left[U\left(x + (H \cdot S)_T\right)\right] = \sup_{X_T \in C(x)} \mathbb{E}[U(X_T)], \quad x \in \text{dom}(U). \tag{68}$$

The Lagrangian now is given by

$$L(\xi_1, \dots, \xi_N, \eta_1, \dots, \eta_M) \tag{69}$$

$$= \sum_{n=1}^{N} p_n U(\xi_n) - \sum_{m=1}^{M} \eta_m \left(\sum_{n=1}^{N} q_n^m \xi_n - x \right) \tag{70}$$

$$= \sum_{n=1}^{N} p_n \left(U(\xi_n) - \sum_{m=1}^{M} \frac{\eta_m q_n^m}{p_n} \xi_n \right) + \sum_{m=1}^{M} \eta_m x, \tag{71}$$

$$\text{where } \ (\xi_1, \dots, \xi_N) \in \operatorname{dom}(U)^N, \ \ (\eta_1, \dots, \eta_M) \in \mathbb{R}_+^M. \tag{72}$$

Writing $y = \eta_1 + \ldots + \eta_M$, $\mu_m = \frac{\eta_m}{y}$, $\mu = (\mu_1, \dots, \mu_m)$ and

$$Q^\mu = \sum_{m=1}^{M} \mu_m Q^m, \tag{73}$$

note that, when $(\eta_1, \dots, \eta_M)$ runs trough $\mathbb{R}_+^M$, the pairs (y, Q^μ) run through $\mathbb{R}_+ \times \mathcal{M}^a(S)$. Hence we may write the Lagrangian as

$$\begin{aligned} &L(\xi_1, \dots, \xi_N, y, Q) = \\ &= \mathbb{E}_{\mathbf{P}}[U(X_T)] - y\left(\mathbb{E}_Q[X_T - x]\right) \\ &= \sum_{n=1}^{N} p_n \left(U(\xi_n) - \frac{y q_n}{p_n} \xi_n \right) + yx, \\ &\text{where } \ \xi_n \in \operatorname{dom}(U), \ \ y > 0, \ \ Q = (q_1, \dots, q_N) \in \mathcal{M}^a(S). \end{aligned} \tag{74}$$

This expression is entirely analogous to (24), the only difference now being that Q runs through the set $\mathcal{M}^a(S)$ instead of being a fixed probability measure. Defining again

$$\Phi(\xi_1, \dots, \xi_n) = \inf_{y>0, Q \in \mathcal{M}^a(S)} L(\xi_1, \dots, \xi_N, y, Q), \tag{75}$$

and

$$\Psi(y, Q) = \sup_{\xi_1, \dots, \xi_N} L(\xi_1, \dots, \xi_N, y, Q), \tag{76}$$

we obtain, just as in the complete case,

$$\sup_{\xi_1, \dots, \xi_N} \Phi(\xi_1, \dots, \xi_N) = u(x), \quad x \in \operatorname{dom}(U), \tag{77}$$

and

$$\Psi(y, Q) = \sum_{n=1}^{N} p_n V\left(\frac{y q_n}{p_n} \right) + yx, \qquad y > 0, \ \ Q \in \mathcal{M}^a(S), \tag{78}$$

where $(q_1, \ldots, q_N)$ denotes the probabilities of $Q \in \mathcal{M}^a(S)$. The minimization of Ψ will be done in two steps: first we fix $y > 0$ and minimize over $\mathcal{M}^a(S)$, i.e.,

$$\Psi(y) := \inf_{Q \in \mathcal{M}^a(S)} \Psi(y, Q), \quad y > 0. \tag{79}$$

For fixed $y > 0$, the continuous function $Q \to \Psi(y, Q)$ attains its minimum on the compact set $\mathcal{M}^a(S)$, and the minimizer $\widehat{Q}(y)$ is unique by the strict convexity of V. Writing $\widehat{Q}(y) = (\widehat{q}_1(y), \ldots, \widehat{q}_N(y))$ for the minimizer, it follows from $V'(0) = -\infty$ that $\widehat{q}_n(y) > 0$, for each $n = 1, \ldots, N$; in other words, $\widehat{Q}(y)$ is an equivalent martingale measure for S.

Defining the dual value function $v(y)$ by

$$v(y) = \inf_{Q \in \mathcal{M}^a(S)} \sum_{n=1}^{N} p_n V\left(y \frac{q_n}{p_n}\right) \tag{80}$$

$$= \sum_{n=1}^{N} p_n V\left(y \frac{\widehat{q}_n(y)}{p_n}\right) \tag{81}$$

we find ourselves in an analogous situation as in the complete case above: defining again $\widehat{y}(x)$ by $v'(\widehat{y}(x)) = -x$ and

$$\widehat{\xi}_n = I\left(\widehat{y}(x) \frac{\widehat{q}_n(y)}{p_n}\right), \tag{82}$$

similar arguments as above apply to show that $(\widehat{\xi}_1, \ldots, \widehat{\xi}_N, \widehat{y}(x), \widehat{Q}(y))$ is the unique saddle-point of the Lagrangian (74) and that the value functions u and v are conjugate.

Let us summarize what we have found in the incomplete case:

Theorem 2.3 (finite Ω, incomplete market). *Let the financial market $S = (S_t)_{t=0}^T$ defined over the finite filtered probability space $(\Omega, \mathcal{F}, (\mathcal{F})_{t=0}^T, \mathbf{P})$ and let $\mathcal{M}^e(S) \neq \emptyset$, and the utility function U satisfies Assumptions 1.2.*

Denote by $u(x)$ and $v(y)$ the value functions

$$u(x) = \sup\nolimits_{X_T \in C(x)} \mathbb{E}[U(X_T)], \qquad x \in \mathrm{dom}(U), \tag{83}$$

$$v(y) = \inf\nolimits_{Q \in \mathcal{M}^a(S)} \mathbb{E}\left[V\left(y \tfrac{dQ}{d\mathbf{P}}\right)\right], \qquad y > 0. \tag{84}$$

We then have:

(i) The value functions $u(x)$ and $v(y)$ are conjugate and u shares the qualitative properties of U listed in Assumption 1.2.
(ii) The optimizers $\widehat{X}_T(x)$ and $\widehat{Q}(y)$ in (83) and (84) exist, are unique, $\widehat{Q}(y) \in \mathcal{M}^e(S)$, and satisfy

$$\widehat{X}_T(x) = I\left(y \frac{d\widehat{Q}(y)}{d\mathbf{P}}\right), \qquad y \frac{d\widehat{Q}(y)}{d\mathbf{P}} = U'(\widehat{X}_T(x)), \tag{85}$$

where $x \in \mathrm{dom}(U)$ *and* $y > 0$ *are related via* $u'(x) = y$ *or, equivalently,* $x = -v'(y)$.

(iii) The following formulae for u' *and* v' *hold true:*

$$u'(x) = \mathbb{E}_{\mathbf{P}}[U'(\widehat{X}_T(x))], \qquad v'(y) = \mathbb{E}_{\widehat{Q}}\left[V'\left(y\frac{d\widehat{Q}(y)}{d\mathbf{P}}\right)\right] \tag{86}$$

$$xu'(x) = \mathbb{E}_{\mathbf{P}}[\widehat{X}_T(x)U'(\widehat{X}_T(x))], \; yv'(y) = \mathbb{E}_{\mathbf{P}}\left[y\frac{d\widehat{Q}(y)}{d\mathbf{P}}V'\left(y\frac{d\widehat{Q}(y)}{d\mathbf{P}}\right)\right]. \tag{87}$$

Remark 2.4. Let us again interpret the formulae (86), (87) for $u'(x)$ similarly as in Remark 2.2 above. In fact, the interpretations of these formulae as well as their derivations remain in the incomplete case exactly the same.

But a new and interesting phenomenon arises when we pass to the variation of the optimal pay-off function $\widehat{X}_T(x)$ by a small unit of an arbitrary pay-off function $f \in L^\infty(\Omega, \mathcal{F}, \mathbf{P})$. Similarly as in (64) we have the formula

$$\mathbb{E}_{\widehat{Q}(y)}[f]u'(x) = \lim_{h \to 0} \frac{\mathbb{E}_{\mathbf{P}}[U(\widehat{X}_T(x) + hf) - U(\widehat{X}_T(x))]}{h}, \tag{88}$$

the only difference being that Q has been replaced by $\widehat{Q}(y)$ (recall that x and y are related via $u'(x) = y$).

The remarkable feature of this formula is that it does not only pertain to variations of the form $f = x + (H \cdot S)_T$, i.e, contingent claims attainable at price x, but to arbitrary contingent claims f, for which — in general — we cannot derive the price from no arbitrage considerations.

The economic interpretation of formula (88) is the following: the pricing rule $f \mapsto \mathbb{E}_{\widehat{Q}(y)}[f]$ yields precisely those prices, at which an economic agent with initial endowment x, utility function U and investing optimally, is indifferent of first order towards adding a (small) unit of the contingent claim f to her portfolio $\widehat{X}_T(x)$.

In fact, one may turn the view around, and this was done by M. Davis [D 97] (compare also the work of L. Foldes [F 90]): one may *define* $\widehat{Q}(y)$ by (88), verify that this indeed is an equivalent martingale measure for S, and interpret this pricing rule as "pricing by marginal utility", which is, of course, a classical and basic paradigm in economics.

Let us give a proof for (88) (under the hypotheses of Theorem 2.3). One possibility, which also has the advantage of a nice economic interpretation, is the idea of introducing "fictitious securities" as developed in [KLSX 91]: fix $x \in \mathrm{dom}(U)$ and $y = u'(x)$ and let $(f^1, \dots, f^k)$ be finitely many elements of $L^\infty(\Omega, \mathcal{F}, \mathbf{P})$ such that the space $K = \{(H \cdot S)_T : H \in \mathcal{H}\}$, the constant function $\mathbf{1}$, and $(f^1, \dots, f^k)$ linearly span $L^\infty(\Omega, \mathcal{F}, \mathbf{P})$. Define the k processes

$$S_t^{d+j} = \mathbb{E}_{\widehat{Q}(y)}[f^j | \mathcal{F}_t], \qquad j = 1, \dots, k, \quad t = 0, \dots, T. \tag{89}$$

Now extend the $\mathbb{R}^{d+1}$-valued process $S = (S^0, \dots, S^d)$ to the $\mathbb{R}^{d+k+1}$-valued process $\overline{S} = (S^0, \dots, S^d, S^{d+1}, \dots, S^{d+k})$ by adding these new coordinates. By (89) we still have that $\overline{S}$ is a martingale under $\widehat{Q}(y)$, which now is the unique probability under which $\overline{S}$ is a martingale, by our choice of $(f^1, \dots, f^k)$.

Hence we find ourselves in the situation of Theorem 2.1. By comparing (54) and (85) we observe that the optimal pay-off function $\widehat{X}_T(x)$ has not changed. Economically speaking this means that in the "completed" market $\overline{S}$ the optimal investment may still be achieved by trading only in the first $d+1$ assets and without touching the "fictitious" securities $S^{d+1}, \dots, S^{d+k}$.

In particular, we now may apply formula (64) to $Q = \widehat{Q}(y)$ to obtain (88).

Finally remark that the pricing rule induced by $\widehat{Q}(y)$ is precisely such that the interpretation of the optimal investment $\widehat{X}_T(x)$ defined in (85) (given in Remark 2.2 in terms of marginal utility and the ratio of Arrow-Debreu prices $\widehat{q}_n(y)$ and probabilities p_n) carries over to the present incomplete setting. The above completion of the market by introducing "fictious securities" allows for an economic interpretation of this fact.

3 The General Case

In the previous section we have analyzed the duality theory of the optimization problem (1) in detail and with full proofs, for the case when the underlying probability space is finite.

We now pass to the question under which conditions the crucial features of the above Theorem 2.3 carry over to the general setting. In particular one is naturally led to ask: under which conditions

- Are the optimizers $\widehat{X}_T(x)$ and $\widehat{Q}(y)$ of the value functions $u(x)$ and $v(y)$ attained?
- Does the basic duality formula

$$U'\left(\widehat{X}_T(x)\right) = \widehat{y}(x)\frac{d\widehat{Q}(\widehat{y}(x))}{d\mathbf{P}} \tag{90}$$

  or, equivalently

$$\widehat{X}_T(x) = I\left(\widehat{y}(x)\frac{d\widehat{Q}(\widehat{y}(x))}{d\mathbf{P}}\right) \tag{91}$$

  hold true?
- Are the value functions $u(x)$ and $v(y)$ conjugate?
- Does the value function $u(x)$ still inherit the qualitative properties of U listed in Assumption 1.2?
- Do the formulae for $u'(x)$ still hold true?

We shall see that we get affirmative answers to these questions under two provisos: firstly, one has to make an appropriate choice of the sets in which X_T and Q are allowed to vary. This choice will be different for case 1, where

$\mathrm{dom}(U) = \mathbb{R}_+$, and case 2, where $\mathrm{dom}(U) = \mathbb{R}$. Secondly, the utility function U has to satisfy — in addition to Assumption 1.2 — a mild regularity condition, namely the property of *"reasonable asymptotic elasticity"*.

The essential message of the theorems below is that, assuming that U has *"reasonable asymptotic elasticity"*, the duality theory works just as well as in the case of finite Ω. Note that we do not have to impose any regularity conditions on the underlying stochastic process S, except for its arbitrage-freeness in the sense made precise by Assumption 1.1. On the other hand, the assumption of reasonable asymptotic elasticity on the utility function U cannot be relaxed, even if we impose very strong assumptions on the process S (e.g., having continuous paths and defining a complete financial market), as we shall see below.

Before passing to the positive results we first analyze the notion of "reasonable asymptotic elasticity" and sketch the announced counterexample.

Definition 3.1. A utility function U satisfying Assumption 1.2 is said to have *"reasonable asymptotic elasticity"* if

$$\limsup_{x\to\infty} \frac{xU'(x)}{U(x)} < 1, \tag{92}$$

and, in case 2 of Assumption 1.2, we also have

$$\liminf_{x\to-\infty} \frac{xU'(x)}{U(x)} > 1. \tag{93}$$

Let us discuss the economic meaning of this notion: as H.-U. Gerber observed, the quantity $\frac{xU'(x)}{U(x)}$ is the elasticity of the function U at x. We are interested in its asymptotic behaviour. It easily follows from Assumption 1.2 that the limits in (92) and (93) are less (resp. bigger) than or equal to one. What does it mean that $\frac{xU'(x)}{U(x)}$ tends to one, for $x \mapsto \infty$? It means that the ratio between the *marginal utility* $U'(x)$ and the *average utility* $\frac{U(x)}{x}$ tends to one. A typical example is a function $U(x)$ which equals $\frac{x}{\ln(x)}$, for x large enough; note however, that in this example Assumption 1.2 is not violated insofar as the marginal utility still decreases to zero for $x \to \infty$, i.e., $\lim_{x\to\infty} U'(x) = 0$.

If the marginal utility $U'(x)$ is approximately equal to the average utility $\frac{U(x)}{x}$ for large x, this means that for an economic agent, modeled by the utility function U, the increase in utility by varying wealth from x to $x+1$, when x is large, is approximately equal to the average of the increase of utility by changing wealth from n to $n+1$, where n runs through $1, 2, \dots, x-1$ (we assume in this argument that x is a large natural number and, w.l.o.g., that $U(1) \approx 0$). We feel that the economic intuition behind decreasing marginal utility suggests that, for large x, the marginal utility $U'(x)$ should be substantially smaller than the average utility $\frac{U(x)}{x}$. Therefore we have denoted a utility function, where the ratio of $U'(x)$ and $\frac{U(x)}{x}$ tends to one, as being "unreasonable". Another justification for this terminology will be the results of Example 3.2 and Theorems 3.4 and 3.5 below.

P. Guasoni observed, that there is a close connection between the asymptotic behaviour of the *elasticity* of U, and the asymptotic behaviour of the *relative risk aversion* associated to U. Recall (see, e.g., [HL88]) that the relative risk aversion of an agent with endowment x, whose preferences are described by the utility function U, equals

$$RRA(U)(x) = -\frac{xU''(x)}{U'(x)}. \tag{94}$$

A formal application of de l'Hôpital's rule yields

$$\lim_{x\mapsto\infty}\frac{xU'(x)}{U(x)} = \lim_{x\mapsto\infty}\frac{U'(x)+xU''(x)}{U'(x)} = 1 - \lim_{x\mapsto\infty}\left(-\frac{xU''(x)}{U'(x)}\right) \tag{95}$$

which suggests that the asymptotic elasticity of U is less than one iff the "*asymptotic relative risk aversion*" is strictly positive.

Turning the above formal argument into a precise statement, one easily proves the following result: if $\lim_{x\mapsto\infty}(-\frac{xU''(x)}{U'(x)})$ exists, then $\lim_{x\mapsto\infty}\frac{xU'(x)}{U(x)}$ exists too, and the former is strictly positive iff the latter is less than one. Hence "*essentially*" these two concepts coincide.

On the other hand, in general (i.e. without assuming that the above limit exists), there is no way to characterize the condition $\lim_{x\mapsto\infty}\sup\frac{xU'(x)}{U(x)} < 1$ in terms of the asymptotic behaviour of $-\frac{xU''(x)}{U'(x)}$, as $x \mapsto \infty$. Firstly, for the second expression to make sense, we have to assume that U is twice differentiable; but, even doing so, does not help, as it is easy to construct examples where $\frac{xU'(x)}{U(x)}$ converges (to 1 or to a number less than 1), while $-\frac{xU''(x)}{U'(x)}$ oscillates wildly in $]0,\infty[$, as $x\mapsto\infty$.

We shall see that the assumption of reasonable asymptotic elasticity is *necessary and sufficient* for several key results in the duality theory of utility maximization to hold true. Hence in order to obtain these sharp results, we cannot reformulate things in terms of the asymptotic relative risk aversion. However, readers which are happy with sufficient conditions, may replace the assumption $\limsup_{x\mapsto\infty}\frac{xU'(x)}{U(x)} < 1$ by the assumption $\liminf_{x\mapsto\infty}(-\frac{xU''(x)}{U'(x)}) > 0$ below: it is easy to verify that the latter assumption implies the former (but not vice versa).

Similar reasoning applies to the asymptotic behaviour of $\frac{xU'(x)}{U(x)}$, as x tends to $-\infty$, in case 2. In this context the typical counter-example is $U(x) \sim x\ln(|x|)$, for $x < x_0$; in this case one finds similarly

$$\lim_{x\to-\infty} U'(x) = \infty, \quad \text{while} \quad \lim_{x\to-\infty}\frac{xU'(x)}{U(x)} = 1. \tag{96}$$

The message of Definition 3.1 above is — roughly speaking — that we want to exclude utility functions U which behave like $U(x) \sim \frac{x}{\ln(x)}$, as $x \to \infty$, or $U(x) \sim x \ln |x|$, as $x \to -\infty$. Similar (but not quite equivalent) notions comparing the behaviour of $U(x)$ with that of power functions in the setting of case 1, were defined and analyzed in [KLSX 91] (see [KS 99], lemma 6.5, for a comparison of these concepts).

We start with a sketch of a counterexample showing the relevance of the notion of asymptotic elasticity in the context of utility maximization: *whenever U fails to have reasonable asymptotic elasticity* the duality theory breaks down in a rather dramatic way. We only state the version of the counterexample where both assumptions (92) and (93) are violated and refer to [KS 99] and [S 00a] for the other cases.

Example 3.2. ([S 00], prop. 3.5) Let U be *any utility function* satisfying Assumption 1.2, case 2 and such that

$$\lim_{x \to -\infty} \frac{xU'(x)}{U(x)} = \lim_{x \to \infty} \frac{xU'(x)}{U(x)} = 1, \tag{97}$$

Then there is an $\mathbb{R}$-valued process $(S_t)_{0 \le t \le T}$ of the form

$$S_t = \exp\left(B_t + \mu_t\right), \tag{98}$$

where $B = (B_t)_{0 \le t \le T}$ is a standard Brownian motion, based on its natural filtered probability space, and μ_t a predictable process, such that the following properties hold true:

(i) $\mathcal{M}^e(S) = \{Q\}$, i.e., S defines a complete financial market.
(ii) The primal value function $u(x)$ fails to be strictly concave and to satisfy $u'(\infty) = 0$, $u'(-\infty) = \infty$ in a rather striking way: $u(x)$ is a straight line of the form $u(x) = c + x$, for some constant $c \in \mathbb{R}$.
(iii) The optimal investment $\widehat{X}_T(x)$ fails to exist, for all $x \in \mathbb{R}$, except for one point $x = x_0$. In particular, for $x \neq x_0$, the formula (91) does not define the optimal investment $\widehat{X}_T(x)$.
(iv) The dual value function v fails to be a finite, smooth, strictly convex function on $\mathbb{R}_+$ in a rather striking way: in fact, $v(1) < \infty$ while $v(y) = \infty$, for all $y > 0$, $y \neq 1$.

We do not give a rigorous proof for these assertions but refer to [S 00, Proposition 3.5], which in turn is a variant of [KS 99, Proposition 5.3].

We shall try to sketch the basic idea underlying the construction of the example, in mathematical as well as economic terms. Arguing mathematically, one starts by translating the assumptions (97) on the utility function U into equivalent properties of the conjugate function V: roughly speaking, the corresponding property of $V(y)$ is, that it increases very rapidly to infinity, as $y \to 0$ and $y \to \infty$ (see [KS 99, Corollary 6.1] and [S 00, Proposition

4.1]). Having isolated this property of V, it is an easy exercise to construct a function $f : [0,1] \to]0,\infty[$, $\mathbb{E}[f] = 1$ such that

$$\mathbb{E}[V(f)] < \infty \text{ while } \mathbb{E}[V(yf)] = \infty, \quad \text{for } y \neq 1, \tag{99}$$

where $\mathbb{E}$ denotes expectation with respect to Lebesgue measure λ. In fact one may find such a function f taking only the values $(y_n)_{n=-\infty}^{\infty}$, for a suitable chosen increasing sequence $(y_n)_{n=-\infty}^{\infty}$, $\lim_{n\to-\infty} y_n = 0$, $\lim_{n\to\infty} y_n = \infty$.

Next we construct a measure Q on the sigma algebra $\mathcal{F} = \mathcal{F}_T$ generated by the Brownian motion $B = (B_t)_{0\le t\le T}$ which is equivalent to Wiener measure $\mathbf{P}$, and such that the distribution of $\frac{dQ}{d\mathbf{P}}$ (under $\mathbf{P}$) equals that of f (under Lebesgue measure λ). There is no uniqueness in this part of the construction, but it is straightforward to find some appropriate measure Q with this property.

By Girsanov's theorem we know that we can find an adapted process $(\mu_t)_{0\le t\le T}$, such that Q is the unique equivalent local martingale measure for the process defined in (98), hence we obtain assertion (i).

This construction makes sure that we obtain property (iv), i.e.

$$v(y) = \mathbb{E}_{\mathbf{P}}\left[V\left(y\frac{dQ}{d\mathbf{P}}\right)\right] = \mathbb{E}_\lambda\left[V(yf)\right] < \infty \text{ iff } y = 1. \tag{100}$$

Once this crucial property is established, most of the assertions made in (ii) and (iii) above easily follow (in fact, for the existence of $\widehat{X}_T(x)$ for precisely one $x = x_0$, some extra care is needed).

Instead of elaborating further on the mathematical details of the construction sketched above, let us try to give an economic interpretation of what is really happening in the above example. This is not easy, but we find it worth trying. We concentrate on the behaviour of U as $x \to \infty$, the case when $x \to -\infty$ being similar.

How is the "unreasonableness" property of the utility function U used to construct the pathologies in the above example? Here is a rough indication of the underlying economic idea: the financial market S is constructed in such a way that one may find positive numbers $(x_n)_{n=1}^{\infty}$, disjoint sets $(A_n)_{n=1}^{\infty}$ in $\mathcal{F}_T$, with $\mathbf{P}[A_n] = p_n$ and $Q[A_n] = q_n$, such that for the contingent claims $x_n \mathbf{1}_{A_n}$ we approximately have

$$\mathbb{E}_Q\left[x_n \mathbf{1}_{A_n}\right] = q_n x_n \approx 1 \tag{101}$$

and

$$\mathbb{E}_{\mathbf{P}}\left[U(x_n)\mathbf{1}_{A_n}\right] = p_n U(x_n) \approx 1. \tag{102}$$

Hence $\frac{q_n}{p_n} \approx \frac{U(x_n)}{x_n}$.

It is easy to construct a complete, continuous market S over the Brownian filtration such that this situation occurs and this is, in fact, what is done in the above "mathematical" argument to define f and Q. We remark in passing that one might just as well construct S as a complete, discrete time model $S=(S_t)_{t=0}^\infty$ over a countable probability space Ω displaying sets A_n and real numbers x_n having the properties listed above. But for esthetical reasons we have prefered to do the construction in terms of an exponential Brownian motion with drift.

We claim that, for any $x \in \mathbb{R}$ and any investment strategy $X_T = x + (H \cdot S)_T$, we can find an investment strategy $\widetilde{X}_T = (x+1) + (\widetilde{H} \cdot S)_T$ such that

$$\mathbb{E}\left[U(\widetilde{X}_T)\right] \approx \mathbb{E}\left[U(X_T)\right] + 1. \tag{103}$$

The above relation should motivate why the value function $u(x)$ becomes a straight line with slope one, at least for x sufficiently large (for the corresponding behaviour of $u(x)$ on the left hand side of $\mathbb{R}$ one has to play in addition a similar game as above with $(x_n)_{n=1}^\infty$ tending to $-\infty$).

To present the idea behind (103), suppose that we have $\mathbb{E}\left[U(X_T)\right] < \infty$, so that $\lim_{n\to\infty} \mathbb{E}[U(X_T)\mathbf{1}_{A_n}] = 0$. Varying our initial endowment from x to $x+1$ C, we may use the additional C to add to the pay-off function X_T the function $x_n \mathbf{1}_{A_n}$, for some large n; by (101) this may be financed (approximately) with the additional C and by (102) this will increase the expected utility (approximately) by 1

$$\begin{aligned}
\mathbb{E}\left[U\left(X_T + x_n \mathbf{1}_{A_n}\right)\right] &\approx \mathbb{E}\left[U(X_T)\mathbf{1}_{\Omega\setminus A_n}\right] + \mathbb{E}\left[U(X_T + x_n)\mathbf{1}_{A_n}\right] \\
&\approx \mathbb{E}\left[U(X_T)\right] + p_n U(x_n) \\
&\approx \mathbb{E}\left[U(X_T)\right] + 1, \qquad (104)
\end{aligned}$$

which was claimed in (103).

The above argument also gives a hint why we cannot expect that the optimal strategy $\widehat{X}_T(x) = x + (\widehat{H} \cdot S)_T$ exists, as one cannot "pass to the limit as $n \to \infty$" in the above reasoning.

Observe that we have not yet used the assumption $\limsup_{x\to\infty} \frac{xU'(x)}{U(x)} = 1$, as it always is possible to construct things in such a way that (101) and (102) hold true (provided only that $\lim_{x\to\infty} U(x) = \infty$, which we assume from now on). How does the "unreasonable asymptotic elasticity" come into play? The point is that we have to do the construction described in (101) and (102) without violating Assumption 1.3, i.e.,

$$u(x) = \sup_{H \in \mathcal{H}} \mathbb{E}\left[U\left(x + (H \cdot S)_T\right)\right] < \infty,$$
$$\text{for some (equivalently, for all) } x \in \mathbb{R}. \tag{105}$$

In order to satisfy Assumption 1.3 we have to make sure that

$$\mathbb{E}\left[\sum_{n=1}^\infty U(\mu_n x_n)\mathbf{1}_{A_n}\right] = \sum_{n=1}^\infty p_n U(\mu_n x_n) \tag{106}$$

remains bounded, when $(\mu_n)_{n=1}^{\infty}$ runs through all convex weights $\mu_n \geq 0$, $\sum_{n=1}^{\infty} \mu_n = 1$, i.e., when we consider all investments into non-negative linear combinations of the contingent claims $x_n \mathbf{1}_{A_n}$, which can be financed with one $\mathbb{C}$.

The message of Example 3.2 is that this is not possible, if and only if $\limsup_{x\to\infty} \frac{xU'(x)}{U(x)} = 1$ (for this part of the construction we only use the asymptotic behaviour of $U(x)$, as $x \to \infty$). To motivate this claim, think for a moment of the "reasonable" case, e.g., $U(x) = \frac{x^\alpha}{\alpha}$, for some $0 < \alpha < 1$, in which case we have $\lim_{x\to\infty} \frac{xU'(x)}{U(x)} = \alpha < 1$. Letting $\mu_n \approx n^{-(1+\epsilon)}$, we get

$$\sum_{n=1}^{\infty} p_n U(\mu_n x_n) \approx \sum_{n=1}^{\infty} n^{-(1+\epsilon)\alpha} p_n U(x_n) \tag{107}$$

$$\approx \sum_{n=1}^{\infty} n^{-(1+\epsilon)\alpha}, \tag{108}$$

which equals infinity if $\epsilon > 0$ is small enough, that $(1+\epsilon)\alpha \leq 1$. This argument indicates that in the case of the power utility $U(x) = \frac{x^\alpha}{\alpha}$ it is impossible to reconcile the validity of (101) and (102) with the requirement (106). On the other hand, it turns out that in the "unreasonable" case, where we have $\lim_{x\to\infty} \frac{xU'(x)}{U(x)} = 1$, we can do the construction in such a way that $U(\mu_n x_n)$ is sufficiently close to $\mu_n U(x_n)$ such that we obtain a uniform bound on the sum in (106).

Let us now stop our attempt at an economic interpretation. We hope that the above informal arguments were of some use for the reader in developing her intuition for the concept of "reasonable asymptotic elasticity" and that she now has some background information to find her way through the corresponding formal arguments in [KS 99] and [S 00].

We now pass to the positive results in the spirit of Theorem 2.1 and Theorem 2.3 above. We first consider the case where U satisfies the Inada conditions (7) and (8), which was studied in [KS 99].

Case 1: $\mathrm{dom}(U) = \mathbb{R}_+$.

The heart of the argument in the proof of Theorem 2.3 (which we now want to extend to the general case) is the applicability of the minimax theorem, which underlies the theory of Lagrange multipliers. We want to extend the applicability of the minimax theorem to the situation. The infinite-dimensional versions of the minimax theorem available in the literature (see, e.g, [ET 76] or [St 85]) are along the following lines: Let $\langle E, F\rangle$ be a pair of locally convex vector spaces in separating duality, $C \subseteq E$, $D \subseteq F$ a pair of convex subsets, and $L(x, y)$ a function defined on $C \times D$, concave in the first and convex in the second variable, having some (semi-)continuity property compatible with the topologies of E and F (which in turn should be compatible with the duality between E and F). If one of the sets C and D is compact and the other is complete, then one may assert the existence of a saddle point $(\widehat{\xi},\widehat{\eta}) \in C \times D$ such that

$$L(\widehat{\xi},\widehat{\eta}) = \sup_{\xi \in C} \inf_{\eta \in D} L(\xi,\eta) = \inf_{\eta \in D} \sup_{\xi \in C} L(\xi,\eta). \tag{109}$$

We try to apply this theorem to the analogue of the Lagrangian encountered in the proof of Theorem 2.3 above. Fixing $x > 0$ and $y > 0$ let us formally write the Lagrangian (74) in the infinite-dimensional setting,

$$L^{x,y}(X_T, Q) = \mathbb{E}_{\mathbf{P}}[U(X_T)] - y(\mathbb{E}_Q[X_T - x]) \tag{110}$$

$$= \mathbb{E}_{\mathbf{P}}\left[U(X_T) - y\tfrac{dQ}{d\mathbf{P}} X_T\right] + yx, \tag{111}$$

where X_T runs through "all" non-negative $\mathcal{F}_T$-measurable functions and Q through the set $\mathcal{M}^a(S)$ of absolutely continuous local martingale measures.

To restrict the set of "all" nonnegative functions to a more amenable one note that $\inf_{y>0, Q \in \mathcal{M}^a(S)} L^{x,y}(X_T, Q) > -\infty$ iff

$$\mathbb{E}_Q[X_T] \leq x, \quad \text{for all} \quad Q \in \mathcal{M}^a(S). \tag{112}$$

Using the basic result on the super-replicability of the contingent claim X_T (see [KQ 95], [J 92], [AS 94], [DS 94], and [DS 98b]), we have — as encountered in the finite dimensional case — that a non-negative $\mathcal{F}_T$-measurable random variable X_T satisfies (112) iff there is an admissible trading strategy H such that

$$X_T \leq x + (H \cdot S)_T. \tag{113}$$

Hence let

$$\begin{aligned} C(x) = \{ X_T \in L^0_+(\Omega, \mathcal{F}_T, \mathbf{P}) : & \\ X_T \leq x + (H \cdot S)_T, & \text{ for some admissible } H \} \end{aligned} \tag{114}$$

and simply write C for $C(1)$ (observe that $C(x) = xC$).

We thus have found a natural set $C(x)$ in which X_T should vary when we are mini-maxing the Lagrangian $L^{x,y}$. Dually, the set $\mathcal{M}^a(S)$ seems to be the natural domain where the measure Q is allowed to vary (in fact, we shall see later, that this set still has to be slightly enlarged). But what are the locally convex vector spaces E and F in separating duality into which C and $\mathcal{M}^a(S)$ are naturally embedded? As regards $\mathcal{M}^a(S)$ the natural choice seems to be $L^1(\mathbf{P})$ (by identifying a measure $Q \in \mathcal{M}^a(S)$ with its Radon-Nikodym derivative $\frac{dQ}{d\mathbf{P}}$); note that $\mathcal{M}^a(S)$ is a *closed* subset of $L^1(\mathbf{P})$, which is good news. On the other hand, there is no reason for C to be contained in $L^\infty(\mathbf{P})$, or even in $L^p(\mathbf{P})$, for any $p > 0$; the natural space in which C is embedded is just $L^0(\Omega, \mathcal{F}_T, \mathbf{P})$, the space of all real-valued $\mathcal{F}_T$-measurable functions endowed with the topology of convergence in probability.

The situation now seems hopeless (if we don't want to impose artificial $\mathbf{P}$-integrability assumptions on X_T and/or $\frac{dQ}{d\mathbf{P}}$), as $L^0(\mathbf{P})$ and $L^1(\mathbf{P})$ are not in any reasonable duality; in fact, $L^0(\mathbf{P})$ is not even a locally convex space, hence there seems to be no hope for a good duality theory, which could serve as a basis for the application of the minimax theorem. But the good news is that the sets C and $\mathcal{M}^a(S)$ are in the *positive orthant* of $L^0(\mathbf{P})$ and $L^1(\mathbf{P})$ respectively; the crucial observation is, that for $f \in L^0_+(\mathbf{P})$ and $g \in L^1_+(\mathbf{P})$, it is possible to well-define

$$\langle f, g \rangle := \mathbb{E}_{\mathbf{P}}[fg] \in [0, \infty]. \tag{115}$$

The spirit here is similar as in the very foundation of Lebesgue integration theory: For positive measurable functions the integral is always defined, but possibly $+\infty$. This does not cause any logical inconsistency.

Similarly the bracket $\langle \cdot, \cdot \rangle$ defined in (115) shares many of the usual properties of a scalar product. The difference is that $\langle f, g \rangle$ now may assume the value $+\infty$ and that the map $(f, g) \mapsto \langle f, g \rangle$ is not continuous on $L^0_+(\mathbf{P}) \times L^1_+(\mathbf{P})$, but only lower semi-continuous (this immediately follows from Fatou's lemma).

At this stage it becomes clear that the role of $L^1_+(\mathbf{P})$ is somewhat artificial, and it is more natural to define (115) in the general setting where f and g are both allowed to vary in $L^0_+(\mathbf{P})$. The pleasant feature of the space $L^0(\mathbf{P})$ in the context of Mathematical Finance is, that it is invariant under the passage to an equivalent measure Q, a property only shared by $L^\infty(\mathbf{P})$, but by no other $L^p(\mathbf{P})$, for $0 < p < \infty$.

We now can turn to the polar relation between the sets C and $\mathcal{M}^a(S)$. By (113) we have, for an element $X_T \in L^0_+(\Omega, \mathcal{F}, \mathbf{P})$,

$$X_T \in C \Leftrightarrow \mathbb{E}_Q[X_T] = \mathbb{E}_{\mathbf{P}}[X_T \tfrac{dQ}{d\mathbf{P}}] \le 1, \quad \text{for} \quad Q \in \mathcal{M}^a(S). \tag{116}$$

Denote by D the closed, convex, solid hull of $\mathcal{M}^a(S)$ in $L^0_+(\mathbf{P})$. It is easy to show (using, e.g., Lemma 3.3 below), that D equals

$$\begin{aligned} D = \{Y_T \in L^0_+(\Omega, \mathcal{F}_T, \mathbf{P}) : \ & \text{there is} \\ & (Q_n)_{n=1}^\infty \in \mathcal{M}^a(S) \text{ s.t. } Y_T \le \lim_{n\to\infty} \tfrac{dQ_n}{d\mathbf{P}}\}, \end{aligned} \tag{117}$$

where the $\lim_{n\to\infty} \frac{dQ_n}{d\mathbf{P}}$ is understood in the sense of almost sure convergence. We have used the letter Y_T for the elements of D to stress the dual relation to the elements X_T in C. In further analogy we write, for $y > 0$, $D(y)$ for yD, so that $D = D(1)$. By (117) and Fatou's lemma we again find that, for $X_T \in L^0_+(\Omega, \mathcal{F}, \mathbf{P})$

$$X_T \in C \Leftrightarrow \mathbb{E}_{\mathbf{P}}[X_T Y_T] \le 1, \quad \text{for} \quad Y_T \in D. \tag{118}$$

Why did we pass to this enlargement D of the set $\mathcal{M}^a(S)$? The reason is that we now obtain a more symmetric relation between C and D: for $Y_T \in$

$L^0_+(\Omega, \mathcal{F}, \mathbf{P})$ we have

$$Y_T \in D \Leftrightarrow \mathbb{E}_{\mathbf{P}}[X_T Y_T] \leq 1, \quad \text{for} \quad X_T \in C. \tag{119}$$

The proof of (119) relies on an adaption of the "bipolar theorem" from the theory of locally convex spaces (see, e.g., [Sch 66]) to the present duality $\langle L^0_+(\mathbf{P}), L^0_+(\mathbf{P}) \rangle$, which was worked out in [BS 99].

Why is it important to define the enlargement D of $\mathcal{M}^a(S)$ in such a way that (119) holds true? After all, $\mathcal{M}^a(S)$ is a nice, convex, closed (w.r.t. the norm of $L^1(\mathbf{P})$) set and we also have that, for $g \in L^1(\mathbf{P})$ *such that* $\mathbb{E}_{\mathbf{P}}[g] = 1$,

$$g \in \mathcal{M}^a(S) \Leftrightarrow \mathbb{E}_{\mathbf{P}}[X_T g] \leq 1, \quad \text{for} \quad X_T \in C. \tag{120}$$

The reason is that, in general, the saddle point $(\widehat{X}_T, \widehat{Q})$ of the Lagrangian will *not* be such that $\widehat{Q}$ is a probability measure; it will only satisfy $\mathbb{E}\left[\frac{d\widehat{Q}}{d\mathbf{P}}\right] \leq 1$, the inequality possibly being strict. But it will turn out that $\widehat{Q}$, which we identify with $\frac{d\widehat{Q}}{d\mathbf{P}}$, is always in D. In fact, the passage from $\mathcal{M}^a(S)$ to D is the *crucial feature* in order to make the duality work in the present setting: we shall see below that even for nice utility functions U, such as the logarithm, and for nice processes, such as a continuous process $(S_t)_{0 \leq t \leq T}$ based on the filtration of two Brownian motions, the above described phenomenon can occur: the saddle point of the Lagrangian leads out of $\mathcal{M}^a(S)$.

The set D can be characterized in several equivalent manners. We have defined D above in the abstract way as the convex, closed, solid hull of $\mathcal{M}^a(S)$ and mentioned the description (117). Equivalently, one may define D as the set of random variables $Y_T \in L^0_+(\Omega, \mathcal{F}, \mathbf{P})$ such that there is a process $(Y_t)_{0 \leq t \leq T}$ starting at $Y_0 = 1$ with $(Y_t X_t)_{0 \leq t \leq T}$ a $\mathbf{P}$-supermartingale, for every non-negative process $(X_t)_{0 \leq t \leq T} = (x + (H \cdot S)_t)_{0 \leq t \leq T}$, where $x > 0$ and H is predictable and S-integrable. This definition was used in [KS 99]. Another equivalent characterization was used in [CSW 00]: Consider the convex, solid hull of $\mathcal{M}^a(S)$, and embed this subset of $L^1(\mathbf{P})$ into the bidual $L^1(\mathbf{P})^{**} = L^\infty(\mathbf{P})^*$; denote by $\overline{\mathcal{M}^a(S)}$ the weak-star closure of the convex solid hull of $\mathcal{M}^a(S)$ in $L^\infty(\mathbf{P})^*$. Each element of $\overline{\mathcal{M}^a(S)}$ may be decomposed into its regular part $\mu^r \in L^1(\mathbf{P})$ and its purely singular part $\mu^s \in L^\infty(\mathbf{P})^*$. It turns out that D equals the set $\{\mu^r \in L^1(\mathbf{P}) : \mu \in \overline{\mathcal{M}^a(S)}\}$, i.e. consists of the regular parts of the elements of $\overline{\mathcal{M}^a(S)}$. This description has the advantage that we may associate to the elements $\mu^r \in D$ a singular part μ^s, and it is this extra information which is crucial when extending the present results to the case of random endowment (see [CSW 00]).

Why are the sets C and D hopeful candidates for the minimax theorem to work out properly for a function L defined on $C \times D$? Both are closed, convex and bounded subsets of $L^0_+(\mathbf{P})$. But recall that we still need some compactness property to be able to localize the mini-maximizers (resp. maxi-minimizers) on C (resp. D). In general, neither C nor D is compact (w.r.t. the topology of convergence in measure), i.e., for a sequence $(f_n)_{n=1}^\infty$ in C (resp. $(g_n)_{n=1}^\infty$ in D) we cannot pass to a subsequence converging in measure. But C and D have a property which is close to compactness and in many applications turns out to serve just as well.

Lemma 3.3. *Let A be a closed, convex, bounded subset of $L^0_+(\Omega, \mathcal{F}, \mathbf{P})$. Then for each sequence $(h_n)_{n=1}^\infty \in A$ there exists a sequence of convex combinations $k_n \in \operatorname{conv}(h_n, h_{n+1}, \dots)$ which converges almost surely to a function $k \in A$.*

This easy lemma (see, e.g., [DS 94, Lemma A.1.1], for a proof) is in the spirit of the celebrated theorem of Komlos [Kom 67], stating that for a bounded sequence $(h_n)_{n=1}^\infty$ in $L^1(\mathbf{P})$ there is a subsequence converging in Cesaro-mean almost surely. The methodology of finding pointwise limits by using convex combinations has turned out to be extremely useful as a surrogate for compactness. For an extensive discussion of more refined versions of the above lemma and their applications to Mathematical Finance we refer to [DS 99].

The application of the above lemma is the following: by passing to convex combinations of optimizing sequences $(f_n)_{n=1}^\infty$ in C (resp. $(g_n)_{n=1}^\infty$ in D), we can always find limits $f \in C$ (resp. $g \in D$) w.r.t. almost sure convergence. Note that the passage to convex combinations does not cost more than passing to a subsequence in the application to convex optimization.

We have now given sufficient motivation to state the central result of [KS 99], which is the generalization of Theorem 2.3 to the semi-martingale setting under Assumption 1.2, case 1, and having reasonable asymptotic elasticity.

Theorem 3.4 ([KS 99], th. 2.2). *Let the semi-martingale $S = (S_t)_{0 \le t \le T}$ and the utility function U satisfy Assumptions 1.1, 1.2 case 1 and 1.3; suppose in addition that U has reasonable asymptotic elasticity. Define*

$$u(x) = \sup_{X_T \in C(x)} \mathbb{E}[U(X_T)], \qquad v(y) = \inf_{Y_T \in D(y)} \mathbb{E}[V(Y_T)]. \tag{121}$$

Then we have:

(i) The value functions $u(x)$ and $v(y)$ are conjugate; they are continuously differentiable, strictly concave (resp. convex) on $]0, \infty[$ and satisfy

$$u'(0) = -v'(0) = \infty, \qquad u'(\infty) = v'(\infty) = 0. \tag{122}$$

(ii) The optimizers $\widehat{X}_T(x)$ and $\widehat{Y}_T(y)$ in (121) exist, are unique and satisfy

$$\widehat{X}_T(x) = I(\widehat{Y}_T(y)), \qquad \widehat{Y}_T(y) = U'(\widehat{X}_T(x)), \tag{123}$$

where $x > 0$, $y > 0$ are related via $u'(x) = y$ or equivalently $x = -v'(y)$.

(iii) We have the following relations between u', v' and $\widehat{X}_T, \widehat{Y}_T$ respectively:

$$u'(x) = \mathbb{E}\left[\frac{\widehat{X}_T(x)U'(\widehat{X}_T(x))}{x}\right], \; x>0, \quad v'(y) = \mathbb{E}\left[\frac{\widehat{Y}_T(y)V'(\widehat{Y}_T(y))}{y}\right], \; y>0. \tag{124}$$

For the proof of the theorem we refer to [KS 99].

We finish the discussion of utility functions satisfying the Inada conditions (7) and (8) by briefly indicating an example, when the dual optimizer $\widehat{Y}_T(y)$ fails to be of the form $\widehat{Y}_T(y) = y\frac{d\widehat{Q}(y)}{d\mathbf{P}}$, for some probability measure $\widehat{Q}(y)$.

It suffices to consider a stock-price process of the form

$$\begin{aligned} S_t &= \left(\exp\left(B_t + \tfrac{t}{2}\right)\right)^{\tau} \\ &= \exp\left(B_{t\wedge\tau} + \tfrac{t\wedge\tau}{2}\right), \quad t \geq 0, \end{aligned} \tag{125}$$

where $(B_t)_{t\geq 0}$ is Brownian motion based on $(\Omega, \mathcal{F}, (\mathcal{F}_t)_{t>0}, \mathbf{P})$ and τ a suitably chosen finite stopping time (to be discussed below) with respect to the filtration $(\mathcal{F}_t)_{t>0}$, after which the process S remains constant.

The usual way to find a risk-neutral measure Q for the process S above is to use Girsanov's formula, which amounts to considering

$$Z_\tau = \exp(-B_\tau - \tfrac{\tau}{2}) \tag{126}$$

as a candidate for the Radon-Nikodym derivative $\frac{dQ}{d\mathbf{P}}$.

It turns out that one may construct τ in such a way that the density process given by Girsanov's theorem

$$Z_t = \exp(-B_{t\wedge\tau} - \tfrac{t\wedge\tau}{2}), t > 0 \tag{127}$$

fails to be a uniformly integrable martingale: Then in particular

$$\mathbb{E}[Z_\tau] < 1. \tag{128}$$

The trick is to choose the filtration $(\mathcal{F}_t)_{t\geq 0}$ to be generated by two independent Brownian motions $(B_t)_{t\geq 0}$ and $(W_t)_{t\geq 0}$. Using the information of *both* $(B_t)_{t\geq 0}$ and $(W_t)_{t\geq 0}$ one may define τ in a suitable way such that (128) holds true and nevertheless we have that $\mathcal{M}^e(S) \neq \emptyset$. In other words, there are equivalent martingale measures Q for the process S, but Girsanov's theorem fails to produce one.

This example is known for quite some time ([DS 98a]) and served as a kind of "universal counterexample" to several questions arising in Mathematical Finance.

How can one use this example in the present context? Consider the logarithmic utility $U(x) = \ln(x)$ and recall that its conjugate function V equals $V(y) = -\ln(y) - 1$. Hence the dual optimization problem — formally — is given by

$$\begin{aligned} \mathbb{E}\left[V\left(y\tfrac{dQ}{d\mathbf{P}}\right)\right] &= \mathbb{E}\left[-\ln\left(y\tfrac{dQ}{d\mathbf{P}}\right) - 1\right] = \\ &= -\mathbb{E}\left[\ln\left(\tfrac{dQ}{d\mathbf{P}}\right)\right] - (\ln(y)+1) \longmapsto \min!, \quad Q \in \mathcal{M}^a(S). \end{aligned} \tag{129}$$

It is well known (see, e.g., the literature on the "numéraire portfolio" [L 90], [J 96], [A 97] and [B 00]), that for a process $(S_t)_{t\geq 0}$ based, e.g., on the filtration generated by an n-dimensional Brownian motion, the martingale measure obtained from applying Girsanov's theorem (which equals the "minimal martingale measure" investigated by Föllmer and Schweizer [FS 91]) is the minimizer for (129), *provided it exists.*

In the present example we have seen that the candidate for the density of the minimal martingale measure Z_τ obtained from a formal application of Girsanov's theorem fails to have full measure; but nevertheless one may show that Z_τ is the optimizer of the dual problem (125), which shows in particular that we have to pass from $\mathcal{M}^a(S)$ to the larger set D to find the dual optimizer in (129).

Passing again to the general setting of Theorem 3.4 one might ask: how severe is the fact that the dual optimizer $\widehat{Y}_T(1)$ may fail to be the density of a probability measure (or that $\mathbb{E}[\widehat{Y}_T(y)] < y$, for $y > 0$, which amounts to the same thing)? In fact, in many respects it does not bother us at all: we still have the basic duality relation between the primal and the dual optimizer displayed in Theorem 3.4 (ii). Even more is true: using the terminology from [KS 99] the product $(\widehat{X}_t(x)\widehat{Y}_t(y))_{0\leq t\leq T}$, where x and y satisfy $u'(x) = y$, is a uniformly integrable martingale. This fact can be interpreted in the following way: by taking the optimal portfolio $(\widehat{X}_t(x))_{0\leq t\leq T}$ as numéraire instead of the original cash account, the pricing rule obtained from the dual optimizer $\widehat{Y}_T(y)$ then is induced by an equivalent martingale measure. We refer to ([KS 99], p. 912) for a thorough discussion of this argument.

Finally we want to draw the attention of the reader that — comparing item (iii) of Theorem 3.4 to the corresponding item of Theorem 2.3 — we only asserted one pair of formulas for $u'(x)$ and $v'(y)$. The reason is that, in general, the formulae (86) do not hold true any more, the reason again being precisely that for the dual optimizer $\widehat{Y}_T(y)$ we may have $\mathbb{E}[\widehat{Y}_T(y)] < y$. Indeed, the validity of $u'(x) = \mathbb{E}[U'(\widehat{X}_T(x))]$ is tantamount to the validity of $y = \mathbb{E}[\widehat{Y}_T(y)]$.

Case 2: $\mathrm{dom}(U) = \mathbb{R}$

We now pass to the case of a utility function U satisfying Assumption 1.2 case 2 which is defined and finitely valued on all of $\mathbb{R}$. The reader should have in mind the exponential utility $U(x) = -e^{-\gamma x}$, for $\gamma > 0$, as the typical example.

We want to obtain a result analogous to Theorem 3.4 also in this setting. Roughly speaking, we get the same theorem, but the sets C and D considered above have to be chosen in a somewhat different way, as the optimal portfolio $\widehat{X}_T$ now may assume negative values too.

Firstly, we have to assume throughout the rest of this section that the semimartingale S is *locally bounded.* The case of non locally bounded processes is not yet understood and waiting for future research.

Next we turn to the question; what is the proper definition of the set $C(x)$ of terminal values X_T dominated by a random variable $x+(H\cdot S)_T$, where H is an "allowed" trading strategy? On the one hand we cannot be too liberal in the choice of "allowed" trading strategies as we have to exclude doubling strategies and similar schemes. We therefore maintain the definition of the value function $u(x)$ unchanged

$$u(x) = \sup_{H\in\mathcal{H}} \mathbb{E}\left[U\left(x+(H\cdot S)_T\right)\right], \quad x\in\mathbb{R}, \tag{130}$$

where we still confine H to run through the set $\mathcal{H}$ of admissible trading strategies, i.e., such that the process $((H\cdot S)_t)_{0\le t\le T}$ is uniformly bounded from below. This notion makes good sense economically as it describes the strategies possible for an agent having a finite credit line.

On the other hand, in general, we have no chance to find the minimizer $\widehat{H}$ in (130) within the set of admissible strategies: already in the classical cases studied by Merton ([M 69] and [M 71]) the optimal solution $x+(\widehat{H}\cdot S)_T$ to (130) is *not* uniformly bounded from below; this random variable typically assumes low values with very small probability, but its essential infimum typically is minus infinity.

In [S 00] the following approach was used to cope with this difficulty: fix the utility function $U:\mathbb{R}\to\mathbb{R}$ and first define the set $C_U^b(x)$ to consist of all random variables G_T dominated by $x+(H\cdot S)_T$, for some *admissible* trading strategy H and such that $\mathbb{E}[U(G_T)]$ makes sense:

$$C_U^b(x) = \{G_T\in L^0(\Omega,\mathcal{F}_T,\mathbf{P}):\ \text{there is } H \text{ admissible s.t.} \tag{131}$$

$$G_T\le x+(H\cdot S)_T \text{ and } \mathbb{E}[|U(G_T)|]<\infty\}. \tag{132}$$

Next we define $C_U(x)$ as the set of $\mathbb{R}\cup\{+\infty\}$-valued random variables X_T such that $U(X_T)$ can be approximated by $U(G_T)$ in the norm of $L^1(\mathbf{P})$, when G_T runs through $C_U^b(x)$:

$$C_U(x) = \{X_T\in L^0(\Omega,\mathcal{F}_T,\mathbf{P};\mathbb{R}\cup\{+\infty\}) : U(X_T) \text{ is in} \tag{133}$$

$$L^1(\mathbf{P})\text{-closure of } \{U(G_T): G_T\in C_U^b(x)\}\}. \tag{134}$$

The optimization problem (130) now reads

$$u(x) = \sup_{X_T\in C_U(x)} \mathbb{E}[U(X_T)], \quad x\in\mathbb{R}. \tag{135}$$

The set $C_U(x)$ was chosen in such a way that the value functions $u(x)$ defined in (130) and (135) coincide; but now we have much better chances to find the maximizer to (135) in the set $C_U(x)$.

Two features of the definition of $C_U(x)$ merit some comment: firstly, we have allowed $X_T\in C_U(x)$ to attain the value $+\infty$; indeed, in the case when $U(\infty)<\infty$ (e.g., the case of exponential utility), this is natural, as the set $\{U(X_T): X_T\in C_U(x)\}$ should equal the $L^1(\mathbf{P})$-closure of the set $\{U(G_T):$

$G_T \in C_U^b(x)\}$. But we shall see that — under appropriate assumptions — the optimizer $\widehat{X}_T$, which we are going to find in $C_U(x)$, will almost surely be finite.

Secondly, the elements X_T of $C_U(x)$ are only *random variables* and, at this stage, they are not related to a *process* of the form $x + (H \cdot S)$. Of course, we finally want to find for each $X_T \in C_U(x)$, or at least for the optimizer $\widehat{X}_T$, a predictable, S-integrable process H having "allowable" properties (in order to exclude doubling strategies) and such that $X_T \leq x + (H \cdot S)_T$. We shall prove later that — under appropriate assumptions — this is possible and give a precise meaning to the word "allowable".

After having specified the proper domain $C_U(x)$ for the primal optimization problem (135), we now pass to the question of finding the proper domain for the dual optimization problem. Here we find a pleasant surprise: contrary to case 1 above, where we had to pass from the set $\mathcal{M}^a(S)$ to its closed, solid hull D, it turns out that, in the present case 2, the dual optimizer always lies in $\mathcal{M}^a(S)$. This fact was first proved by F. Bellini and M. Frittelli ([BF 00]).

We now can state the main result of [S 00]:

Theorem 3.5. *[S 00, Theorem 2.2] Let the locally bounded semi-martingale $S = (S_t)_{0 \leq t \leq T}$ and the utility function U satisfy Assumptions 1.1, 1.2 case 2 and 1.3; suppose in addition that U has reasonable asymptotic elasticity. Define*

$$u(x) = \sup_{X_T \in C_U(x)} \mathbb{E}[U(X_T)], \qquad v(y) = \inf_{Q \in \mathcal{M}^a(S)} \mathbb{E}\left[V\left(y\tfrac{dQ}{d\mathbf{P}}\right)\right]. \tag{136}$$

Then we have:

(i) The value functions $u(x)$ and $v(y)$ are conjugate; they are continuously differentiable, strictly concave (resp. convex) on $\mathbb{R}$ (resp. on $]0, \infty[$) and satisfy

$$u'(-\infty) = -v'(0) = v'(\infty) = \infty, \qquad u'(\infty) = 0. \tag{137}$$

(ii) The optimizers $\widehat{X}_T(x)$ and $\widehat{Q}(y)$ in (136) exist, are unique and satisfy

$$\widehat{X}_T(x) = I\left(y\frac{d\widehat{Q}(y)}{d\mathbf{P}}\right), \qquad y\frac{d\widehat{Q}(y)}{d\mathbf{P}} = U'(\widehat{X}_T(x)), \tag{138}$$

where $x \in \mathbb{R}$ and $y > 0$ are related via $u'(x) = y$ or equivalently $x = -v'(y)$.

(iii) We have the following relations between u', v' and $\widehat{X}, \widehat{Q}$ respectively:

$$u'(x) = \mathbb{E}_\mathbf{P}[U'(\widehat{X}_T(x))], \qquad v'(y) = \mathbb{E}_{\widehat{Q}}\left[V'\left(y\tfrac{d\widehat{Q}(y)}{d\mathbf{P}}\right)\right] \tag{139}$$

$$xu'(x) = \mathbb{E}_\mathbf{P}[\widehat{X}_T(x)U'(\widehat{X}_T(x))], \quad yv'(y) = \mathbb{E}_\mathbf{P}\left[y\tfrac{d\widehat{Q}(y)}{d\mathbf{P}}V'\left(y\tfrac{d\widehat{Q}(y)}{d\mathbf{P}}\right)\right]. \tag{140}$$

(iv) *If* $\widehat{Q}(y) \in \mathcal{M}^e(S)$ *and* $x = -v'(y)$, *then* $\widehat{X}_T(x)$ *equals the terminal value of a process of the form* $\widehat{X}_t(x) = x + (H \cdot S)_t$, *where* H *is predictable and* S*-integrable, and such that* $\widehat{X}$ *is a uniformly integrable martingale under* $\widehat{Q}(y)$.

We refer to [S 00] for a proof of this theorem and further related results. We cannot go into the technicalities here, but a few comments on the proof of the above theorem are in order: the technique is to reduce case 2 to case 1 by approximating the utility function $U : \mathbb{R} \to \mathbb{R}$ by a sequence $(U^{(n)})_{n=1}^{\infty}$ of utility functions $U^{(n)} : \mathbb{R} \to \mathbb{R} \cup \{-\infty\}$ such that $U^{(n)}$ coincides with U on $[-n, \infty[$ and equals $-\infty$ on $]-\infty, -(n+1)]$. For fixed initial endowment $x \in \mathbb{R}$, we then apply Theorem 3.4 to find for each $U^{(n)}$ the saddle-point $(\widehat{X}_T^{(n)}(x), \widehat{Y}_T^{(n)}(\widehat{y}_n)) \in C_U^b(x) \times D(\widehat{y}_n)$; finally we show that this sequence converges to some $(\widehat{X}_T(x), \widehat{y}\widehat{Q}_T) \in C_U(x) \times \widehat{y}\mathcal{M}^a(S)$, which then is shown to be the saddle-point for the present problem. The details of this construction are rather technical and lengthy (see [S 00]).

We have assumed in item (iv) that $\widehat{Q}(y)$ is equivalent to $\mathbf{P}$ and left open the case when $\widehat{Q}(y)$ is only absolutely continuous to $\mathbf{P}$. F. Bellini and M. Frittelli have observed ([BF 00]) that, in the case $U(\infty) = \infty$ (or, equivalently, $V(0) = \infty$), it follows from (136) that $\widehat{Q}(y)$ is equivalent to $\mathbf{P}$. But there are also other important cases where we can assert that $\widehat{Q}(y)$ is equivalent to $\mathbf{P}$: for example, for of the exponential utility $U(x) = -e^{-\gamma x}$, in which case the dual optimization becomes the problem of finding $\widehat{Q} \in \mathcal{M}^a(S)$ minimizing the relative entropy with respect $\mathbf{P}$, it follows from the work of Csiszar [C 75] (compare also [R 84], [F 00], [GR 00]) that the dual optimizer $\widehat{Q}(y)$ is equivalent to $\mathbf{P}$, provided only that there is at least one $Q \in \mathcal{M}^e(S)$ with finite relative entropy.

Under the condition $\widehat{Q}(y) \in \mathcal{M}^e(S)$, item (iv) tells us that the optimizer $\widehat{X}_T \in C_U(x)$ is almost surely finite and equals the terminal value of a process $x + (H \cdot S)$, which is a uniformly integrable martingale under $\widehat{Q}(y)$; this property qualifies H to be a "allowable", as it certainly excludes doubling strategies and related schemes. One may turn the point of view around and take this as the *definition* of the "allowable" trading strategies; this was done in [DGRSSS 00] for the case of exponential utility, where this approach is thoroughly studied and some other definitions of "allowable" trading strategies, over which the primal problem may be optimized, are also investigated. Further results on these lines were obtained in [KaS 00] for the case of exponential utility, and in [S 00b] for general utility functions.

We finish this survey with a brief account on the recent literature related to maximizing expected utility in financial markets. There are many aspects going beyond the basic problem surveyed above. We can only give a very brief indication on the many interesting papers and hope to have provided the reader with some introductory motivation to study this literature.

G. Zitkovic [Z 00] has analyzed the problem of optimizing expected utility of consumption during the time interval $[0,T]$. He obtained a similar result as Theorem 3.4 above, provided the utility functions $U_{t,\omega}$, which in this setting may depend on $t \in [0,T]$ and $\omega \in \Omega$ in an $\mathcal{F}_t$-measurable way, satisfy the reasonable elasticity condition in a uniform way.

Results related to the duality theory of utility maximization and notably to the dual optimizer $\widehat{Q} \in \mathcal{M}^e(S)$ were obtained in [F 00], [K 00], [XY 00], [GK 00], [GR 00] and [BF 00].

Utility maximization under transaction costs was investigated, e.g., in [HN 89], [CK 96], [CW 00] and [DPT 00]; in the latter two papers the phenomenon arising in Theorem 3.4 is of crucial importance: for the dual optimizer one has to perform a similar enlargement as the passage from $\mathcal{M}^a(S)$ to D encountered in Theorem 3.4 above.

The theme of random endowment, which is intimately related to the concept of utility based hedging of contingent claims is treated in [KJ 98], [KR 00], [CSW 00], [D 00], [JS 00], [CH 00], and in the context of minimizing expected shortfall, which leads to non-smooth utility functions, in [C 00] and [FL 00]. Nonsmooth utility functions also come up in a natural way in [DPT 00] and in [L 00].

References

[A 97] P. Artzner, (1997), *On the numeraire portfolio.* Mathematics of Derivative Securities, M. Dempster and S. Pliska, eds., Cambridge University Press, pp. 53–60.

[AS 94] J.P. Ansel, C. Stricker, (1994), *Couverture des actifs contingents et prix maximum.* Ann. Inst. Henri Poincaré, Vol. 30, pp. 303–315.

[B 00] D. Becherer, (2000), *The numeraire portfolio for unbounded semimartingales.* preprint, TU Berlin.

[BF 00] F. Bellini, M. Frittelli, (2000), *On the existence of minimax martingale measures.* preprint.

[BS 99] W. Brannath, W. Schachermayer, (1999), *A Bipolar Theorem for Subsets of $L^0_+(\Omega,\mathcal{F},P)$.* Séminaire de Probabilités, Vol. XXXIII, pp. 349–354.

[C 75] I. Csiszar, (1975), *I-Divergence Geometry of Probability Distributions and Minimization Problems.* Annals of Probability, Vol. 3, No. 1, pp. 146–158.

[C 00] J. Cvitanic, (2000), *Minimizing expected loss of hedging in incomplete and constrained markets.* Preprint Columbia University, New York.

[CH 00] P. Collin-Dufresne, J.-N. Huggonnier, (2000), *Utility-based pricing of contingent claims subject to counterparty credit risk.* Working paper GSIA & Department of Mathematics, Carnegie Mellon University.

[CH 89] J.C. Cox, C.F. Huang, (1989), *Optimal consumption and portfolio policies when asset prices follow a diffusion process.* J. Economic Theory, Vol. 49, pp. 33–83.

[CH 91] J.C. Cox, C.F. Huang, (1991), *A variational problem arising in financial economics.* J. Math. Econ., Vol. 20, pp. 465–487.

[CK 96] J. Cvitanic, I. Karatzas, (1996), *Hedging and portfolio optimization under transaction costs: A martingale approach.*, Mathematical Finance 6, pp. 133–165.

[CW 00] J. Cvitanic, H. Wang, (2000), *On optimal terminal wealth under transaction costs.* Preprint.

[CSW 00] J. Cvitanic, H. Wang, W. Schachermayer, (2000), *Utility Maximization in Incomplete Markets with Random Endowment.* preprint (12 pages), to appear in Finance and Stochastics.

[D 97] M. Davis, (1997), *Option pricing in incomplete markets.* Mathematics of Derivative Securities, eds. M.A.H. Dempster and S.R. Pliska, Cambridge University Press, pp. 216–226.

[D 00] M. Davis, (2000), *Optimal hedging with basis risk.* Preprint of the TU Vienna.

[DGRSSS 00] F. Delbaen, P. Grandits, T. Rheinländer, D. Samperi, M. Schweizer, C. Stricker, (2000), *Exponential hedging and entropic penalties.* preprint.

[DPT 00] G. Deelstra, H. Pham, N. Touzi, (2000), *Dual formulation of the utility maximisation problem under transaction costs.* Preprint of ENSAE and CREST.

[DS 94] F. Delbaen, W. Schachermayer, (1994), *A General Version of the Fundamental Theorem of Asset Pricing.* Math. Annalen, Vol. 300, pp. 463–520.

[DS 95] F. Delbaen, W. Schachermayer, (1995), *The No-Arbitrage Property under a change of numéraire.* Stochastics and Stochastic Reports, Vol. 53, pp. 213–226.

[DS 98b] F. Delbaen, W. Schachermayer, (1998), *A Simple Counter-example to Several Problems in the Theory of Asset Pricing, which arises in many incomplete markets.* Mathematical Finance, Vol. 8, pp. 1–12.

[DS 98a] F. Delbaen, W. Schachermayer, (1998), *The Fundamental Theorem of Asset Pricing for Unbounded Stochastic Processes.* Mathematische Annalen, Vol. 312, pp. 215–250.

[DS 99] F. Delbaen, W. Schachermayer, (1999), *A Compactness Principle for Bounded Sequences of Martingales with Applications.* Proceedings of the Seminar of Stochastic Analysis, Random Fields and Applications, Progress in Probability, Vol. 45, pp. 137–173.

[E 80] M. Emery, (1980), *Compensation de processus à variation finie non localement intégrables.* Séminaire de Probabilités XIV, Springer Lecture Notes in Mathematics, Vol. 784, pp. 152–160.

[ET 76] I. Ekeland, R. Temam, (1976), *Convex Analysis and Variational Problems.* North Holland.

[F 00] M. Frittelli, (2000), *The minimal entropy martingale measure and the valuation problem in incomplete markets.* Mathematical Finance, Vol. 10, pp. 39–52.

[F 90] L.P. Foldes, (1990), *Conditions for optimality in the infinite-horizon portfolio-cum-savings problem with semimartingale investments.* Stochastics and Stochastics Report, Vol. 29, pp. 133–171.

[FL 00] H. Föllmer, P. Leukert, (2000), *Efficient Hedging: Cost versus Shortfall Risk.* Finance and Stochastics, Vol. 4, No. 2, pp. 117–146.

[FS 91] H. Föllmer, M. Schweizer, (1991), *Hedging of contingent claims under incomplete information.* Applied Stochastic Analysis, Stochastic Monographs, M.H.A. Davis and R.J. Elliott, eds., Gordon and Breach, London New York, Vol. 5, pp. 389–414.

[GK 00] T. Goll, J. Kallsen, (2000), *Optimal portfolios for logarithmic utility.* Stochastic Processes and Their Applications, Vol. 89, pp. 31–48.

[GR 00] T. Goll, L. Rüschendorf, (2000), *Minimax and minimal distance martingale measures and their relationship to portfolio optimization.* Preprint of the Universität Freiburg, Germany.

[HL88] C.-F. Huang, R.H. Litzenberger, (1988), *Foundations for Fiancial Economics.* North-Holland Publishing Co. New York.

[HN 89] S.D. Hodges, A. Neuberger, (1989), *Optimal replication of contingent claims under transaction costs.* Review of Futures Markets, Vol. 8, pp. 222–239.

[HP 81] J.M. Harrison, S.R. Pliska, (1981), *Martingales and Stochastic intefrals in the theory of continuous trading.* Stoch. Proc. & Appl., Vol. 11, pp. 215–260.

[HP 91] H. He, N.D. Pearson, (1991), *Consumption and Portfolio Policies with Incomplete Markets and Short-Sale Constraints: The Finite-Dimensional Case.* Mathematical Finance, Vol. 1, pp. 1–10.

[HP 91a] H. He, N.D. Pearson, (1991), *Consumption and Portfolio Policies with Incomplete Markets and Short-Sale Constraints: The Infinite-Dimensional Case.* Journal of Economic Theory, Vol. 54, pp. 239–250.

[J 92] S.D. Jacka, (1992), *A martingale representation result and an application to incomplete financial markets.* Mathematical Finance, Vol. 2, pp. 239–250.

[J 96] B.E. Johnson, (1996), *The pricing property of the optimal growth portfolio: extensions and applications.* preprint, Department of Engineering-Economic-Systems, Stanford University.

[JS 00] M. Jonsson, K.R. Sircar, (2000), *Partial hedging in a stochastic volatility environment.* Preprint of the Princeton University, Dept. of Operation Research and Financial Engineering.

[K 00] J. Kallsen, (2000), *Optimal portfolios for exponential Lévy processes* Mathematical Methods of Operation Research, Vol. 51, No. 3, pp. 357–374.

[KaS 00] Yu.M. Kabanov, C. Stricker, (2000), *On the optimal portfolio for the exponential utility maximization: remarks to the six-author paper.* Preprint.

[KJ 98] N. El Karoui, M. Jeanblanc, (1998), *Optimization of consumptions with labor income.* Finance and Stochastics, Vol. 4, pp. 409–440.

[KLS 87] I. Karatzas, J.P. Lehoczky, S.E. Shreve, (1987), *Optimal portfolio and consumption decisions for a "small investo" on a finite horizon.* SIAM Journal of Control and Optimization, Vol. 25, pp. 1557–1586.

[KLSX 91] I. Karatzas, J.P. Lehoczky, S.E. Shreve, G.L. Xu, (1991), *Martingale and duality methods for utility maximization in an incomplete market.* SIAM Journal of Control and Optimization, Vol. 29, pp. 702–730.

[KQ 95] N. El Karoui, M.-C. Quenez, (1995), *Dynamic programming and pricing of contingent claims in an incomplete market.* SIAM J. Control Optim., Vol. 33, pp. 29–66.

[KR 00] N. El Karoui, R. Rouge, (2000), *Pricing via utility maximization and entropy.* Preprint, to appear in Mathematical Finance.

[Kom 67] J. Komlos, (1967), *A generalization of a problem of Steinhaus.* Acta Math. Sci. Hung., Vol. 18, pp. 217–229.

[KS 99] D. Kramkov, W. Schachermayer, (1999), *The Asymptotic Elasticity of Utility Functions and Optimal Investment in Incomplete Markets.* Annals of Applied Probability, Vol. 9, No. 3, pp. 904–950.

[L 00] P. Lakner, (2000), *Portfolio Optimization with an Insurance Constraint.* Preprint of the NYU, Dept. of Statistics and Operation Research.

[L 90] J.B. Long, (1990), *The numeraire portfolio.* Journal of Financial Economics, Vol. 26, pp. 29–69.

[M 69] R.C. Merton, (1969), *Lifetime portfolio selection under uncertainty: the continuous-time model.* Rev. Econom. Statist., Vol. 51, pp. 247–257.

[M 71] R.C. Merton, (1971), *Optimum consumption and portfolio rules in a continuous-time model.* J. Econom. Theory, Vol. 3, pp. 373–413.

[M 90] R.C. Merton, (1990), *Continuous-Time Finance.* Basil Blackwell, Oxford.

[P 86] S.R. Pliska, (1986), *A stochastic calculus model of continuous trading: optimal portfolios.* Math. Oper. Res., Vol. 11, pp. 371–382.

[R 70] R.T. Rockafellar, (1970), *Convex Analysis.* Princeton University Press, Princeton, New Jersey.

[R 84] L. Rüschendorf, (1984), *On the minimum discrimination information theorem.* Statistics & Decisions Supplement Issue, Vol. 1, pp. 263–283.

[S 69] P.A. Samuelson, (1969), *Lifetime portfolio selection by dynamic stochastic programming.* Rev. Econom. Statist., Vol. 51, pp. 239–246.

[S 00] W. Schachermayer, (1999), *Optimal Investment in Incomplete Markets when Wealth may Become Negative.* Preprint (45 pages), to appear in Annals of Applied Probability.

[S 00a] W. Schachermayer, (2000), *Introduction to the Mathematics of Financial Markets.* Notes on the St. Flour summer school 2000, preprint, to appear in Springer Lecture Notes.

[S 00b] W. Schachermayer, (2000), *How Potential Investments may Change the Optimal Portfolio for the Exponential Utility.* preprint.

[Sch 66] H.H. Schäfer, (1966), *Topological Vector Spaces.* Graduate Texts in Mathematics.

[St 85] H. Strasser, (1985), *Mathematical theory of statistics: statistical experiments and asymptotic decision theory.* De Gruyter studies in mathematics, Vol. 7.

[XY 00] J. Xia, J. Yan, (2000), *Martingale measure method for expected utility maximisation and valuation in incomplete markets.* Preprint.

[Z 00] G. Zitkovic, (2000), *Maximization of utility of consumption in incomplete semimartingale markets.* In preparation.

[Z 00a] G. Zitkovic, (2000), *A filtered version of the Bipolar Theorem of Brannath and Schachermayer.* Preprint of the Dept. of Statistics, Columbia University, NY.

Evaluating Investments in Disruptive Technologies

Eduardo S. Schwartz[1] and Carlos Zozaya-Gorostiza[2]

[1] University of California, Los Angeles, Anderson Graduate School of Management, Los Angeles, CA. 90095-1481,
eduardo.schwartz@anderson.ucla.edu

[2] Instituto Tecnológico Autónomo de México,
Río Hondo #1, México D.F. 01000, México,
zozaya@itam.mx

Abstract. A computer simulation model for the valuation of investments in disruptive technologies is developed. Based on the conceptual framework proposed by Christensen (1997) for explaining the *Innovator's Dilemma* phenomenon, an investment project is divided into two sequential phases representing the evolution of the disruptive technology from an emerging to a mainstream market. In each of these phases, development costs and net commercialization cash flows are modeled using various stochastic processes that interact with each other. As a result, the initial estimate on the value of the project is continuously updated to reflect the stochastic changes of these variables. An example illustrates the usefulness of the model for understanding the effects of cash flow and cost volatilities in the value of a disruptive technology investment.

1 Introduction

Innovation, defined as the creation of something new or as the adoption of an idea or behavior that is new for the organization that adopts it, has become the fundamental process to promote the growth of an organization. After decades of cost reductions and down-sizing, organizations have understood that they have to innovate their administrative processes, the products and services they deliver, and the technology they use in order to obtain a differentiation that allows them to compete successfully (Tushman and O'Reilly 1997). Jonash and Sommerlatte (1999) have found strong evidence of an innovation premium: the top 20% companies of *Fortune's* ratings on innovation enjoy double the shareholder returns of the other companies in their industries.

In spite of its importance, most executives have serious doubts about how they should invest in a process that traditionally has been full of uncertainty and high costs. Emerging technologies do not fit well with the traditional project evaluation methods used in most organizations and projects involving these technologies are often ignored or considered a second priority. The situation is even more dramatic in the case of *disruptive technologies* whose performance is worse than that of existing technologies according to prevailing value networks (Christensen and Rosenbloom 1995). Consider, for instance, what happened when early PCs first appeared in the market. As pointed out by Christensen and Overdof (2000), early PCs were not powerful enough to run the computing applications of existing mainframes and minicomputers. PCs had other attributes, such as their low cost, that enabled new applications for the personal user to emerge; however, in their earlier stages, PCs did not address the next-generation needs of the leading customers of mainframes and minicomputers. Therefore, leading manufacturers of minicomputers such as Digital Equipment Corporation gave higher priority to projects related to enhancing existing technologies than to those involving the development of the emerging disruptive technology.

Several frameworks have been proposed for the valuation of investment projects under uncertainty including the use of real options (Amram and Kulatilaka 1999, Dixit and Pindyck 1994, Schwartz and Zozaya-Gorostiza 2000), decision trees (Tipping, Zeffren 1995), scenario-analysis (Jovanovic 1999) and Monte Carlo simulation (Boer 1999). Each of these frameworks provides the decision maker with additional elements for evaluating an investment decision in comparison with traditional valuation methods. Real options, for instance, allow to value the flexibility of an investment project in terms of postponing the decision to invest, suspend (or resume) the execution of a project or abandon it completely (Brennan and Schwartz 1985), as well as to model the growth opportunity acquired when investing in a new technology (Schwartz and Moon 2000a).

Computer simulation, and its application in the analysis of multiple scenarios, has proven to be useful in modeling complex phenomena for which analytical solutions are difficult to obtain. Schwartz and Moon (2000b) developed a simulation model for the valuation of Internet companies and found that the pricing of Internet stocks may be rational given high enough growth rates and volatilities in the corresponding cash flows. Bers, Lynn and Spurling (1999) note that, despite the power of scenario analysis as a planning tool, there have been few published reports of its successful application to business planning. They developed a Monte Carlo simulation model designed to assist technical professionals in developing business plans for emerging technologies in emerging markets under different scenarios.

In this paper we develop a computer simulation model for the valuation of investments in disruptive technologies. Based on the conceptual framework proposed by Christensen (1997) for explaining the *Innovator's Dilemma* phenomenon, an investment project is divided into two sequential phases representing the evolution of the disruptive technology from an emerging to a mainstream market. In each of these phases, development costs and net commercialization cash flows are modeled using various stochastic processes that interact with each other. As a result, the initial estimate on the value of the project is continuously updated to reflect the stochastic changes of these variables.

The next section describes some characteristics of disruptive technologies as a background for subsequent discussion. Section 3 describes Christensen's conceptual model of the evolution of disruptive technologies over time. Based on this conceptualization, Section 4 describes the proposed valuation model for investments in disruptive technologies. First, the stochastic processes associated with the evolution of development costs and commercialization cash flows are presented. Then, the manner in which these processes are used to compute the expected value of the project is discussed. Section 5 provides the reader with a pseudo-code of the overall simulation process. Section 6 illustrates the application of our model for the valuation of a hypothetical project involving the development of a disruptive optical technology for data storage. Finally, Section 7 discusses some possible extensions to the proposed model and provides some conclusions of our work.

2 Disruptive Technologies and Change

In his book entitled "The Innovator's Dilemma", Chistensen (1997) explains why leading companies fail when they are confronted with certain types of market and technological changes:

- There is a strategically important distinction between a *sustaining* and a *disruptive* technology. *Sustaining* technologies are new technologies that foster improved product performance. They can be discontinuous radical or incremental. In contrast *disruptive* technologies result in *worse* product performance, at least in the near term.
- When faced with disruptive technologies, good management practices (planning better, working harder, becoming customer driven and taking a long-time perspective) do not work and may lead to failure. Even though most disruptive technologies are originated in the leading firms, they are often neglected and their creators have to continue their development in a new firm.
- The pace of technological progress can outstrip what markets need. Leading firms are generally aggressive and innovative in sustaining technologies but they were unable to confront a *downward* vision. They often "overshoot" their mainstream markets by giving their customers more than what they need or are willing to pay for, but they fail to look at new technologies that might bring a new value proposition provided by disruptive technologies. As a result, when disruptive technologies are able to satisfy the minimum performance required by the low end of the market, the new value proposition comes into play and incumbent firms find that they are unable to react successfully.
- Customers and financial structures of successful companies affect the sorts of investments that appear to be attractive to them, relative to certain types of entering firms. Leading companies often invest aggressively in technological innovation. However, they often disregard projects involving disrupting technologies because they are less attractive than those

of sustaining technologies. Investing aggressively in disruptive technologies is not a rational financial decision according to traditional project evaluation techniques because it promises lower margins (i.e., products are generally simpler and cheaper) in emerging and insignificant markets. Furthermore, the most profitable customers do not want the disruptive technology products.

3 Conceptual Model

In order to develop a formal model for the valuation of investments in disruptive technologies, it is necessary to understand the decision making traps in which a successful organization falls as well as the trajectories followed by established and disruptive technologies. Christensen summarizes the pattern of decisions regarding disruptive technological change as follows (Christensen 1997, p. 43):

a) Disruptive technologies are first developed within established firms. These innovations are the result of many factors including creativity, serendipity, and entrepreneurial spirit of the technical personnel in the leading firms. Instead of being sponsored by senior management, these initiatives generally use bootleg resources.
b) Marketing personnel tests the reactions from their lead customers. The most profitable customers are unwilling or unable to benefit from the disruptive technology new capabilities because they are concerned with the current performance measures of the established technologies.
c) Established firms step up the pace of sustaining technological development. In order to satisfy (and exceed the expectations of) their customers, firms develop sustaining technologies even at a faster rate than the corresponding market needs.
d) New companies are formed, and markets for disruptive technologies are found by trial and error. The lack of attention and support to disruptive technologies causes some of the technical personnel to leave and create new companies to commercialize the disruptive technologies.
e) The entrants move up and enter the mainstream market. The disruptive technologies normally improve at very fast rates and get to a point in which their performance is enough to satisfy the needs of the mainstream market with respect to the current measures of performance. In addition, the new technologies provide better value along other performance dimensions so customers are willing to switch.
f) Established firms belatedly jump on the bandwagon to defend their customer base. However, their organizational structures and processes difficult them to react effectively.

Figure 1 shows the intersecting trajectories of the performance demanded by mainstream and emerging markets and that supplied by current and emerging technologies. As pointed out by Christensen, what matters is not whether the established technology is improving at increasing or decreasing rates, but whether the disruptive technology is improving from below along a trajectory that will ultimately intersect with what the market needs ($t = \tau$).

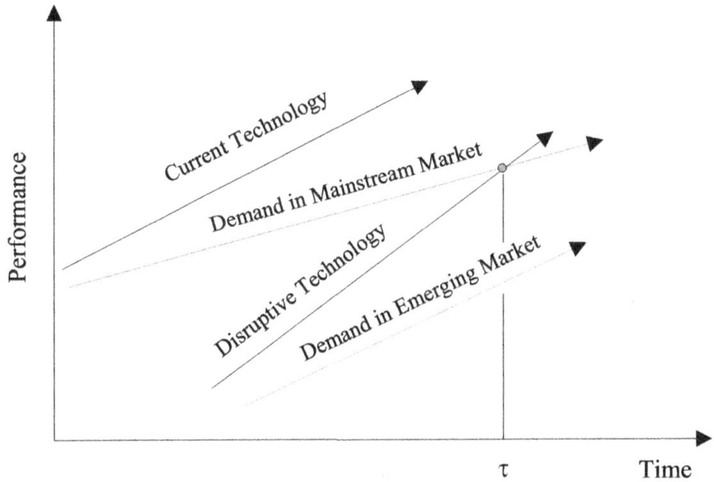

Fig. 1. Intersecting Trajectories of Performance Demanded and Supplied by Technologies (adapted from (Christensen 1997))

4 Valuation Model

In this section we explain the elements used in our model for the valuation of investments in disruptive technologies. Following the conceptual description presented in the previous section, our model assumes that a firm that invests in disruptive technologies goes through two sequential phases (see Figure 2):

- *Emerging Market Phase.* This phase has two stages: one in which the technology is being developed and tested in the emerging market, and a second stage in which it is commercialized only in this market. In this second stage, the organization continues to invest in the technology to enhance its performance according to several attributes including those valued by the mainstream market. The costs and times for developing the technology for any of the markets as well as the cash flows that will be received when the technology is ready for commercialization are uncertain.
- *Mainstream Market Phase.* Once the technology has achieved an acceptable performance level for the mainstream market, the organization moves up and starts competing for the most profitable customers.

At the beginning of the project, the organization will have an initial estimate of the development costs and cash flows associated with each phase, but these values will be updated periodically as time moves forward. Furthermore, there will be a probability of failure in each of the phases to represent the cases in which the development and/or commercialization of the technology are unsuccessful and the project is permanently abandoned.

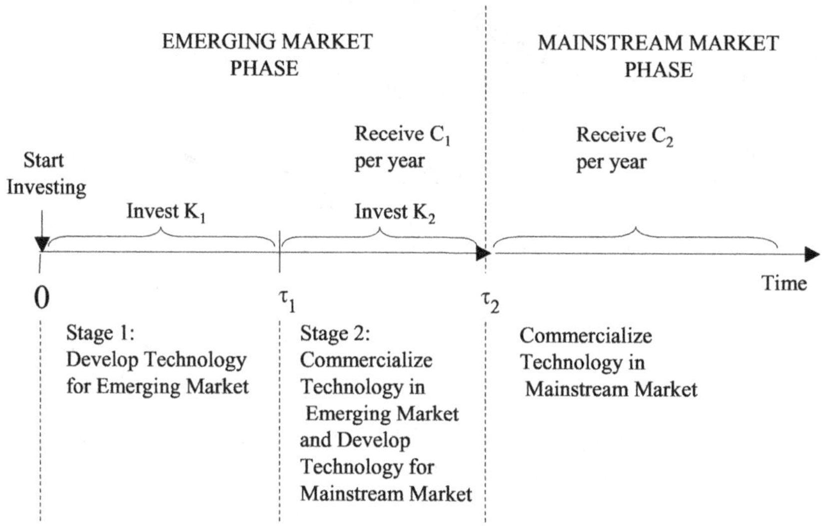

Fig. 2. Phases of a Disruptive Technology Project

As an initial approximation, our valuation model accounts for uncertainty in development costs and cash flow rates using three stochastic processes associated, respectively, with the development costs of Stages 1 and 2 of the Emerging Phase and with the evolution of net commercialization cash flows, as explained below.

4.1 Development Costs

In Pindyck's model (1993) for investment under uncertainty, the expected cost of completion of a project $K(t)$ is assumed to follow a controlled diffusion process given by the following expression:

$$dK = -I dt + g(I, K) dz \tag{1}$$

where I is the rate of investment, dz is an increment to a Wiener process that might or might not be correlated with the economy and the stock market, and g is a function such that $g_I \geq 0$, $g_{II} \leq 0$, and $g_K \geq 0$. According to this model, the expected cost to completion declines with ongoing investment but also changes stochastically. If $g_I > 0$, the effect of the stochastic component increases as we invest more in the project; if $g_{II} < 0$, the changes in these increasing effects will be smaller as we proceed forward; finally, if $g_K > 0$, the effect of the stochastic component decreases when the expected cost of completion is reduced. Pindyck also shows that letting $g(I, K) = \beta(IK)^{1/2}$ provides a formulation that complies with the following conditions that make economic sense: a) an increase in the expected cost of an investment reduces its value; b) the instantaneous variance of $\mathrm{d}K$ is bounded for all finite K and approaches to zero as $K \to 0$; and c) if the firms invests at a maximum rate I_m until the project is complete, K is indeed the expected cost to completion.

Following Pindyck's reasoning, our model assumes that the completion costs of developing a technology evolve according to the following stochastic process:

$$\mathrm{d}K = -I\mathrm{d}t + \beta(IK)^{1/2}\mathrm{d}z \tag{2}$$

where $\mathrm{d}z$ is an increment to a Wiener process that is uncorrelated with the economy, β is the instant standard deviation of the proportional changes in K (i.e., the volatility of the costs) and I is the corresponding investment rate.

In our model an organization is assumed to keep investing in a technology until its performance is adequate for the target market (i.e., the technology "catches up" with the market) unless a disaster occurs and the development project is aborted. Also, completion costs include both the costs required for improving the disruptive technology and the launching costs required prior to the commercialization of the technology. Examples of launching costs are those associated with advertising the technology as well as with the development of the appropriate distribution channels.

After a technology is ready to be commercialized (i.e., completion costs of initial development are zero), the organization incurs in sales, operating and other types of technology enhancement costs. In our model, however, these post-launch costs are not modeled separately but considered within the *net* cash flows of the commercialization process.

For each simulation run, Equation (2) is used to compute the development cost for Stages 1 and 2 of the *Emerging Phase* as follows:

- *Stage 1: Development of Technology for the Emerging Market.* First, the new completion costs for Stage 1 of the Emerging Phase is obtained from the completion cost of the previous time step using the following expression:

$$K_{1,t+1} = K_{1,t} - I_1\Delta t + \beta(I_1 K_{1,t})^{1/2}\sqrt{\Delta t}\varepsilon_1 \ , \tag{3}$$

  where Δt is time interval from t to $t + 1$ and ε_1 is a $N(0, 1)$ standard normal. The total cost for Stage 1 is updated by adding the corresponding investment over time:

$$TotCost_{1,t+1} = TotCost_{1,t} + I_1\Delta t \ . \tag{4}$$

At the end of Stage 1, this value is then used to obtain an estimate of the expected completion costs of Stage 2 of the Emerging Phase:

$$K_{2,\tau_1} = TotCost_{1,\tau_1} * CostRatio \,, \tag{5}$$

where *CostRatio* is an input parameter representing the ratio of total development costs for the Emerging and Mainstream markets.

- *Stage 2: Development of Technology for the Mainstream Market.* Once Stage 2 has been started, the completion cost for this stage is updated using Eq. 2 as follows:

$$K_{2,t+1} = K_{2,t} - I_2 \Delta t + \beta (I_2 K_{2,t})^{1/2} \sqrt{\Delta t} \varepsilon_1 \,. \tag{6}$$

In our model, no risk premium for costs is considered[1]. Therefore, the development costs in any of the stages can be discounted using the risk-free rate without further adjustment for risk.

4.2 Commercialization Cash Flows

The net cash flow rates received from commercializing a technology in a particular market are assumed to behave according to the following expression:

$$\mathrm{d}C = \alpha C \mathrm{d}t + \phi C \mathrm{d}x \,, \tag{7}$$

where $\mathrm{d}x$ is an increment to a Wiener process that is uncorrelated with the economy but that might be correlated with the costs of developing the technology, and ϕ is the instantaneous standard deviation of the proportional changes in C (i.e., the volatility of the cash flows). The term $(\alpha C \mathrm{d}t)$ describes the expected change in cash flows over time. A positive α might be used for modeling situations in which cash flows increase during the commercialization of the disruptive technology due, for instance, to a major penetration in the market. In contrast, a negative α might be used for situations in which cash flows decrease due to a reduction in prices triggered by more intense competition.

Our model considers that there is a risk-premium associated with the cash flows. Therefore, the true process of cash flows described by Eq. (7) is transformed into the following risk-adjusted process:

$$\mathrm{d}C = (\alpha - \eta_C) C \mathrm{d}t + \phi C \mathrm{d}x \tag{8}$$

where η_C represents the risk-premium for cash flow values. Once this adjustment has been made, risk-adjusted cash flows can be discounted using the risk-free rate for valuating the project under a risk-neutral scenario.

The model uses one stochastic process for representing the evolution of the net cash flows of the emerging market phase. This evolution is used to

[1] That is, we assume that development costs are uncorrelated with aggregate wealth.

obtain an estimate of the expected net cash flows of the mainstream phase. Once the first phase has been completed and the technology is ready for being commercialized in the mainstream market, the expected cash flow of the mainstream phase is obtained as a multiple of the final cash flow from the first phase.

We allow the stochastic changes in the net cash flows to be correlated with the stochastic changes in the development costs:

$$\mathrm{d}x\mathrm{d}z = \rho\mathrm{d}t \tag{9}$$

A positive ρ could represent, for instance, that higher development costs will lead to higher commercialization benefits because the customers are willing to pay more for a better product. In contrast, a negative ρ could represent, for instance, that the inability to control the costs of developing the technology are associated with an inability to commercialize the technology in an adequate manner.

For each time step of the simulation run, Eq. (8) is used to update the cash flows received during the commercialization of the disruptive technology in the emerging market as follows:

$$C_{1,t+1} = C_{1,t} * \mathrm{e}^{[(\alpha_1 - \eta_C - \frac{1}{2}\phi^2)\Delta t + \phi\sqrt{\Delta t}\varepsilon_2]}\,, \tag{10}$$

where Δt is time interval from t to $t+1$, α_1 is the drift of cash flows in the emerging market, η_C is the risk-premium for cash flow values, ϕ is the volatility of cash flows and ε_2 is a $N(0,1)$ standard normal which has a correlation of ρ with ε_1. During Stage 1, $C_{1,t}$ represents the cash flow that the organization *expects* to receive once this stage is completed. During Stage 2 of the Emerging Phase, $C_{1,t}$ represents the *actual* cash flows that the organization is receiving when the technology is being further developed for the mainstream market.

At the end of the Emerging Phase, the initial cash flow of the Mainstream Phase is computed as a multiple of the final cash flow of the first phase:

$$C_{2,\tau_2} = C_{1,\tau_1} * \mathit{CashRatio}\,, \tag{11}$$

where *CashRatio* is an input parameter of the simulation representing the increase in cash flows that the organization would get if it is able to commercialize the disruptive technology in the mainstream market.

Finally, the following equation is used to update the cash flows received during the commercialization of the disruptive technology in the mainstream market:

$$C_{2,t+1} = C_{2,t} * \mathrm{e}^{[(\alpha_2 - \eta_C - \frac{1}{2}\phi^2)\Delta t + \phi\sqrt{\Delta t}\varepsilon_2]}\,, \tag{12}$$

where Δt is time interval from t to $t+1$, α_2 is the drift of cash flows in the mainstream market, η_C is the risk-premium for cash flow values, ϕ is the volatility of cash flows and ε_2 is a $N(0,1)$ standard normal.

4.3 Value of the Project

The *Value* of the investment project (V) is computed by subtracting the present value of the development costs ($PVCost$) from the present value of the future cash flows ($PVCash$). This assumes that the organization will invest in the development of the disruptive technology immediately if the value of the project is positive:

$$V = PVCash - PVCost\,. \tag{13}$$

In the next section we consider the possibility of abandoning the project once it has been started; however, in this section we assume that investment will proceed without interruptions unless a disaster occurs and the project is permanently interrupted.

For the purpose of comparisons, we will compute the value of the project for two scenarios: one in which only the emerging phase is considered (V_e) and another in which both phases are included. (V_T). This will allow us to measure the disruptive impact of the technology in the mainstream market. As we will see later on, there are many projects which are unprofitable if only the emerging phase is considered, but that are very profitable when both phases are included.

The expected present value of the development costs for entering the emerging market is obtained by integrating Eq. (2) during the interval from $t = 0$ to $t = \tau_1$[2]:

$$PVCost_{\text{Stage 1}} = E_o\left[\int_0^{\tau_1} I_1 \mathrm{e}^{-(r_f+\lambda_1)t}\mathrm{d}t\right] \tag{14}$$

where I_1 is the rate of investment during Stage 1 of the Emerging Phase, r_f is the risk-free rate, τ_1 is a random variable representing the end of Stage 1, and λ_1 is a constant rate of failure for this stage assuming that failures follow a Poisson process. Similarly, the expected present value of the development costs for entering the mainstream market is obtained by integrating Eq. (2) during the interval from $t = \tau_1$ to $t = \tau_2$ and by discounting the result from $t = \tau_1$ to time zero:

$$PVCost_{\text{Stage 2}} = E_o\left[\left(\int_{\tau_1}^{\tau_2} I_2 \mathrm{e}^{-(r_f+\lambda_2)(t-\tau_1)}\mathrm{d}t\right) * \mathrm{e}^{-(r_f+\lambda_1)\tau_1}\right], \tag{15}$$

where I_2 is the rate of investment during Stage 2 of the Emerging Phase, r_f is the risk-free rate, τ_2 is a random variable representing the end of Stage 2, τ_1 is a random variable representing the end of Stage 1, and λ_1 and λ_2 are

[2] As explained in Schwartz and Moon (2000a), if failures of a particular project follow a Poisson distribution with mean rate of failure per unit time λ, the cash flows of the project should be discounted at a rate equal to the risk-free rate plus λ when computing their present value under a risk-neutral scenario.

respectively the constant rates of failure for the two developmental stages of the Emerging Phase.

The present value of the future cash flows from commercialization of the disruptive technology in the emerging market is obtained by integrating Eq. (8) from the time in which the technology is ready to be commercialized in this market ($t = \tau_1$) to the time in which this technology has reached the end of its useful life for this market ($t = T_1$) and by discounting the result from $t = \tau_1$ to time zero:

$$PVCash_{\text{Emerging}} = E_o\left(\left[\int_{\tau_1}^{T_1} C_{1,t}\mathrm{e}^{-(r_f-\alpha_1^*)(t-\tau_1)}\mathrm{d}t\right]\mathrm{e}^{-(r_f+\lambda_1)\tau_1}\right), \quad (16)$$

where α_1^* is the risk-adjusted growth rate of the net cash flows obtained by subtracting a risk-premium due to cash flow uncertainty, η_c, from the rate of cash flow growth α_1 in the emerging market.

Similarly, the expected present value of future cash flows from continuing the development and commercialization of the disruptive technology into the mainstream market is obtained by discounting the cash flows of Eq. (8) for the second stage of development and adding the result to the present value of the commercialization cash flows from the time in which the technology is ready to be commercialized in this market ($t = \tau_2$) to the time in which the technology has reached the end of its useful life for this market ($t = T_2$):

$$PVCash_{\text{Mainstream}} = E_o\left[\begin{array}{c}\left(\int_{\tau_1}^{\tau_2} C_{1,t}\mathrm{e}^{\alpha_1^*(t-\tau_1)}\mathrm{e}^{-(r_f+\lambda_2)(t-\tau_1)}\mathrm{d}t\right)\mathrm{e}^{-(r_f+\lambda_1)\tau_1}\\ +\left(\int_{\tau_2}^{T_2} C_{2,t}\mathrm{e}^{\alpha_2^*(t-\tau_2)}\mathrm{e}^{-r_f(t-\tau_2)}\mathrm{d}t\right)\mathrm{e}^{-(r_f+\lambda_2)(\tau_2-\tau_1)}\\ \times\mathrm{e}^{-(r_f+\lambda_1)\tau_1}\end{array}\right]. \quad (17)$$

Equations (14) through (17) cannot be solved analytically because they involve random variables τ_2 and τ_1 in the integration limits. Therefore, we use a simulation approach to solve the problem. For each simulation run, we proceed by iteratively by adding the benefits and subtracting the costs at each point in time to the value of the project V_t. The formula for updating V_t depends on the phase and stage of the simulation as follows:

$$V_{t+1} = V_t - I_1\Delta tPVF_{1,t+1} \quad \text{if } 0 \le t \le \tau_1(\text{Stage 1}) \quad (18)$$

$$V_{t+1} = V_t + (C_{1,t} - I_2)\Delta tPVF_{2,t+1} \quad \text{if } \tau_1 \le t \le \tau_2(\text{Stage 2}) \quad (19)$$

$$V_{t+1} = V_t + C_{2,t}\Delta tPVF_{2,t+1} \quad \text{if } \tau_2 \le t \le T_2(\text{Mainstream Phase}), \quad (20)$$

where $PVF_{1,t+1}$ and $PVF_{2,t+1}$ represent the present value discount factors and are obtained as follows:

$$PVF_{1,0} = 1.0 \quad (21)$$

$$PVF_{1,t+1} = PVF_{1,t} * \mathrm{e}^{-(r_f+\lambda_1)\Delta t} \quad \text{if } 0 \le t \le \tau_1(\text{Stage 1}) \quad (22)$$

$$PVF_{2,t+1} = PVF_{2,t} * \mathrm{e}^{-(r_f+\lambda_2)\Delta t} \quad \text{if } \tau_1 \le t \le \tau_2(\text{Stage 2}). \quad (23)$$

For each simulation run, a final value of the project V_T^i is obtained and used to compute the expected value of random variable V_T.

4.4 Abandonment Value

In our simulation model we allow an organization to abandon the investment project once it has been started whenever the future of the project does not seem promising. For instance, if a project is having substantially higher development costs than originally expected or cash flows are lower than expected, it might be convenient to permanently abandon it. For this purpose, we estimate the Net Present Value that we expect to receive from the project based on the current value of the completion costs ($K_{1,t}$ and $K_{2,t}$) and cash flows ($C_{1,t}$ and $C_{2,t}$) at each time step of a simulation run, assuming that there is no uncertainty.

Given that the major benefits from a disruptive technology come from moving up into the mainstream market, the value of the project when both phases are included will always be higher than the value of the project when the technology is only commercialized in the emerging market. Therefore, our decision to abandon or not the project will depend upon the present value of the remaining development costs and cash flows including both phases.

During Stage 1, the present value of the remaining development costs at each time step t is estimated based on the current value of the completion costs $K_{1,t}$ and $K_{2,0}$ by setting $\tau_1 = K_{1,t}/I_1$ and $\tau_2 = K_{1,t}/I_1 + K_{2,0}/I_2$ in Eqs. (14) and (15) and solving the corresponding integrals:

$$PVCost_t = \frac{I_1}{r_f + \lambda_2}\left(1 - \mathrm{e}^{-(r_f+\lambda_1)\frac{K_{1,t}}{I_1}}\right) + \frac{I_2}{r_f + \lambda_2}\left(1 - \mathrm{e}^{-(r_f+\lambda_2)\frac{K_{2,0}}{I_2}}\right) \times \mathrm{e}^{-(r_f+\lambda_1)\frac{K_{1,t}}{I_1}} \quad \text{for } t < \tau_1 \,. \tag{24}$$

Similarly, the present value of the remaining development costs at each time step of Stage 2 is estimated as follows:

$$PVCost_t = \frac{I_2}{r_f + \lambda_2}\left(1 - \mathrm{e}^{-(r_f+\lambda_2)\frac{K_{2,t}}{I_2}}\right) \quad \text{for } \tau_1 \le t \le \tau_2 \,. \tag{25}$$

Once Stage 2 has been completed, there are no more development costs and $PVCost_t$ becomes zero since commercialization cost are already included in the corresponding net cash flows.

With respect to the present value of the cash flows, they are also estimated at each time step t by computing the length of each Stage in terms of its remaining development costs and investment rates and substituting the results in Eqs. (16) and (17). Letting T_1 and T_2 be large in comparison to τ_1 and τ_2

we obtain the following expressions for $PVCash_t$:

$$PVCash_t = \begin{cases} \frac{C_{1,t}}{r_f+\lambda_2-\alpha_1^*} \mathrm{e}^{-(r_f+\lambda_1-\alpha_1^*)\frac{K_{1,t}}{I_1}} \left(1 - \mathrm{e}^{-(r_f+\lambda_2-\alpha_1^*)\frac{K_{2,0}}{I_2}}\right) \\ + \frac{CashRatio*C_{1,t}}{r_f-\alpha_2^*} \left(\mathrm{e}^{-(r_f+\lambda_1-\alpha_1^*)\frac{K_{1,t}}{I_1}}\right) \mathrm{e}^{-(r_f+\lambda_2-\alpha_1^*)\frac{K_{2,0}}{I_2}} \end{cases} \tag{26}$$

for $t < \tau_1$

$$PVCash_t = \begin{cases} \frac{C_{1,t}}{r_f+\lambda_2-\alpha_1^*} \left(1 - \mathrm{e}^{-(r_f+\lambda_2-\alpha_1^*)\frac{K_{2,t}}{I_2}}\right) \\ + \frac{CashRatio*C_{1,t}}{r_f-\alpha_2^*} \mathrm{e}^{-(r_f+\lambda_2-\alpha_1^*)\frac{K_{2,t}}{I_2}} \end{cases} \tag{27}$$

for $\tau_1 \leq t \leq \tau_2$

$$PVCash_t = \frac{C_{2,t}}{r_f - \alpha_2^*} \quad \text{for } \tau_2 < t \tag{28}$$

The model allows the decision maker to simulate the effects of different *abandonment policies*. A project can be never abandoned, abandoned as soon as the NPV at a particular time step t becomes zero (i.e., the traditional NPV criteria) or abandoned as soon as the difference between the present value of the benefits minus a percentage of the present value of the costs becomes zero. For this purpose, we compute an *abandonment value* using the following expression:

$$AbandonmentValue_t = PVCash_t - \gamma * PVCost_t \ , \tag{29}$$

where γ is a number between 0 and 1. Note that when γ is 1 the abandonment value equals the NPV of the project. Also, setting this parameter to zero will cause never to abandon a project since the abandonment value will never become zero.

Abandoning a project implies foregoing any opportunity of things to improve in the future. Even if a decision to abandon the project is taken at a moment in which the NPV becomes substantially negative, this decision might be incorrect. As will be shown in the example, different abandonment policies have different effects on the value of the project. Therefore, we will refer to "optimal abandonment policy" to the abandonment criteria (i.e., the value of γ) that maximizes the value of the project for a given set of input parameters.

Note, however, that the above abandonment strategy is not optimal. Even if we search for the value of γ which maximizes the value of the project, the strategy would be sub-optimal. As it is well known from the vast literature on American options, the optimal exercise strategy is state dependant and varies depending upon the time to maturity. In our situation, time to maturity is random and the problem is strongly path dependant since costs and cash flows in subsequent stages depend on the realization of the state variables of the preceding stages. In these circumstances, there is not a known optimal exercise strategy.

5 Simulation Process

Figure 3 shows the pseudo-code of the simulation process. The value of an investment project is obtained as the average of the values obtained over a certain number of simulation runs. On each simulation run, a realization of the stochastic processes representing the development costs and the cash flows associated with the project is obtained by taking as a starting point the corresponding initial expected values of these variables in Equations (2) and (7) updating these initial estimations over time.

```
For each simulation run (i)
  Initialize K1,0, and C1,0 to their initial estimates
  Compute K2,0 and C2,0 using Eqs. (5) and (11) with TotCost1,0=0
  For each time slice (t)
    -- Emerging Phase
    While K1,t is positive DO (Stage 1)
        Determine the abandonment value using Eqs. (24), (26) and (29)
        IF the abandonment value is less than or equal to zero THEN
            Flag the project as abandoned
            Increase the counter of abandoned projects
        Update K1,t,TotCost1,t and C1,t using Eqs. (3), (4) and (10)
        Update Vt using Eq.(18)
    Compute K2,τ1 using Eq. (5)
    While K2,t is positive DO (Stage 2)
        Determine the abandonment value using Eqs. (25), (27) and (29)
        IF the abandonment value is less than or equal to zero THEN
            Flag the project as abandoned
            Increase the counter of abandoned projects
        Update K2,t and C2,t using Eqs. (6) and (12)
        Update Vt using Eq. (19)
    Compute C2,τ2 using Eq. (11)
    -- Mainstream Phase
    Update Vt using Eq. (20)
Compute the Mean and the Std. Dev. of VT
Compute other Statistics of the simulation
```

Fig. 3. Pseudo-code of the Simulation Process

Figure 4 shows an example of a realization of the simulation using one hundred periods per year. The vertical dotted line represents the point in time in which the disruptive technology is starting to be commercialized in the emerging market. In this realization, Stage 1 of the Emerging Phase is finished at the end of period 240 (approximately 2 years and 5 months). At this point, the system computes the value of the project when only the Emerging Phase is considered. Then, the system continues the evolution of the stochastic processes representing the cash flows received in the emerging market ($C_{1,t}$) and the development cost of Stage 2 of the Emerging Phase ($K_{2,t}$) until the technology is ready for being commercialized into the mainstream market (end of the Emerging Phase). At this point, the system estimates the expected cash flow in the mainstream market using the current value of the cash flow and the corresponding multiplying factor.

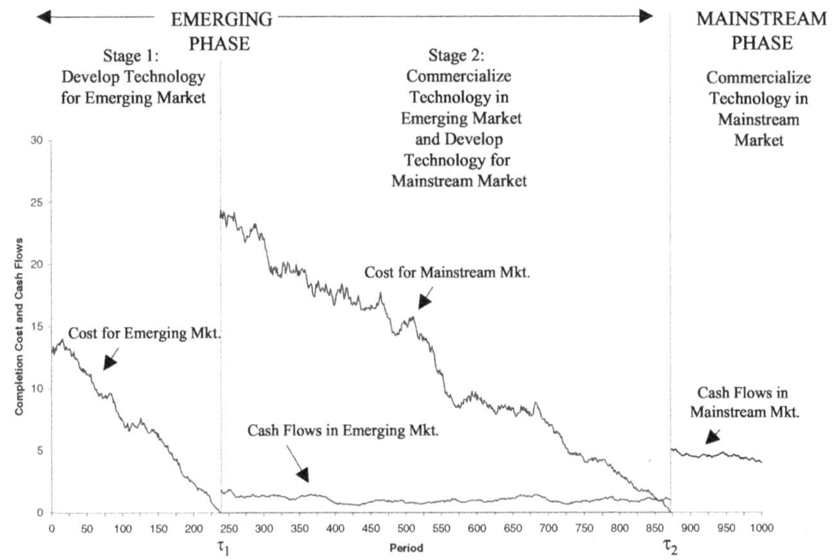

Fig. 4. Evolution of Completion Costs and Cash Flows in a Simulation Run

6 Example and Results

Suppose that a group of technicians of *High Tech Corp.* (HTC) has just developed a prototype of a technology that allows them to store and retrieve data from a crystal using optical technology. The potential market for this type of application has not yet been identified; however, they foresee that data storage devices using this technology will be less subject to failure than conventional magnetic storage devices because they would have no moving parts. Furthermore, the data stored in the crystal may be retrieved in parallel instead of being read in a sequential manner, leading to faster access times. However, the most profitable customers of the company have said that they are not interested in this type of technology because its cost per megabyte stored is substantially higher than the cost of magnetic storage devices.

At the moment of the assessment, the group of technicians is uncertain about the costs that will be needed in order find a market for this technology and starting its commercialization. Also, they are uncertain about the time in which the technology will be ready for launching since all they have at this moment is a prototype device. However, they believe that this technology might be potentially disruptive to some of the data storage technologies currently being used by the organization to supply the needs of its current customers, and that some other markets such as the military or the airlines might be interested in having safer and faster devices. Therefore, they decide to prepare a presentation to senior managers with some preliminary figures, in the understanding that these assumptions will be tested during the development of the project. Table 1 shows the preliminary figures used for the initial valuation of the project (Base Case)

After running the simulation for 100,000 times, the results shown in Table 2 were obtained for the case where no abandonment is allowed ($\gamma = 0$). The Value of the project is \$3.452 million when the technology is commercialized both in the emerging and in the mainstream market. In contrast, the Value of the project is negative (−\$4.707 million) if the organization decides only to develop (and commercialize) the technology for the emerging market. These results indicate that the main value of the disruptive technology developed by HTC will come precisely from the first-mover advantages it will provide to the company in the mainstream market of storage devices.

In order to study the impact of the cash flow and cost volatilities in the value of the investment project, we performed sensitivity analysis on various input parameters. Simulations were run for two scenarios depending on whether the organization has the option of abandoning the project once it has been started or not. For each simulation in which abandonment was allowed, we identified which *abandonment policy* was "optimal" by varying parameter γ from 0 to 1 in increments of 0.1. The results of this exercise are discussed below.

Effects of Different Abandonment Policies

The value of a project with an abandonment option is dependent upon the policy used for deciding whether to continue or not with the development of the disruptive technology. Figure 5 shows the expected value of the project for various abandonment policies represented by parameter γ. The policy of no abandonment occurs when parameter γ is zero. In this example, not abandoning the project gives better results than the policy in which the project is abandoned as soon as its expected NPV (ignoring sunk costs) becomes negative ($\gamma = 1$). The best abandonment policy for our base case occurs when the company decides to abandon the project as soon as the present value of the benefits minus 70% of the costs ($\gamma = 0.7$) is less than zero.

The value of the *option* to abandon can be estimated as the difference between the value with abandonment and the value without abandonment. As the figure shows, the option value can be an important component of the total value of the project, even if our pseudo-optimal abandonment strategy is followed. For instance, when $\gamma = 0.7$ the option value (\$3.986 – \$3.452 = \$0.534 million) represents 13.4% of the project value.

Table 1. Parameters for the Example of High-Tech Corp. (Base Case)

Initial Expected Completion Costs and Cash Flow Values		
Completion Cost for Developing and Launching the Technology in Emerging Market	$K_1(0)$	13 million
Annual Net Cash Flow from Commercializing the Technology in Emerging Market	$C_1(0)$	1 million
Cost Ratio (Total Completion Cost of Stage 2/ Total Completion Cost of Stage 1)	CostRatio	2 times
Cash Flow Ratio (Cash Flow in Mainstream Market/ Cash Flow in Emerging Market)	CashRatio	5 times
Volatility, Drift and Correlation Parameters		
Volatility of Developmental Costs	β	0.3
Volatility of Cash Flows	ϕ	0.2
Drift of Cash Flows in Emerging Market	α_1	0.04
Drift of Cash Flows in Mainstream Market	α_2	0.10
Correlation of Costs and Cash Flow Changes	ρ	0.0
Investment Parameters		
Investment Rate during the Development of the Technology for Emerging Market	I_1	5 million per year
Investment Rate during the Development of the Technology for Mainstream Market	I_2	5 million per year
Other Parameters		
Risk-free rate	r_f	0.06
Risk premium on Cash Flow Value	η_c	0.08
Catastrophe Probability Rate for Developing and Launching the Technology in Emerging Market	λ_1	0.3
Catastrophe Probability Rate for Developing and Launching the Technology in Mainstream Market	λ_2	0.1
Fraction of Costs Subtracted from Benefits when computing the Abandonment Value	γ	From 0 to 1

Table 2. Results for the Example of *High-Tech* Corp (Base Case) with no abandonment-millions

	Mean (μ)	Std Dev of Mean (σ_μ)
Value of the Project (Total)	\$3.452	\$0.043
Value of the Project (only Emerging Market)	–\$4.707	\$0.007

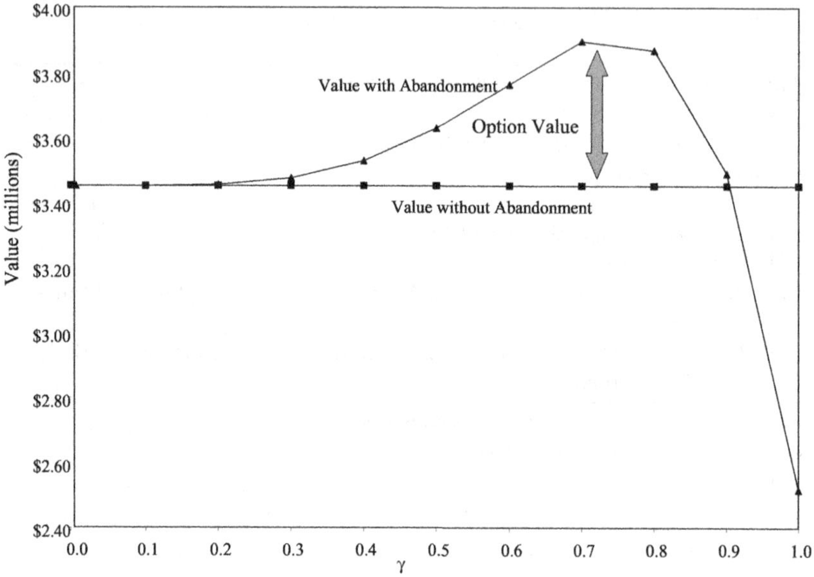

Fig. 5. Effects of Different Abandonment Policies in the Value of the Project (Base Case)

As discussed earlier, the optimal abandonment policy is dependent upon the value of the input parameters. Figure 6 shows, for instance, the appropriateness of different abandonment policies for different values of the cost volatility β when all the other parameters remain constant. When $\beta = 0.1$, optimal abandonment occurs when γ is set to 0.9. In contrast, when $\beta = 0.5$, optimal abandonment occurs when γ is set to 0.4. These results indicate that projects with high volatilities should be abandoned only when the NPV becomes sufficiently negative. Note that in both cases, abandoning the project as soon as its NPV becomes negative (i.e. the traditional criteria represented by $\gamma = 1.0$) gives worse results than not abandoning the project at all.

Effects of Cost Volatility

Figure 7 shows the value of the project with and without abandonment for different cost volatilities. The value of the project is higher when costs are more volatile. This is a consequence of the value being a convex function of costs and, due to Jensen's inequality (Dixit and Pindyck 1994, p. 49), the value of the expected costs is greater than the expected value of the project. For the case with abandonment, the option value also increases with volatility. Note that the value of the project with optimal abandonment is always higher than the value of the project without abandonment. This is true since the decision

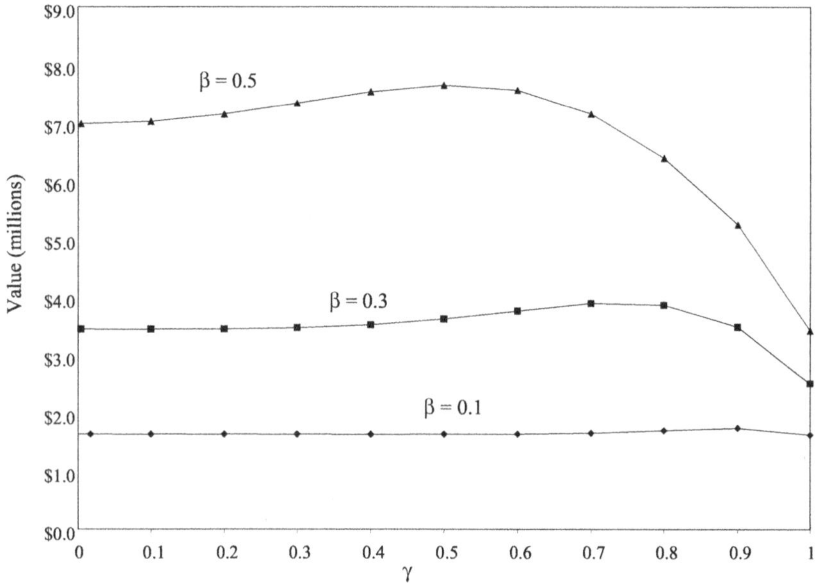

Fig. 6. Effects of Different Abandonment Policies in the Value of the Project for Various Cost Volatilities

maker can always decide not to abandon the project by setting $\gamma = 0$ whenever this policy happens to give better results.

Effects of Cash Flow Volatility

Figure 8 shows the value of the project with and without abandonment for different cash flow volatilities. In contrast with cost volatility, cash flow volatility does not impact the value of the project when no abandonment is allowed. This occurs because the value of the project without abandonment is a linear function of the cash flows and, therefore, the volatility has no impact on the overall expected value of the project. Note, however, that the value of the project with optimal abandonment increases monotonically when more volatility is present. This is a consequence of the asymmetry associated with the abandonment option: those projects that are positively affected by the higher volatility will be continued but those that are negatively affected will be abandoned.

Combined Effects of Cost and Cash Flow Volatilities

Since the value of the project without abandonment does not change with respect to the volatility of the cash flows but increases with respect to the

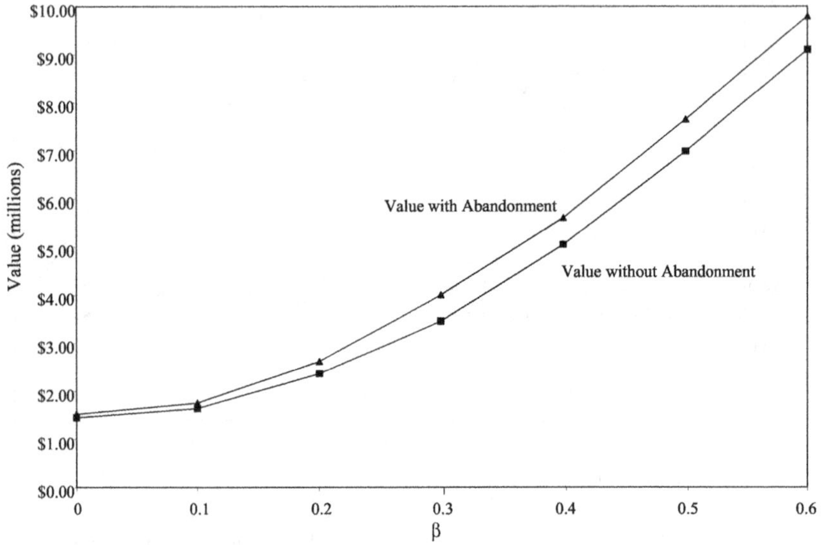

Fig. 7. Effects of Cost Volatility in the Value of the Project

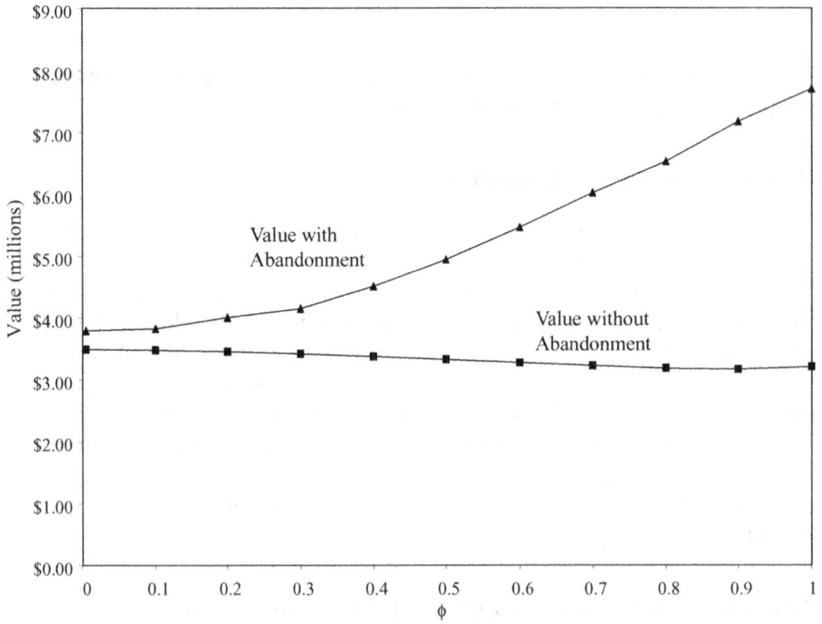

Fig. 8. Effect of Cash Flow Volatility in the Value of the Project

volatility of the development costs, the combined effect of both types of volatilities will be to increase the overall value of the project. Similarly, since the value of the project with abandonment is always higher when cash flow or cost volatility increases, the combined effect of this type of uncertainty will be to increase the value of the project when abandonment is allowed.

Effects of Costs – Cash Flows Correlations

Figure 9 shows the effects of the correlation between completion costs and benefit cash flows on the value of the project. When these two variables are positively correlated, an unexpected increase in cash flows will be partially offset by an unexpected increase in costs. Conversely, when these variables are negatively correlated, an unexpected increase in cash flows will be associated with an unexpected decrease in costs and vice versa. Therefore, a positive correlation partially neutralizes the effect of uncertainty and a negative correlation enhances it. Since the value of the project with or without abandonment is higher when more volatility is present, a positive correlation will impact negatively the value of the project and vice versa.

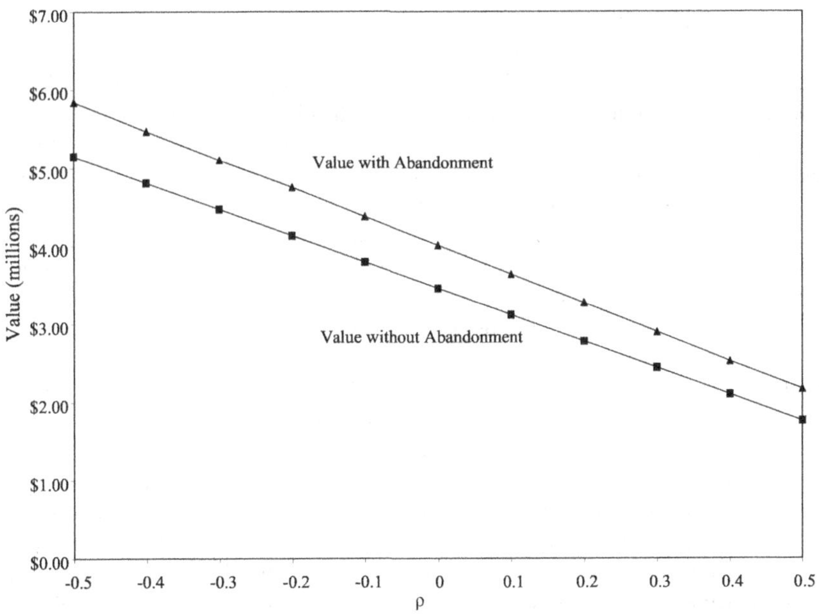

Fig. 9. Effects of Cost – Cash Flows Correlation in the Value of the Project

7 Summary and Conclusions

In this paper we developed a computer simulation model for the valuation of investments in disruptive technologies. Based on the conceptual framework proposed by Christensen (1997) for explaining the Innovator's Dilemma phenomenon, an investment project is divided into two sequential phases representing the evolution of the disruptive technology from an emerging to a mainstream market. The model incorporates the effects of the uncertainty associated with the development costs, with the possibility that a catastrophic event causes the permanent abandonment of the project, and with the time required to complete each stage of the project. Also, the model accounts for the uncertainty associated with the net cash flows that an organization expects to obtain once the technology is commercialized in each of the markets and allows modeling the interaction between these cash flows and the development costs of each stage.

The application of the model to a hypothetical example provided some insights of the effects that cash flow and cost volatilities have on the overall value of this type of projects. When abandonment is allowed, higher cash flow or cost volatilities increase the value of the investment project. When no abandonment is possible, only cost volatilities have an effect in the expected value of the project. Also, a negative correlation between cash flow and cost changes increases the value of the project and vice versa.

We also discussed how to implement a pseudo-optimal policy for deciding whether to abandon or not a project once it has been started. For each set of the input parameters, an "optimal" percentage of the present value of the remaining completion costs is subtracted from the present value of the expected cash flows to determine an abandonment value. As shown in the example, the traditional criteria of stopping a project as soon as NPV becomes negative leads to a lower expected value of the project than the value obtained by other abandonment policies when cash flow or cost volatilities are considered.

The model described in the paper constitutes an attempt to develop a more formal methodology for evaluating investments in disruptive technology projects that, by their own nature, are characterized by a high degree of uncertainty in all their relevant variables. We believe that such an attempt is not in contraposition with the principles of disruptive innovation that Christensen recommends for dealing with this type of technologies. Modeling such a project does not mean that a company should not be cautious about keeping alive those ideas that their mainstream customers do not currently want. Neither does it mean that an organization will solve its growth needs by only developing such projects. Organizations need to treat projects of disruptive technologies as a new endeavor that might require the creation of new business processes and values that correspond to the size of the emerging market. However, they also need new tools that help the decision maker to account for the high uncertainty associated with this phenomenon.

Our model can be used to measure the effects of the various assumptions related to a particular scenario, including those related to the volatilities of cash flows and development costs, the probability of failure in each stage of development, the maximum investment rates that the organization can invest, the size of the jump in cash flows that occur when an organization is successful in moving up into the mainstream market, the risk-adjusted growth rates of the commercialization cash flows in each phase, and the risk-free rate. It also provides us with information related to the additional value coming from the flexibility that an organization has for abandoning the project once it has been started that traditional valuation tools such as the NPV method do not take into consideration. Therefore, the model can be used as a valuable input for other decision-making frameworks, such as discovery-driven planning (McGrath and MacMillan 1995), that help to identify what needs to be learned during the course of the project in order to solve for the uncertainties that characterize the phenomenon.

References

Amram, M. and Kulatilaka, N. (1999) *Real Options: Managing Strategic Investment in an Uncertain World*, Boston, MA.: Harvard Business School Press

Bers, J.A., Lynn, G.S., and Spurling, C. (1999) *A Computer Simulation Model for Emerging Technology Business Planning and Forecasting.* International Journal of Technology Management. Vol. 18, No. 1/2, 31–45

Boer, F.P. (1999) *The Valuation of Technology.* Wiley Operations Management Series, New York, NY.: John Wiley & Sons, Inc., 404

Brennan, M.J. and Schwartz, E.S. (1985) *Evaluating Natural Resource Investments.* Journal of Business. Vol. 58, No. 2, 135–157

Christensen, C.M. (1997) *The Innovator's Dilemma*: Harvard Business School Press

Christensen, C.M. and Overdorf, M. (2000) *Meeting the Challenge of Disruptive Change.* Harvard Business Review, No. March-April, 67–76

Christensen, C.M. and Rosenbloom, R.S. (1995) *Explaining the Attacker's Advantage: Technological Paradigms, Organizational Dynamics, and the Value Network.* Research Policy. Vol. 24, 233–257

Dixit, A.K. and Pindyck, R.S. (1994) *Investment under Uncertainty*, New Jersey: Princeton University Press

Jonash, R.S. and Sommerlatte, T. (1999) *The Innovation Premium*: Perseus Books

Jovanovic, P. (1999) *Application of Sensitivity Analysis in Investment Project Evaluation Under Uncertainty and Risk.* International Journal of Project Management. Vol. 17, No. 4, 217–222

McGrath, R.G. and MacMillan, I.C. (1995) *Discovery-Driven Planning.* Harvard Business Review, No. July-August, 4–12

Pindyck, R.S. (1993) *Investments of Uncertain Cost.* Journal of Financial Economics. Vol. 34, 53–76

Schwartz, E.S. and Moon, M. (2000a) *Evaluating Research and Development Investments*, in *Project Flexibility, Agency, and Competition*, Brennan, M.J. and Trigeorgis, L., Editors, Oxford University Press: New York, 85–106

Schwartz, E.S. and Moon, M. (2000b) *Rational Pricing of Internet Companies.* Financial Analysts Journal, Vol. 56, No. 3, 62–75

Schwartz, E.S. and Zozaya-Gorostiza, C. (2000) *Valuation of Information Technology Investments as Real Options*, Anderson Graduate School of Management, University of California at Los Angeles: Los Angeles, CA., 1–37

Tipping, J.W., Zeffren, E., and Fusfeld, A.R. (1995) *Assessing the Value of Your Technology.* Research Technology Management, No. September- October, 22–39

Tushman, M.L. and O'Reilly, C.A. (1997) *Winning Through Innovation: A Practical Guide to Leading Organizational Change and Renewal*, Boston, MA.: Harvard Business School Press

Quickest Detection Problems in the Technical Analysis of the Financial Data

Albert N. Shiryaev

Steklov Mathematical Institute, Gubkina str. 8, 117966, Moscow, Russia

1 Introduction. The Statement of Some Problems of the "Technical Analysis"

Suppose that we are observing a random process $X = (X_t)$ on an interval $[0, T]$. The objects θ and τ introduced below are essential throughout the paper:

θ – a *parameter* or a *random variable*; this is the time at which the observed process $X = (X_t)_{t \geq 0}$ changes its *character* of behaviour or its probability *characteristics*;

τ – a *stopping (Markov) time* which serves as the time of "alarm"; it warns of the coming of the time θ.

Let us consider a particular example, where θ is the time at which the process X attains its maximum over $[0, T]$. Figure 1 shows that θ is a time at which the process X changes the *character* of behaviour. Namely, the tendency of "growth towards the maximum" is replaced by the tendency of "decrease from the maximum".

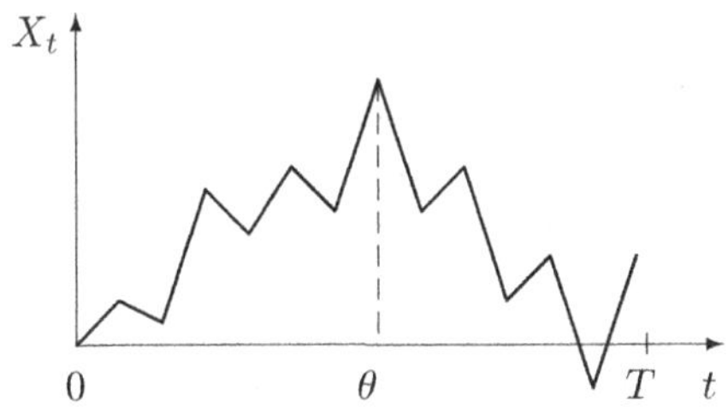

Figure 1

The following setting of the detection (or prediction) problem for θ is rather natural for the "technical analysis" of the financial data.

Problem I. Let $X = (X_t)_{t \leq T}$ be a stochastic process (prices, indexes,...) observed on $[0, T]$. Let $\theta = \theta(\omega)$ be such that

$$X_\theta = \sup_{t \in [0,T]} X_t.$$

We want to find a *stopping time* $\tau^* = \tau^*(\omega)$ such that

$$X_{\tau^*} \sim X_\theta$$

or

$$\tau^* \sim \theta,$$

for example, in the following senses:

$$\inf_\tau \mathsf{E}(X_\theta - X_\tau)^p = \mathsf{E}[X_\theta - X_{\tau^*}]^p,$$
$$\inf_\tau \mathsf{E}(\theta - \tau)^p = \mathsf{E}|\theta - \tau^*|^p.$$

(The other criteria are:

$$\inf_\tau \mathsf{E}G(X_\theta - X_\tau),$$
$$\inf_\tau \mathsf{E}G\left(\frac{X_\tau}{X_\theta} - 1\right), \ldots$$

for some "performance" functions $G = G(x)$).

We will consider below the case where the process X is a *standard linear Brownian motion* B. From the viewpoint of the modern mathematical finance, this model (due to L. Bachelier) is too idealized. However, we will further see that even in this (relatively simple) case the solution of the corresponding optimization problem is rather nontrivial. On the other hand, the solution for the case of a Brownian motion gives a way to solve this problem for more general cases.

Problem II is related to the models for which θ is the time the observed process changes its *probability characteristics*. In connection with the technical analysis of the financial data, it is of interest to consider the schemes in which θ is interpreted as the time of the appearance of an *arbitrage* ("transition from a martingale to a submartingale", for example) or as the time of the appearance of a *change-point.*

We will concentrate our attention on the model in which the observed process X has the following form (see Figure 2):

$$X_t = r(t-\theta)^+ + \sigma B_t,$$

i.e.

$$dX_t = \begin{cases} \sigma\, dB_t, & t < \theta, \\ r\, dt + \sigma\, dB_t, & t \geq \theta. \end{cases}$$

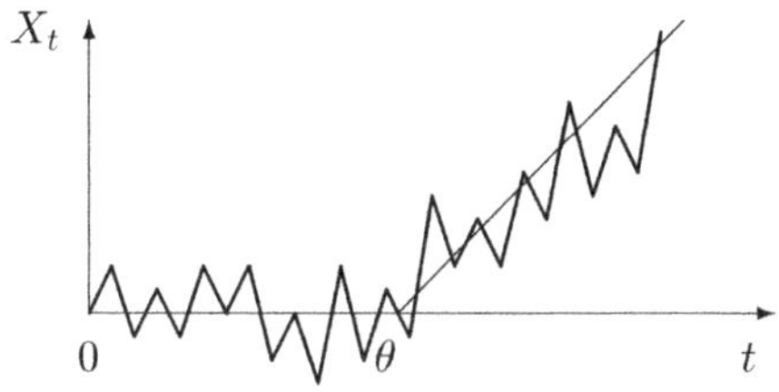

Figure 2

One can associate the following two events with a time τ (recall that τ is interpreted as the time of "alarm"):

$$\{\tau < \theta\} \text{ and } \{\tau \geq \theta\}.$$

The first event ($\{\tau < \theta\}$) corresponds to the "false alarm": the alarm time τ comes before the time θ. The second event ($\{\tau \geq \theta\}$) corresponds to the case where the alarm is raised in due time, i.e. after the time θ.

The above reasoning leads to several formulations of the quickest detection problem for the time θ. These formulations are given below.

Suppose that $\theta = \theta(\omega)$ is a *random variable* ($\theta \geq 0$). Then the first formulation (Variant **A**) of the *quickest detection problem* for θ is as follows. (This formulation can be called *conditionally-extremal*).

Variant A. For a given $\alpha \in (0,1)$, find a stopping time τ_α^* such that

$$\mathbb{A}(\alpha) = \inf_{\tau \in \mathfrak{M}_\alpha} \mathsf{E}(\tau - \theta \,|\, \tau \geq \theta) = \mathsf{E}(\tau_\alpha^* - \theta \,|\, \tau_\alpha^* \geq \theta),$$

where

$$\mathfrak{M}_\alpha = \{\tau : \mathsf{P}(\tau < \theta) \leq \alpha\}.$$

Variant B. For a given $c > 0$, find

$$\mathbb{B}(c) = \inf_\tau \big\{\mathsf{P}(\tau < \theta) + c\,\mathsf{E}(\tau - \theta)^+\big\}$$

and the corresponding optimal (Bayes) stopping time. (Note that

$$\mathsf{E}(\tau - \theta)^+ = \mathsf{E}(\tau - \theta \,|\, \tau \geq \theta)\,\mathsf{P}(\tau \geq \theta)\ ).$$

In the following Variants **C** and **D**, θ is an *unknown parameter*.

Variant C. Find

$$\mathbb{C}(T) = \inf_{\tau \in \mathfrak{M}^T} \sup_\theta \operatorname*{ess\,sup}_\omega \mathsf{E}_\theta\big((\tau - \theta)^+ \,|\, \mathcal{F}_\theta\big)(\omega)$$

and the corresponding optimal stopping time. Here,

$$\mathfrak{M}^T = \{\tau : \mathsf{E}_\infty \tau = T\},$$
$$\mathsf{P}_\theta(\,\cdot\,) = \mathrm{Law}(\,\cdot\,|\theta).$$

Variant D. Find

$$\mathbb{D}(T) = \inf_{\tau \in \mathfrak{M}^T} \sup_\theta \mathsf{E}_\theta(\tau - \theta \,|\, \tau \geq \theta)$$

and the corresponding optimal stopping time.

The following Variant **E** of the quickest detection problem is interesting due to the unusual assumptions made on the nature of θ and on the character of the observation procedure.

First, the assumption that θ is a random variable (or an unknown parameter) is replaced by the assumption that θ appears after the *stationary regime of the observations* is established. (Of course, the time θ is preceded by a long period of observations that contains many alarms of the appearance of "change-points"). Second, we suppose that the observation procedure in Variant **E** is *multistage*.

Informally, Variant **E** (to be more precise, its particular case; a more general formulation, Variant **E**$'$, is given in Subsection 9 of Section 3 below) is formulated as follows.

Let $\Psi_t = \Psi(t; X_s,\, s \leq t)$ be a functional of the observations with $\Psi_0 = 0$, $t \geq 0$. Suppose that the alarm of the appearance of a "change-point" is based on the observation of Ψ and the alarm procedure has the following form.

We observe the process $(X_t)_{t\geq 0}$ and use it to construct a process $\Psi = (\Psi_t)_{t\geq 0}$. Once this process has reached a level $a > 0$, we raise an alarm of the appearance of a "change-point". Let us call this time τ_1. After this time, the process Ψ is returned to zero, i.e. for $t > \tau_1$, we observe the process $\Psi(t - \tau_1; X_s - X_{\tau_1},\, \tau_1 < s \leq t)$. The next alarm is raised at a time $\tau_1 + \tau_2$. In a similar way, this procedure (let us call it δ) is repeated after the time $\tau_1 + \tau_2$ with the alarm raised at a time $\tau_1 + \tau_2 + \tau_3$ and so on.

Let us denote the process constructed above (which is a "renewal process") by $\Psi^\delta = (\Psi_t^\delta)_{t\geq 0}$ (see Figure 3). We will suppose that this process has a limit distribution $F^\delta(\psi) = \lim_{t\to\infty} \mathsf{P}_\infty(\Psi_t^\delta \leq \psi)$, where P_∞ denotes the distribution of the process X under the assumption that there is no change-point (i.e. for the case where $\theta = \infty$).

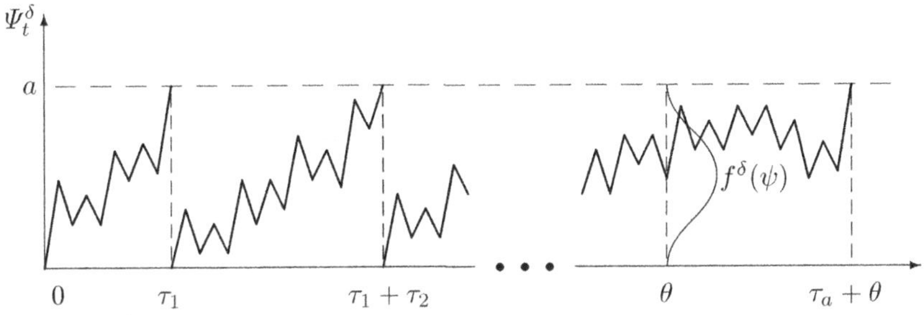

Figure 3. (The function $f^\delta = f^\delta(\psi)$ is the density of the stationary distribution $F^\delta = F^\delta(\psi)$)

Let the mean time between two false alarms for the above observation procedure (it is determined by a functional Ψ and a level $a > 0$) be equal to T, i.e. $\mathsf{E}_\infty \tau_i = T$ $(i = 1, 2, \dots)$. Here, E_∞ is the expectation taken with respect to the measure P_∞. Then the mean time of the delay after the appearance of θ is given by

$$R^\delta(T) = \int_0^a (\mathsf{E}_0^\psi \tau_a)\, F^\delta(d\psi),$$

where $\mathsf{E}_0^\psi \tau_a$ is the expectation of the hitting time of the level a by the process Ψ^δ under the assumption that $\Psi_0^\delta = \psi$. Here, E_0 is the expectation with respect to the measure P_0.

The formulation of the quickest detection problem for the multistage observations and the stationary regime is as follows:

Variant E. Find among all the methods determined by a pair (Ψ, a) the infimum

$$\mathbb{E}(T) = \inf_{\{\Psi, a\}} R^\delta(T)$$

assuming that the mean time of delay between two false alarms is equal to $T > 0$.

Our aim is to describe the optimal or asymptotically optimal methods for Variants **A**, **B**, **C**, **D**, **E**, **E′**.

2 Problem I

1. Let $X = B$, where $B = (B_t)_{0 \le t \le 1}$ is a standard linear Brownian motion on $[0, 1]$. For $p > 0$, $p \neq 1$, set

$$V_*^{(p)} = \inf_\tau \mathsf{E}\Big(\max_{0 \le s \le 1} B_s - B_\tau\Big)^p. \tag{2.1}$$

Here, the infimum is taken over all $(\mathcal{F}_t^B)$-stopping times τ, where $(\mathcal{F}_t^B)$ is the natural filtration of B and $0 \le \tau \le 1$.

In what follows, we will denote $V_*^{(2)}$ simply as V_*. Note that, for our stopping times τ, $\mathsf{E}B_\tau = 0$. On the other hand, $\mathsf{E}\max_{0\le s\le 1} B_s = \sqrt{\frac{2}{\pi}}$. Thus, from the statistical point of view, B_τ is a *biased* estimate for $\max_{0\le s\le 1} B_s$. Therefore, one could consider the following value instead of V_*:

$$\widetilde{V}_* = \inf_{\{\tau,\, a\in\mathbb{R}\}} \mathsf{E}\Big(\max_{0\le s\le 1} B_s - (a + B_\tau)\Big)^2. \tag{2.2}$$

However, it is clear that $\widetilde{V}_* = V_* - \frac{2}{\pi}$ and the "optimal" a is equal to

$$a_* = \mathsf{E}\max_{0\le s\le 1} B_s = \sqrt{\frac{2}{\pi}}.$$

Thus, the solution of problem (2.2) is reduced to the solution of problem (2.1) with $p = 2$.

2. The following solution of problem (2.1) with $p = 2$ was obtained in the paper [4] by S.E. Graversen, G. Peskir and the author.

Theorem 1. *Suppose that $p > 0$, $p \neq 1$. Then there exists a number $z_p > 0$ such that the optimal stopping time $\tau_*^{(p)}$ is given by*

$$\tau_*^{(p)} = \inf\big\{t \le 1 : S_t - B_t \ge z_p\sqrt{1-t}\big\}.$$

Here, $S_t = \max_{0\le u\le t} B_u$. For $p = 2$, $z_2 = z_* = 1.12\ldots$ is the unique solution of the equation

$$4\Phi(z_*) - 2z_*\varphi(z_*) - 3 = 0,$$

where

$$\varphi(z) = \frac{1}{\sqrt{2\pi}}\exp\Big(-\frac{z^2}{2}\Big), \quad \Phi(z) = \int_{-\infty}^{z} \varphi(u)\,du.$$

Furthermore,

$$V_* = 2\Phi(z_*) - 1 = 0.73\ldots$$

The following theorem (see [4], [11]) yields some probability characteristics of the distribution of the optimal time τ_*.

Theorem 2 (some distributional properties of the optimal stopping time τ_*; the case $p = 2$). *For $\tau_* = \tau_*^{(2)}$, we have*

$$\mathsf{E}\tau_* = \frac{z_*^2}{1+z_*^2} = 0.55\ldots$$

$$\mathsf{D}\tau_* = \frac{2z_*^4}{(1+z_*^2)(3+6z_*^2+z_*^4)} = 0.05\ldots$$

$$\mathsf{E}(1-\tau_*)^\mu = \frac{1}{M\big(-\mu, \frac{1}{2}; -\frac{z_*^2}{2}\big)}, \quad \mu > 0,$$

where $M(a,b;x)$ is the Kummer confluent hypergeometric function defined by

$$M(a,b;x) = 1 + \frac{a}{b}x + \frac{a(a+1)}{b(b+1)}\frac{x^2}{2!} + \dots$$

It follows from the properties of the Kummer functions (see [1]) that, for $\mu = n \in \mathbb{N}$, we have

$$\mathsf{E}(1-\tau_*)^n = \frac{He_{2n}(0)}{He_{2n}(iz_*)},$$

where

$$He_n(x) = (-1)^n e^{x^2/2}\frac{d^n}{dx^n}(e^{-x^2/2}).$$

Hence, for $\nu > 0$,

$$e^{\nu}\mathsf{E}\exp(-\nu\tau_*) = \sum_{n=0}^{\infty}\frac{\nu^n}{n!}\frac{He_{2n}(0)}{He_{2n}(iz_*)},$$

and therefore,

$$\mathsf{E}\exp(-\nu\tau_*) = \sum_{n=0}^{\infty}\frac{e^{-\nu}\nu^n}{n!}\frac{He_{2n}(0)}{He_{2n}(iz_*)}.$$

3. *Sketch of the proof of Theorem 1.* Let $G = G(x)$ be a function such that $\mathsf{E}|G(S_1 - B_t)| < \infty$. Using the strong Markov property, for any stopping time τ with values in $[0,1]$, we can write

$$\begin{aligned}
\mathsf{E}\big(G(S_1 - B_\tau)|\mathcal{F}_\tau^B\big) &= \mathsf{E}\Big(G\Big(\max_{0\le u\le\tau} B_u \vee \max_{\tau<u\le 1} B_u - B_t\Big)\Big|\,\mathcal{F}_\tau^B\Big)\\
&= \mathsf{E}\Big(G\Big(s \vee \Big\{\max_{0\le r\le 1-t} B_r + x\Big\} - x\Big)\Big)\Big|_{x=B_\tau,\, s=S_\tau,\, t=\tau}\\
&= \mathsf{E}\,G\big(\max(s,\eta+x) - x\big)\Big|_{x=B_\tau,\, s=S_\tau,\, t=\tau},
\end{aligned}$$

where $\eta = \max_{0\le r\le 1-t} B_r$ has the distribution

$$dF_\eta(t,y) = 2\varphi\Big(\frac{y}{\sqrt{1-t}}\Big)\frac{dy}{\sqrt{1-t}}.$$

Thus,

$$\begin{aligned}
&\mathsf{E}\big(G(S_1 - B_\tau)\,|\,\mathcal{F}_\tau^B\big)\\
&= G(s-x)\,F_\eta(t,s-x) + \int_{s-x}^{\infty} G(y)\,dF_\eta(t,y)\\
&= G(s-x) + \int_{s-x}^{\infty}[G(y) - G(s-x)]\,dF_\eta(t,y)
\end{aligned}$$

with $x = B_\tau$, $s = S_\tau$, $t = \tau$.

Using the above representation for $\mathsf{E}\big(G(S_1 - B_\tau) \,|\, \mathcal{F}_\tau^B\big)$, we get

$$V_* = \inf_\tau \mathsf{E} G(S_1 - B_\tau) = \inf_\tau \mathsf{E}\bigg\{ G(S_\tau - B_\tau)\Big(2\Phi\Big(\frac{S_\tau - B_\tau}{\sqrt{1-\tau}}\Big) - 1\Big)$$
$$+ 2\int_{\frac{S_\tau - B_\tau}{\sqrt{1-\tau}}}^{\infty} G(z\sqrt{1-\tau})\,\varphi(z)\,dz \bigg\}.$$

Note that, for $G(x) = x^p$ $(x \geq 0)$, we have (see also [11])

$$V_* = \inf_\tau \mathsf{E}\bigg\{ (1-\tau)^{p/2} \bigg[\Big(\frac{S_\tau - B_\tau}{\sqrt{1-\tau}}\Big)^p \Big(2\Phi\Big(\frac{S_\tau - B_\tau}{\sqrt{1-\tau}}\Big) - 1\Big)$$
$$+ 2\int_{\frac{S_\tau - B_\tau}{\sqrt{1-\tau}}}^{\infty} z^p\,\varphi(z)\,dz \bigg]\bigg\}.$$

Since $\mathrm{Law}(S - B) = \mathrm{Law}(|B|)$ (see [13, Ch. VI, (2.3)]), we get for $G(x) = x^p$

$$V_* = \inf_\tau \mathsf{E}\bigg\{ (1-\tau)^{p/2} \bigg[\Big(\frac{|B_\tau|}{\sqrt{1-\tau}}\Big)^p \Big(2\Phi\Big(\frac{|B_\tau|}{\sqrt{1-\tau}}\Big) - 1\Big)$$
$$+ 2\int_{\frac{|B_\tau|}{\sqrt{1-\tau}}}^{\infty} z^p\,\varphi(z)\,dz \bigg]\bigg\} = \inf_\tau \mathsf{E}\Big\{ (1-\tau)^{p/2} H_p\Big(\frac{|B_\tau|}{\sqrt{1-\tau}}\Big)\Big\}$$

with

$$H_p(z) = z^p + 2p\int_z^\infty u^{p-1}\,(1 - \Phi(u))\,du.$$

Let us introduce the "new" time $s \in [0, \infty)$ defined by

$$e^{-2s} = 1 - t$$

($t \in [0, 1)$ is the "old" time). Set

$$Z_s \equiv e^s B_{1-e^{-2s}} = \frac{B_t}{\sqrt{1-t}}.$$

By Itô's formula, we have

$$dZ_s = Z_s\,ds + \sqrt{2}\,d\beta_s,$$

where

$$\beta_s = \frac{1}{\sqrt{2}}\int_0^{1-e^{-2s}} \frac{dB_u}{\sqrt{1-u}}, \quad s \geq 0$$

is a standard Brownian motion. Let $Z = (Z_s)_{s \geq 0}$ be a Markov process with

$$dZ_s = Z_s\,ds + \sqrt{2}\,d\beta_s, \quad Z_0 = z \in \mathbb{R}.$$

Set

$$V_*(z) = \inf_{\sigma} \mathsf{E}_z e^{-p\sigma} H_p(|Z_\sigma|). \tag{2.3}$$

Then

$$V_* = V_*(0).$$

In order to solve problem (2.3), we consider the corresponding *Stefan (free boundary) problem* (see [17] for more details):

$$\begin{aligned} &L_Z V(z) = p\,V(z) && z \in (-z_p, z_p),\\ &V(\pm z_p) = H_p(z_p) && (\textit{instantaneous stopping}),\\ &V'(\pm z_p) = \pm H_p'(z_p) && (\textit{smooth fit}) \end{aligned}$$

with $L_Z = z\frac{d}{dz} + \frac{d^2}{dz^2}$.

The equation $L_Z V(z) = pV(z)$ is

$$V''(z) + z\,V'(z) = p\,V(z)$$

with a general solution (see [1])

$$V(z) = c_1\, e^{-z^2/2} M\left(\frac{p+1}{2}, \frac{1}{2}; \frac{z^2}{2}\right) + c_2\, z\, e^{-z^2/2} M\left(\frac{p+2}{2}, \frac{3}{2}; \frac{z^2}{2}\right),$$

where

$$M(a, b; x) = 1 + \frac{a}{b}x + \frac{a(a+1)}{b(b+1)}\frac{x^2}{2!} + \dots$$

is the *Kummer confluent hypergeometric function.*

As the function $V_*(z)$ should be even, we have $c_2 = 0$. Using the boundary conditions, we get

$$c_1 = e^{z_p^2/2} \frac{H_p(z_p)}{M\left(\frac{p+1}{2}, \frac{1}{2}; \frac{z_p^2}{2}\right)},$$

where z_p is the strictly positive solution of the equation

$$\frac{H_p'(z)}{H_p(z)} + z = (p+1)\, z\, \frac{M\left(\frac{p+3}{2}, \frac{3}{2}; \frac{z^2}{2}\right)}{M\left(\frac{p+1}{2}, \frac{3}{2}; \frac{z^2}{2}\right)}$$

with

$$H_p = z^p + 2p \int_z^\infty u^{p-1}\,(1 - \Phi(u))\,du.$$

Thus, we have the following candidate for the *value function* $V_*(z)$ (see Figure 4):

$$V(z) = \begin{cases} H_p(z_p)\exp\left(\frac{z_p^2 - z^2}{2}\right) \dfrac{M\left(\frac{p+1}{2}, \frac{1}{2}; \frac{z^2}{2}\right)}{M\left(\frac{p+1}{2}, \frac{1}{2}; \frac{z_p^2}{2}\right)}, & |z| \le z_p,\\ H_p(|z|), & |z| \ge z_p. \end{cases}$$

The candidate for the *optimal stopping time* $\sigma_*^{(p)}$ is

$$\sigma_{z_p} = \inf\{s \geq 0 : |z_s| \geq z_p\}.$$

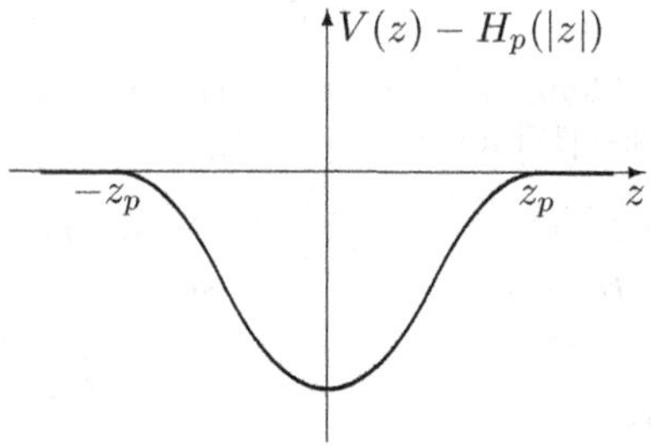

Figure 4

The usual technique of the "verification theorems" (see, for example, [19, p. 756]), together with the Itô-Tanaka-Meyer formula, shows that these candidates are indeed correct.

The "old" time is given by $t = 1 - e^{-2s}$. As $Z_t = \frac{B_t}{\sqrt{1-t}}$ and $\mathrm{Law}(B - S) = \mathrm{Law}(|B|)$, we deduce from the expression

$$\sigma_{z_p} = \inf\{s \geq 0 : |Z_s| \geq z_p\}$$

that in our initial problem

$$V_*^{(p)} = \inf_\tau \mathsf{E}\Big(\max_{0\leq t\leq 1} B_t - B_\tau\Big)^p$$

the optimal stopping time has the form

$$\tau_*^{(p)} = \inf\{t \leq 1 : S_t - B_t \geq z_p\sqrt{1-t}\}.$$

For example, if $p = 2$, we have

$$z_2 = 1.12\ldots \text{ and } \mathsf{E}\tau_*^{(2)} = 0.55\ldots,\ \mathsf{D}\tau_*^{(2)} = 0.05\ldots$$

For the case $p = \frac{1}{2}$, we have

$$z_{\frac{1}{2}} = 0.96\ldots \text{ and } \mathsf{E}\tau_*^{(\frac{1}{2})} = 0.48\ldots,\ \mathsf{D}\tau_*^{(\frac{1}{2})} = 0.05\ldots$$

4. *Distributional properties of* $\tau_*^{(p)}$. Since

$$e^{-2\sigma_*^{(p)}} = 1 - \tau_*^{(p)},$$

we have

$$\mathsf{E}\big(1 - \tau_*^{(p)}\big)^{\lambda/2} = \mathsf{E}e^{-\lambda\sigma_*^{(p)}}.$$

Set $f(z) = \mathsf{E}_z e^{-\lambda \sigma_*^{(p)}}$. Then

$$L_Z f(z) = \lambda f(z), \quad |z| \le z_p,$$

and consequently,

$$f(z) = \begin{cases} \exp\Bigl(\frac{z_p^2 - z^2}{2}\Bigr) \dfrac{M\bigl(\frac{\lambda+1}{2}, \frac{1}{2}; \frac{z^2}{2}\bigr)}{M\bigl(\frac{\lambda+1}{2}, \frac{1}{2}; \frac{z_p^2}{2}\bigr)}, & |z| \le z_p, \\ 1, & |z| \ge z_p. \end{cases}$$

Thus,

$$\mathsf{E}\bigl(1 - \tau_*^{(p)}\bigr)^{\lambda/2} = e^{z_p^2/2} \frac{1}{M\Bigl(\frac{\lambda+1}{2}, \frac{1}{2}; \frac{z_p^2}{2}\Bigr)}.$$

As

$$e^{-z^2/2} M\Bigl(a + \frac{1}{2}, \frac{1}{2}; \frac{z^2}{2}\Bigr) = M\Bigl(-a, \frac{1}{2}; -\frac{z^2}{2}\Bigr),$$

we get

$$\mathsf{E}\bigl(1 - \tau_*^{(p)}\bigr)^{\lambda/2} = \frac{1}{M\Bigl(-\frac{\lambda}{2}, \frac{1}{2}; -\frac{z_p^2}{2}\Bigr)}.$$

Hence, for $\mu > 0$,

$$\mathsf{E}\bigl(1 - \tau_*^{(p)}\bigr)^{\mu} = \frac{1}{M\Bigl(-\mu, \frac{1}{2}; -\frac{z_p^2}{2}\Bigr)}.$$

If $\mu = n \in \mathbb{N}$, then

$$M\Bigl(-n, \frac{1}{2}; -\frac{z_p^2}{2}\Bigr) = \frac{He_{2n}(iz_p)}{He_{2n}(0)},$$

where

$$He_n(x) = (-1)^n e^{x^2/2} \frac{d^n}{dx^n} \bigl(e^{-x^2/2}\bigr).$$

Therefore,

$$\mathsf{E}\bigl(1 - \tau_*^{(p)}\bigr)^{n} = \frac{He_{2n}(0)}{He_{2n}(iz_p)}.$$

Consequently,

$$\mathsf{E} \exp\bigl(-\nu \tau_*^{(p)}\bigr) = \sum_{n=0}^{\infty} \frac{e^{-\nu} \nu^n}{n!} \frac{He_{2n}(0)}{He_{2n}(iz_p)}.$$

Remark 1. If we consider the time interval $[0, T]$ instead of $[0, 1]$, then

$$V_*^{(p)}(T) = V_*^{(p)}(1) \cdot T^{p/2}$$

and

$$\tau_*^{(p)}(T) = \inf\bigl\{t \le T : S_t - B_t \ge z_p \sqrt{T - t}\bigr\}.$$

Remark 2. For

$$V_0^{(2)} = \inf_{0\le t\le 1} \mathsf{E}\Big(\max_{0\le u\le 1} B_u - B_t\Big)^2,$$

the optimal stopping time is $t_0 = \frac{1}{2}$ and

$$V_0^{(2)} = \frac{1}{\pi} + \frac{1}{2} = 0.81\ldots$$

($V_*^{(2)} = 0.73\ldots$).

For

$$\widetilde{V}_0^{(2)} = \inf_{\{0\le t\le 1,\, a\in\mathbb{R}\}} \mathsf{E}\Big(\max_{0\le u\le 1} B_u - (B_t + a)\Big)^2,$$

we have

$$\widetilde{V}_0^{(2)} = V_0^{(2)} - \frac{2}{\pi} = 0.18\ldots$$

On the other hand,

$$\widetilde{V}_*^{(2)} = V_*^{(2)} - \frac{2}{\pi} = 0.09\ldots$$

5. Consider now some problems of the type $\inf_\tau \mathsf{E}|\tau - \theta|$ for $X = B$, where $B = (B_t)_{0\le t\le 1}$. Set

$$K_1 = \inf_\tau \mathsf{E}|\tau - \theta| \ \Big(= \inf_\tau \mathsf{E}\big[(\theta-\tau)^+ + (\tau-\theta)^+\big]\Big),$$

$$K_1(c) = \inf_\tau \mathsf{E}\big[(\theta-\tau)^+ + c\,(\tau-\theta)^+\big],$$

$$K_2(c) = \inf_\tau \mathsf{E}\big[(1-\tau)\, I(\tau<\theta) + c\,(\tau-\theta)^+\big],$$

$$K_3(c) = \inf_\tau \mathsf{E}\big[I(\tau<\theta) + c\,(\tau-\theta)^+\big] = \inf_\tau \big[\mathsf{P}(\tau<\theta) + c\,\mathsf{E}(\tau-\theta)^+\big].$$

As $t \in [0,1]$, we have

$$K_1(c) \le K_2(c) \le K_3(c).$$

Set $\pi_t = \mathsf{P}(\theta \le t \,|\, \mathcal{F}_t^B)$. Then

$$K_1(c) = \inf_\tau \mathsf{E}\Big[\int_\tau^1 (1-\pi_s)\, ds + c\int_0^\tau \pi_s\, ds\Big],$$

$$K_2(c) = \inf_\tau \mathsf{E}\Big[(1-\tau)(1-\pi_\tau) + c\int_0^\tau \pi_s\, ds\Big],$$

$$K_3(c) = \inf_\tau \mathsf{E}\Big[(1-\pi_\tau) + c\int_0^\tau \pi_s\, ds\Big].$$

It is easy to show that, for $X = B$,

$$\pi_t = 2\Phi\Big(\frac{S_t - B_t}{\sqrt{1-t}}\Big) - 1,$$

where $S_t = \max_{0\le u\le t} B_u$. Set $F(x) = 2\Phi(x) - 1$. Then $1 - F = 2(1 - \Phi)$ and

$$K_1(c) = \inf_\tau \mathsf{E}\bigg[\int_\tau^1 \bigg(1 - F\bigg(\frac{|B_u|}{\sqrt{1-u}}\bigg)\bigg)\,du + c\int_0^\tau F\bigg(\frac{|B_u|}{\sqrt{1-u}}\bigg)\,du\bigg], \qquad (2.4)$$

$$K_2(c) = \inf_\tau \mathsf{E}\bigg[(1-\tau)\bigg(1 - F\bigg(\frac{|B_\tau|}{\sqrt{1-\tau}}\bigg)\bigg) + c\int_0^\tau F\bigg(\frac{|B_u|}{\sqrt{1-u}}\bigg)\,du\bigg], \qquad (2.5)$$

$$K_3(c) = \inf_\tau \mathsf{E}\bigg[\bigg(1 - F\bigg(\frac{|B_\tau|}{\sqrt{1-\tau}}\bigg)\bigg) + c\int_0^\tau F\bigg(\frac{|B_u|}{\sqrt{1-u}}\bigg)\,du\bigg]. \qquad (2.6)$$

Remark 3. The process $\Phi(\frac{B_u}{\sqrt{1-u}})$ is a martingale; the process $\Phi(\frac{|B_u|}{\sqrt{1-u}})$ is a submartingale.

Remark 4. We have

$$\pi_t = \mathsf{P}(\theta \le t \,|\, \mathcal{F}_t^B) = 2\Phi\bigg(\frac{S_t - B_t}{\sqrt{1-t}}\bigg) - 1 \overset{\text{law}}{=} 2\Phi\bigg(\frac{|B_t|}{\sqrt{1-t}}\bigg) - 1.$$

Hence,

$$\mathsf{P}(\theta \le t) = \mathsf{E}\pi_t = 2\mathsf{E}\Phi\bigg(\frac{|B_t|}{\sqrt{1-t}}\bigg) - 1, \quad t \in [0,1].$$

In view of the equality

$$\mathsf{E}\Phi\bigg(\frac{|B_t|}{\sqrt{1-t}}\bigg) = 1 - \frac{1}{\pi}\arctan\sqrt{\frac{1-t}{t}},$$

we get

$$\begin{aligned}\mathsf{P}(\theta \le t) &= 1 - \frac{2}{\pi}\arctan\sqrt{\frac{1-t}{t}} = 1 - \frac{2}{\pi}\arccos\sqrt{t}\\ &= 1 - \frac{2}{\pi}\Big(\frac{\pi}{2} - \arcsin\sqrt{t}\Big) = \frac{2}{\pi}\arcsin\sqrt{t}.\end{aligned}$$

This is the "Arcsine law".

It is hard to find the function $K_1(c)$ and, in particular, the value $K_1 = K_1(1)$. This is due to the nonstandard form of the minimizing functional in the formulation of the optimal stopping problem (2.4). It has been found, however, that problem (2.5) can be solved by the same method as used to solve problem (2.1). M.A. Urusov (Department of Probability Theory, Moscow State University) showed that

$$\begin{aligned}K_2(c) &= \inf_\tau \mathsf{E}\big[(1-\tau)\,I(\tau < \theta) + c\,(\tau - \theta)^+\big]\\ &= \inf_\tau \mathsf{E}\bigg[(1-\tau)\bigg(1 - F\bigg(\frac{|B_\tau|}{\sqrt{1-\tau}}\bigg)\bigg) + c\int_0^\tau F\bigg(\frac{|B_u|}{\sqrt{1-u}}\bigg)\,du\bigg]\\ &= \inf_\sigma \mathsf{E}\bigg[e^{-2\sigma}(1 - F(|Z_\sigma|)) + 2c\int_0^\sigma e^{-2u}F(|Z_u|)\,du\bigg],\end{aligned}$$

where $Z_s = e^s B_{1-e^{-2s}}$, $1-\tau = e^{-2\sigma}$, $\tau \in [0,1]$, $\sigma \in [0,\infty)$.

Thus, for the criterion $K_2(c)$, the formulation of the corresponding optimal stopping problem takes the form: find

$$K_2(c) = \inf_{\sigma} \mathsf{E}\left[e^{-2\sigma}(1 - F(|Z_\sigma|)) + 2c\int_0^\sigma e^{-2u} F(|Z_u|)\,du\right] \tag{2.7}$$

and the corresponding optimal stopping time.

This problem is solved similarly to problem (2.3). It has been found that the solutions of problems (2.3) and (2.7) (and consequently, the solutions of (2.1) and (2.5)) are qualitatively similar. Namely, the following statement is true: for problem (2.5), the optimal stopping time is

$$\tau_*(c) = \inf\{t \le 1 : S_t - B_t = z_*(c)\sqrt{1-t}\},$$

where $z_*(c)$ is the solution of the equation

$$(c+1)\,z\,(2\Phi(z) - 1) - ((c+1)z^2 + 1)\,\varphi(z) - z = 0.$$

We have

$$z_*(c) \sim \begin{cases} 1/\sqrt{c}, & c \uparrow \infty, \\ \sqrt{2\,|\log c|}, & c \downarrow 0. \end{cases}$$

3 Problem II

1. Suppose that we observe (see Section 1) a continuous-time random process $X = (X_t)_{t\ge 0}$ defined as

$$X_t = r(t-\theta)^+ + \sigma B_t, \tag{3.8}$$

i.e.

$$dX_t = \begin{cases} \sigma\,dB_t, & t < \theta, \\ r\,dt + \sigma\,dB_t, & t \ge \theta, \end{cases}$$

where θ is the time of the appearance of the arbitrage possibility and $B = (B_t)_{t\ge 0}$ is a standard Brownian motion.

We will need the following notation:

$$\mathsf{P}_\theta = \mathrm{Law}(X\,|\,\theta),$$

$$L_t = \frac{d\mathsf{P}_0}{d\mathsf{P}_\infty}(t, X),$$

where

$$\frac{d\mathsf{P}_\theta}{d\mathsf{P}_\infty}(t, X) = \frac{d(\mathsf{P}_\theta\,|\,\mathcal{F}_t^X)}{d(\mathsf{P}_\infty\,|\,\mathcal{F}_t^X)}, \qquad \theta \in [0,\infty].$$

For our model, we have

$$L_t = \frac{d\mathsf{P}_0}{d\mathsf{P}_\infty}(t, X) = \frac{d\mathsf{P}_0}{d\mathsf{P}_t}(t, X),$$

$$\frac{d\mathsf{P}_\theta}{d\mathsf{P}_\infty}(t, X) = \frac{d\mathsf{P}_\theta}{d\mathsf{P}_t}(t, X) = \frac{L_t}{L_\theta}, \quad \theta \le t.$$

The following two statistics $\gamma = (\gamma_t)_{t\ge0}$ and $\psi = (\psi_t)_{t\ge0}$, defined as

$$\gamma_t = \max_{\theta\le t} \frac{L_t}{L_\theta}$$

and

$$\psi_t = \int_0^t \frac{L_t}{L_\theta}\, d\theta$$

are essential in all the subsequent considerations.

We call $\gamma = (\gamma_t)_{t\ge0}$ the *exponential CUSUM* (cumulative sum) process or the exponential *CUSUM-statistics*. For model (3.8), we have

$$L_t = e^{H_t}, \quad H_t = \frac{r}{\sigma^2} X_t - \frac{r^2}{2\sigma^2} t$$

and

$$\gamma_t = \exp\Big\{ H_t - \min_{\theta\le t} H_\theta \Big\}. \tag{3.9}$$

In the statistical literature (see, for example, [12], [20]), the statistics $\psi = (\psi_t)_{t\ge0}$ is called the *Shiryaev-Roberts statistics.*

For model (3.8), we have

$$\psi_t = \int_0^t e^{H_t - H_\theta}\, d\theta.$$

Remark 1. Let us consider the discrete-time case. We assume that the disorder (the change-point) can appear at the times $\theta = 0, 1 \ldots$. If $\theta = 0$, then the observed sequence $x_0, x_1, \ldots$ is a sequence of independent identically distributed random variables with the distribution density $f_1(x)$. If $\theta \ge 1$, then the observed sequence $x_0, \ldots, x_{\theta-1}, x_\theta, x_{\theta+1}, \ldots$ is again a sequence of independent random variables such that $x_0, \ldots, x_{\theta-1}$ have the distribution density $f_0(x)$ and $x_\theta, \ldots, x_{\theta+1}$ have the distribution density $f_1(x)$.

In this case,

$$L_n = \prod_{i=0}^n \frac{f_1(x)}{f_0(x)} = \exp\Big\{ \sum_{i=0}^n \log \frac{f_1(x_i)}{f_0(x_i)} \Big\}.$$

Let

$$S_n = \sum_{i=0}^n \log \frac{f_1(x_i)}{f_0(x_i)}, \quad n \ge 0.$$

Then $L_n = e^{S_n}$ and (compare with (3.9))

$$\gamma_n = \max_{0\le\theta\le n} \frac{L_n}{L_\theta} = \max_{0\le\theta\le n} \exp\{S_n - S_\theta\} = \exp\Big\{S_n - \min_{0\le\theta\le n} S_\theta\Big\}. \tag{3.10}$$

Set

$$\widetilde{S}_n = S_n - \min_{0\le\theta\le n} S_\theta.$$

Obviously, $\widetilde{S}_0 = 0$ and, for $n \ge 1$, we have

$$\widetilde{S}_n = \max\Big\{0, \widetilde{S}_{n-1} + \log\frac{f_1(x_n)}{f_0(x_n)}\Big\}.$$

In the papers [9], [11], this recurrent relation was used to define the *cumulative sum* (CUSUM) process $\widetilde{S} = (\widetilde{S}_n)_{n\ge0}$. This, together with representation (3.9), explains the above-mentioned name "exponential CUSUM-process" for the statistics $\gamma = (\gamma_t)_{t\ge0}$. Here, γ_t has the form $\gamma_t = e^{\widetilde{H}_t}$ with $\widetilde{H}_t = H_t - \min_{0\le\theta\le t} H_\theta$ (compare with (3.10)).

The following statistics $\psi = (\psi_n)_{n\in\mathbb{N}}$ serves as a discrete-time analog of the statistics $\psi = (\psi_t)_{t\ge0}$:

$$\psi_n = \sum_{\theta=1}^{n} \frac{L_n}{L_{\theta-1}}.$$

The statistics ψ defined by the above equality satisfies the recurrent relation

$$\psi_n = (1+\psi_{n-1})\frac{f_1(x_n)}{f_0(x_n)}, \quad n \ge 1$$

with $\psi_0 = 0$.

Remark 2. Instead of the process $H_t - \min_{\theta\le t} H_\theta$ that appears in (3.9), one can consider the process $\max_{\theta\le t} H_\theta - H_t$.

If $H_t = B_t$, then, by P. Lévy's theorem,

$$\mathrm{Law}\Big(\max_{\theta\le t} B_\theta - B_t;\, t\ge 0\Big) = \mathrm{Law}(|B_t|;\, t\ge0).$$

Due to this equality, the process $\max B - B$ is often called the *reflected* Brownian motion.

If $H_t = B_t^\mu$ with $B_t^\mu = \mu t + B_t$, then (see [5])

$$\mathrm{Law}\Big(\max_{\theta\le t} B_\theta^\mu - B_t^\mu;\, t\ge 0\Big) = \mathrm{Law}\big(|Y_t^\mu|;\, t\ge0\big),$$

where $Y^\mu = (Y_t^\mu)_{t\ge0}$ is the so-called "bang-bang process" defined as the solution of the stochastic differential equation

$$dY_t^\mu = -\mu \operatorname{sgn} Y_t^\mu\, dt + dB_t, \quad Y_0^\mu = 0.$$

2. We will now turn to the quickest detection problem in Variants **A** and **B**. First of all, we will make an assumption on the distribution of the random variable $\theta = \theta(\omega)$ (recall that θ corresponds to the "change-point" in the observed process (3.8)).

Let us assume that θ takes values in $[0, \infty)$ and it has the following distribution:

$$\mathsf{P}(\theta = 0) = \pi,$$

where $\pi \in [0, 1]$, and

$$\mathsf{P}(\theta \geq t \,|\, \theta > 0) = e^{-\lambda t},$$

where $\lambda > 0$ is a known constant.

Let

$$\pi_t = \mathsf{P}(\theta \leq t \,|\, \mathcal{F}_t^X)$$

be the posterior probability (constructed through the observation of the process X) that the change-point has appeared within the time-interval $[0, t]$ (here, $\mathcal{F}_t^X = \sigma(X_s;\, s \leq t)$).

Set

$$\varphi_t = \frac{\pi_t}{1 - \pi_t}.$$

Applying the Bayes formula, we get

$$\begin{aligned}\varphi_t = \frac{\mathsf{P}(\theta \leq t \,|\, \mathcal{F}_t^X)}{\mathsf{P}(\theta > t \,|\, \mathcal{F}_t^X)} &= \frac{\pi}{1-\pi} e^{\lambda t} \frac{d\mathsf{P}_0}{d\mathsf{P}_\infty}(t, X) + e^{\lambda t} \int_0^t \frac{d\mathsf{P}_\theta}{d\mathsf{P}_\infty}(t, X)\, \lambda e^{-\lambda\theta}\, d\theta \\ &= \frac{\pi}{1-\pi} e^{\lambda t} L_t + \lambda e^{\lambda t} \int_0^t \frac{L_t}{L_\theta}\, e^{-\lambda t}\, d\theta.\end{aligned} \tag{3.11}$$

Set

$$U_t = e^{\lambda t} L_t \frac{\pi}{1 - \pi},$$

$$V_t = \lambda e^{\lambda t} \int_0^t \frac{L_t}{L_\theta}\, e^{-\lambda\theta}\, d\theta.$$

Applying Itô's formula and taking the equality $dL_t = \frac{r}{\sigma^2} L_t dX_t$ into account, we get

$$\begin{aligned} dU_t &= \lambda\, U_t\, dt + \frac{r}{\sigma^2}\, U_t\, dX_t, \quad U_0 = \frac{\pi}{1-\pi}, \\ dV_t &= \lambda(1 + V_t)\, dt + \frac{r}{\sigma^2}\, V_t\, dX_t, \quad V_0 = 0. \end{aligned}$$

Therefore, the process $\varphi_t = U_t + V_t$ satisfies the equation (see [15], [16])

$$d\varphi_t = \lambda(1 + \varphi_t)\, dt + \frac{r}{\sigma^2}\, \varphi_t\, dX_t, \quad \varphi_0 = \frac{\pi}{1 - \pi}.$$

Applying once again Itô's formula, we conclude that the posterior probability $(\pi_t)_{t\ge 0}$ satisfies the following stochastic differential equation (see [15], [16]):

$$d\pi_t = \Big(\lambda - \frac{r^2}{\sigma^2}\pi_t^2\Big)\,(1-\pi_t)\,dt + \frac{r}{\sigma^2}\,\pi_t(1-\pi_t)\,dX_t, \quad \pi_0 = \pi. \tag{3.12}$$

Let us denote φ_t by $\varphi_t(\lambda)$ in order to emphasize the dependence of φ_t on λ. Applying (3.11) with $\pi = 0$, we get

$$\frac{\varphi_t(\lambda)}{\lambda} = e^{\lambda t}\int_0^t \frac{L_t}{L_\theta}\,e^{-\lambda\theta}\,d\theta.$$

Consequently,

$$\lim_{\lambda\to 0}\frac{\varphi_t(\lambda)}{\lambda} = \psi_t\Big(= \int_0^t \frac{L_t}{L_\theta}\,d\theta\Big), \tag{3.13}$$

and the statistics ψ satisfies the following stochastic differential equation (it was obtained by the author in [15], [16]):

$$d\psi_t = dt + \frac{r}{\sigma^2}\,\psi_t\,dX_t, \quad \psi_0 = 0. \tag{3.14}$$

3. We will first solve the quickest detection problem in the Bayes formulation (**Variant B**). Set

$$\mathbb{B}(c;\pi) = \inf_\tau\big\{\mathsf{P}_\pi(\tau<\theta) + c\mathsf{E}_\pi(\tau-\theta)^+\big\}, \tag{3.15}$$

where the subscript π indicates that the prior probability of the event $\{\theta=0\}$ equals π.

Using (3.15), we get

$$\mathbb{B}(c;\pi) = \inf_\tau \mathsf{E}_\pi\Big\{(1-\pi_\tau) + c\int_0^\tau \pi_s\,ds\Big\}\ \big(=\rho_*(\pi)\big). \tag{3.16}$$

It is useful to introduce the *innovation* representation of the process X:

$$dX_t = r\pi_t\,dt + \sigma\,d\overline{B}_t,$$

where $\overline{B} = (\overline{B}_t)_{t\ge 0}$ is again a Brownian motion (with respect to the filtration $(\mathcal{F}_t^X)_{t\ge 0}$). See [6] for details.

In view of this representation, stochastic differential equation (3.12) takes the following form:

$$d\pi_t = \lambda(1-\pi_t)\,dt + \frac{r}{\sigma}\,\pi_t(1-\pi_t)\,d\overline{B}_t. \tag{3.17}$$

Consequently, the infinitesimal generator of the diffusion process $(\pi_t)_{t\ge 0}$ has the form

$$L = a(\pi)\,\frac{d}{d\pi} + \frac{1}{2}\,b^2(\pi)\,\frac{d^2}{d\pi^2}, \tag{3.18}$$

where

$$a(\pi) = \lambda(1-\pi),$$
$$b^2(\pi) = \left(\frac{r}{\sigma}\right)^2 \pi^2(1-\pi)^2.$$

Remark 3. It follows from (3.17) that

$$\int_0^t \pi_s\,ds = \frac{\pi_0 - \pi_t}{\lambda} + \frac{1}{\lambda}\frac{r}{\sigma}\int_0^t \pi_s(1-\pi_s)\,d\overline{B}_s + t.$$

Consequently, at least for τ satisfying the inequality $\mathsf{E}_\pi \tau < \infty$, we derive from (3.16) that $\mathbb{B}(c;\pi)$ can be represented as

$$\mathbb{B}(c;\pi) = \inf_\tau \mathsf{E}_\pi \Big\{ \Big(1 + \frac{c}{\lambda}\pi\Big) - \Big(1+\frac{c}{\lambda}\Big)\pi_\tau + c\tau \Big\} \ (= \rho_*(\pi)). \tag{3.19}$$

Following the standard scheme of solving the quickest detection problems of the type (3.16) or (3.19) (see 16) consider the corresponding free-boundary (Stefan) problem (see Figure 5):

$$\begin{aligned}
&L\rho(\pi) = -c\pi, \quad \pi \in [0,B), && && (3.20)\\
&\rho(B) = 1 - B, \quad \pi \in [B,1], && \text{(instantaneous stopping)}, && (3.21)\\
&\rho'(B) = -1 && \text{(smooth fit)}, && (3.22)\\
&\rho'(0) = 0, && && (3.23)
\end{aligned}$$

where L is the infinitesimal generator defined by (3.18).

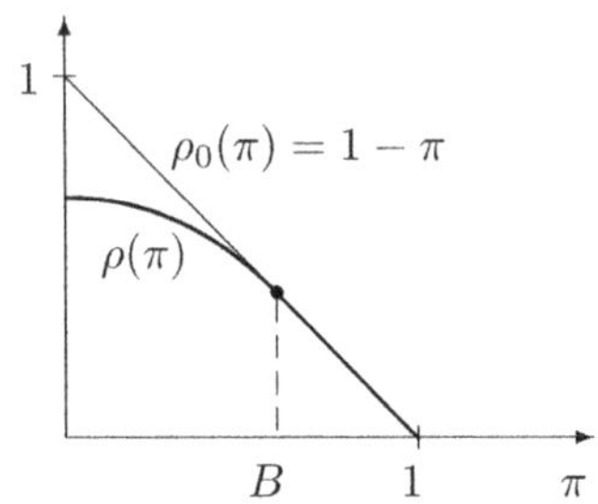

Figure 5. The line $\rho_0(\pi) = 1 - \pi$ corresponds to the risk of the instantaneous stopping

A general solution of equation (3.20) includes two arbitrary constants. An additional unknown constant is the point B. Thus, we have *three* unknown parameters that can be found using conditions (3.21) for $\pi = B$, (3.22) and (3.23).

These considerations lead to the following solution of problems (3.20)–(3.23):

$$\rho(\pi) = \begin{cases} (1 - B_*) - \int_\pi^{B_*} y_*(x)\,dx, & \pi \in [0, B_*], \\ 1 - \pi, & \pi \in [B_*, 1] \end{cases}$$

with

$$y_*(x) = -C \int_0^x e^{-\Lambda[G(x)-G(y)]} \frac{dy}{y\,(1-y)^2},$$

$$G(y) = \log \frac{y}{1-y} - \frac{1}{y},$$

$$\Lambda = \lambda \Big/ \frac{r^2}{2\sigma^2}, \quad C = c \Big/ \frac{r^2}{2\sigma^2}.$$

The parameter B_* is defined as the root of the equation

$$C \int_0^{B_*} e^{-\Lambda[G(B_*)-G(y)]} \frac{dy}{y\,(1-y)^2} = 1. \tag{3.24}$$

(For more details, see [15]–[17]).

The standard technique of the "verification theorems" (see, for example, [19, p. 756]) shows that the obtained solution $\rho(\pi)$ is equal to $\rho_*(\pi)$ and the time $\tau_* = \tau_*(B_*)$ with

$$\tau_*(B_*) = \inf\{t : \pi_t \geq B_*\}$$

is optimal for any $0 \leq \pi \leq 1$. Namely, for any $\theta = \theta(\omega)$ with $\mathsf{P}(\theta = 0) = \pi$ and $\mathsf{P}(\theta \geq t \,|\, \theta > 0) = e^{-\lambda t}$ (here, $\pi \in [0, 1]$ and $\lambda > 0$), we have

$$\rho_*(\pi) = \mathsf{E}_\pi \Big\{ (1 - \pi_{\tau_*}) + c \int_0^{\tau_*} \pi_s\,ds \Big\}$$

and

$$\mathbb{B}(c; \pi) = \mathsf{P}_\pi(\tau_* < \theta) + c\mathsf{E}_\pi(\tau_* - \theta)^+.$$

4. In order to solve the conditionally-extremal problem in **Variant A**, we use the method of the Lagrange multipliers and the obtained solution for Variant **B**.

Let $B_* = B_*(\lambda; c)$ be the value defined from equation (3.24) and $\mathfrak{M}_\alpha = \{\tau : \mathsf{P}_\pi(\tau < \theta) \leq \alpha\}$, where π is a *fixed* value in $[0, 1]$ and α is a constant that defines the upper boundary for the probability of the false alarm $\mathsf{P}_\pi(\tau < \theta)$.

One can show that there exists a value c_α such that

$$B_*(\lambda, c_\alpha) = 1 - \alpha.$$

Then the stopping time $\tau_*(B_*) = \tau_*(B_*(\lambda, c_\alpha))$ belongs to $\mathfrak{M}_\alpha$ and we have

$$\begin{aligned}\mathbb{B}(c_\alpha, \pi) &= \inf_\tau \{\mathsf{P}_\pi(\tau < \theta) + c_\alpha \mathsf{E}_\pi(\tau - \theta)^+\} \\ &= \mathsf{P}_\pi\big(\tau_*(B_*) < \theta\big) + c_\alpha \mathsf{E}_\pi\big(\tau_*(B_*) - \theta\big)^+ \\ &= \mathsf{E}_\pi\big(1 - \pi_{\tau_*(B_*)}\big) + c_\alpha \mathsf{E}_\pi\big(\tau_*(B_*) - \theta\big)^+ \\ &= \alpha + c_\alpha \mathsf{E}_\pi\big(\tau_*(B_*) - \theta \,\big|\, \tau_*(B_*) \geq \theta\big)(1-\alpha).\end{aligned}$$

Since the time $\tau_*(B_*(\lambda, c_\lambda))$ is optimal for Variant **B**, this time is also optimal in the class $\mathfrak{M}_\alpha$. Furthermore, the mean delay in discovering the change point, i.e.

$$\mathbb{R}(\alpha, \lambda) = \mathsf{E}_\pi\big(\tau_*(B_*) - \theta \,\big|\, \tau_*(B_*) \geq \theta\big),$$

is given by the following formula (see [16], [17]):

$$\mathbb{R}(\alpha, \lambda) = \frac{\int_0^{1-\alpha}\left[\int_0^x e^{-\frac{\lambda}{\nu}[G(x)-G(y)]}\,\frac{dy}{y\,(1-y)^2}\right]dx}{(1-\alpha)\,\nu}$$

with $\nu = \frac{r^2}{2\sigma^2}$.

Now, let $\lambda \to 0$, $\alpha \to 1$ in such a way that

$$\frac{1-\alpha}{\lambda} \to T, \tag{3.25}$$

where $T > 0$ is a constant. Then

$$\mathbb{R}(T) \equiv \lim \mathbb{R}(\alpha, \lambda) = \frac{1}{\nu}\left\{e^b(-Ei(-b)) - 1 + b\int_0^\infty e^{-bz}\,\frac{\log(1+z)}{z}\,dz\right\},$$

where

$$b = \frac{1}{\nu T}, \qquad -Ei(-y) = \int_y^\infty \frac{e^{-z}}{z}\,dz. \tag{3.26}$$

Using the above formula, we get for $\nu = 1$:

$$\mathbb{R}(T) = \begin{cases}\log T - 1 - \mathbb{C} + O(\frac{1}{T}), & T \to \infty, \\ \frac{T}{2} + O(T^2), & T \to 0,\end{cases} \tag{3.27}$$

where $\mathbb{C} = 0.577\ldots$ is the Euler constant.

Let us mention the following important property in connection with the passage to the limit $\lambda \to 0$, $\alpha \to 1$ (together with condition (3.25)). Suppose that $\pi = 0$ and $\mathsf{P}(\theta \geq t) = e^{-\lambda t}$. Let $(a, b) \in (A, B)$. Then

$$\mathsf{P}\big(\theta \in (a, b) \,|\, \theta \in (A, B)\big) \longrightarrow \frac{b-a}{B-A}.$$

In other words, using the limit procedure as $\lambda \to \infty$, we get from the exponential distribution on $[0,\infty)$ a *generalized distribution* on $[0,\infty)$ that is *conditionally uniform* in the following sense: the conditional distribution of θ under the condition that θ appears within (A,B) is the uniform distribution on (A,B).

This assumption on the distribution of θ is rather natural in the case where *there is no information* on the distribution of θ. Thus, the use of the exponential distribution, combined with the passage to the limit as $\lambda \to 0$, can be regarded as a useful *technical means* for the study of the schemes with the conditionally uniform distribution.

5. The above reasoning shows that the time $\tau_\alpha^* = \tau_*(B_*(\lambda, c_\alpha))$, i.e. the time

$$\tau_\alpha^* = \inf\{T : \pi_t \geq 1-\alpha\}, \tag{3.28}$$

is optimal in the class $\mathfrak{M}_\alpha$:

$$\inf_{\tau\in\mathfrak{M}_\alpha} \mathsf{E}(\tau-\theta\,|\,\tau\geq\theta) = \mathsf{E}(\tau_\alpha^*-\theta\,|\,\tau_\alpha^*\geq\theta).$$

Using the above notation $\varphi_t = \frac{\pi_t}{1-\pi_t}$, we can express (3.28) as

$$\tau_\alpha^* = \inf\Big\{t : \frac{\varphi_t(\lambda)}{\lambda} \geq \frac{1-\alpha}{\alpha\lambda}\Big\}.$$

For $\psi_t = \lim_{\lambda\to\infty} \frac{\varphi_t(\lambda)}{\lambda}$ (see (3.13), (3.14)), we find that, in the case $\pi = 0$,

$$d\psi_t = dt + \frac{r^2}{\sigma^2}\,\psi_t\,dX_t, \quad \psi_0 = 0$$

or, equivalently,

$$\psi_t = t + \frac{r}{\sigma^2}\int_0^t \psi_s\,dX_s.$$

If $\theta = \infty$, then

$$\psi_t = t + \frac{r}{\sigma}\int_0^t \psi_s\,dB_s.$$

Passing to the limit $\lambda \to 0$, $\alpha \to 1$ in such a way that

$$\frac{1-\alpha}{\lambda} \to T,$$

we find that

$$\tau_\alpha^* \to \tau^*(T) = \inf\{t : \psi_t \geq T\}.$$

Consequently,

$$\psi_{\tau^*(T)} = \tau^*(T) + \frac{r}{\sigma}\int_0^{\tau^*(T)} \psi_s\,dB_s.$$

It follows that

$$T = \mathsf{E}_\infty \psi_{\tau^*(T)} = \mathsf{E}_\infty \tau^*(T).$$

In other words, the constant T has a simple intuitive meaning: it is the *mean time up to the false alarm*. The value

$$\mathbb{R}(T) = \lim_{\{\lambda\to 0,\alpha\to 1,\frac{1-\alpha}{\lambda}\to T\}} \mathbb{R}(\alpha,\lambda)$$

is the *mean delay time* in discovering the change-point (arbitrage). It follows from (3.27) that, for large T, this time has order $\log T$ while, for small T, this time has order $T/2$ (for the case $\nu = \frac{r^2}{2\sigma^2} = 1$).

6. We will now turn to the parametric *minimax* formulations of the quickest detection problems. In these formulations, θ is supposed to be a parameter with values in $[0,\infty)$.

The formulation in **Variant C** is to find

$$\mathbb{C}(T) = \inf_{\tau\in\mathfrak{M}^T} \sup_\theta \operatorname*{esssup}_\omega \mathsf{E}_\theta\big((\tau-\theta)^+ \mid \mathcal{F}_\theta\big)(\omega)$$

and the corresponding optimal stopping time. Here,

$$\mathfrak{M}^T = \{\tau : \mathsf{E}_\infty \tau = T\}$$

and $\mathcal{F}_\theta = \mathcal{F}_\theta^X = \sigma(X_s;\, s\le\theta)$.

For $\tau\in\mathfrak{M}^T$, we set

$$C(\tau) = \sup_\theta \operatorname*{esssup}_\omega \mathsf{E}_\theta\big((\tau-\theta)^+ \mid \mathcal{F}_\theta\big)(\omega).$$

The main idea in finding the optimal stopping time in Variant **C** is to give a *lower* estimate for $C(\tau)$ (τ belongs to $\mathfrak{M}^T$).

The author showed in [18] that the following estimate is true (see details below in Subsection 7):

$$C(\tau) \ge \frac{\mathsf{E}_\infty \int_0^\tau \gamma_t\,dt}{\mathsf{E}_\infty \gamma_\tau}, \tag{3.29}$$

where $\gamma = (\gamma_t)_{t\ge 0}$ is the exponential CUSUM-statistics introduced above:

$$\gamma_t = \sup_{\theta\le t} \frac{L_t}{L_\theta}.$$

Here,

$$L_t = \frac{d\mathsf{P}_0}{d\mathsf{P}_t}(t,X) = \exp\Big\{\frac{r}{\sigma^2}X_t - \frac{r^2}{2\sigma^2}t\Big\}.$$

Suppose that $\nu = \frac{r^2}{2\sigma^2} = 1$ and set

$$\tau^*(B) = \inf\{t : \gamma_t \ge B\}.$$

Using the Markov property of the process $\gamma = (\gamma_t)_{t\ge 0}$, one can show that

$$C(\tau^*(B)) = \mathsf{E}_0\tau^*(B)$$

(the proof of this property is given in Subsection 7 below).

Let us now find $B = B_T$ such that

$$\tau^*(B) \in \mathfrak{M}^T = \{\tau : \mathsf{E}_\infty\tau = T\}.$$

Using the standard method of the differential backward Kolmogorov equation and taking equality (3.34) (it is given below) into account, we deduce that

$$\mathsf{E}_\infty\tau^*(B) = B - 1 - \log B,$$

and consequently, B_T is defined from the equation

$$B_T - 1 - \log B_T = T.$$

By the Markov property of the process γ, we get by analogy with $\mathsf{E}_\infty\tau^*(B)$,

$$\mathsf{E}_0\tau^*(B_T) = \frac{B_T\log B_T + 1 - B_T}{B_T}. \tag{3.30}$$

Using the inequalities

$$\mathsf{E}_0\tau^*(B_T) \ge C(\tau^*(B_T)),$$

$$C(\tau) \ge \frac{\mathsf{E}_\infty\int_0^\tau \gamma_t\,dt}{\mathsf{E}_\infty\gamma_\tau},$$

we get

$$\begin{aligned}\mathsf{E}_0\tau^*(B_T) &\ge C(\tau^*(B_T)) \ge \inf_{\tau\in\mathfrak{M}^T} C(\tau) \ge \frac{\inf\limits_{\tau\in\mathfrak{M}^T}\mathsf{E}_\infty\int_0^\tau \gamma_t\,dt}{\sup\limits_{\tau\in\mathfrak{M}^T}\mathsf{E}_\infty\gamma_\tau}\\ &= \frac{\inf\limits_{\tau\in\mathfrak{M}^T}\mathsf{E}_\infty\int_0^\tau \gamma_t\,dt}{B_T} = \frac{\mathsf{E}_\infty\int_0^{\tau^*(B_T)}\gamma_t\,dt}{B_T} = \frac{B_T\log B_T + 1 - B_T}{B_T}.\end{aligned} \tag{3.31}$$

From (3.30) and (3.31), it follows that $\tau^*(B_T)$ is the *optimal* stopping time in the class $\mathfrak{M}^T = \{\tau : \mathsf{E}_\infty\tau = T\}$:

$$\mathbb{C}(T) = C(\tau^*(B_T))$$

and

$$\begin{aligned}\mathbb{C}(T) &= \frac{B_T\log B_T + 1 - B_T}{B_T}\\ &\sim \begin{cases}\log T - 1 + O(\frac{1}{T}), & T\to\infty,\\ \frac{T}{2} + O(T^2), & T\to 0.\end{cases}\end{aligned}$$

Thus, the exponential CUSUM-process $\gamma = (\gamma_t)_{t\ge 0}$ defined as

$$\gamma_t = \sup_{\theta \le t} \frac{L_t}{L_\theta}$$

is the *optimal* statistics in Variant **C**.

It is interesting to note that, for Variant **C**, the statistics $\psi = (\psi_t)_{t\ge 0}$ is *asymptotically optimal.*

Indeed, take

$$\psi_t = \int_0^t \frac{L_t}{L_\theta}\, d\theta$$

and set

$$\sigma_T^* = \inf\{t : \psi_t \ge T\}.$$

Then $\mathsf{E}_\infty \sigma_T^* = T$. Using (3.14) and (3.8), we find that

$$\mathsf{E}_0 \sigma_T^* = e^b(-Ei(-b)), \quad b = \frac{1}{T}$$

(the function Ei is defined in (3.26)). By the arguments similar to those used to obtain (3.31) (see also Section 7 below), we can show that

$$\mathsf{E}_0 \sigma_T^* \ge C(\sigma_T^*) \ge \frac{1}{T}\,\mathsf{E}_\infty \int_0^{\sigma_T^*} \psi_s\, ds.$$

For $b \to 0$, we have

$$e^b(-Ei(-b)) = -\mathbb{C} - \log b + O(b),$$

where $\mathbb{C} = 0.577\ldots$ is the Euler constant. Thus, for $T \to \infty$, we have

$$\mathsf{E}_0 \sigma_T^* = \log T - \mathbb{C} + O\Big(\frac{1}{T}\Big). \tag{3.32}$$

It can also be shown that

$$\frac{1}{T}\mathsf{E}_\infty \int_0^{\sigma_T^*} \psi_s\, ds = \log T - \mathbb{C} - 1 + O\Big(\frac{1}{T}\Big). \tag{3.33}$$

Consequently,

$$\log T - \mathbb{C} + O\Big(\frac{1}{T}\Big) \ge C(\sigma_T^*) \ge \log T - \mathbb{C} - 1 + O\Big(\frac{1}{T}\Big).$$

On the other hand, as we have already seen,

$$\mathbb{C}(T) = \log T - 1 + O\Big(\frac{1}{T}\Big).$$

Thus, the statistics $\psi = (\psi_t)_{t\ge 0}$ is *asymptotically* ($T \to \infty$) optimal in Variant **C**.

7. We now turn to the proof of the basic inequality (3.29) that was used to prove that the exponential CUSUM-statistics is optimal in Variant **C**.

Set

$$C_\theta(\tau;\omega) = \mathsf{E}_\theta\big((\tau-\theta)^+ \,|\, \mathcal{F}_\theta\big)(\omega).$$

We have

$$\gamma_t = \sup_{\theta\le t} \frac{L_t}{L_\theta} = \frac{L_t}{\inf_{\theta\le t} L_\theta} = \frac{L_t}{N_t},$$

$$d\gamma_t = d\Big(\frac{L_t}{N_t}\Big) = \frac{dL_t}{N_t} - \frac{L_t\,dN_t}{(N_t)^2} = \frac{r}{\sigma^2}\,\gamma_t\,dX_t - \gamma_t\,\frac{dN_t}{N_t}.$$

Note that $\gamma_t \ge 1$ and N changes its values only on the set $\{(t,\omega) : \gamma_t = 1\}$. Hence,

$$d\gamma_t = \frac{r}{\sigma^2}\,\gamma_t\,dX_t - \gamma_t\, I(\gamma_t = 1)\,\frac{dN_t}{N_t}.$$

Denote

$$H_t = -\int_0^t \gamma_s\, I(\gamma_s = 1)\,\frac{dN_s}{N_s}.$$

Then

$$d\gamma_t = dH_t + \frac{r}{\sigma^2}\,\gamma_t\,dX_t, \quad \gamma_0 = 1, \tag{3.34}$$

and therefore,

$$\gamma_t = L_t\Big[1 + \int_0^t \frac{dH_\theta}{L_\theta}\Big].$$

We have

$$\begin{aligned} C(\tau)\,\mathsf{E}_\infty H_\tau &= \mathsf{E}_\infty\big(C(\tau)H_\tau\big) = \mathsf{E}_\infty\Big(C(\tau)\int_0^\infty I(\theta\le\tau(\omega))\,dH_\theta(\omega)\Big) \\ &\ge \mathsf{E}_\infty \int_0^\infty C_\theta(\tau;\omega)\, I(\theta\le\tau(\omega))\,dH_\theta(\omega), \end{aligned}$$

where $C_\theta(\tau;\omega)$ is defined above and $C(\tau)$ is defined in Subsection 6.

Since

$$(\tau-\theta)^+ = \int_\theta^\infty I(u\le\tau)\,du,$$

we get

$$\begin{aligned} C_\theta(\tau;\omega) &= \int_\theta^\infty \mathsf{E}_\theta\big(I(u\le\tau)\,|\,\mathcal{F}_\theta\big)(\omega)\,du \\ &= \int_\theta^\infty \mathsf{E}_\infty\Big(\frac{L_u}{L_\theta}\, I(u\le\tau)\,\Big|\,\mathcal{F}_\theta\Big)(\omega)\,du = \mathsf{E}_\infty\Big(\int_\theta^\tau \frac{L_u}{L_\theta}\,du\,\Big|\,\mathcal{F}_\theta\Big)(\omega). \end{aligned}$$

We used here the fact that $\xi = I(u \le \tau)$ is $\mathcal{F}_u$-measurable and

$$\mathsf{E}_\theta(\xi \,|\, \mathcal{F}_\theta) = \mathsf{E}_\infty\Big(\xi\, \frac{L_u}{L_\theta} \,\Big|\, \mathcal{F}_\theta\Big).$$

As a result,

$$\begin{aligned} C(\tau)\,\mathsf{E}_\infty H_\tau &\ge \mathsf{E}_\infty \int_0^\infty I(\theta \le \tau(\omega))\,\mathsf{E}_\infty\Big(\int_\theta^\tau \frac{L_u}{L_\theta}\,du \,\Big|\, \mathcal{F}_\theta\Big)\,dH_\theta \\ &= \mathsf{E}_\infty \int_0^\tau \mathsf{E}_\infty\Big(\int_\theta^\tau \frac{L_u}{L_\theta}\,du \,\Big|\, \mathcal{F}_\theta\Big)\,dH_\theta. \end{aligned}$$

Set

$$\widetilde{H}_\theta = \int_0^\theta \frac{dH_s}{L_s}, \qquad \xi_\theta = \int_0^\theta L_s\,ds.$$

Then

$$\begin{aligned} &C(\tau)\,\mathsf{E}_\infty H_\tau \ge \mathsf{E}_\infty \int_0^\tau \mathsf{E}_\infty\Big(\int_\theta^\tau \frac{L_u}{L_\theta}\,du \,\Big|\, \mathcal{F}_\theta\Big)\,dH_\theta \\ &= \mathsf{E}_\infty \int_0^\tau \mathsf{E}_\infty\Big(\int_\theta^\tau L_u\,du \,\Big|\, \mathcal{F}_\theta\Big)\,d\widetilde{H}_\theta \\ &= \mathsf{E}_\infty \int_0^\tau \mathsf{E}_\infty\Big(\int_0^\tau L_u\,du \,\Big|\, \mathcal{F}_\theta\Big)\,d\widetilde{H}_\theta - \mathsf{E}_\infty \int_0^\tau \mathsf{E}_\infty\Big(\int_0^\theta L_u\,du \,\Big|\, \mathcal{F}_\theta\Big)\,d\widetilde{H}_\theta \\ &= \mathsf{E}_\infty \int_0^\tau \mathsf{E}_\infty(\xi_\tau \,|\, \mathcal{F}_\theta)\,d\widetilde{H}_\theta - \mathsf{E}_\infty \int_0^\tau \xi_\theta\,d\widetilde{H}_\theta. \end{aligned}$$

The process $M_\theta = \mathsf{E}_\infty(\xi_\tau \,|\, \mathcal{F}_\theta)$ is a P_∞-martingale and

$$\mathsf{E}_\infty \int_0^\tau M_\theta\,d\widetilde{H}_\theta = \mathsf{E}_\infty M_\tau \widetilde{H}_\tau - \mathsf{E}_\infty M_0 \widetilde{H}_0 = \mathsf{E}_\infty \xi_\tau \widetilde{H}_\tau.$$

Thus,

$$\begin{aligned} C(\tau)\,\mathsf{E}_\infty H_\tau &\ge \mathsf{E}_\infty \xi_\tau \widetilde{H}_\tau - \mathsf{E}_\infty \int_0^\tau \xi_\theta\,d\widetilde{H}_\theta = \mathsf{E}_\infty \int_0^\tau \widetilde{H}_\theta\,d\xi_\theta \\ &= \mathsf{E}_\infty \int_0^\tau \widetilde{H}_\theta\,L_\theta\,d\theta = \mathsf{E}_\infty \int_0^\tau L_\theta\Big(\int_0^\theta \frac{dH_s}{L_s}\Big)\,d\theta \\ &= \mathsf{E}_\infty \int_0^\tau \Big[\int_0^\theta \frac{L_\theta}{L_s}\,dH_s\Big]\,d\theta = \mathsf{E}_\infty \int_0^\tau [\gamma_\theta - L_\theta]\,d\theta \\ &= \mathsf{E}_\infty \int_0^\tau \gamma_\theta\,d\theta - \mathsf{E}_\infty \int_0^\tau L_\theta\,d\theta. \end{aligned}$$

We have

$$\gamma_t = 1 + H_t + \int_0^t \frac{r}{\sigma^2}\,\gamma_s\,dX_s.$$

Hence,

$$\mathsf{E}_\infty \gamma_\tau = 1 + \mathsf{E}_\infty H_\tau,$$

and, from the obtained inequality

$$C(\tau)\,\mathsf{E}_\infty H_\tau \geq \mathsf{E}_\infty \int_0^\tau \gamma_\theta\, d\theta - \mathsf{E}_\infty \int_0^\tau L_\theta\, d\theta,$$

we get

$$C(\tau)[\mathsf{E}_\infty \gamma_\tau - 1] \geq \mathsf{E}_\infty \int_0^\tau \gamma_\theta\, d\theta - \mathsf{E}_\infty \int_0^\tau L_\theta\, d\theta. \tag{3.35}$$

Now, note that, at least for bounded stopping times τ,

$$C(\tau) = \sup_\theta \operatorname*{esssup}_\omega \mathsf{E}_\theta\big((\tau-\theta)^+ \,|\, \mathcal{F}_\theta\big)(\omega) \geq \mathsf{E}_0\tau = \mathsf{E}_\infty(\tau L_\tau) = \mathsf{E}_\infty \int_0^\tau L_\theta\, d\theta \tag{3.36}$$

as $d(tL_t) = L_t\,dt + t\,dL_t$. In the general case of finite τ, we have

$$C(\tau) \geq \mathsf{E}_0\tau \geq \mathsf{E}_0(\tau\wedge N)\, L_{\tau\wedge N} = \mathsf{E}_\infty \int_0^{\tau\wedge N} L_\theta\, d\theta \uparrow \mathsf{E}_\infty \int_0^\tau L_\theta\, d\theta.$$

From (3.35) and (3.36), we get

$$C(\tau)\,\mathsf{E}_\infty \gamma_\tau \geq \mathsf{E}_\infty \int_0^\tau \gamma_\theta\, d\theta.$$

Thus,

$$C(\tau) \geq \frac{\mathsf{E}_\infty \int_0^\tau \gamma_\theta\, d\theta}{\mathsf{E}_\infty \gamma_\tau}$$

and

$$\mathbb{C}(T) = \inf_{\tau\in\mathfrak{M}^T} C(\tau) \geq \frac{\inf_{\tau\in\mathfrak{M}^T} \mathsf{E}_\infty \int_0^\tau \gamma_\theta\, d\theta}{\sup_{\tau\in\mathfrak{M}^T} \mathsf{E}_\infty \gamma_\tau},$$

where

$$\mathfrak{M}^T = \{\tau : \mathsf{E}_\infty\tau = T\}.$$

We will now give a proof of the property

$$\mathsf{E}_0\tau^*(B) = C(\tau^*(B)),$$

where

$$\tau^*(B) = \inf\{t : \gamma_t \geq B\},$$

$$C(\tau^*(B)) = \sup_\theta \operatorname*{esssup}_\omega \mathsf{E}_\theta\big((\tau^*(B)-\theta)^+ \,|\, \mathcal{F}_\theta\big)(\omega).$$

On the set $\{\omega : \tau^*(B) \le \theta\}$, we have:

$$\mathsf{E}_\theta\big((\tau^*(B) - \theta)^+ \,|\, \mathcal{F}_\theta\big)(\omega) = 0.$$

Consider the set

$$\{\omega : \tau^*(B) > \theta\} = \{\omega : \gamma_u < B,\ u \le \theta\}.$$

By the Markov property of the process $\gamma = (\gamma_t)$, we have on the set

$$\{\omega : \tau^*(B) > \theta\} \cap \{\gamma_\theta = x\}$$

that

$$\mathsf{E}_\theta\big((\tau^*(B) - \theta)^+ \,|\, \mathcal{F}_\theta\big)(\omega) = f(x),$$

i.e. this conditional expectation is the function only of the value x of γ_θ. It is clear that $\max_{1\le x\le B} f(x) = f(1)$ and

$$f(1) = \mathsf{E}_0\tau^*(B).$$

Hence,

$$C(\tau^*(B)) = \mathsf{E}_0\tau^*(B).$$

8. Let us now consider the quickest detection problem in **Variant D**. One should find

$$\mathbb{D}(T) = \inf_{\tau\in\mathfrak{M}^T} \sup_\theta \mathsf{E}_\theta(\tau - \theta \,|\, \tau \ge \theta)$$

and the optimal stopping time in the class $\mathfrak{M}^T$.

We will follow the same scheme as that used in Variant **C**. Let us first prove that

$$\mathsf{E}_0\sigma_T^* \ge \mathbb{D}(T) \ge \inf_{\tau\in\mathfrak{M}^T} \frac{\mathsf{E}_\infty \int_0^\tau \psi_s\, ds}{T}, \tag{3.37}$$

where

$$\sigma_T^* = \inf\{t \ge 0 : \psi_t \ge T\}.$$

We have

$$\begin{aligned}
D_\theta(\tau) &= \mathsf{E}_\theta(\tau - \theta \,|\, \tau \ge \theta) = \mathsf{E}_\theta\big((\tau - \theta)^+ \,|\, \tau \ge \theta\big) \\
&= \int_\theta^\infty \mathsf{E}_\theta\big(I(u \le \tau) \,|\, \tau \ge \theta\big)\, du = \int_\theta^\infty \frac{\mathsf{E}_\theta I(u \le \tau)}{\mathsf{E}_\theta I(\theta \le \tau)}\, du \\
&= \int_\theta^\infty \frac{\mathsf{E}_\theta I(u \le \tau)}{\mathsf{E}_\infty I(\theta \le \tau)}\, du = \int_\theta^\infty \frac{\mathsf{E}_\infty\big(\frac{L_u}{L_\theta} I(u \le \tau)\big)}{\mathsf{E}_\infty I(\theta \le \tau)}\, du.
\end{aligned}$$

Thus,

$$\mathbb{D}(T)\,\mathsf{E}_\infty I(\tau \le \theta) \ge D_\theta(\tau)\,\mathsf{E}_\infty I(\theta \le \tau)$$
$$= \int_\theta^\infty \mathsf{E}_\infty\left(\frac{L_u}{L_\theta} I(u \le \tau)\right) du = \mathsf{E}_\infty \int_\theta^\tau \frac{L_u}{L_\theta}\, du.$$

Integrating on θ (from 0 to ∞), we get

$$\mathbb{D}(T)\,\mathsf{E}_\infty \tau \ge \mathsf{E}_\infty \int_0^\infty \left(\int_\theta^\tau \frac{L_u}{L_\theta}\, du\right) d\theta$$
$$= \mathsf{E}_\infty \int_0^\tau \left(\int_0^\theta \frac{L_\theta}{L_u}\, du\right) d\theta = \mathsf{E}_\infty \int_0^\tau \psi_\theta\, d\theta.$$

As a result,

$$\begin{aligned}\mathsf{E}_0 \sigma_T^* &= \sup_\theta \mathsf{E}_\theta(\sigma_T^* - \theta \,|\, \sigma_T^* \ge \theta) \ge \inf_{\tau \in \mathfrak{M}^T} \sup_\theta \mathsf{E}_\theta(\tau - \theta \,|\, \tau \ge \theta) \\ &\ge \inf_{\tau \in \mathfrak{M}^T} \frac{\mathsf{E}_\infty \int_0^\tau \psi_s\, ds}{\mathsf{E}_\infty \tau} = \inf_{\tau \in \mathfrak{M}^T} \frac{\mathsf{E}_\infty \int_0^\tau \psi_s\, ds}{T},\end{aligned} \tag{3.38}$$

where $\mathfrak{M}^T = \{\tau : \mathsf{E}_\infty \tau = T\}$.

Similarly to Variant **C**, we have here

$$\sup_\theta \mathsf{E}_\theta(\sigma_T^* - \theta \,|\, \sigma_T^* \ge \theta) = \mathsf{E}_0 \sigma_T^*.$$

This, together with (3.38), leads to the desired inequalities (3.37).

It has already been mentioned in Subsection 6 that, for large T, formulas (3.32) and (3.33) are true. Combining these formulas with inequalities (3.37), we deduce that the statistics $\psi = (\psi_t)_{t \ge 0}$ is asymptotically $(T \to \infty)$ optimal.

9. Variant E. In Section 1, we considered a particular case of the observation procedure for which the observations are multistage, i.e. the observations are continued after each alarm. It is also important that the change-point should be preceded by a long period of observations and a stationary regime of observations should be established within this period. The change-point can appear only after this stationary regime has been established.

The analysis of the observation procedure that was described in Section 1 shows that any observation procedure is eventually described by a sequence of stopping times $\delta = (\tau_1, \tau_2, \dots)$ such that $\tau_1, \tau_2, \dots$ are independent identically distributed random variables with respect to the measure P_∞. For this procedure, the alarms are raised at the times $\tau_1, \tau_1 + \tau_2, \dots$:

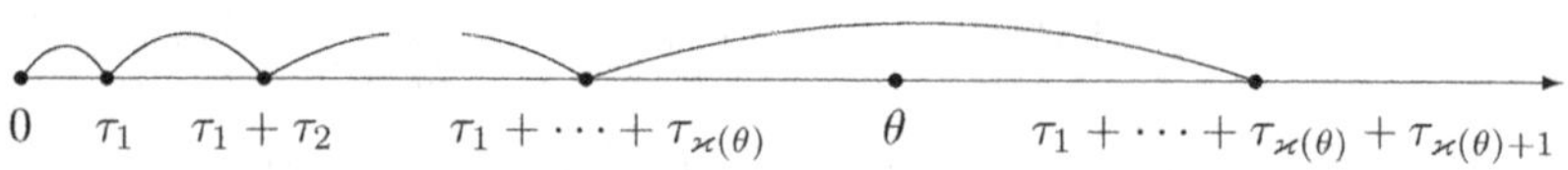

Figure 6

For $\theta > 0$, define $\varkappa(\theta)$ (it takes values in $0, 1, \dots$) from the inequality

$$\tau_0 + \tau_1 + \cdots + \tau_{\varkappa(\theta)} < \theta \le \tau_0 + \tau_1 + \cdots + \tau_{\varkappa(\theta)} + \tau_{\varkappa(\theta)+1},$$

where $\tau_0 = 0$.

For a fixed θ and an observation procedure $\delta = (\tau_1, \tau_2, \dots)$, the mean delay time in discovering the time θ equals

$$R_\theta^\delta(T) = \mathsf{E}_\theta(\tau_1 + \cdots + \tau_{\varkappa(\theta)+1} - \theta)$$

(we assume here that $\mathsf{E}_\infty \tau_i = T > 0$).

Let

$$F_\theta^\delta(u) = \mathsf{P}_\theta\big(\theta - (\tau_1 + \cdots + \tau_{\varkappa(\theta)}) \le u\big).$$

Then

$$\begin{aligned} R_\theta^\delta(T) &= \int_0^\infty \mathsf{E}_\theta\big(\tau_1 + \cdots + \tau_{\varkappa(\theta)+1} - \theta \,|\, \theta - (\tau_1 + \cdots + \tau_{\varkappa(\theta)}) = u\big)\, dF_\theta^\delta(u) \\ &= \int_0^\infty \mathsf{E}_u(\tau_1 - u \,|\, \tau_1 > u)\, dF_\theta^\delta(u). \end{aligned}$$

We will suppose that the observation procedure $\delta = (\tau_1, \tau_2, \dots)$ is such that the distribution of $\mathrm{Law}(\tau_i | \mathsf{P}_\infty)$ is nonlattice. Then the general renewal theory guarantees that there exists a limit distribution

$$F_\infty^\delta(u) = w\text{-}\lim_{\theta \to \infty} F_\theta^\delta(u).$$

By the well-known *Basic Renewal Theorem*,

$$F_\infty^\delta(u) = \frac{1}{T} \int_0^u (1 - F(x))\, dx,$$

where $F(x) = \mathsf{P}_\infty(\tau_1 \le x)$.

Suppose now that the method δ is such that, for $\theta \to \infty$,

$$\begin{aligned} R_\theta^\delta(T) &= \int_0^\infty \mathsf{E}_u(\tau_1 - u | \tau_1 \ge u)\, dF_\theta^\delta(u) \\ &\to \int_0^\infty \mathsf{E}_u(\tau_1 - u | \tau_1 \ge u)\, dF_\infty^\delta(u) \\ &= \frac{1}{T} \int_0^\infty \mathsf{E}_u(\tau_1 - u | \tau_1 \ge u)\, \mathsf{P}_\infty(\tau_1 \ge u)\, du \\ &= \frac{1}{T} \int_0^\infty \mathsf{E}_u(\tau_1 - u | \tau_1 \ge u)\, \mathsf{P}_u(\tau_1 \ge u)\, du \\ &= \frac{1}{T} \int_0^\infty \mathsf{E}_u(\tau_1 - u)^+\, du \; \big(= R^\delta(T)\big). \end{aligned}$$

For the Bayes setting with Law$(\theta) = \exp(\lambda)$, we have

$$\mathsf{E}(\tau - \theta)^+ = \lambda \int_0^\infty e^{-\lambda u}\, \mathsf{E}_u(\tau_1 - u)^+\, du.$$

Fix $T > 0$ and let $\lambda \to 0$, $\alpha \to 1$ in such a way that $\frac{1-\alpha}{\alpha\lambda} = T$. Then

$$\begin{aligned}\frac{1}{T}\int_0^\infty \mathsf{E}_u(\tau_1 - u)^+\, du &= \\ &= \lim_{\{\lambda\to 0,\, \alpha\to 1,\, \frac{1-\alpha}{\alpha\lambda}=T\}} \frac{\lambda\int_0^\infty e^{-\lambda u}\,\mathsf{E}_u(\tau_1 - u)^+\, du}{(1-\alpha)/\alpha} \\ &= \lim_{\{\lambda\to 0,\, \alpha\to 1,\, \frac{1-\alpha}{\alpha\lambda}=T\}} \frac{\mathsf{E}(\tau_1 - \theta)^+}{1-\alpha} \\ &= \lim_{\{\lambda\to 0,\, \alpha\to 1,\, \frac{1-\alpha}{\alpha\lambda}=T\}} \mathsf{E}(\tau_1 - \theta \mid \tau_1 \geq \theta).\end{aligned}$$

From these formulas and the asymptotic ($\lambda \to 0$, $\alpha \to 1$, $\frac{1-\alpha}{\lambda} \to T$) optimality of the statistics $\psi = (\psi_t)_{t\geq 0}$, we conclude that this statistics is *optimal* for **Variant E′**: find

$$\mathbb{E}'(T) = \inf_{\tau\in\mathfrak{M}^T} \frac{1}{T}\int_0^\infty \mathsf{E}_\theta(\tau - \theta)^+ d\theta. \tag{3.39}$$

Note that, in this formulation, the parameter θ can be regarded as a *generalized* random variable with the "uniform distribution on $[0, \infty)$". Formulation (3.39) deals only with one stopping time (in other words, with *one stage* of observations). However, this formulation is directly related to the multistage procedure described above because $\mathbb{E}'(T) = \lim_{\theta\to\infty} R_\theta^\delta(T)$.

10. Table 1 sums up the results described above. These results are related to the optimality and the asymptotic optimality of the statistics $(\pi_t)_{t\geq 0}$, $(\gamma_t)_{t\geq 0}$ and $(\psi_t)_{t\geq 0}$ in various formulations of the quickest detection problem of the change-point (or arbitrage).

Table 1. Optimal and asymptotically optimal statistics for Variants **A**–**E′**

θ	Variants	Optimality of statistics
r.v. $\exp(\lambda)$	**A:** $\inf\limits_{\tau\in\mathfrak{M}_\alpha} \mathsf{E}(\tau-\theta \mid \tau\geq\theta)$	$\pi=(\pi_t)$ is optimal $\pi_t=\mathsf{P}(\theta\leq t \mid \mathcal{F}_t^X)$
r.v. $\exp(\lambda)$	**B:** $\inf\limits_{\tau}\bigl\{\mathsf{P}(\tau<\theta)+c\,\mathsf{E}(\tau-\theta)^+\bigr\}$	$\pi=(\pi_t)$ is optimal
parameter $\theta\in\mathbb{R}_+$	**C:** $\inf\limits_{\tau\in\mathfrak{M}^T}\sup\limits_{\theta}\operatorname*{ess\,sup}\limits_{\omega}\mathsf{E}_\theta\bigl((\tau-\theta)^+ \mid \mathcal{F}_\theta\bigr)$	$\gamma=(\gamma_t)$ is optimal $\gamma_t=\sup\limits_{\theta\leq t}\dfrac{L_t}{L_\theta}$ (exponential CUSUM)
parameter $\theta\in\mathbb{R}_+$	**D:** $\inf\limits_{\tau\in\mathfrak{M}^T}\sup\limits_{\theta}\mathsf{E}_\theta(\tau-\theta \mid \tau\geq\theta)$	$\psi=(\psi_t)$ is asymptotically $(T\to\infty)$ optimal $\gamma=(\gamma_t)$ is asymptotically $(T\to\infty)$ optimal
θ is a generalized r.v.	**E′:** $\inf\limits_{\tau\in\mathfrak{M}^T}\dfrac{1}{T}\displaystyle\int_0^\infty \mathsf{E}_\theta(\tau-\theta)^+d\theta$	$\psi=(\psi_t)$ is optimal $\psi_t=\displaystyle\int_0^t\frac{L_t}{L_\theta}\,d\theta$

$$\mathfrak{M}_\alpha=\{\tau:\mathsf{P}(\tau\leq\theta)\leq\alpha\},$$

$$\mathfrak{M}^T=\{\tau:\mathsf{E}_\infty\tau=T\}.$$

4 Some Comments

1. The researchers working in the field of finance as well as the market operators can be grouped, regarding their approach to the analysis of the dynamics of prices, as follows:

(1) "fundamentalists";
(2) "quantitative analysts";
(3) "technicians".

"Fundamentalists" make their decisions by looking at the state of the "economy at large" or by analyzing some of its sectors. The development prospects are of particular interest to them. The basis of their analysis is the assumption that the actions of the market operators are "rational". The second group ("quantitative analysts") emerged in the 1950s as the followers of L. Bachelier. This group is closer to the "fundamentalists" than to the "technicians" due to the fact that the "quantitative analysts" attach more significance to the rational aspects of the investors' behaviour than to the tones of the market.

It should be pointed out that the theoretical basis for groups (1) and (2) is currently stronger compared to that of group (3). We believe that Problems **I** and **II** considered in this paper are directly related to the technical analysis of the financial data and to the decision-making procedures. (For Problems **I** and **II**, the decisions were connected with discovering the time θ that defines the change of the character of the observed process). Let us also mention our paper [3] that is related to the theoretical foundations of the technical analysis.

2. The formulations of the quickest detection problem (Problem **II**) have a long history. Variants **A** and **B** were considered by the author in [14], [15], [16], [17]. The formulation of Variant **C** was proposed in [7]. In this paper, the asymptotic optimality of the statistics $\gamma = (\gamma_n)_{n \geq 0}$ was proved for the discrete-time case. The proof of the asymptotic optimality of $\gamma = (\gamma_t)_{t \geq 0}$ in the continuous-time case was given in our paper [18]. In particular, this paper contains the lower estimate (3.29) that is essential in the proof of the asymptotic optimality. For the discrete-time case, the corresponding estimate was proved in the paper [8].

The formulation of Variant **D** was considered in many papers. See, for instance, [12] and the collection of works [2].

It seems that our approach to the proof of the asymptotic optimality of the statistics $\psi = (\psi_t)_{t \geq 0}$ and $\gamma = (\gamma_t)_{t \geq 0}$ (based on the lower estimate (3.37)) has not previously been considered.

Finally, the optimality of the statistics $\psi = (\psi_t)_{t \geq 0}$ in Variants **E** and **E′** was proved in the author's papers [15] and [16].

References

1. M. Abramowicz, I.A. Stegun: *Handbook of mathematical functions*. Dover, NY 1968.

2. E. Carlstein, H.-G. Müller, D. Siegmund (eds): *Change-point problems.* Institute of Mathematical Statistics. Lecture Notes Monograph. Ser. 23, 1994.
3. R. Douady, A.N. Shiryaev, M. Yor: On the probability characteristics of "falls" for the standard Brownian motion. Theory of Probability and its Applications **44** (1999) No. 1 3–13.
4. S.E. Graversen, G. Peskir, A.N. Shiryaev: Stopping Brownian motion without anticipation as close as possible to its ultimate maximum. Theory of Probability and its Applications **45** (2000) No. 1 125–136.
5. S.E. Graversen, A.N. Shiryaev: An extension of P. Lévy's distributional properties to the case of a Brownian motion with a drift. Bernoulli **6** (2000) No. 4 615–620.
6. R.S. Liptser, A.N. Shiryaev: *Statistics of random processes, I and II.* Springer-Verlag 1977 and 1978.
7. G. Lorden. Procedures for reacting to a change of distributions. The Annals of Mathematical Statistics **42** (1971) No. 6 1897–1908.
8. G.V. Moustakides: Optimal stopping times for detecting changes in distributions. The Annals of Statistics **14** (1986) No. 4 1379–1387.
9. E.S. Page: Continuous inspection schemes. Biometrika **41** (1954) 100–115.
10. E.S. Page: A test for a change in a parameter occurring at an unknown point. Biometrika **42** (1955) 523–527.
11. J.L. Pedersen: Optimal prediction of the ultimate maximum of Brownian motion. In thesis "Some results on optimal stopping and Skorokhod embedding with applications". University of Aarhus, January, 2000.
12. M. Pollak, D. Siegmund: A diffusion process and its application to detecting a change in the drift of Brownian motion. Biometrika **72** (1985) 267–280.
13. D. Revuz, M. Yor: *Continuous martingales and Brownian motion.* Springer-Verlag 1998.
14. A.N. Shiryaev: Detection of spontaneously occurring effects. Dokl. Acad. Nauk SSSR **138** (1961) 799–801.
15. A.N. Shiryaev: The problem of quickest detection of the destruction of a stationary regime. Dokl. Acad. Nauk. SSSR **138** (1961) 1039–1042.
16. A.N. Shiryaev: On optimum methods in the quickest detection problems. Theory of Probability and its Applications **8** (1963) 22–46.
17. A.N. Shiryaev: *Optimal stopping rules.* Springer-Verlag 1978.
18. A.N. Shiryaev: Minimax optimality of the method of Cumulative Sums (CUSUM) in the case of continuous time. Russian Mathematical Surveys **51** (1996) 750–751.
19. A.N. Shiryaev: *Essentials of stochastic finance.* World Scientific 1999.
20. D. Siegmund: *Sequential analysis.* Springer-Verlag 1985.

Springer Finance

Springer Finance is a programme of books aimed at students, academics and practitioners working on increasingly technical approaches to the analysis of financial markets. It aims to cover a variety of topics, not only mathematical finance but foreign exchanges, term structure, risk management, portfolio theory, equity derivatives, and financial economics.

Credit Risk Valuation: Methods, Models, and Application
Ammann, M.
ISBN 3-540-67805-0 (2001)

Credit Risk: Modeling, Valuation and Hedging
T.R. Bielecki and M. Rutkowski
ISBN 3-540-67593-0 (2001)

Risk-Neutral Valuation: Pricing and Hedging of Finance Derivatives
N.H. Bingham and R. Kiesel
ISBN 1-85233-001-5 (1998)

Interest Rate Models – Theory and Practice
D. Brigo and Fabio Mercurio
ISBN 3-540-41772-9 (2001)

Visual Explorations in Finance with Self-Organizing Maps
G. Deboeck and T. Kohonen (Editors)
ISBN 3-540-76266-3 (1998)

Mathematics of Financial Markets
R. J. Elliott and P. E. Kopp
ISBN 0-387-98533-0 (1999)

Mathematical Models of Financial Derivatives
Y.-K. Kwok
ISBN 981-3083-25-5 (1998)

Efficient Methods for Valuing Interest Rate Derivatives
A. Pelsser
ISBN 1-85233-304-9 (2000)

Exponential Functionals of Brownian Motion and Related Processes
M. Yor
ISBN 3-540-65943-9 (2001)

Made in the USA
Monee, IL
20 March 2025

14337598R00295